GENETICS, GENOMICS AND BREEDING OF STONE FRUITS

Genetics, Genomics and Breeding of Crop Plants

Series Editor

Chittaranjan Kole
Department of Genetics and Biochemistry
Clemson University
Clemson, SC
USA

Books in this Series:

Published or in Press:

- Jinguo Hu, Gerald Seiler & Chittaranjan Kole: *Sunflower*
- Kristin D. Bilyeu, Milind B. Ratnaparkhe & Chittaranjan Kole: *Soybean*
- Robert Henry & Chittaranjan Kole: *Sugarcane*
- Kevin Folta & Chittaranjan Kole: *Berries*
- Jan Sadowsky & Chittaranjan Kole: *Vegetable Brassicas*
- James M. Bradeen & Chittaranjan Kole: *Potato*
- C.P. Joshi, Stephen DiFazio & Chittaranjan Kole: *Poplar*
- Anne-Françoise Adam-Blondon, José M. Martínez-Zapater & Chittaranjan Kole: *Grapes*
- Christophe Plomion, Jean Bousquet & Chittaranjan Kole: *Conifers*
- Dave Edwards, Jacqueline Batley, Isobel Parkin & Chittaranjan Kole: *Oilseed Brassicas*
- Marcelino Pérez de la Vega, Ana María Torres, José Ignacio Cubero & Chittaranjan Kole: *Cool Season Grain Legumes*
- Yi-Hong Wang, Tusar Kanti Behera & Chittaranjan Kole: *Cucurbits*
- Chittaranjan Kole & Albert G. Abbott: *Stone Fruits*
- Barbara E. Liedl, Joanne A. Labate, John R. Stommel, Ann Slade & Chittaranjan Kole: *Tomato*

GENETICS, GENOMICS AND BREEDING OF STONE FRUITS

Editors

Chittaranjan Kole
Department of Genetics and Biochemistry
Clemson University
Clemson, SC
USA

Albert G. Abbott
Department of Genetics and Biochemistry
Clemson University
Clemson, SC
USA

CRC Press
Taylor & Francis Group
Boca Raton London New York

CRC Press is an imprint of the
Taylor & Francis Group, an **informa** business
A SCIENCE PUBLISHERS BOOK

CRC Press
Taylor & Francis Group
6000 Broken Sound Parkway NW, Suite 300
Boca Raton, FL 33487-2742

The photographs of the stone fruits were kindly provided by Dr. Thomas Gradziel, University of California, Davis, USA (almond); Dr. Tomas Necas, Mendel University at Brno, Czech Republic (apricot); Dr. Daniel Sargent, Istituto Agrario S. Michele all'Adige, Italy (Cherry); Dr. Herman Silva, Universidad de Chile, Santiago, Chile (peach); and Dr. Teresa Barreneche, INRA UREF, Villenaved'OrnonCédex, France (plum).

Printed in the United States of America on acid-free paper
Version Date: 20120327

International Standard Book Number: 978-1-57808-801-0 (Hardback)

Library of Congress Cataloging-in-Publication Data

Genetics, genomics and breeding of stone fruits / editors: Chittaranjan Kole, Albert G. Abbott. -- 1st ed.
p. cm. -- (Genetics, genomics and breeding of crop plants)
ISBN 978-1-57808-801-0 (hardcover)
1. Stone fruit--Genetics. 2. Stone fruit--Genome mapping. 3. Stone fruit--Breeding. I. Kole, Chittaranjan. II. Abbott, Albert G. III. Series: Genetics, genomics and breeding of crop plants.
SB378.G46 2012
634--dc23

2012018333

Visit the Taylor & Francis Web site at
http://www.taylorandfrancis.com

CRC Press Web site at
http://www.crcpress.com

Science Publishers Web site at
http://www.scipub.net

Preface to the Series

Genetics, genomics and breeding has emerged as three overlapping and complimentary disciplines for comprehensive and fine-scale analysis of plant genomes and their precise and rapid improvement. While genetics and plant breeding have contributed enormously towards several new concepts and strategies for elucidation of plant genes and genomes as well as development of a huge number of crop varieties with desirable traits, genomics has depicted the chemical nature of genes, gene products and genomes and also provided additional resources for crop improvement.

In today's world, teaching, research, funding, regulation and utilization of plant genetics, genomics and breeding essentially require thorough understanding of their components including classical, biochemical, cytological and molecular genetics; and traditional, molecular, transgenic and genomics-assisted breeding. There are several book volumes and reviews available that cover individually or in combination of a few of these components for the major plants or plant groups; and also on the concepts and strategies for these individual components with examples drawn mainly from the major plants. Therefore, we planned to fill an existing gap with individual book volumes dedicated to the leading crop and model plants with comprehensive deliberations on all the classical, advanced and modern concepts of depiction and improvement of genomes. The success stories and limitations in the different plant species, crop or model, must vary; however, we have tried to include a more or less general outline of the contents of the chapters of the volumes to maintain uniformity as far as possible.

Often genetics, genomics and plant breeding and particularly their complimentary and supplementary disciplines are studied and practiced by people who do not have, and reasonably so, the basic understanding of biology of the plants for which they are contributing. A general description of the plants and their botany would surely instill more interest among them on the plant species they are working for and therefore we presented lucid details on the economic and/or academic importance of the plant(s); historical information on geographical origin and distribution; botanical origin and evolution; available germplasms and gene pools, and genetic and cytogenetic stocks as genetic, genomic and breeding resources; and basic information on taxonomy, habit, habitat, morphology, karyotype, ploidy level and genome size, etc.

Classical genetics and traditional breeding have contributed enormously even by employing the phenotype-to-genotype approach. We included detailed descriptions on these classical efforts such as genetic mapping using morphological, cytological and isozyme markers; and achievements of conventional breeding for desirable and against undesirable traits. Employment of the *in vitro* culture techniques such as micro- and megaspore culture, and somatic mutation and hybridization, has also been enumerated. In addition, an assessment of the achievements and limitations of the basic genetics and conventional breeding efforts has been presented.

It is a hard truth that in many instances we depend too much on a few advanced technologies, we are trained in, for creating and using novel or alien genes but forget the infinite wealth of desirable genes in the indigenous cultivars and wild allied species besides the available germplasms in national and international institutes or centers. Exploring as broad as possible natural genetic diversity not only provides information on availability of target donor genes but also on genetically divergent genotypes, botanical varieties, subspecies, species and even genera to be used as potential parents in crosses to realize optimum genetic polymorphism required for mapping and breeding. Genetic divergence has been evaluated using the available tools at a particular point of time. We included discussions on phenotype-based strategies employing morphological markers, genotype-based strategies employing molecular markers; the statistical procedures utilized; their utilities for evaluation of genetic divergence among genotypes, local landraces, species and genera; and also on the effects of breeding pedigrees and geographical locations on the degree of genetic diversity.

Association mapping using molecular markers is a recent strategy to utilize the natural genetic variability to detect marker-trait association and to validate the genomic locations of genes, particularly those controlling the quantitative traits. Association mapping has been employed effectively in genetic studies in human and other animal models and those have inspired the plant scientists to take advantage of this tool. We included examples of its use and implication in some of the volumes that devote to the plants for which this technique has been successfully employed for assessment of the degree of linkage disequilibrium related to a particular gene or genome, and for germplasm enhancement.

Genetic linkage mapping using molecular markers have been discussed in many books, reviews and book series. However, in this series, genetic mapping has been discussed at length with more elaborations and examples on diverse markers including the anonymous type 2 markers such as RFLPs, RAPDs, AFLPs, etc. and the gene-specific type 1 markers such as EST-SSRs, SNPs, etc.; various mapping populations including F_2, backcross, recombinant inbred, doubled haploid, near-isogenic and pseudotestcross; computer software including MapMaker, JoinMap, etc. used; and different types of genetic maps including preliminary, high-resolution, high-density, saturated, reference, consensus and integrated developed so far.

Mapping of simply inherited traits and quantitative traits controlled by oligogenes and polygenes, respectively has been deliberated in the earlier literature crop-wise or crop group-wise. However, more detailed information on mapping or tagging oligogenes by linkage mapping or bulked segregant analysis, mapping polygenes by QTL analysis, and different computer software employed such as MapMaker, JoinMap, QTL Cartographer, Map Manager, etc. for these purposes have been discussed at more depth in the present volumes.

The strategies and achievements of marker-assisted or molecular breeding have been discussed in a few books and reviews earlier. However, those mostly deliberated on the general aspects with examples drawn mainly from major plants. In this series, we included comprehensive descriptions on the use of molecular markers for germplasm characterization, detection and maintenance of distinctiveness, uniformity and stability of genotypes, introgression and pyramiding of genes. We have also included elucidations on the strategies and achievements of transgenic breeding for developing genotypes particularly with resistance to herbicide, biotic and abiotic stresses; for biofuel production, biopharming, phytoremediation; and also for producing resources for functional genomics.

A number of desirable genes and QTLs have been cloned in plants since 1992 and 2000, respectively using different strategies, mainly positional cloning and transposon tagging. We included enumeration of these and other strategies for isolation of genes and QTLs, testing of their expression and their effective utilization in the relevant volumes.

Physical maps and integrated physical-genetic maps are now available in most of the leading crop and model plants owing mainly to the BAC, YAC, EST and cDNA libraries. Similar libraries and other required genomic resources have also been developed for the remaining crops. We have devoted a section on the library development and sequencing of these resources; detection, validation and utilization of gene-based molecular markers; and impact of new generation sequencing technologies on structural genomics.

As mentioned earlier, whole genome sequencing has been completed in one model plant (Arabidopsis) and several economic plants (e.g., rice, poplar, peach, papaya, grapes, soybean and sorghum, etc. to name a few) and is progressing in an array of model and economic plants. Advent of massively parallel DNA sequencing using 454-pyrosequencing, Solexa Genome Analyzer, SOLiD system, Heliscope and SMRT have facilitated whole genome sequencing in many other plants more rapidly, cheaply and precisely. We have included extensive coverage on the level (national or international) of collaboration and the strategies and status of whole genome sequencing in plants for which sequencing efforts have been completed or are progressing currently. We have also included critical assessment of the impact of these genome initiatives in the respective volumes.

Comparative genome mapping based on molecular markers and map positions of genes and QTLs practiced during the last two decades of the last century provided answers to many basic questions related to evolution, origin and phylogenetic relationship of close plant taxa. Enrichment of genomic resources has reinforced the study of genome homology and synteny of genes among plants not only in the same family but also of taxonomically distant families. Comparative genomics is not only delivering answers to the questions of academic interest but also providing many candidate genes for plant genetic improvement.

The 'central dogma' enunciated in 1958 provided a simple picture of gene function—gene to mRNA to transcripts to proteins (enzymes) to metabolites. The enormous amount of information generated on characterization of transcripts, proteins and metabolites now have led to the emergence of individual disciplines including functional genomics, transcriptomics, proteomics and metabolomics. Although all of them ultimately strengthen the analysis and improvement of a genome, they deserve individual deliberations for each plant species. For example, microarrays, SAGE, MPSS for transcriptome analysis; and 2D gel electrophoresis, MALDI, NMR, MS for proteomics and metabolomics studies require elaboration. Besides transcriptome, proteome or metabolome QTL mapping and application of transcriptomics, proteomics and metabolomics in genomics-assisted breeding are frontier fields now. We included discussions on them in the relevant volumes.

The databases for storage, search and utilization on the genomes, genes, gene products and their sequences are growing enormously in each second and they require robust bioinformatics tools plant-wise and purpose-wise. We included a section on databases on the gene and genomes, gene expression, comparative genomes, molecular marker and genetic maps, protein and metabolomes, and their integration.

Notwithstanding the progress made so far, each crop or model plant species requires more pragmatic retrospect. For the model plants we need to answer how much they have been utilized to answer the basic questions of genetics and genomics as compared to other wild and domesticated species. For the economic plants we need to answer as to whether they have been genetically tailored perfectly for expanded geographical regions and current requirements for green fuel, plant-based bioproducts and for improvements of ecology and environment. These futuristic explanations have been addressed finally in the volumes.

We are aware of exclusions of some plants for which we have comprehensive compilations on genetics, genomics and breeding in hard copy or digital format and also some other plants which will have enough achievements to claim for individual book volume only in distant future. However, we feel satisfied that we could present comprehensive deliberations on genetics, genomics and breeding of 30 model and economic plants, and their groups in a few cases, in this series. I personally feel also happy that I could work with many internationally celebrated scientists who edited the book volumes on the leading plants and plant groups and included chapters authored by many scientists reputed globally for their contributions on the concerned plant or plant group.

We paid serious attention to reviewing, revising and updating of the manuscripts of all the chapters of this book series, but some technical and formatting mistakes will remain for sure. As the series editor, I take complete responsibility for all these mistakes and will look forward to the readers for corrections of these mistakes and also for their suggestions for further improvement of the volumes and the series so that future editions can serve better the purposes of the students, scientists, industries, and the society of this and future generations.

Science publishers, Inc. has been serving the requirements of science and society for a long time with publications of books devoted to advanced concepts, strategies, tools, methodologies and achievements of various science disciplines. Myself as the editor and also on behalf of the volume editors, chapter authors and the ultimate beneficiaries of the volumes take this opportunity to acknowledge the publisher for presenting these books that could be useful for teaching, research and extension of genetics, genomics and breeding.

Chittaranjan Kole

Preface to the Volume

Plants of the Rosaceae comprise many of the most important specialty crops worldwide. They are grown both for their fruits (peaches, plums, apples, cherries, strawberries, raspberries and others), for their lumber (black cherry) and for their ornamental value (roses). Collectively, it is one of the most important temperate region plant families. Within this broadly diverse plant family, the genus *Prunus* is comprised of a small number of tree species that produce drupe fruits (peaches, apricots, almonds, plums, cherries) containing a highly lignified seed capsule referred to as the "stone". These stone fruit species have been bred and grown by humankind for thousands of years and today are significant agricultural crops in many local economies worldwide providing important components to healthy diets. However, like most crop species, their growth, fruit production and sustainability are influenced by many different biotic and abiotic factors. Stone fruit breeding programs are continually confronted with the need to find genetic solutions to ever-changing problems posed by disease and pests (e.g., viruses, fungi, insects) and the ever-changing environmental landscape (e.g., drought, global warming, cold temperatures, marginalization of lands).

Until recently, the genetic understanding of characters important to stone fruit agriculture lagged well behind that of the large commodity crops such as maize, rice, soybean, etc. This was primarily due to the lack of significant public investment and the refractory nature of the plants for achieving high-resolution genetic manipulations (juvenility periods in years, requirements for large amounts of labor and space). However, with the shift in the marketplace focus in many countries from large-scale commodity crops to smaller specialty crops of increased nutritional value, the development of genetic resources in the stone fruit crops came to the forefront. In the last 8–10 years, significant investment by the worldwide agricultural community and the application of state of the art molecular genetics technologies, have both substantially contributed to the progress in development of the genetic and genomic resources for key stone fruit species. These resources are enabling discovery and marking of many key genetic loci that ultimately will contribute to advancing our understanding of the basic biology of these important tree fruit species, as well as, contribute to sustained breeding improvement of these important crops. It is the intent of this volume to highlight key achievements, both past and present, in the advancement of the stone fruit breeding, genetics and genomics that are paving the future road for stone fruit agriculture.

This volume presents 14 chapters contributed by 37 eminent scientists from seven countries including Argentina, Chile, France, Germany, Italy, Spain and USA.

We dedicate this work to all the young scientists (undergraduate students, graduate students, post-doctoral associates and other researchers) whose work has been and continues to be the foundation for our understanding of the genetics of the stone fruit species.

Contents

List of Contributors

Albert G. Abbott
Department of Genetics and Biochemistry, Clemson University, 116 Jordan Hall, Clemson, SC 29634, USA.
Tel: +1-864-656-3060
Email: *aalbert@clemson.edu*

María José Aranzana
IRTA, Centre de Recerca en Agrigenòmica CSIC-IRTA-UAB-UB, Campus UAB, Bellaterra (Cerdanyola del Vallès), 08193, Barcelona, Spain.
Tel: +34-935-636-600
Email: *mariajose.aranzana@irta.cat*

Pere Arús
IRTA, Centre de Recerca en Agrigenòmica CSIC-IRTA-UAB-UB, Campus UAB, Bellaterra (Cerdanyola del Vallès), 08193 Barcelona, Spain.
Tel:+34-935-636-600
Email: *Pere.Arus@irta.cat*

Maria Luisa Badenes
Instituto Valenciano de Investigaciones Agrarias, Apartado Oficial 46113 Moncada, Valencia, Spain.
Tel: +34-963424049
Email: *badenes_mlu@gva.es*

Teresa Barreneche
INRA, UR419, Unité de Recherches sur les Espèces Fruitières, BP 81, F-33140 Villenave d'Ornon, France.
Tel: +33-557-12 2-4 49
Email: *teresa.barreneche@bordeaux.inra.fr*

Claudio Bonghi
Department of Environmental Agronomy and Crop Science, University of Padova, Agripolis, Legnaro (PD) 35020, Italy.
Tel: +39-049-8272844
Email: *claudio.bonghi@unipd.it*

Stefano Cagnin
Department of Biology, University of Padova, Via U. Bassi 58/B, 35121 Padova, Italy and CRIBI Biotechnology Centre, University of Padova, Via U. Bassi 58/B, 35121 Padova, Italy.
Tel: +39-049-8276162
Email: *stefano.cagnin@unipd.it*

Michel Claverie
CIRAD, UMR DAP, Avenue Agropolis, TA A-96/03, F–34398 Montpellier, Cedex 5, France.
Tel: +33-4-67-53-81
Email: *michel.claverie@cirad.fr*

Maria Teresa Dettori
CRA—Fruit Tree Research Center, via di Fioranello 52, 00134 Roma, Italy.
Tel: +39-06-79348144
Email: *mariateresa.dettori@entecra.it*

Amit Dhingra
Department of Horticulture and Landscape Architecture and Molecular Plant Sciences Graduate Program, Washington State University, PO BOX 646414, Pullman, WA 99164, USA.
Tel: +1-509-335-3625
Email: *adhingra@wsu.edu*

Elisabeth Dirlewanger
INRA, UR419, Unité de Recherches sur les Espèces Fruitières (UREF), B.P. 81, F-33140 Villenave d'Ornon, France.
Tel: +33-5-57-12-24-61
Email: *dirlewan@bordeaux.inra.fr*

María F. Drincovich
Center of Photosynthetic and Biochemical Studies (CEFOBI), Faculty of Pharmaceutical and Biochemical Sciences, Rosario National University, Suipacha 531, Rosario, Argentina.
Tel: +54-341-4371955 ext. 104
Email: *drincovich@cefobi-conicet.gov.ar*

Daniel Esmenjaud
Equipe Interactions Plantes-Nématodes, INRA, Centre de recherche Provence-Alpes-Côte d'Azur, UMR ISA (Institut Sophia Agrobiotech), INRA/CNRS/Université de Nice, 400 route des Chappes, BP 167, 06903 Sophia Antipolis Cedex, France.
Tel: +33-4-92-38-64-60

Email: *daniel.esmenjaud@sophia.inra.fr*

Claudio Forcato
CRIBI Biotechnology Centre, University of Padova, Via U. Bassi 58/B, 35121 Padova, Italy.
Tel: +39-049-8276165
Email: *claudio.forcato@unipd.it*

Thomas M. Gradziel
Department of Plant Sciences, University of California-Davis, 1 Shields Avenue, Davis, CA 95616, USA.
Phone +1-530 752-1575
Email: *tmgradziel@ucdavis.edu*

Renate Horn
Institute of Biological Sciences, Department of Plant Genetics, University of Rostock, Albert-Einstein-Str. 3, 18051 Rostock, Germany.
Tel: +49-381-498-6170
Email: *renate.horn@uni-rostock.de*

Werner Howad
IRTA, Centre de Recerca en Agrigenòmica CSIC-IRTA-UAB-UB, Campus UAB, Bellaterra (Cerdanyola del Vallès), 08193 Barcelona, Spain.
Tel:+34-935-636-600
Email: *werner.howad@irta.cat*

Eudald Illa
IRTA, Centre de Recerca en Agrigenòmica CSIC-IRTA-UAB-UB, Campus UAB, Bellaterra (Cerdanyola del Vallès), 08193 Barcelona, Spain.
Tel: +34-935-636-600
Email: *eudald.illa@gmail.com*

Sook Jung
Department of Horticulture and Landscape, Washington State University, Pullman, WA 99164, USA.
Tel: +1-509-335-2774
Email: *sook_jung@wsu.edu*

Tyson Koepke
Department of Horticulture and Landscape Architecture and Molecular Plant Sciences Graduate Program, Washington State University, PO BOX 646414, Pullman, WA 99164, USA.
Tel: +1-509-335-6586
Email: *tkoepke@wsu.edu*

Chittaranjan Kole
Department of Genetics and Biochemistry, Clemson University, 109 Jordan Hall, Clemson, SC 29634, USA.
Tel: +1-864-656-5188
Email: *ckole@clemson.edu*

María V. Lara
Center of Photosynthetic and Biochemical Studies (CEFOBI), Faculty of Pharmaceutical and Biochemical Sciences, Rosario National University, Suipacha 531, Rosario, Argentina.
Tel: +54-341-4371955 ext. 104
Email: *lara@cefobi-conicet.gov.ar*

Dorrie Main
Department of Horticulture and Landscape, Washington State University, Pullman, WA 99164, USA.
Tel: +1-509-335-2774
Email: *dorrie@bioinfo.wsu.edu*

Jonathan Maldonado
Millennium Nucleus in Plant Cell Biotechnology and Functional Genomics & Bioinformatic Lab, Departamento de Producción Agrícola, Facultad de Ciencias Agronómicas, Universidad de Chile, Av. Santa Rosa 11315, 8820808 La Pintana, Casilla, 1004 Santiago, Chile.
Tel: +56-2-9785860 ext. 222
Email: *jomaldon@gmail.com*

Sabrina Micali
CRA—Fruit Tree Research Center, via di Fioranello 52, 00134 Roma, Italy.
Tel: +39-06-79348150
Email: *sabrina.micali@entecra.it*

Bode A. Olukolu
Department of Plant Pathology, North Carolina State University, 112 Derieux Place, Raleigh, NC 27607, USA.
Tel: +1-919-515-7376
Email: *baolukol@ncsu.edu*

Daniel Potter
Department of Plant Sciences, Mail Stop 2, University of California, 1 Shields Avenue, Davis, CA 95616, USA.
Tel: +1-530-754-6141
Email: *dpotter@ucdavis.edu*

Loreto Prat
Millennium Nucleus in Plant Cell Biotechnology and Functional Genomics & Bioinformatic Lab, Departamento de Producción Agrícola, Facultad de Ciencias Agronómicas, Universidad de Chile, Av. Santa Rosa 11315, 8820808 La Pintana, Casilla 1004, Santiago, Chile.
Tel: +56-2-9785723 ext. 203
Email: *mlprat@uchile.cl*

Osama Sajer
Institute of Biological Sciences, Department of Plant Genetics, University of Rostock, Albert-Einstein-Str. 3, 18051 Rostock, Germany.
Tel: +49-381-498-6170
Email: *osama.sajer@hotmail.de*

Daniel James Sargent
Fondazione Edmund Mach Istituto Agrario S. Michele all'Adige, Via E. Mach, 1 38010 S. Michele all'Adige, Italy.
Tel: +39-0461- 615536
Email: *dan.sargent@iasma.it*

Herman Silva
Millennium Nucleus in Plant Cell Biotechnology and Functional Genomics & Bioinformatic Lab, Departamento de Producción Agrícola, Facultad de Ciencias Agronómicas, Universidad de Chile, Av. Santa Rosa 11315, 8820808 La Pintana, Casilla 1004, Santiago, Chile.
Tel: +56-2-9785726
Email: *hesilva@uchile.cl*

Jose Miguel Soriano
Instituto Valenciano de Investigaciones Agrarias, Apartado Oficial 46113 Moncada, Valencia, Spain.
Tel: +34-963424000
Email: *soriano_jmi@gva.es*

Bryon Sosinski
Department of Horticultural Science, North Carolina State University, Raleigh, NC, USA.
Tel: +1-919-513-3883
Email: *bryon_sosinski@ncsu.edu*

Chinnathambi Srinivasan
USDA-ARS Appalachian Fruit Research Station, 2217 Wiltshire Road, Kearneysville, WV 25430, USA.
Tel: +1-304-725-3451 x 316
Email: *chinnathambi.srinivasan@ars.usda.gov*

Livio Trainotti
Department of Biology, University of Padova, Via U. Bassi 58/B, 35121 Padova, Italy.
Tel: +39-049-8276292
Email: *livio.trainotti@unipd.it*

Elisa Vendramin
CRA—Fruit Tree Research Center, via di Fioranello 52, 00134 Roma, Italy.
Tel: +39-06-79348193
Email: *elisa.vendramin@entecra.it*

Ignazio Verde
CRA—Fruit Tree Research Center, via di Fioranello 52, 00134 Roma, Italy.
Tel:+39-06-79348183
Email: *ignazio.verde@entecra.it*

Abbreviations

1-DE	One-dimensional electrophoresis
1-MCP	1-Methylcyclopropene
2-DE	Two-dimensional electrophoresis
2D-PAGE	Two-dimensional-PAGE
ABA	Abscisic acid
ABI3	*ABSCISIC ACID INSENSITIVE3*
ACC	1-Aminocyclopropane-1-carboxylic acid
ACO	1-Aminocyclopropane-1-carboxylate oxidase
AFLP	Amplified fragment length polymorphism
ANOVA	Analysis of variance
BAC	Bacterial artificial chromosome
BC	Backcross
BD	Blooming date
BIBAC	Binary-BAC
BIND	Biomolecular Interaction Network Database
BLAST	Basic local alignment search tool
BLAT	BLAST-like alignment tool
BN-PAGE	Blue-native-PAGE
BSA	Bulked segregant analysis
BTL	Binary trait loci
C×A	Contender × Ambra
CAPS	Cleaved amplified polymorphic sequence
CC	Coil-coiled
cDNA	Complementary-DNA
CDS	Coding sequence
CEMS	Capillary electrophoresis-MS
CG	Candidate gene
CI	Chilling injury
CIM	Composite interval mapping
cM	CentiMorgan
CNL	Coil-coiled-NBS-LRR
CP	Coat protein
CR	Chilling requirement
CUGI	Clemson University Genomics Institute
DAH	Days after harvest

DAM	Dormancy-associated MADS-box
DAS	Double-antibody sandwich
DD	Differential display
DDBJ	DNA Data Bank of Japan
DIGE	Difference in gel electrophoresis
DRS	Direct RNA sequencing
DUS	Distinctness, uniformity and stability
ECPGR	European Cooperative Programme for Crop Genetic Resources
EFA	EST frequency analysis
eIF	Eukaryotic translation initiation factor
ELISA	Enzyme-linked immunosorbent assay
EMBL	European Molecular Biology Laboratory
ENA	European Nucleotide Archive
endoPG	Endopolygalacturonase
EPDB	European *Prunus* Database
eQTL	Expression QTL
ESFY	European stone fruit yellows
ESI-MS/MS	Electrospray ionization tandem-MS
EST	Expressed sequence tag
EVG	Evergrowing locus
F×T	Fiesta × Totem
F_1	First filial generation
F_2	Second filial generation
FAO	Food and Agriculture Organization
FG	*Fragaria* linkage group
FPC	Fingerprinted contig
FRT-seq	Flowcell reverse-transcription sequencing
FV×FN	*Fragaria vesca* × *Fragaria nubicola*
GBS	Genotyping-by-sequencing
GDR	Genomic Database for Rosaceae
GEO	Gene expression omnibus
GFP	Green fluorescence protein
GM	Genetic modification
GMO	Genetically-modified organism
GO	Gene ontology
GRIN	Germplasm Resources Information Network
GSA	Genome scanning approach
GUS	ß-Glucuronidase
GWAS	Genome-wide association study
GxE	Genetic x environment interaction
HC	High cot (sequencing)
HICF	High-information content fingerprinting

HLR	Hypersensitive-like reaction
HR	Heat requirement
HSP	Heat shock protein
HT	Heat treatment
IEF	Isoelectric focusing
IGA	Istituto di Genomica Applicata
Ihp	Intron-hairpin
ihpRNA	Intron-hairpin-RNA
InDels	Insertion/deletion
INRA	Institut National de la Recherche Agronomique
IPG	Immobilized pH gradient
IPGI	International Peach Genome Initiative
IT	Information technology
ITS	Internal transcribed spacer
JA	Jasmonate
JGI	Joint Genome Institute
KEGG	Kyoto Encyclopedia of Genes and Genomes
KOG	Eukaryotic orthologous groups
LC	Liquid chromatography
LD	Linkage disequilibrium
LecPK	Lectin protein kinase
LG	Linkage group
LOX	Lipoxygenase
LQ-RNA-seq	Low quantity RNA sequencing
LRR	Leucine-rich repeat
LTR	Long terminal repeat
MAB	Marker-assisted breeding
MAF	Minimal allele frequency
MALDI-TOF MS	Matrix-assisted laser desorption-ionization time-of-flight MS
MAS	Marker-assisted selection
MASCP	Multinational Arabidopsis Steering Committee Proteomics Subcommittee
MBC	Map-based cloning
MDR	Multiple disease resistance genes
MF	Methylation filtration
MG	*Malus* linkage group
MIM	Multiple interval mapping
MPK7	*MITOGEN-ACTIVATED PROTEIN KINASE7*
MPSS	Massively parallel signature sequencing
MQM	Multiple QTL mapping
MS	Mass spectrometry
MT	Metallothionein

MTL	Mendelian trait loci
MTP	Minimal tiling path
NBS	Nucleotide binding site
NCBI	National Center for Biotechnology Information
NCED	9-cis-epoxycarotenoid dioxygenase
NGS	Next-generation sequencing
NIAS	National Institute of Agrobiological Sciences
NIL	Near isogenic lines
NMR	Nuclear magnetic resonance
NPGS	National Plant Germplasm System
NPT	Neomycinephosphotransferase
ORF	Open reading frame
Overgo	Overlapping oligonucleotide
P×F (*Malus*)	Prima × Fiesta
P×F (*Prunus*)	*P. ferganensis* × IF7310828
P3DB	Plant Protein Phosphorylation Data Base
PA×PN	*P. avium* × *P. nipponica*
PAGE	Polyacrylamide gel electrophoresis
PANTHER	Protein analysis through evolutionary relationships
PCA	Principal component analysis
PCR	Polymerase chain reaction
PDC	Pyruvate decarboxylase
PDS	Phytoenedesaturase
PFAM	Protein Family Database
PG	Polygalacturonase
PG	*Prunus* linkage group
PGDD	Plant Genome Duplication Database
PL	Post-LRR
PME	Pectin methylesterase
PMF	Peptide mass fingerprinting
PPO	Polyphenol oxidase
PPV	Plum pox virus
PR	Pathogen-related
PTGS	Post-transcriptional gene silencing
PTSL	Peach tree short life
qRT-PCR	Quantitative RT-PCR
QTA	Quantitative trait allele
QTL	Quantitative trait loci
qTOF	Quadrupole/time-of-flight
RACE	Rapid amplification of cDNA ends
RAPD	Random amplified polymorphic DNA
RCSB	Resource for Studying Biological Macromolecules
RefSeq	NCBI Reference Sequence

RFLP	Restriction fragment length polymorphism
RGA	Resistance gene analog
RIL	Recombinant inbred line
RKN	Root-knot nematode
RNAseq	RNA sequencing
ROS	Reactive oxygen species
ROSCOS	Rosaceae conserved orthologous set
RRL	Reduced representation library
RT-PCR	Reverse transcriptase-PCR
SAGE	Serial analysis of gene expression
SALAD	Surveyed contained motif alignment diagram and associating dendrogram
SC	Self-compatible
SCAR	Sequence characterized amplified region
SCOP	Structural Classification of the Protein Database
SDFR	Single dose restriction fragment
SDS	Sodium dodecyl sulfate
SFB	S-locus F-box
SGN	Solanaceae Genomics Network
SGS	Second-generation sequencing
SI	Self-incompatibility
SI	Self-incompatible
SIM	Simple interval mapping
siRNA	Small interfering RNA
SLR	Slow ripening
SMRT	Single-molecule real time
smsDGE	Single-molecule sequencing-direct gene expression
SNP	Single nucleotide polymorphism
SQR-RNA seq	Short quantitative random RNA sequencing
SRA	Short read archives
SSH	Suppression subtractive hybridization
SSR	Simple sequence repeat
ST	Standard
STS	Sequence tagged site
SUBA	Arabidopsis Subcellular Database
T×E	Texas × Earlygold
TA	Titratable acid
TAC	Transformation-competent artificial chromosome
TAIL	Thermal asymmetric interlaced
TAIR	The Arabidopsis Information Resource
TDF	Transcript-derived fragments
TGS	Third-generation sequencing

TIR	Toll/interleukin receptor
TNL	Toll/interleukin receptor-NBS-LLR
TrEMBL	Translated European Molecular Biology Laboratory Nucleotide Sequence Database
tSMS	True Single Molecule Sequencing
UniProt	Universal Protein Resource
UP	Upright
USRosEXEC	US Rosaceae Genomics, Genetics and Breeding Executive Committee
UTR	Untranslated region
WGS	Whole-genome shotgun
WTSS	Whole-transcriptome shotgun sequencing
wwPDB	Worldwide Protein Data Bank
XET	Xyloglucanendotransglycosylase
YAC	Yeast artificial chromosome

1

Basic Information on the Stone Fruit Crops

Daniel Potter

ABSTRACT

The stone fruits are species of the large widely distributed genus *Prunus* (Rosaceae); they include apricots, cherries, peaches, nectarines, and plums, cultivated primarily for their edible fleshy mesocarps, and almonds, cultivated primarily for their edible seeds. *Prunus* comprises about 230 species and includes, besides the stone fruits, many taxa cultivated as ornamentals and many wild species of local economic importance, as well as wild crop relatives that have been useful in research and breeding programs for the major crops. The stone fruit species are native to temperate regions of the northern hemisphere, but *Prunus* also includes about 35 species native to the Old World tropics and 25 native to the New World tropics. *Prunus* species are trees and shrubs with typically showy 5-merous flowers with a single carpel that matures into a drupe with a fleshy mesocarp and a hard endocarp enclosing a singe seed. The major stone fruit crop species all originated in Asia or Europe but they have been widely disseminated by people throughout the world and major producers are now found on all continents except Antarctica. Nutritionally, stone fruits are good sources of vitamins and minerals, and there is increasing interest in their potential value as nutraceuticals due to the presence of phenolic compounds with antioxidant properties. Because of the large number of cultivated species and their economic importance and the diversity and widespread distribution of wild species, *Prunus* has been the subject of extensive basic and applied research, and significant germplasm and genomics resources are available.

Keywords: germplasm resources, infrageneric classification, morphology, phylogeny, taxonomy

Department of Plant Sciences Mail Stop 2, University of California, 1 Shields Avenue, Davis, CA 95616, USA; e-mail: *dpotter@ucdavis.edu*

1.1 Introduction

Stone fruits—almonds, apricots, cherries, peaches and nectarines, and plums (Table 1-1)—belong to the large cosmopolitan genus *Prunus* L., which in turn is classified, along with strawberries, raspberries, blackberries, apples, pears, and roses, in the large and diverse family Rosaceae. All of the major fruit crop species in *Prunus* originated in temperate regions of the Old World, in areas from eastern Europe to eastern Asia, but are now cultivated throughout the world and are of enormous economic importance. Each category of stone fruits also includes several minor crops and/or wild species, many of which have been important to some degree in the genetic improvement of the main crop species. In addition to the stone fruits, the genus includes a significant number of species cultivated as ornamentals, several that are valued for their wood, and at least one wild species (*P. africana*) of global economic importance due to its medicinal properties, as well as others that have traditional local medicinal uses in their native regions. This chapter presents a brief description of the botany of the stone fruit species, including their morphology and systematics, an overview of their status as world crops, including their origins and histories, data on their nutritional value and current economic significance, a review of topics of research interest and a summary of available germplasm for *Prunus*.

Table 1-1 Species of *Prunus* cultivated for edible fruits and seeds.

Species	Common Name	Ploidy	Native Region
	Almond		
P. dulcis (Miller) D.A. Webb	Almond	2×	Central Asia
	Apricots		
P. armeniaca L.	Apricot	2×	Asia
	Cherries		
P. avium L.	Sweet cherry	2×	Southwestern Asia
P. cerasus L.	Sour cherry	4×	Southwestern Asia
P. fruticosa Pall.	Ground cherry	4×	Central Europe to Siberia
	Peaches		
P. persica (L.) Batsch	Peach, Nectarine	2×	China
	Plums		
P. domestica L. subsp. *domestica*	European plum	6×	Europe
P. domestica L. subsp. *insititia* (L.) C. K. Schneid.	Bullace	6×	Europe
P. spinosa L.	Sloe; blackthorn	4×	Europe
P. cerasifera Ehrh.	Cherry-plum	2×	Western-Central Asia
P. salicina Lindl.	Japanese plum	2×	Japan
P. americana Marshall	American plum	2×	Eastern North America

1.2 Botany

1.2.1 General Description

Prunus, a genus of about 230 species (Table 1-2), is distributed throughout the world, with the greatest native diversity in Asia. The group is generally considered to have a primarily north temperate distribution, and indeed roughly two thirds to three quarters of the described species are native to temperate regions in Asia, Europe, or North America, but there is significant representation of the genus in tropical regions of both the Old (Southeast Asia) and New (Central and South America and the Caribbean) Worlds and the diversity of the genus in some of those areas is still poorly known. As might be expected given the large size and widespread distribution of the genus, *Prunus* also exhibits considerable ecological diversity, with different species occurring from lowlands to alpine zones and in wet and dry forests as well as deserts.

Species of *Prunus* are woody plants, ranging from small shrubs to very large trees, which may be either deciduous or evergreen. The leaves are alternate and simple, pinnately veined, and usually glabrous but pubescent in a few species; stipules are present but fall early. The leaf margins range from serrate to entire; the leaf teeth are generally glandular, and glands also occur on the petioles or leaf blade bases of most species. The inflorescence ranges from a solitary flower to an umbel-like cluster or a raceme. The flowers are radially symmetric and have well-developed hypanthia, as is characteristic of all Rosaceae; there are five sepals and five petals that vary in color from white to pink or red. The flowers are usually perfect, with ten or more stamens and a single, unicarpellate pistil with a superior ovary that matures into a drupe. The mesocarp varies from fleshy and juicy, as in the most of the stone fruit species and their wild relatives, to dry and/or leathery; the endocarp (stone) is hard and contains a single seed. As in many other Rosaceae, the primary transport carbohydrate in species of *Prunus* is the sugar alcohol sorbitol (Zimmermann and Ziegler 1975; Moing et al. 1997), and the plants produce cyanogenic glycosides that result in a characteristic acrid odor when the plant is wounded. These compounds also make the seeds of many species toxic, but the seeds of some species—most notably, the cultivated almond—are edible and highly prized for their flavor and nutritional value. As in several other genera of Rosaceae, polyploidy and interspecific hybridization are common.

1.2.2 Taxonomy and Phylogeny

Position in Rosaceae. Based on its base chromosome number ($x = 8$) and fruit type (drupe), both of which are quite rare in Rosaceae, *Prunus* has often

Table 1-2 Summary of Rehder's (1940) infrageneric classification of *Prunus* L. and modifications.

Infrageneric Taxon	Common Name(s)	Approximate Number of Species	Distribution
Subgenus *Prunus* (*Prunophora* Focke in Rehder)	plums, apricots	45	north temperate
Section *Prunus* (*Euprunus* Koehne in Rehder)	Eurasian plums	12	Europe, Asia
Section *Prunocerasus* Koehne	North American plums	20	North America
Section *Armeniaca*	apricots	10	Europe, Asia
Section *Penarmeniaca* Mason[1]	desert peach, desert apricot	2	North America
Section *Piloprunus* Mason[1]	wild peach	1	North America
Subgenus *Amygdalus* (L.) Focke	peaches, almonds	25	Europe, Asia
Section *Amygdalus* (*Euamygdalus* Schneid. in Rehder)	peaches, almonds	22	Asia
Section *Chamaeamygdalus* Focke		3	Europe, Asia
Subgenus *Cerasus* Pers.	cherries	55	north temperate
Section *Microcerasus* Webb.		12	north temperate
Section *Pseudocerasus* Koehne	flowering cherries	20	Asia
Section *Lobopetalum* Koehne		5	Asia
Section *Cerasus* (*Eucerasus* Koehne in Rehder)		3	Europe, Asia
Section *Mahaleb* Focke		5	north temperate
Section *Phyllocerasus* Koehne		5	Asia
Section *Phyllomahaleb* Koehne		5	Asia
Subgenus *Padus* (Moench) Koehne	bird-cherries	25	north temperate
Subgenus *Laurocerasus* Koehne	laurel-cherries	75	global
Section *Laurocerasus*[2]		15	Old World
Section *Mesopygeum* (Koehne) Kalkm.[2]		35	Old World tropics
Unnamed section-primarily tropical America[2]		25	New World
Subgenus *Emplectocladus* (Torr.) Sargent[1]	desert almond	5	North America

[1]following Mason (1913)
[2]following Kalkman (1964)

been classified in a subfamily (Prunoideae or Amygdaloideae) with the few other genera in the family that exhibit one or both of those characters. Two of these other genera—*Pygeum* Gaertn. and *Maddenia* Hook. f. & Thomson—have been shown to be nested within *Prunus* in recent molecular phylogenetic analyses of the genus (e.g., Wen et al. 2008) and the species of both have been taxonomically transferred to *Prunus* (Kalkman 1965; Chin et al. 2010). In contrast, molecular phylogenetic studies of Rosaceae (e.g., Potter et al. 2007) have shown that the other genera sometimes classified in Amygdaloideae—*Exochorda* Lindl., *Oemleria* Rchb., and *Prinsepia* Royle —form a well-supported clade, but that clade is not the sister-group to *Prunus*. Consequently, in the most recent infrafamilial classification of Rosaceae, Potter et al. (2007) included *Exochorda, Prinsepia,* and *Oemleria* in tribe Osmaronieae and placed *Prunus* by itself in tribe Amygdaleae; both tribes were included within an expanded subfamily Spiraeoideae, though recent changes to the International Code of Nomenclature for Algae, Fungi and Plants dictate that the name Amygdaloideae should be used for this group.

Infrageneric Classification and Relationships. In addition to the former separation of the genera *Pygeum* and *Maddenia,* discussed above, the rest of *Prunus* has historically been divided into anywhere from two to six genera by various botanists (Bortiri et al. 2001). Molecular phylogenetic studies, however, do not provide support for these divisions, and instead indicate that a broad circumscription of *Prunus* (e.g., Rehder 1940) is most appropriate.

Rehder (1940) divided *Prunus* into five subgenera, several of which he further divided into sections (Table 1-2). Three of Rehder's subgenera correspond to broad categories of stone fruits. His subgenus *Prunophora* (correctly, subgenus *Prunus*), characterized by sulcate fruits with bloomy (glaucous) surfaces and stems with solitary axillary buds and no terminal buds, includes plums and apricots (section *Armeniaca*), the latter characterized by usually pubescent fruits, the former separated into Eurasian (section *Euprunus* or *Prunus*) and North American (section *Prunocerasus*) subgroups. Species of subgenus *Amygdalus* (peaches and almonds) also have sulcate fruits but their stems have three axillary buds and terminal buds are present. In subgenus *Cerasus*—cherries—the fruits are neither sulcate nor bloomy; Rehder divided this subgenus into seven sections.

These first three subgenera have flowers that are either solitary or borne in umbel-like clusters, while the remaining two have flowers in elongated racemes, with the peduncles leafy in subgenus *Padus* and leafless in subgenus *Laurocerasus*. No major fruit crop species is included in either of the last two subgenera, though the fruits of *P. serotina* are consumed in Mexico (Fresnedo-Ramírez et al. 2011). Most of the species of subgenus *Laurocerasus* bear fruits with thin, dry to leathery mesocarps (Kalkman 1965).

Mason (1913) erected the subgenus *Emplectocladus* to accommodate several species native to arid regions of western North America (e.g., *Prunus fasciculata* (Torr.) A. Gray, desert almond) that have the unique character within *Prunus* of being dioecious. Rehder (1940) included *P. fasciculata* in subgenus *Amygdalus*, but molecular phylogenetic studies (e.g., Chin et al. 2010) resolve *Emplectocladus* and *Amygdalus* as distinct lineages.

Molecular phylogenetic studies of *Prunus* conducted over the last decade (Bortiri et al. 2001; Lee and Wen 2001; Shaw and Small 2005; Bortiri et al. 2006; Wen et al. 2008; Chin et al. 2010) have resulted in several conclusions about relationships across the genus (summarized diagrammatically in Fig. 1-1). Neither subgenus *Padus* nor subgenus *Laurocerasus*, as formerly circumscribed, is monophyletic, but those two subgenera and the species formerly classified in *Maddenia* together form a well-supported clade, within which several geographically distinct lineages are resolved. Most of subgenus *Cerasus* forms a strongly supported clade. Members of the remaining subgenera together form a moderately well-supported monophyletic group, within which subgenus *Emplectocladus* is sister to a strongly supported clade of subgenera *Prunus* and *Amygdalus*, each of which is moderately well-supported with some modifications from Rehder's (1940) treatment (specifically, each of sections *Microcerasus* and *Chamaeamygdalus* is more closely related to members of subgenus *Prunus* than to members

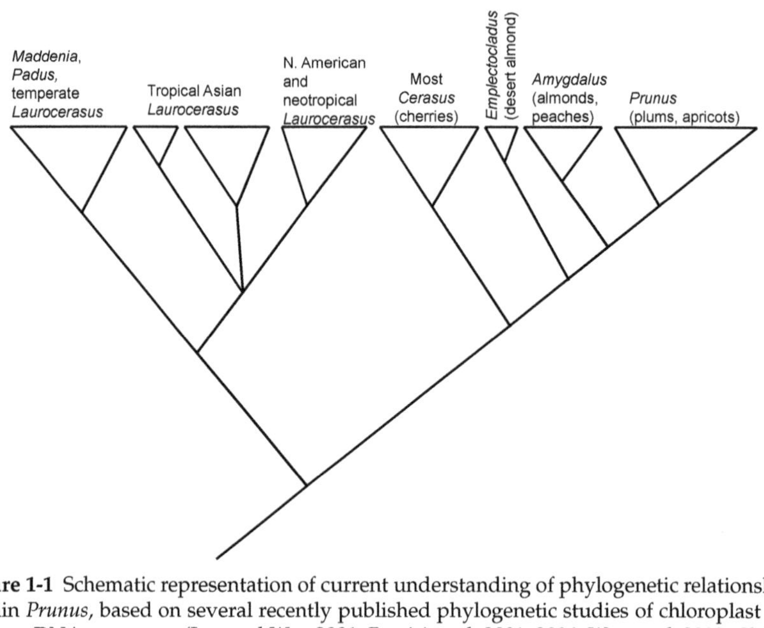

Figure 1-1 Schematic representation of current understanding of phylogenetic relationships within *Prunus*, based on several recently published phylogenetic studies of chloroplast and nuclear DNA sequences (Lee and Wen 2001; Bortiri et al. 2001, 2006; Wen et al. 2008; Chin et al. 2010). Subgeneric names refer to those listed in Table 1-2.

of the respective subgenus in which Rehder originally classified it). The position of the *Cerasus* group has varied among studies to date. Analyses based on chloroplast DNA markers (e.g., Bortiri et al. 2001; Wen et al. 2008) and combined data from nuclear and chloroplast DNA and morphology (Bortiri et al. 2006) have supported a sister relationship between the *Cerasus* and the *Amygdalus-Prunus* clade, as in Fig. 1-1, while analyses based on nuclear internal transcribed spacer (ITS) sequences alone (e.g., Bortiri et al. 2001; Lee and Wen 2001; Chin et al. 2010) have instead supported a sister relationship between the *Cerasus* and the *Padus-Laurocerasus* clades. These intriguing results suggest the possibility that hybridizations were involved in the early diversification of the genus. Moreover, the groups resolved in molecular phylogenetic analyses coupled with the observation that many of the characters that have been important in previous taxonomic treatments of *Prunus* exhibit considerable homoplasy within the genus (Bortiri et al. 2006) highlight the need for a revised infrageneric classification for this very important group.

1.3 Stone Fruits as Global Crops

1.3.1 Histories and Origins

Species of *Prunus* have been cultivated and highly appreciated for their edible fruits by people in Asia and Europe for thousands of years and the major cultivated species have been widely disseminated by people across the globe over many centuries. In addition, the fruits of wild species have been locally important in Asia, Europe, and North America. Pinpointing the precise geographic origins of cultivated *Prunus* species can be extremely challenging for two primary reasons. First, as a result of their long history of cultivation and human-mediated dispersal, stone fruit species have had numerous opportunities to become naturalized over broad geographic ranges, rendering it impossible in many cases to distinguish between ancestrally wild populations and those that have become established as a result of escape from cultivation. Second, interspecific hybridizations have played a role in the development of many of the cultivated varieties, such that each of them is derived from multiple progenitors. In most cases, only broad areas of origin can be designated with certainty (Table 1-1). Nonetheless, there has been considerable interest in tracking the origins and dissemination of the stone fruit species, and many insights have been gained.

Candolle (1886) provided very thorough accounts of understanding of the origins and histories of the major cultivated *Prunus* species. Based on a combination of botanical, archaeological, and historical evidence, he deduced the origin and history of spread of each species. Most of the conclusions he reached have been upheld and in some cases refined by more recent analyses of molecular genetic data.

Cultivated almond (*Prunus dulcis*) is believed to have originated in Western-Central Asia, more precisely in the area that is now Iran, where the species now generally accepted to be its wild progenitor, *P. fenzliana* Fritsch, is native, but hybridizations with other wild species were also important in the development of modern cultivated almond (Zeinalabedini et al. 2010). Apricot (*P. armeniaca*) provides an excellent example of a case in which determining the origin of the species was hampered by the fact that it was widely disseminated by people early in human history. The Latin name reflects the belief by early botanists, and maintained by some into the 20th century (Faust et al. 1998), that the species was native to Armenia, where it was cultivated as early as the first century AD (Ledbetter 2008), but a Chinese origin is now generally accepted (Candolle 1886; Faust et al. 1998).

Sweet cherry, *Prunus avium*, is considered to be native to an area of western Asia south of the Black and Caspian Seas, including modern Armenia, Iran, Iraq, and Syria (Candolle 1886; Vavilov 1951; Iezzoni 2008) and likely became naturalized throughout a large area in Asia and Europe in prehistoric times (Candolle 1886). The tetraploid sour cherry, *P. cerasus*, comes from the same region and is believed to have originated from hybridization between diploid *P. avium* and another tetraploid species, *P. fruticosa* (Oldén and Nybom 1968).

Peach (*P. persica*), like apricot, originated in China but has a misleading Latin name, reflecting its early dissemination from China to Persia, from where it was subsequently introduced to Greece and Rome in the first or second century BC (Candolle 1886; Faust and Timon 1995). Nectarines, glabrous-skinned varieties of *P. persica*, also originated in China (Faust and Timon 1995).

The major cultivated plum species are European plum, *P. domestica*, and Japanese plum, *P. salacina*. The latter species is in fact native to China but extensively cultivated in Japan (Rehder 1940), while the former, a hexaploid species whose geographic and genealogical origins have been somewhat controversial, is now generally believed to be derived from the European tetraploid species *P. spinosa*, which in turn is an allopolyploid that originated from hybridization between the diploid *P. cerasifera* and another species (Okie and Hancock 2008).

1.3.2 Current Production and Monetary Value

Today, the stone fruits are major global crops. The top 20 countries in the world for area harvested, quantity produced, and yield of each major category of stone fruits (almonds, apricots, cherries, peaches and nectarines,

Table 1-3 Production of stone fruits by various countries in 2009 (values from FAO 2011).

Top 20 Countries: Area Harvested	Area Harvested (Ha)	Top 20 Countries: Production	Production (Tonnes)	Top 20 Countries: Yield	Yield (Hg/Ha)
Almonds					
Spain	650,000	USA	1,162,200	UAR	212,500
USA	287,330	Spain	282,100	Lebanon	47,656
Tunisia	190,000	Iran	128,464	USA	40,448
Morocco	147,696	Italy	113,700	Afghanistan	39,154
Iran	140,108	Morocco	104,115	Israel	34,594
Italy	80,300	Syria	97,002	Turkey	32,185
Libya	50,070	Tunisia	60,000	Jordan	32,044
Syria	42,400	Turkey	54,844	Greece	30,193
Algeria	39,313	Algeria	47,393	Kazakhstan	29,903
Portugal	38,444	Greece	44,273	Uzbekistan	29,344
Turkey	17,040	Afghanistan	43,183	Chile	28,985
Australia	15,800	China	35,000	China	28,000
Greece	14,663	Lebanon	30,500	Pakistan	24,000
China	12,500	Pakistan	26,400	Syria	22,877
Afghanistan	11,029	Libya	25,000	Kyrgyzstan	22,823
Pakistan	11,000	Chile	20,000	Argentina	19,440
Chile	6,900	Australia	18,957	Croatia	18,350
Lebanon	6,400	Uzbekistan	18,610	Macedonia	16,612
Uzbekistan	6,342	Portugal	12,454	Occ. Palestinian Terr.	16,456
Tajikistan	4,651	Israel	11,762	Italy	14,159
World Total	1,796,475	World Total	2,361,676	World Total	13,146

Table 1-3 contd....

Table 1-3 contd....

Top 20 Countries: Area Harvested	Area Harvested (Ha)	Top 20 Countries: Production	Production (Tonnes)	Top 20 Countries: Yield	Yield (Hg/Ha)
Apricots					
Turkey	59,000	Turkey	695,364	Austria	477,690
Iran	53,107	Iran	397,749	Slovenia	192,500
Uzbekistan	42,000	Uzbekistan	290,000	Malta	175,000
Algeria	34,119	Italy	233,600	Turkmenistan	149,754
Pakistan	30,206	Algeria	202,806	Greece	141,362
China	22,349	Pakistan	193,936	Switzerland	141,255
Italy	18,400	France	190,382	France	135,822
Spain	18,000	Morocco	122,798	Italy	126,956
Russia	18,000	Ukraine	115,800	Romania	126,062
Japan	18,000	Japan	115,200	Ukraine	125,869
Egypt	15,000	Egypt	100,000	USA	124,009
France	14,017	Syria	98,913	Turkey	117,858
Syria	13,563	Spain	97,100	Morocco	107,963
Morocco	11,374	China	89,890	Argentina	106,234
Tajikistan	9,800	Greece	77,000	South Africa	98,444
Ukraine	9,200	USA	61,980	Canada	92,222
Afghanistan	8,170	Russia	53,000	Portugal	88,626
Kyrgyzstan	8,000	Afghanistan	46,895	Chile	88,180
Tunisia	7,900	South Africa	46,269	Azerbaijan	84,227
Bulgaria	7,200	Lebanon	35,251	Macedonia	80,163
World Total	503,808	World Total	3,728,083	World Total	73,998

Cherries					
Turkey	37,900	Turkey	417,694	Slovenia	359,181
USA	34,524	USA	390,000	Switzerland	219,098
Italy	30,000	Iran	225,000	USA	112,964
Iran	29,935	Italy	116,200	Austria	112,133
Russia	27,000	Spain	96,400	Turkey	110,209
Spain	26,000	Syria	78,289	Armenia	107,142
France	14,640	Russia	69,000	Canada	102,743
Syria	13,073	Romania	67,874	Romania	99,144
Ukraine	12,600	Uzbekistan	67,000	Uzbekistan	83,750
Chile	12,500	Chile	56,000	Iran	75,162
Bulgaria	12,500	France	53,577	Australia	72,899
Poland	10,625	Ukraine	53,000	Germany	72,542
Greece	10,475	Poland	50,505	Israel	71,174
Serbia	9,108	Greece	48,051	Peru	69,736
Lebanon	9,039	Germany	39,463	Mexico	66,506
Uzbekistan	8,000	Lebanon	34,662	Kazakhstan	65,000
China	7,000	Austria	30,276	Syria	59,886
Romania	6,846	Serbia	29,228	Occ. Palestinian Terr.	54,795
Portugal	6,258	China	27,000	Denmark	53,809
Bosnia	5,595	Japan	18,000	Albania	51,766
World Total	381,482	World Total	2196,537	World Total	57,579

Table 1-3 contd....

Table 1-3 contd....

Top 20 Countries: Area Harvested	Area Harvested (Ha)	Top 20 Countries: Production	Production (Tonnes)	Top 20 Countries: Yield	Yield (Hg/Ha)
Peaches and Nectarines					
China	705,783	China	10,170,038	Austria	455,515
Italy	92,700	Italy	1,692,500	Israel	364,327
Egypt	82,000	USA	1,197,670	France	238,372
Spain	72,000	Spain	1,191,300	Malta	223,833
USA	60,716	Greece	734,000	Venezuela	200,000
Iran	49,586	Turkey	547,219	USA	197,257
Mexico	40,000	Egypt	425,000	Turkey	196,135
Greece	38,849	Iran	396,059	Slovenia	195,481
India	35,531	Chile	388,000	Greece	188,936
Turkey	27,900	France	347,476	Chile	184,761
Argentina	24,710	Argentina	290,774	Italy	182,578
Chile	21,000	India	237,931	Spain	165,458
Australia	20,917	Brazil	216,236	Morocco	161,504
North Korea	20,244	Mexico	200,000	South Africa	157,808
Brazil	19,043	South Korea	190,000	Switzerland	155,000
Algeria	17,750	South Africa	157,808	South Korea	146,153
Tunisia	16,500	Japan	150,700	China	144,095
Pakistan	15,349	Algeria	147,400	Turkmenistan	142,576
France	14,577	North Korea	147,095	Japan	137,000
South Korea	13,000	Tunisia	118,000	Colombia	136,309
World Total	1,568,447	World Total	20,315,245	World Total	129,525

Plums and Sloes					
China	1,662,873	China	5,372,899	Austria	2,964,132
Serbia	201,230	Serbia	662,631	UK	286,582
Bosnia	75,000	USA	561,366	Turkmenistan	208,333
Romania	74,688	Romania	533,691	Armenia	208,000
Russia	58,000	Chile	296,000	Switzerland	200,293
USA	37,955	Turkey	245,782	Netherlands	198,013
India	24,737	France	243,746	Kenya	178,579
Moldova	22,000	Spain	227,800	Germany	161,230
Poland	21,044	India	196,900	Chile	160,000
Ukraine	19,900	Italy	194,100	USA	147,903
Spain	19,500	Iran	167,429	Italy	138,642
Turkey	19,400	Bosnia	155,767	France	130,582
Croatia	19,000	Ukraine	136,700	Turkey	126,691
Iran	18,689	Russia	125,000	Lebanon	117,867
France	18,666	Poland	120,718	Spain	116,820
Chile	18,500	Argentina	105,868	Syria	112,359
Argentina	18,038	Belarus	80,914	South Korea	112,068
Bulgaria	17,000	Algeria	74,406	Portugal	107,002
Algeria	15,049	Germany	73,102	Israel	105,464
Mexico	14,389	Morocco	72,038	Belarus	101,205
World Total	2,534,873	World Total	10,776,232	World Total	42,512

plums) are listed in Table 1-3. The major stone fruit-producing areas in the world (Table 1-3) include countries on all continents except Antarctica. Yields vary considerably by country and commodity, and, for all commodities, the country with the highest yield is not among the top 20 for production. In nearly all cases, the highest-ranked country for area harvested was also highest-ranked for quantity produced (the one exception was almonds, for which Spain had more than twice as much area harvested but produced only about 25% as much as the USA; this large difference in yield appears in several previous years' data as well).

Based on 2008 values (FAO 2011), countries' ranks for monetary value of production of each commodity closely track those for the quantity of the commodity produced, with occasional exceptions (e.g., for apricots, Uzbekistan was third in the world for quantity but fourth for value, for cherries, Spain was sixth in the world for quantity produced, but eighth for value, and for plums and sloes, Turkey was fifth for quantity but sixth for value) and where discrepancies did occur, they involved cases with relatively small differences between the successive ranks in each category. In 2008, the total monetary values for the top 20 producers of stone fruits were about $2.50 billion for almonds, $1.20 billion for apricots, $1.40 billion for cherries, $6.10 billion for peaches and nectarines, and $2.17 billion for plums and sloes, making the total value of world production of *Prunus* fruit crops well over $13 billion (FAO 2011).

1.3.3 Nutritional Content

The stone fruits have long been highly esteemed for their delicious flavors; they are also excellent sources of vitamins and minerals (Table 1-4). Due to the presence of high levels of vitamins as well as anthocyanins and other phenolics with antioxidant, antimicrobial, and anti-inflammatory properties, considerable attention has been called to the potential value of fruits of *Prunus* species as nutraceuticals. Numerous studies have measured the levels of these compounds and their antioxidant activities in various species of *Prunus*, including both plants found in the wild (e.g., *P. armeniaca*: Jablonska-Rys et al. 2009; *P. avium* and *P. cerasus*: Serteser et al. 2008; *P. spinosa*: Fraternale et al. 2009) and those in cultivation (e.g., *P. armeniaca*: Drogoudi et al. 2008; Leccese et al. 2008; Davarynejad et al. 2010; Hegedus et al. 2010; Leccese et al. 2010; Schmitzer et al. 2011; *P. avium*: Usenik et al. 2008; Ahn et al. 2009; Faniadis et al. 2010; Gonzalez-Gomez et al. 2010; *P. cerasus*: Blando et al. 2004; *P. domestiica*: Kayano et al. 2004; Lombardi-Boccia et al. 2004, Vasantha Rupasinghe et al. 2006; Rop et al. 2009; Slimestad et al. 2009; Voca et al. 2009; *P. dulcis*: Mishra et al. 2010; P. *persica*: Drogoudi and Tsipoulidis 2007; Cantin et al. 2009; *P. salicina*: Lozano et al. 2009). These studies have demonstrated high antioxidant activity in several of the stone

Table 1-4 Nutritional composition of stone fruits (USDA-ARS 2011b). Values given are per 100g raw fruit or nut (edible portions only). The common names correspond to the species listed in Table 1-1, but the values for plums are not separated by particular species (the scientific name listed is *Prunus* spp.).

Nutrient	Almonds	Apricots	Cherries, sour	Cherries, sweet	Peaches	Plums
Water, g	4.70	86.35	86.13	82.25	88.87	87.23
Energy, kcal	575	48	50	63	39	46
Protein, g	21.22	1.40	1.00	1.06	0.91	0.70
Fat, g	49.42	0.39	0.30	0.20	0.25	0.28
Carbohydrate, g	21.67	11.12	12.18	16.01	9.54	11.42
Dietary fiber, g	12.2	2.0	1.6	2.1	1.5	1.4
Sugars, g	3.89	9.24	8.49	12.82	8.39	9.92
Ca, mg	264	13	16	13	6	6
Fe, mg	3.72	0.39	0.32	0.36	0.25	0.17
Mg, mg	268	10	9	11	9	7
P, mg	484	23	15	21	20	16
K, mg	705	259	173	222	190	157
Na, mg	1	1	3	0	0	0
Zn, mg	3.08	0.20	0.10	0.07	0.17	0.10
Cu, mg	0.996	0.078	0.104	0.06	0.068	0.057
Mn, mg	2.285	0.077	0.112	0.07	0.061	0.052
Se, µg	2.5	0.1	0.0	0.0	0.1	0
Vitamin C, mg	0.00	10.0	10.0	7.0	6.6	9.5
Thiamin, mg	0.211	0.030	0.03	0.027	0.024	0.028
Riboflavin, mg	1.014	0.040	0.04	0.033	0.031	0.026
Niacin, mg	3.385	0.600	0.40	0.154	0.806	0.417
Pantothenic acid, mg	0.469	0.240	0.143	0.199	0.153	0.135
Vitamin B-6, mg	0.143	0.054	0.044	0.049	0.025	0.029
Folate, µg	50	9	8	4	4	5
Choline, mg	52.1	2.8	6.1	6.1	6.1	1.9
Vitamin A, RAE, µg	0.00	96	64	3	16	17
Carotene, beta, µg	1	1094	770	38	162	190
Carotene, alpha, µg	0	19	0	0	0	0
Cryptoxanthin, beta, µg	0	104	0	0	67	35
Vitamin A, IU	1	1926	1283	64	326	345
Lutein + zeaxanthin, µg	1	89	85	85	91	73
Vitamin E (alpha-tocopherol), mg	26.22	0.89	0.07	0.07	0.73	0.26
Vitamin K (phylloquinone), µg	0.00	3.3	2.1	2.1	2.6	6.4

fruit species and also shown that phenolic content and antioxidant activity can vary significantly within species by genotype (cultivar), harvest year, and cultural practices. Taken together with current interest in the health benefits of compounds with high antioxidant activities, these results suggest that selection of varieties that produce high levels of particular phenolic compounds in particular horticultural systems may become an important target for stone fruit breeding programs (e.g., Cevallos-Casals et al. 2006).

Over the last several decades, nutritionists have increasingly emphasized the health benefits of fruits and vegetables and recommended inclusion of plant-derived sources of proteins and oils, such as nuts, in the diet. This is highly evident, for example, in MyPlate (http://www.choosemyplate.gov/), the new dietary guidelines recently released by the USDA's Center for Nutrition Policy and Promotion. It is therefore likely that the importance of *Prunus* species as sources of human nutrition will only increase in the coming years.

1.4 *Prunus* Research and Resources

1.4.1 Research on Wild Species

Due to the large number of crop species of tremendous economic importance and the abundance and widespread distribution of wild species, which range from immediate progenitors of the domesticates to distant relatives, the genus *Prunus* has been the focus of a tremendous amount of research aimed at understanding genetic diversity and natural variation among and within species and using that variation for genetic improvement of cultivated taxa. There is also a wealth of germplasm of both wild and cultivated species available from collections and repositories throughout the world.

Potter (2011) reviewed the diversity of wild crop relatives in *Prunus* and their importance as resources for breeding and genomic studies of the domesticated species. Historically, wild species have been of considerable importance in the breeding of plums, where wild species have contributed to crop improvement efforts in both Japanese (*P. salacina*) and European (*P. domestica*) plums (Okie and Hancock 2008), but much less so in sweet (*P. avium*) and sour (*P. cerasus*) cherries (Iezzoni 2008), apricots (*P. armeniaca*; Ledbetter 2008), and almonds (Gradziel 2009) and least of all in peaches (Hancock et al. 2008). More recently, however, there has been considerable interest in the use of wild species as sources of desirable traits, including pest and disease resistance (e.g., Foulongne et al. 2003) and fruit quality (e.g., Quilot et al. 2004) characteristics, and modern genetic and genomic approaches can facilitate efficient transfer of genes conferring desirable traits while avoiding the introgression of undesirable traits that can occur when wild species are used as parents in traditional breeding programs.

Wild species, and wild populations of cultivated species, of *Prunus* have also been the subjects of numerous evolutionary and population genetic studies (reviewed in Potter 2011). The interest in these plants is based not only on their relationships to very economically important ornamentals and stone fruit crops, but also on the fact that the genus includes some species that are threatened, rare, or endangered and others that have become invasive outside of their native ranges, and also provides excellent opportunities for the study of historical biogeographic patterns and the evolution of biologically interesting traits such as gametophytic self-incompatibility. The diversity of wild and cultivated species in *Prunus* makes it an outstanding system for the integration of applied and fundamental research across a range of disciplines.

1.4.2 Genomics Research

Over the last two decades, significant advances have been made in genomics research in Rosaceae in general and *Prunus* in particular (Shulaev et al. 2008). An international consortium of Rosaceae researchers has established on-line resources, including the Genome Database for Rosaceae (GDR, http://www.rosaceae.org), a web-based repository of curated genetic and genomic data for species in the family (Jung et al. 2008), and the website for RosBREED (http://www.rosbreed.org/), a multi-institutional international project whose overarching objective is to facilitate integration of the use of genomic tools with traditional breeding approaches (Iezzoni et al. 2010). Because it is a diploid species with a relatively small genome size, peach has emerged as a model system for Rosaceae genomics research, and the complete peach genome was made available by the International Peach Genome Initiative (IPGI) in April, 2010 (GDR 2011).

1.4.3 Germplasm Resources

There are numerous *Prunus* germplasm collections housed at various institutions in multiple countries, which include many accessions of all of the major cultivated species and a broad diversity of the wild relatives. Members of subgenus *Laurocerasus*, especially the tropical species, are least well represented. Potter (2011) reported that a search of Biodiversity International's Biodiversity Directory of Germplasm Collections in 2009 returned over 60,000 accessions of *Prunus* species, held at 168 institutions. That database is not currently available so no updated figures can be provided here.

In the US, the majority of *Prunus* collections are held at three locations within the USDA germplasm system. According to information available through GRIN, the Germplasm Resources Information Network (USDA-ARS 2011a), the largest of the three, the National Clonal Germplasm Repository at Davis, currently has 1036 *Prunus* accessions available for distribution. Ninety-nine of these are classified as almonds, representing *P. dulcis* (73 accessions) and 11 other species (16 accessions), as well as eight accessions unidentified to species and two listed as hybrids. One hundred and sixty-eight of the accessions are classified as apricots, representing *P. armeniaca* (156 accessions), *P. mume* (seven accessions), three other species (four accessions), and one accession unidentified to species. One hundred and forty-six accessions are classified as cherries, representing *P. avium* (68 accessions), *P. cerasus* (28 accessions), 16 other species (32 accessions) including several classified in subgenera *Padus* and *Laurocerasus*, 13 accessions unidentified to species, and five hybrid accessions. Two hundred and eighty accessions are classified as peaches, representing *P. persica* (256 accessions), four other species (13 accessions), eight accessions unidentified to species and three listed as hybrids. Finally, 313 accessions are listed as plums, representing *P. domestica* (113 accessions), *P. salicina* (48 accessions), *P. cerasifera* (42 accessions), *P. americana* (six accessions), *P. spinosa* (six accessions), and 12 other species (24 accessions), 43 accessions unidentified to species, and 31 listed as hybrids; and 30 accessions are listed as plumcots (all hybrids). The Geneva Repository lists 115 accessions, the vast majority of which represent *P. cerasus* (91 accessions), followed by *P. fruticosa* (13 accessions). The National Arboretum has 80 accessions of *Prunus*, dominated by ornamental flowering cherry species, including *P. sargentii* Rehd. (35 accessions), *P. nipponica* Matsum. (seven accessions), and *P. serrulata* Lindl. (five accessions).

References

Ahn S-M, Ryu H-Y, Kang D-K, Jung I-C, Sohn H-Y (2009) Antimicrobial and antioxidant activity of the fruit of *Prunus avium* L. Kor J Microbiol Biotechnol 37: 371–376.

Blando F, Gerardi C, Nicoletti I (2004) Sour cherry (*Prunus cerasus* L) anthocyanins as ingredients for functional foods. J Biomed Biotechnol (5): 253–258.

Bortiri E, Oh S-H, Jiang J, Baggett S, Granger A, Weeks C, Buckingham M, Potter D, Parfitt D (2001) Phylogeny and systematics of *Prunus* (Rosaceae) as determined by sequence analysis of ITS and the chloroplast *trnL-trnF* spacer DNA. Syst Bot 26: 797–807.

Bortiri E, Vanden Heuvel B, Potter D (2006) Phylogenetic analysis of morphology in *Prunus* reveals extensive homoplasy. Plant Syst Evol 259: 53–71.

Candolle A de (1886) Origin of Cultivated Plants, 2nd edn, reprinted in 1959. Hafner Publishing, New York.

Cantin CM, Moreno MA, Gogorcena Y (2009) Evaluation of the antioxidant capacity, phenolic compounds, and vitamin C content of different peach and nectarine [*Prunus persica* (L.) Batsch] breeding progenies. J Agri Food Chem 57: 4586–4592.

Cevallos-Casals BA, Byrne D, Okie WR, Cisneros-Zevallos L (2006) Selecting new peach and plum genotypes rich in phenolic compounds and enhanced functional properties. Food Chem 96: 273–280.

Chin SW, Wen J, Johnson G, Potter D (2010) Merging *Maddenia* with the morphologically diverse *Prunus* (Rosaceae). Bot J Linn Soc 164: 236–245.

Davarynejad G, Khorshidi S, Nyeki J, Szabo Z, Gal-Remennyik J (2010) Antioxidant capacity, chemical composition and physical properties of some apricot (*Prunus armeniaca* L.) cultivars. HortEnviron Biotechnol 51: 477–482.

Drogoudi PD, Tsipoulidis CG (2007) Effects of cultivar and rootstock on the antioxidant content and physical characters of clingstone peaches. Sci Hort 115: 34–39.

Drogoudi PD, Vemmos S, Pantelidis G, Petri E, Tzoutzoukou C, Karayiannis I (2008) Physical characters and antioxidant, sugar, and mineral nutrient contents in fruit from 29 Apricot (*Prunus armeniaca* L.) cultivars and hybrids. J Agri Food Chem 56: 10754–10760.

Faniadis D, Drogoudi PD, Vasilakakis M (2010) Effects of cultivar, orchard elevation, and storage on fruit quality characters of sweet cherry (*Prunus avium* L.). Sci Hort 125: 301–304.

FAO (2011) FAOSTAT: http://faostat.fao.org/site/291/default.aspx (Cited 28 Sept 2011).

Faust M, Surányi D, Nyujtó F (1998) Origin and dissemination of apricot. Hort Rev 22: 225–266.

Faust M, Timon B (1995) Origin and dissemination of peach. Hort Rev 17: 331–379.

Foulongne M, Pascal T, Pfeiffer F, Kervella J (2003) QTLs for powdery mildew resistance in peach X *Prunus davidiana* crosses: consistency across generations and environments. Mol Breed 12: 33–50.

Fraternale D, Giamperi L, Bucchini A, Ricci D (2009) Antioxidant activity of *Prunus spinosa* L. fruit juice. Ital J Food Sci 3: 337–346.

Fresnedo-Ramírez J, Segura S, Muratalla-Lúa A (2011) Morphovariability of capulín (*Prunus serotina* Ehrh.) in the central-western region of Mexico from a plant genetic resources perspective. Genet Resour Crop Evol 58: 481–495.

GDR (2011) Peach genome: http://www.rosaceae.org/peach/genome (cited 28 Sept 2011).

Gonzalez-Gomez D, Lozano M, Fernandez-Leon MF, Bernalte MJ, Ayuso MC, Rodriguez AB (2010) Sweet cherry phytochemicals: identification and characterization by HPLC-DAD/ESI-MS in six sweet-cherry cultivars grown in Valle del Jerte (Spain). J Food Compos Analys 23: 533–539.

Gradziel TM (2009) Almond (*Prunus dulcis*) breeding. In: Jain SM, Priyadarshan PM (eds) Breeding Plantation Tree Crops: Temperate Species. Springer, New York, pp 1–31.

Hancock JF, Scorza R, Lobos GA (2008) Peaches. In: Hancock JF (ed) Temperate Fruit Crop Breeding: Germplasm to Genomics. Springer, Heidelberg, pp 265–298.

Hegedus A, Engel R, Abranko L, Balogh E, Blazovics A, Herman R, Halasz J, Ercisli S, Pedryc A, Stefanovits-Banyai E (2010) Antioxidant and antiradical capacities in apricot (*Prunus armeniaca* L.) fruits: variations from genotypes, years, and analytical methods. J Food Sci 75: C722–C730.

Iezzoni A, Weebadde C, Luby J, Yue C, van de Weg E, Fazio G, Main D, Peace CP, Bassil NV, McFerson J (2010) RosBREED: enabling marker-assisted breeding in Rosaceae. Acta Hort 859: 389–394.

Iezzoni AF (2008) Cherries. In: Hancock JF (ed) Temperate Fruit Crop Breeding: Germplasm to Genomics. Springer, Berlin, Heidelberg, New York, pp 151–176.

Jablonska-Rys E, Zalewska-Korona M, Kalbarczyk J (2009) Antioxidant capacity, ascorbic acid and phenolics content in wild edible fruits. J Fruit Ornament Plant Res 17: 115–120.

Jung S, Staton M, Lee T, Blenda A, Svancara R, Abbott A, Main D (2008) GDR (Genome Database for Rosaceae): integrated web-database for Rosaceae genomics and genetics data. Nucl Acids Res 36: D1034–D1040.

Kalkman C (1965) The Old World species of *Prunus* subg. *Laurocerasus* including those formerly referred to *Pygeum*. Blumea 13: 1–174.

Kayano S, Kikuzaki H, Yamada NF, Aoki A, Kasamatsu K, Yamasaki Y, Ikami T, Suzuki T, Mitani T, Nakatani N (2004) Antioxidant properties of prunes (*Prunus domestica* L.) and their constituents. Biofactors 21: 309–313.

Leccese A, Bartolini S, Viti R (2008) Total antioxidant capacity and phenolics content in fresh apricots. Acta Aliment 37: 65–76.

Leccese A, Bureau S, Reich M, Renard MGCC, Audergon J-M, Mennone C, Bartolini S, Viti, R (2010) Pomological and nutraceutical properties in apricot fruit: cultivation systems and cold storage fruit management. Plant Foods Human Nutr 65: 112–120.

Ledbetter CA (2008) Apricots. In: Hancock JF (ed) Temperate Fruit Crop Breeding: Germplasm to Genomics. Springer, Berlin, Heidelberg New York, pp 39–82.

Lee S, Wen J (2001) A phylogenetic analysis of *Prunus* and the Amygdaloideae (Rosaceae) using ITS sequences of nuclear ribosomal DNA. Am J Bot 88: 150–160.

Lombardi-Boccia G, Lucarini M, Lanzi S, Aguzzi A, Cappelloni M (2004) Nutrients and antioxidant molecules in yellow plums (*Prunus domestica* L.) from conventional and organic productions: a comparative study. J Agri Food Chem 52: 90–94.

Lozano M, Carmen Vidal-Aragon M, Teresa Hernandez M, Concepcion Ayuso M, Josefa Bernalte M, Garcia J, Velardo B (2009) Physicochemical and nutritional properties and volatile constituents of six Japanese plum (*Prunus salicina* Lindl.) cultivars. Eur Food Res Technol 228: 403–410.

Mason SC (1913) The pubescent-fruited species of *Prunus* of the southwestern states. J Agri Res 1: 147–179.

Mishra N, Dubey A, Mishra R, Barik N (2010) Study on antioxidant activity of common dry fruits. Food Chem Toxicol 48: 3316–3320.

Moing A, Langlois N, Svanella L, Zanetto A, Gaudillere JP (1997) Variability in sorbitol: sucrose ratio in mature leaves of different *Prunus* species. J Am Soc Hort Sci 122: 83–90.

Okie WR, Hancock JF (2008) Plums. In: Hancock JF (ed) Temperate Fruit Crop Breeding: Germplasm to Genomics. Springer, Berlin, Heidelberg, New York, pp 337–358.

Oldén EJ, Nybom N (1968) On the origin of *Prunus cerasus* L. Hereditas 59: 327–345.

Potter D (2011) *Prunus*. In: Kole C (ed) Wild Crop Relatives: Genomic and Breeding Resources. Vol: Temperate Fruits. Springer, Heidelberg, Dordrecht, London, New York, pp 129–145.

Potter D, Eriksson T, Evans RC, Oh S-H, Smedmark J, Morgan DR, Kerr M, Robertson KR, Arsenault M, Dickinson TA, Campbell CS (2007) Phylogeny and classification of Rosaceae. Plant Syst Evol 266: 5–43.

Quilot B, Wu BH, Kervella J, Génard M, Foulongne M, Moreau K (2004) QTL analysis of quality traits in an advanced backcross between *Prunus persica* cultivars and the wild relative species *P. davidiana*. Theor Appl Genet 109: 884–897.

Rehder A (1940) Manual of Cultivated Trees and Shrubs Hardy in North America Exclusive of the Subtropical and Warmer Temperate Regions. Dioscorides Press, Portland, Oregon.

Rop O, Jurikova T, Mlcek J, Kramarova D, Sengee Z (2009) Antioxidant activity and selected nutritional values of plums (*Prunus domestica* L.) typical of the White Carpathian Mountains. Sci Hort 122: 545–549.

Schmitzer V, Slatnar A, Mikulic-Petkovsek M, Veberic R, Krska B, Stampar F (2011) Comparative study of primary and secondary metabolites in apricot (*Prunus armeniaca* L.) cultivars. J Sci Food Agri 91: 860–866 .

Serteser A, Kargioglu M, Gok V, Bagci Y, Ozcan MM, Arslan D (2008) Determination of antioxidant effects of some plant species wild growing in Turkey. Int J Food Sci Nutr 59: 643–651.

Shaw J, Small RL (2005) Chloroplast DNA phylogeny and phylogeography of the North American plums (*Prunus* subgenus *Prunus* section *Prunocerasus*, Rosaceae). Am J Bot 92: 2011–2030.

Shulaev V, Korban SS, Sosinski B, Abbott AG, Aldwinckle HS, Folta KM, Iezzoni A, Main D, Arús P, Dandekar AM, Lewers K, Brown SK, Davis TM, Gardiner SE, Potter D, Veilleux RE (2008) Multiple models for Rosaceae genomics. Plant Physiol 147: 985–1003.

Slimestad R, Vangdal E, Bredo C (2009) Analysis of phenolic compounds in six Norwegian plum cultivars (*Prunus domestica* L.). J Agri Food Chem 57: 11370–11375.
USDA-ARS (2011a) Germplasm Resources Information Network—(GRIN): http://www.ars-grin.gov/ (Cited 28 Sept 2011).
USDA (2011b) USDA National Nutrient Database for Standard Reference: http://www.nal.usda.gov/fnic/foodcomp/search/ (Cited 28 Sept 2011).
Usenik V, Fabcic J, Stampar F (2008) Sugars, organic acids, phenolic composition and antioxidant activity of sweet cherry (*Prunus avium* L.). Food Chem 107: 185–192.
Vasantha Rupasinghe HP, Jayasankar S, Lay W (2006) Variation in total phenolics and antioxidant capacity among European plum genotypes Sci Hort 108: 243–246.
Vavilov NI (1951) The origin, variation, immunity and breeding of cultivated plants. Ronald, New York.
Voca S, Galic A, Sindrak Z, Dobricevic N, Pliestic S, Druzic J (2009) Chemical composition and antioxidant capacity of three plum cultivars. Agri Conspec Sci 74: 273–276.
Wen J, Berggren ST, Lee C-H, Ickert-Bond S, Yi T-S, Yoo K-O, Xie L, Shaw J, Potter D (2008) Phylogenetic inferences in *Prunus* (Rosaceae) using chloroplast ndhF and nuclear ribosomal ITS sequences. J Syst Evol 46: 322–332.
Zeinalabedini M, Khayam-Nekoui M, Grigorian V, Gradziel TM, Martínez-Gómez P (2010) The origin and dissemination of the cultivated almond as determined by nuclear and chloroplast SSR marker analysis. Sci Hort 125: 593–601.
Zimmermann MH, Ziegler H (1975) List of sugars and sugar alcohols in sieve-tube exudates. In: Zimmermann MH, Milburn JA (eds) Transport in Plants I. Phloem Transport. Springer, Heidelberg, pp 480–502.

2

Traditional Genetics and Breeding

Thomas M. Gradziel

ABSTRACT

Commercial success of a new cultivar is determined not only by improved performance in a specific area, but also a consistently superior performance for the wide range of required agronomic traits. This is particularly relevant for stone fruits where orchards are expected to be productive for ten or more years and where failed cultivars cannot be readily plowed under and replanted as with cereal and vegetable crops. Stone fruits also differ from most agronomic crops in that they are not seed propagated but clonally propagated. Clonal propagation combined with traditional clonal selection strategies remains one of the most effective methods for capturing the fullest range of breeding potential, including additive, dominance, epistatic, epigenetic and genomic interactions for crop improvement. Clonal propagation also confers a near-immortality to stone fruit cultivars, permitting extended time periods of tens to hundreds of years for continued genetic improvements through bud-sport mutation or further, rare, favorable recombinations. Consequently, continued cultivar improvement using marker-assisted selection and associated molecular based techniques may be limited by their inherently reductionistic and additive gene emphasis. A major barrier to further breeding progress remains a poor understanding of non-additive genetic as well as epigenetic mechanisms and plant development. The capacity of clone-based cultivars to capture the fullest range of these beneficial interactions for crop improvement provides both a prerequisite and unique opportunity for stone fruit researchers to evolve beyond the current additive-gene emphasis, but would require/stimulate significant parallel progress in our understanding of the basics underlying developmental and inheritance mechanisms at the epigenetic and genomic as well as genetic level. An even greater challenge/opportunity is the shift from the current focus on single trait

Department of Plant Sciences, University of California-Davis, 1 Shields Avenue, Davis, CA 95616, USA; e-mail: *tmgradziel@ucdavis.edu*

improvement to the achievement of a more holistic cultivar breeding with concurrent management/advancement of the multitude of traits required for commercial success.

Keywords: cultivar, clone, non-additive, epigenetic, inter-locus heterosis, MAS, MAB

2.1 Introduction

Historically, the term 'breed' referred to a type of domesticated animal such as the Clydesdale horse that has been selected for specific phenotypes or well-defined traits. The term 'breeding', in turn, refers to the selection of parent combination to achieve the desired phenotype in subsequent offspring. Just as natural selection can result in the gradual evolution of individuals and populations towards greater fitness within the selecting environment, human selection of parental combinations and resultant progeny can result in pronounced phenotype changes in individuals and populations, which can occur relatively rapidly depending upon the intensity of selection. A primary objective of most breeding approaches is to maximize the desired response to selection. In plant systems, the goal of breeding is also the development of an improved phenotype which is often referred to as a 'variety', or more specifically a 'cultivar' (derived from 'cultivated variety') to distinguish it from the more broadly defined 'botanical variety'. Because most perennial, woody plants, including all stone fruits, can be asexually propagated, a typical cultivar is usually a single genotype which may be the result of selection over a very large number of years and/or from a very large population of progeny. For example, virtually all commercial sweet orange (*Citrus sinensis*) plantings are essentially asexual propagations of a single ancestral genotype (Soost and Roose 1996). Chance mutations leading to improved phenotypes (improved flavor, sweetness, color, later maturity, etc.) among the millions of otherwise clonally identical trees cultivated over the past several hundred years have been discovered and, if found to be true-to-type following asexual propagation, are often propagated and distributed as new orange cultivars such as the Washington Navel and Valencia cultivars (Nicolosi et al. 2000). Recent evidence indicates that fig (*Ficus carica*) has been cultivated for over 11,000 years (Kislev et al. 2006) supporting a very early domestication of fruit crops. In addition, many modern fruit cultivars have been cultivated continuously for hundreds to thousands of years since their initial selection (Janick 2002, 2006), presumably derived from the leading cultivars of their day. The capacity of asexual propagation to essentially capture these rare, horticulturally elite genotypes and, in addition, allow their continued improvement through the accumulation of desirable sports or mutations, offered considerable advantages over early breeding efforts with cereals and other seed-propagated crops. This is because propagation by seed inevitably results in a risk reshuffling of genes resulting in genetically and so phenotypically variable progeny. In contrast, modern cultivar breeding methods including marker-assisted selection (MAS), and marker-assisted breeding (MAB), have proven more effective in crops which are seed propagated as compared to those which are vegetatively propagated. This is due in part to the long seed-to-seed breeding cycle time as well as the relatively low population size restrictions of vegetatively propagated tree crops. However, both molecular-based and traditional analysis of genetic variability are demonstrating that clonal propagation combined with traditional clonal selection strategies is among the most effective methods for capturing the full range of genetic potential, including additive, dominance, and epistatic interactions, for crop improvement (Luby and Shaw 2001; Janick 2006; Ortiz et al. 2006). These and ongoing studies are also shedding light on fundamental reasons why MAS and associated molecular strategies as currently applied, are less effective in tree crop cultivar development.

2.2 Genetic Analysis as a Basis for Applied Breeding

Early breeders were generally aware that the characteristics or phenotypes of progeny from a specific set of parents were determined by the environmental conditions during their development as well as by factors inherited from parents (Burbank and Hall 1927). The only way to determine a given individual's breeding potential, however, was through experience; that is by keeping track of the general breeding value for each individual parent as well as the specific value of each individual parental combination. Such trial and error approaches required both extensive experience as well as a good understanding of various environmental effects on the final phenotype since the final trait heritability was determined by the proportion of the total phenotypic variability that was due to parental (genetic) contribution relative to the variability from environmental causes. Breeding was largely reactive since the extent of heritability of a specific trait from a specific parental combination had to first be developed empirically and then, if desired, reproduced on a larger scale. More proactive and analytical approaches to cultivar breeding resulted from the discovery in the early to mid-1900s, that genes coded by unique DNA sequences were the factors controlling heredity, and the rediscovery of Mendel's research showing that genes are often inherited in predictable patterns. An example of the proactive breeding potential of Mendelian analysis is apparent in the classical single gene (1:2:1) and two gene (9:3:3:1) ratios expected in heterozygous diploid crosses (Figs. 2-1 and 2-2). With sufficient knowledge of the inheritance for the trait of interest and the genetic composition of

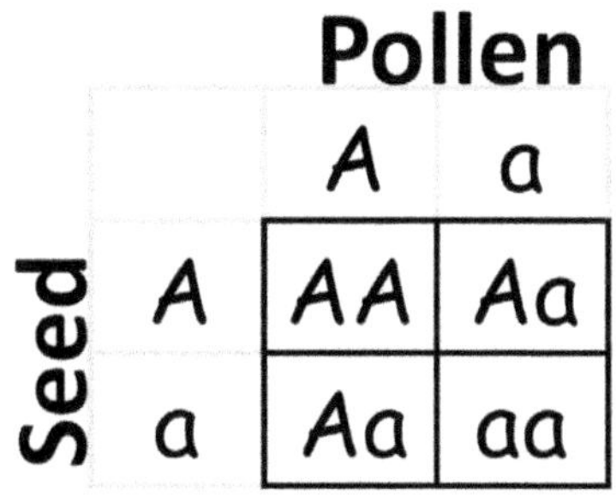

Figure 2-1 Punnett square diagram showing predicted gamete as well as progeny genotypes and their probabilities from a cross between two diploid plants heterozygous at locus A.

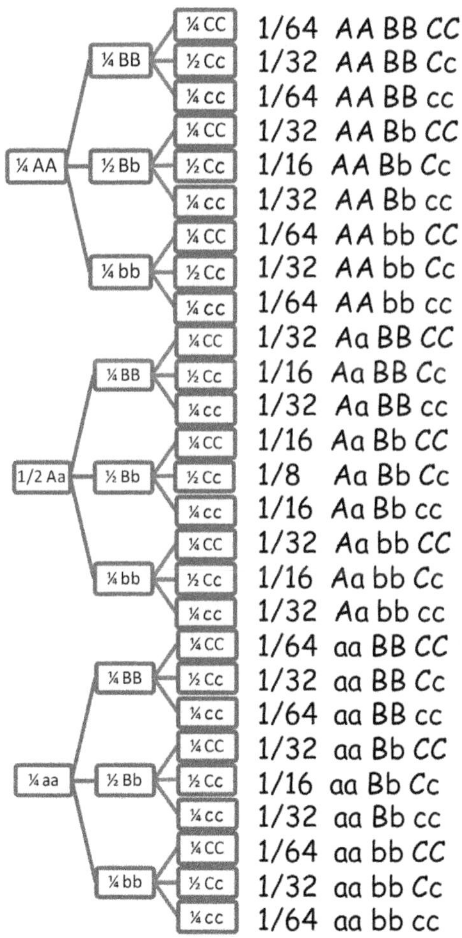

Figure 2-2 Tree diagram showing genotypes and their probabilities predicted from a cross between two diploid plants heterozygous at unlinked loci *A/a*, *B/b* and *C/c*.

the parents, the breeder could accurately predict the proportion of progeny expected to inherit the traits (and thus determine the minimum number of progeny required to obtain individuals possessing the desired traits). Similarly, by analyzing segregation ratios of progeny from known crosses, the breeder could sometimes deduce both the genetic control for the trait as well as the genetic composition of the parents. The major Mendelian or qualitative genes identified through this process for peach, the stone fruit with the most extensive genetic database, are summarized in Table 2-1.

The reductionist approach made possible through Mendelian analysis remains the foundation for the genetic manipulation of most major or *qualitative* genes. Similarly, the recognition that genetic contributions could be isolated and then recombined in a largely additive manner forms the basis for most molecular marker approaches, including both MAS and MAB. As segregation ratios become more complex, however, the ability to discriminate the diminishing individual genetic effect from environmental effect becomes limiting (Figs. 2-2 and 2-3) so that for traits controlled by three or more genes, an analysis based on statistical probabilities is usually required to help discriminate genetic from environmental background effects. In such *quantitative* genetic analysis, the variation in phenotypic expression is partitioned into environmental and genetic components where genes are generally assumed to be independent in action and alleles contribute equal and additive effects to final phenotype. Heritability (H) in this narrow sense can then be defined by the ratio of additive genetic variance [V_G] to total variance (genetic [V_G] + environmental [V_E] + genetic by environment interaction [V_{GxE}]) resulting in the formula:

$$\text{Heritability (H)} = V_G / (V_G + V_E + V_{GxE})$$

Published heritability estimates for apricot and peach are presented in Tables 2-2 and 2-3. Traditional breeding methods by necessity targeted those alleles whose heritability was large enough to be differentiated from background environmental variance. Effective molecular markers combined with advanced statistical analysis techniques now offer the opportunity for more accurate discrimination among individual additive genetic from environmental effects, resulting in the opportunity for more efficient, incremental genetic improvement. For this reason MAS has been particularly successful in the genetic improvement of self-pollinating crops such as most cereals and vegetables since most advanced selections have been inbred to near homozygosity. In out-crossed crops, however, high levels of heterozygosity exist, with additional and often exploitable genetic contributions resulting from interactions within individual loci (dominance), among different loci (epistasis and other genetic interactions) and even between genomes (as in interspecies hybrid vigor). The relative

Table 2-1 Qualitative traits in peach (modified from Monet and Bassi 2008).

Phenotype and symbol	Genotype[a]	Note	Reference
Tree			
Broomy (columnar, pillar) (*Br*)	*br/br*	Incomplete dominance; phenotype is upright when *Br* is heterozygous with the alleles for the standard, dwarf, compact or weeping growth habits	Scorza et al. (1989, 2002); Yamazaki et al. (1987); Chaparro et al. (1994)
Upright (*Up*)	*Br/br*	See: Columnar	Scorza et al. (1989)
	br/pl		
	br/dw		
	br/Ct		
Arching (*Ar*)	*Brbr/plpl*	Upright weeping; similar to the *Up*, but with a distinct curvature of the 1-year-old shoots;	Werner and Chaparro (2005)
Bushy (*Bu*)	*bu*1/*bu*1 *bu*2/*bu*2	from F2 or backcross progenies of columnar (*Br*) × weeping (*Pl*) crosses; *Br* is epistatic to *Pl* *Bu*1 and *Bu*2 are independent.	Lammerts (1945)
Compact (*Ct*)	*Ct*/-		Mehlenbacher and Scorza (1986)
Dwarf (*Dw*)	*dw/dw*	Short internode (<10 mm)	Lammerts (1945)
	*dw*2/*dw*2	Very dwarf	Hansche (1988)
	*dw*3/*dw*3	Extremely dwarf, thin stem	Chaparro et al. (1994)
Semi-dwarf (*N*)	*n/n*	Incomplete dominance	Monet and Salesses (1975)

Table 2-1 contd....

Table 2-1 contd....

Phenotype and symbol	Genotype[a]	Note	Reference
Weeping (*Pl*)	*pl/pl*	Incomplete dominance, featuring open, intermediate canopy when heterozygous (Pl/pl) (Bassi and Rizzo 2000). Pl from pleureur (to weep, in French)	Monet et al. (1988)
Standard	*Dw/-* *Br/Br*	This growth habit results from the allelic status of any of these known genotypes	
	Pl/Pl		
	ct/ct		
Graft incompatibility with	*I1/-*	Two dominant genes; incompatibility found only	Salesses and Al-Kai (1985)
Damas 1869 plum (*I*)	*I2/-*	in some nectarines	
Corky triangle (*T*)	*t/t*	Epidermic suberification at bud base in 1-year-old shoot	Monet and Bastard (1982)
Evergreen (*Evg*)	*evg/evg*	Also called evergrowing (Bielenberg et al. 2004) terminal buds do not go dormant.	Lammerts (1945); Rodriguez et al. (1994)
Anthocyanin deficiency (*An*)	*an/an*	Pale pink flowers	Monet (1967)
Anthocyaninless (*W*)	*w/w*	White flowers; no red anywhere	Lammerts (1945)
	w^v/w^v	w^v is unstable producing variegated flowers (peppermint)	Lammerts (1945); Chaparro et al. (1995)
Resistance to root-knot nematode, *Meloidogyne javanica* (*Mj*)	*Mj1/-* *Mj2/-*	*Mj*1 and *Mj*2 are independent	Lownsberry and Thomson (1959) Sharpe et al. (1970); Lu et al. (2000)
Resistance to root-knot nematode, *Meloidogyne incognita* (*Mi*)	*Mi/-*		Weinberger et al. (1943); Lu *et al.* (2000)
Resistant to both species	*Mij/*		

Green aphid resistance (*Rm*)	*Rm*1/-		Monet and Massonié (1994)
Leaf			
Foliar glands (*E*) Reniform Globose Eglandular	 *E/E* *E/e* *e/e*	Incomplete dominance High susceptibility to powdery mildew; serrate leaf margin	Connors (1921)
Redleaf (*Gr*)	*Gr/-*	Red incompletely dominant over green in leaves and fruit skin ground colour.	Blake (1937)
Albinism (*C*)	*c/c*	Plant does not survive	Bailey and French (1932)
Wavy-leaf (*Wa*)	*wa/wa*		Scott and Cullinan (1942)
Willow-leaf (*Wa2*) Crinkle leaf (*CL*)	*wa2/wa2* *cl/cl*	Associated with very oblate fruit shape	Chaparro et al. (1994) Ledbetter (1996)
Flower			
Non-showy (Sh)	*Sh/-*		Connors (1920); Bailey and French (1942); Lammerts (1945)
Large size (*L*)	*L/-*		Connors (1920); Lammerts (1945);
Double (*D1*)	*d1/d1*	More than five petals; often incompletely dominant; number of extra petals controlled by one or two recessive genes	Bailey and French (1949); Lammerts (1945)
Fewer extra petals (*Dm1*, *Dm2*)	*dm1/dm1* *dm2/dm2*	*Dm1* and *Dm2* are independent and additive	Lammerts (1945); Yamazaki et al. (1987)
Dark pink petal (*P*)	*P/-*		Lammerts (1945)

Table 2-1 contd....

Table 2-1 contd....

Phenotype and symbol	Genotype [a]	Note	Reference
Red petal (*R*)	*r/r*		Lammerts (1945); Chaparro et al. (1995)
Male sterility (*Ps*)	*ps/ps*	Sometimes has some viable pollen (from 'J.H. Hale')	Connors (1926); Blake and Connors (1936); Scott and Weinberger (1944)
Male sterility (*Ps*2)	*ps*2/*ps*2	(from 'White Glory')	Blake (1932); Werner and Creller (1997)
Fruit			
Slow ripening (*Sr*)	*sr/sr*		Ramming (1991)
Saucer (flat) shape (*S*)	*S/-*	*S/S* is lethal (Guo et al. 2002)	Lesley (1940)
Aborting fruit (*Af*)b	*af/af*		Dirlewanger et al. (2006); Blake (1932); Werner et al. (1998)
Blood red flesh (*Bf*)	*Bf/-*	Pigment appears in immature fruit and main leaf vein; often smaller trees	Blake (1932); Werner et al. (1998)
Rough skin (*Rs*)	*rs/rs*	Matte skin surface; glabrous flower buds	Okie and Prince (1982); Okie (1988b)
Glabrous skin (nectarine) (*G*)	*g/g*	Fuzzless	Blake (1932); Blake and Connors (1936)
Full red skin (*Fr*)	*fr/fr*	Only on fruit	Beckman and Sherman (2003)
Highlighter (*H*)	*h/h*	Red colour suppression on fruit skin	Beckman et al. (2005)

White flesh (*Y*)	*Y*/-	Also affects calyx cup and leaf colour	Connors (1920)
Flesh texture/pit adherence (*F*)c Melting freestone	*Fl*		Bailey and French (1932, 1949); Monet (1989); Peace et al. (2005)
Melting clingstone	*f/f* *f/f1* *f/fn*		Peace et al. (2005)
Nonmelting clingstone	*f1/f1* *f1/n* *n/n*		Peace et al. (2005)
Stony Hard flesh (*Hd*)*d*	*hd/hd* *hdhd/F* *hdhd/f1f1*	 Stony hard, melting Stony hard, nonmelting	Yoshida (1976); Scorza and Sherman (1996) Haji et al. (2005) Haji et al. (2005)
Low-acid flesh (*D*)	*D*/-	*D* for *douce* (sweet in French)	Monet (1979)
Sweet kernel (Sk)	*Sk/sk*		Werner and Creller (1997)

[a]Traits determined by traditional Mendelian analysis prior to advent of molecular analysis.
[b]The 'aborting fruit' has been reported as a recessive trait causing the abortion of all fruits within 2 months after full bloom; it is still not clear whether the 'aborting fruit' phenotype is regulated from the same locus or from a novel gene (*Af*).
[c]Four alleles at the same locus controlling both flesh texture (endopolygalacturonase enzyme expression) and pit adherence; the fourth, null allele (*n*), has the
same effect as the *f1* allele (non-melting clingstone) (Peace et al. 2005) both traditional Mendelian as well as molecular markers used in determination.
[d]The independent inheritance of this trait was demonstrated, also suggesting an epistatic influence on the *F* locus, since the stony-hard, melting (*hdhd*/*f*-)
phenotype is induced to soften when exogenous ethylene is applied (Haji et al. 2005).

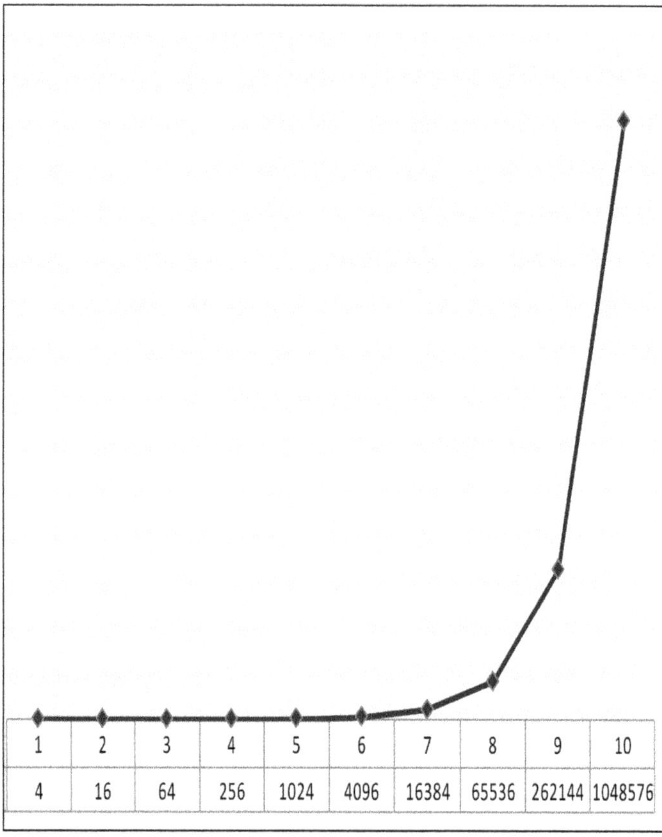

Figure 2-3 Plot showing the minimum diploid population size (Y-axis and bottom row) predicted by Mendelian analysis for obtaining specific homozygous genotypes at increasing numbers of independent loci (X-axis and top row).

importance of these different genetic components among different crops can be better appreciated by comparing the breeding strategies found most effective in their genetic improvement.

2.3 Crop Breeding Strategies

2.3.1 Basic Approaches

In the century since the genetic basis of inheritance was rediscovered and exploited for crop improvement, a large number of diverse breeding strategies have evolved (Simmonds and Smartt 1999). Most, however, are based on four fundamental approaches: *Inbreeding, Hybridization, Synthetics* and *Cloning* (Fig. 2-4). *Inbreeding* and *Hybridization* are commonly used for crops which are self-pollinating and so tolerant of inbreeding.

Table 2-2 Correlation coefficients between GCA and parent phenotype, correlation coefficients between F_1 progeny mean and midparent performance, and heritability coefficients with their standard deviation for 11 apricot characters (from Couranjou 1995).

	Coefficient of correlation *(r)*		Heritability and its standard deviation	
	Between GCA and parent phenotype	Between progeny mean and midparent	h^2	*S*
Flowering date	0.951**	0.886**	0.936	0.135
Maturity date	0.938**	0.834**	0.935	0.172
Yield	0.888*	0.831**	0.784	0.145
Fruit size	0.887*	0.753**	0.615	0.149
Background color	0.921**	0.835**	0.538	0.098
Overall color	0.765	0.519*	0.294	0.134
Flesh color	0.972**	0.880**	0.588	0.088
Firmness	0.878*	0.845**	0.564	0.099
Flavor	0.826*	0.633*	0.769	0.261
Aroma	0.912**	0.831**	0.603	0.112
Juiciness	0.944**	0.871**	0.915	0.143

**P=0.01, *P=0.05

Table 2-3 Reported heritability estimates for quantitative traits in peach (from Monet and Bassi 2008).

	Trait Heritability	Reference
Full bloom	0.39	Hansche et al. (1972)
Amount of ripening	0.38	Hansche et al. (1972)
Ripening date	0.84	Hansche et al. (1972)
Crop	0.08	Hansche et al. (1972)
Fruit length	0.31	Hansche et al. (1972)
Fruit cheek	0.26	Hansche et al. (1972)
Fruit suture	0.29	Hansche et al. (1972)
Fruit firmness	0.13	Hansche et al. (1972)
Fruit acidity	0.19	Hansche et al. (1972)
Soluble solids	0.01	Hansche et al. (1972)
Juvenility (flower number)	0.16	Hansche and Boyton (1986a)
Intensity of browning	0.35	Hansche and Boyton (1986b)

Synthetics and *Clones* are more often used in cross-pollinating crop species where self-pollination may result in reduced fitness, including inbreeding depression.

2.3.2 *Inbreeding*

Inbreeding typically involves the recurrent inbreeding of populations which are thus more genetically homogeneous than would occur with random mating. Ranging from recurrent selection to the development of inbred pure lines, this strategy is characterized by the selection of transgressive phenotypes in the F_2 to F_7 generations. Inbreeding drives individual loci

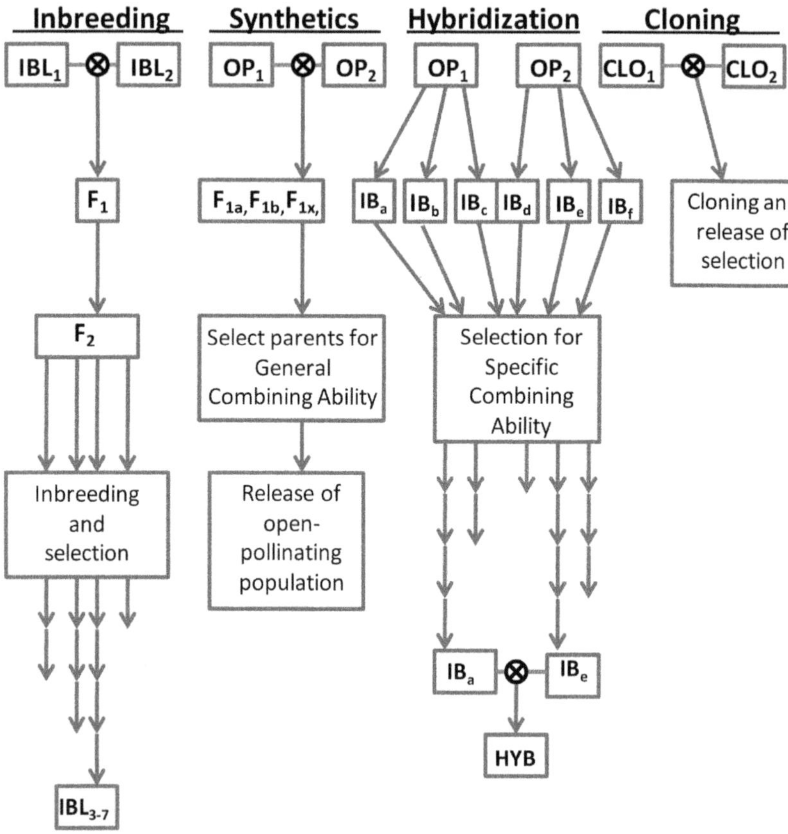

Figure 2-4 Four basic breeding strategies for cultivar development. (The horizontal length of individual boxes roughly reflects genetic variability while the number of vertical arrows approximate number of breeding cycles). In addition to genetic improvement, the capability to generate true-breeding grower seed/propagules is a major determinant of breeding strategy. (Modified from Simmonds and Smartt 1999).

towards homozygosity and so primarily targets additive genetic effects. The increasing level of homozygosity with each inbred generation is a distinct advantage in seed-propagated crops as individuals in advanced inbred lines will be more homogenous and so consequently more true-breeding in seed provided to growers.

2.3.3 Hybridization

Hybridization involves the development of hybrids between inbred parental lines which have been carefully selected for their specific combining ability (typically heterotic vigor or heterosis). Resulting hybrid progeny are genetically uniform (homogenous) yet can be highly heterozygous and so capable of exploiting additive, dominance and epistasis interactions. However, the full exploitation of these genetic effects is limited by the tedious parental combining-ability testing required for each desired inbred line combination (Duvick 2009).

2.3.4 *Synthetics*

Where *Hybridization* involves the selection of inbred parent pairs based on their specific combining ability (as determined by previous assessment of progeny performance), Synthetics involve the selection of a number of genotypes for good general combining ability (i.e., moderate to good heterosis recovered in all possible crossing combinations). While capable of exploiting additive, dominance and epistatic genetic effects, synthetics are generally less efficient in accumulating additive genetic effects than inbreeding with recurrent selection and less efficient than hybridization at capturing dominance and epistasis effects since the realized genetic gain is the average of the many potential hybrids and so difficult to optimize. Because heterosis can be partially maintained in growers' fields through continued natural outcrossing, synthetics have proven particularly useful in perennial forage crops such as alfalfa where annual re-seeding is important.

2.3.5 *Cloning*

Cloning depends upon the capability for asexual or vegetative propagation of the cultivar from the breeding program to the grower's field, and is thus common in perennial, woody crops such as the stone fruit. It usually involves an initial hybridization between two distinct parents, but may also involve self-pollination in genotypes were inbreeding depression is not limiting. Unlike *Inbreeding* and *Hybridization*, there is typically no pre-breeding requirement (i.e., development of early generation inbred lines, etc.) in cloning and, because selected genotypes can be asexually propagated, all genetic potential is essentially captured for the grower without the risk of the often regressive, genetic recombination associated with foundation seed increase for seed propagated crops. Consequently, Cloning can fully capture additive, dominance and epistasis effects in cultivars which then remain true-to-type in subsequent vegetative propagations. The level of genetic gain is limited only by the quality of the breeding parents and the size of the progeny population. Cloning of interspecies hybrids has also been shown to be very effective for the breeding of vigorous and often disease resistant rootstocks for many perennial tree crop cultivars (Rom and Carlson 1987; Kester et al. 2001). Vegetative growth vigor in interspecies hybrids, which is sometimes termed 'luxuriance' to distinguish it from intraspecies hybrid vigor or heterosis (Dobzhansky 1952), can often transgress well beyond that of even a highly-vigorous parent, and appears to involve both gene-gene and even genome-genome interaction.

2.4 Genetic Components of Stone Fruit Fitness

Because breeding strategies differentially exploit the genetic components contributing to final cultivar fitness, the approach ultimately converged upon by crop breeders can often be informative concerning the genetic components critical to that crop. While recurrent mass selection is still commonly used by forest tree breeders (Simmonds and Smartt 1999) and was extensively employed in Europe in the early 1900s for certain low-value stone fruits such as almond (Gradziel 2011), virtually all modern stone fruit breeding programs currently employ versions of the *Cloning* strategy. Early studies by Scorza and Pooler (1993) had suggested the possibility that hybrid vigor or heterosis was exploitable in peach if inbred parent lines could be properly developed though later results of a long-term peach *Inbreeding* and *Hybridization* study by Monet's program in France were considered unsuccessful (Monet and Bassi 2008). Most commercially successful stone fruit cultivars appear to result from an initial hybridization between two distinct parents rather than the more convenient self-pollination, even in crops such as peach and sour cherry which are naturally self-pollinated (Hesse 1975; Yamane et al. 2001) and may show no discernible inbreeding depression (Monet and Bassi 2008). Results from a survey on cultivar origins (hybridization, selfing or sport mutation) for the different stone fruit cultivars having parentage reported in the Brooks and Olmo Register of Fruit and Nut Varieties (Anon 1997) are presented in Fig. 2-5. Even for nectarine, fresh-market peach, and processing peach, where self-pollination is the natural mating system, a hybrid origin clearly dominates despite the relatively large numbers of cultivars evaluated (162, 540, and 22 respectively). Similar trends are seen for apricot, where selfing is readily achieved for a number of major cultivars. No cultivars originating from self-pollination were observed in plums despite self-pollination being common in several plum species including prominent cultivars (Okie and Weinberger 1996) but this may also be a reflection of the prevalence of a more exotic, even interspecies hybrid origins of some plum cultivars. An earlier, comprehensive study by Blazek (1996) on origins of major prune (Table 2-4) and Japanese plum (Table 2-5) cultivars developed during the previous 50 years also document the predominance of hybrid origins, often between parents that were not closely related. Similarly, no cultivars originating from self-pollination were observed for sweet cherry as might be expected in this mostly outcrossing species (Fig. 2-5). Little information was

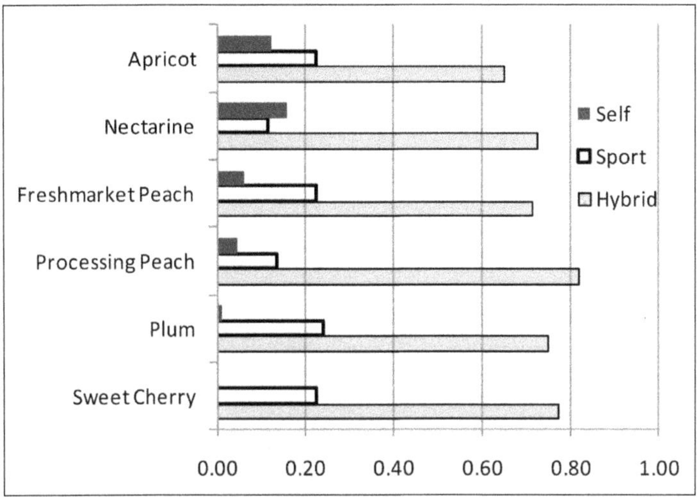

Figure 2-5 Results from a survey on cultivar origins (hybridization, selfing or sport mutation) for the different stone fruit cultivars having parentage reported in the Brooks and Olmo Register of Fruit and Nut Varieties (Anon 1997) showing predominance of hybrid origins as well as importance of origins from mutation.

available for sour cherry cultivars, which are self-fertile, though a Brown et al. (1996) summary of important world cultivars showed only 4 of the 35 listed cultivars which had resulted from self-pollination, all of which were sister lines from the German variety 'Schattenmorelle'. Results from a recent survey of origins for the major (annual production exceeding 500,000 cases) peach and plum cultivars currently grown in California (CDFA 2011) also shows a clear predominance of a hybrid over self-pollination origin (Table 2-6) despite the common breeding practice of selfing which is much less tedious. Hybridization in stone fruits often requires careful emasculation of the seed parent flower (see Fig. 2-6) followed by hand-pollination using previously collected and processed pollen. [For species specific techniques, see reviews by Bailey and Hough (1975) for apricot, Brown et al. (1996) for cherry, Okie and Weinberger (1996) for plum, and Hesse (1975) for peach]. Self-pollination, in contrast involves simply bagging the flowering branch to exclude outside pollen transfer by visiting insects, or merely allowing the flowers to open-pollinate and then using molecular markers to rogue-out the occasional out-cross. [Under field conditions, insect cross-pollination has been shown to be relatively common for plum and peach with out-crossing proportions of 30% or higher reported by Parfitt et al. (1985)].

The observed fitness of hybrids relative to self-pollinations is consistent with the out-breeding nature of most stone fruits where deleterious recessive alleles would be expected to accumulate. Hybridization would encourage greater heterozygosity at these vulnerable loci, where a dominant allele

Table 2-4 Origins of major prune cultivars developed during the previous 50 years. (from Blazek 1996).

Variety	Parentage	Number of new cultivars	% of the Total (190)
Agen	?	13	7
Althanova (Reine Claude Althans)	Green Gage o.p.	12	6
Anna Spath	?	5	3
Auerbacher	?	3	2
Čačanská najbolja	Wangenheimer × Pozegaca	5	3
Czar	Prinz Engelbert × Early Rivers	3	2
Common Prune (Pozegaca)	?	16	8
Early Rivers	?	11	6
Grand Duke	?	7	4
Hall	Golden Drop × Grand Duke	3	2
Imperial Epineuse	?	3	2
Iroquois	Italian Prune × Hall	3	2
Italian Prune	?	13	7
Liivi Kollane Munaploom	?	4	2
Monfort	?	6	3
Ortenauer (Borsumer)	?	4	2
President	?	10	5
Ruth Gerstetter	Czar × Bonne de Bry	10	5
Stanley	Agen × Grand Duke	14	7
Tartu Kaunitar	?	3	2
Tuleu Gras	?	14	7
Tuleu Timpuriu	Tuleu Gras × Peach Plum	4	2
Valor	Imperial Epineuse × Grand Duke	4	2
Victoria	?	3	2
Wangenheimer	?	3	2
Wilhelmine Spath	?	7	4
Unknown origin		33	17
Open-pollinated cultivar		22	12
Bud-sport mutation		4	2

would mask expression of deleterious recessive alleles. An example in peach would be the homozygous recessive eglandular genotype (see in Table 2-1) which is associated with high susceptibility to powdery mildew (*Sphaerotheca pannosa* (Wallr:Fr.) disease (Hesse 1975). Both the homozygous dominant genotype (EE, reniform leaf gland) as well as the heterozygous genotype (Ee, globose leaf gland) show resistance. At certain loci, the heterozygote

Table 2-5 Origins of major Japanese plum cultivars developed during the previous 50 years (from Blazek 1996).

Cultivars	Parentage	Number of new cultivars	% of the total (223)
Angelo	(Queen Ann seedl.) o.p.	3	1
Burbank	?	13	6
Burmosa	Burbank × Formosa	3	1
Casselman	mutant of Santa Rosa	3	1
Eldorado	?	5	2
Eldorado seedling	Eldorado o.p.	4	2
Friar	Gaviota × Nubiana	13	6
Gaviota	?	5	2
Golden King	?	4	2
Grand Rosa	(Eldorado seedl.) o.p.	3	1
King David	Late Santa Rosa × Queen Ann	3	1
Laroda	Gaviota × Santa Rosa	9	4
Mariposa	?	3	1
Methley	?	8	4
Nubiana	Gaviota × Eldorado	3	1
Ozark Premier	Burbank × Methly	5	2
Queen Ann	Gaviota × Eldorado	6	3
Queen Rosa	Queen Ann × Sana Rosa	5	2
Red Beauty	(El Dorado × Burmosa) o.p.	8	4
Santa Rosa	?	15	7
Showtime	Santa Rosa o.p.	4	2
Wickson	Burbank × Kelsey	8	4
Unknown origin		59	26
Open-pollinated cultivar		62	28
Bud-sport mutation		20	9

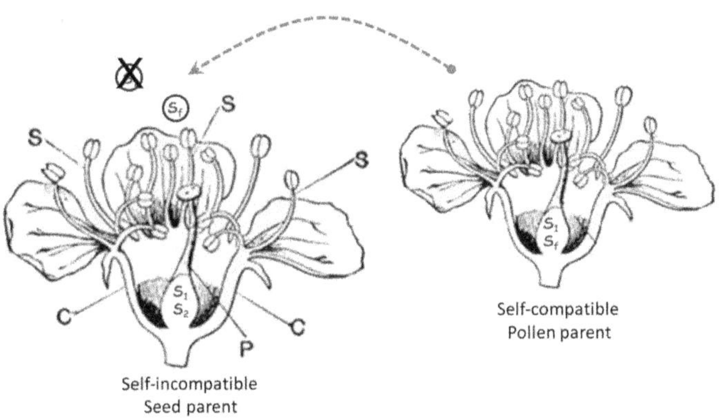

Figure 2-6 Example of a MAB crossing strategy for self-incompatible species such as sweet cherry and almond which ensures that all progeny will inherit the S_f allele and so be self-compatible since all donor pollen not possessing the S_f allele will possess the S_1 pollen which is cross-incompatible on the S_1S_2 seed parent (Gradziel et al. 2001). (S-stamen, P- pistil, C-floral cup. Note that for a controlled cross onto a self-compatible flower as seed parent, the seed parent anthers are removed before pollen dehiscence by pinching and removing the top-half of the floral cup—and so all attached anthers and petals).

Table 2-6 Breeding strategy origins California peach and plum cultivars showing high commercial success as indicated by a harvest of over 500,000 cases in 2010 (CDFA 2011). ('?' -indicates origin implied but not clearly stated in patent description).

Peach	Origin
August Fire	budsport
August Red	OP?
Autumn Flame	hybrid
Brittney Lane	hybrid
Crimson Lady	hybrid
Diamond Bright	hybrid
Elegant Lady	hybrid
Honey Blaze	F2
O'Henry	hybrid
Rich Lady	Amparo OP?
Ruby Diamond	hybrid
Spring Bright	hybrid
Spring Snow	F3?
Summer Bright	F2?
Summer Fire	hybrid
Super Rich	hybrid
Plum	**Origin**
Blackamber	Friar hybrid
Fortune	hybrid
Friar	hybrid
Angeleno	Queen Ann OP?

may also show a fitness advantage over either homozygote, presumably because the greater allelic diversity confers greater overall fitness in differing environments. This situation, sometimes called heterozygote advantage would further encourage hybridization over selfing. A possible example of heterozygote advantage may be the previously described leaf gland loci in peach where the homozygous dominant (EE) phenotype has been associated with greater susceptibility to leaf-curl (*Taphrina deformans* (Burk.)) disease but the heterozygote (Ee) shows relatively greater resistance to both leaf-curl and powdery mildew diseases (Hesse 1975).

Such improved hybrid fitness, which may involve beneficial interactions at the intra-locus, inter-locus and even inter-genomic level (in interspecies hybrids and introgression lines), would confer significant crop performance advantages (Anderson 1949; Abbott 1992; Hancock 2003), particularly in the extensive year by site replicated trials common in stone fruit evaluations. Heterosis may be involved but improved fitness could independently result from the accumulation of other beneficial genetic, inter-locus and genomic interactions. For example, recent reports indicate that a large proportion of the improved fitness observed in maize hybrids result from favorable inter-locus interactions rather than intra-locus based heterosis (Gore et al. 2009; McMullen et al. 2009). While commonly associated with greater vegetative vigor, hybrids often also show improved harvest index (Barber 1970; Duvick 2005, 2009), which in the predominantly spur-bearing stone fruit crops might be expected to confer improved marketable fruit size (i.e. more fruit above the marketable threshold size) as well as increased final yields. Since much of this improved fitness appears to involve non-additive effects, a significant advantage of hybrids over selfs in final stone fruit cultivar performance would seriously undermine the breeding value of MAS as the associated statistical procedures identify primarily additive genetic effects.

2.5 Stone Fruit Breeding Objectives

2.5.1 *Genetic Improvement* vs. *Cultivar Development*

Breeding objectives can be divided into two major categories: genetic improvement and cultivar development. Genetic improvement, sometimes referred to as pre-breeding, typically has a well-defined, focused goal such as improved disease resistance within locally adapted genetic background. In contrast, success at cultivar development is usually indicated by sizable commercial plantings over the long production time required for commercially profitability. For example, a successful processing peach cultivar is expected to have an average production of over 7 metric tons per hectare of marketable fruit and an orchard-life expectancy of at least 20 years in order to be commercially profitable (Gradziel and McCaa 2008). Cultivar success, then, is rarely determined by superior performance in one or a few traits, but rather is determined by the absence of deficiencies for the large number of fruit and tree characteristics required for commercial viability. The need in stone fruit crops to simultaneously optimize a large number of essential traits remains the greatest challenge to breeding strategies including the use of MAS and other molecular-based techniques.

In genetic improvement, the specific strategy utilized for trait manipulation will depend on the nature of genetic control. Genetic control is traditionally classified into three groups: monogenic, oligogenetic, and polygenic, each of which has unique opportunities and limitations.

2.5.2 *Monogenic Traits*

In a monogenic trait, the controlling gene will segregate in a classic single gene Mendelian ratio (Figs. 2-1 and 2-2) which can be readily manipulated provided that gene expressivity and penetrance are sufficiently high to be discernible from environmental background. Dihybrid (2 genes) ratios are also simple enough to also be considered within this group. MAS is particularly effective in the identification and manipulation of monogenic traits with low expressivity and penetrance.

A unique advantage of *clone*-based breeding methods is the ability to accumulate desirable single gene mutations (sometimes referred to as point mutations). Naturally occurring mutations are often identified as bud-sports (novel phenotypes originating from a single bud) which while typically occurring with relatively low probabilities, become increasingly likely with larger planting size and time periods. In addition to such spontaneous mutations, mutations can be induced through the use of various carcinogens (see Brown et al. 1996; Scorza and Sherman 1996). Desirable mutations in an established cultivar have the advantage of providing a discrete improvement in an otherwise well-established genotype, (i.e., a cultivar whose cultural management has already been well worked out), making them very desirable commercially. The commercial value of cultivars originating from bud-sports is well documented by the large numbers listed in Fig. 2-5 and Tables 2-4, 2-5 and 2-6.

For several reported bud sports, however, molecular analysis has shown the actual origin was from sexual crosses (Dangl et al. 2009). Often the cultivar was bud-propagated onto rootstocks of a same species but the scion bud failed and a rootstock bud grew instead. Where the rootstock phenotype is clearly different, it is recognized as an 'escape' and rebudded. But where it is similar enough, it is frequently mistaken as a bud sport, which, if of sufficiently good quality may be propagated as a new cultivar.

Examples of beneficial bud-sports include changes in harvest date (Ramming 1991), fruit quality (Callahan et al. 2004), tree structure (Gradziel and Beres 1993) and change from a pubescent to glabrous fruit skin type (Okie and Prince 1982). Induced mutations, while rarer, can be also be valuable, as in the induction of self-compatibility and compact, spur-type bearing habit in sweet cherry (Brown et al. 1996). While bud sports and induced mutations typically are limited and discrete genetic changes, the risk of negative pleiotropic effects requires careful field evaluations of these altered genotypes before commercial release as cultivars (Marcotrigiano and Gradziel 1997). Sometimes apparent pleiotropic effects are the result of closely linked genes rather than a secondary effect of the primary gene mutation. An example is the common association of the nectarine trait with reduced fruit size which plagued early breeding efforts. Large sized nectarine genotypes were eventually recovered after extensive breeding efforts to break the relatively tight linkage of nectarine with small fruit size (Hesse 1975).

Genetically-modified organisms (GMOs) are in many ways similar to induced mutations as they provide the opportunity to introduce a discrete new trait to an otherwise well established genetic background, but also run the risk of introducing undesirable characteristics (either through pleiotropy or multiple insertion events) and so also require extensive field testing before release as an improved cultivar.

Traits controlled by 1 to 2 gene(s) can also be readily transferred to locally adapted genetic backgrounds through recurrent selection, as might occur with mass-selection, selfing/inbreeding, and backcrossing. Because of the longer generation time and smaller progeny population sizes typical of stone fruit crops, recurrent backcrossing is often utilized as it allows a more efficient concurrent improvement in both recurrent population and targeted traits. An example of recurrent backcrossing program to transfer *Monilinia* fruit rot resistance from the Brazilian cultivar *Bolinha* to California peach breeding lines is presented in Fig. 2-7. Hybridization is also a very efficient strategy for recombining multiple monogenic traits when controlled by dominant genes. In this approach, which is used extensively in vegetable crop breeding but only rarely and fruit crop improvement, the two parents are selected for the both presence of complementary dominant genes as well as their specific combining ability (Scorza and Pooler 1993; Duvick 2009).

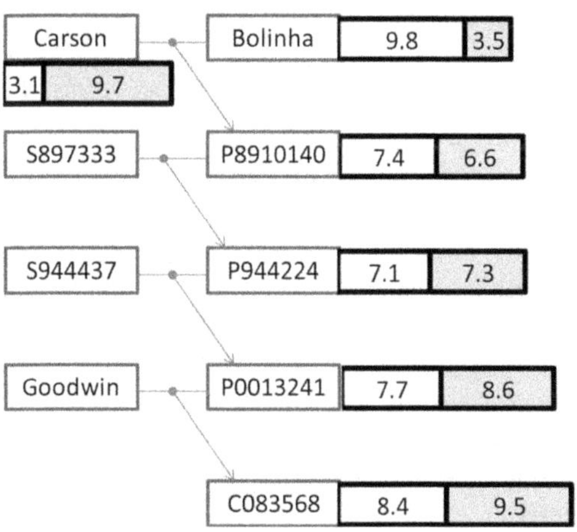

Figure 2-7 Results of a recurrent backcrossing program to transfer *Monilinia* fruit rot resistance from the Brazilian cultivar *Bolinha* to California peach breeding lines. Horizontal bar-graph shows changes in resistance score [unshaded; 10= high] and flesh color quality [shaded] with backcross generation.

2.5.3 Oligogenic Traits

For oligogenic traits, which are controlled by a relatively few genes, the expected Mendelian segregation ratios become increasingly complex and so increasingly difficult to distinguish from background environmental variance (Fig. 2-2). MAS and associated molecular marker strategies should be particularly effective for oligogenic manipulation provided the number of genes remains relatively low. As the number of controlling genes increases, the additive value of individual genes diminishes as does its final breeding value. More significantly, as the number of genes contributing additive affect to a certain trait such as crop yield increases, the population size required to ensure that individuals will be present that possess all or even most of the desired genes becomes prohibitively large (Fig. 2-3) even if effective molecular markers were identified for all targeted genes. In these situations, molecular markers can be employed to identify parents homozygous for some of the desired alleles, which could then be fixed in subsequent progeny populations. By such sequential and recurrent selection/fixation, additional targeted loci can be concentrated in the progeny populations though many of the multitude of other genes required for commercial success are often lost from the recurrent breeding population in the process.

Given the long seed-to-seed generation time and limited progeny population sizes typical of stone fruit, this molecular-based recurrent selection approach has proven more effective in cereal and vegetable improvement relative to fruit crop improvement. However, the improved understanding of the genetic control of targeted traits made possible by molecular analysis may be of considerable value to the breeder and may lead to novel breeding strategies. For example, some of the earliest application of MAB in stone fruit improvement was the development of molecular markers which could allow the identification/selection of individuals possessing desired cross-compatibility genotype (*S*-alleles, including S_f, the self-compatibility allele) at the seedling stage and so eliminate timely process of tree growth and field screenings (Tao et al 1997; Yamane et al. 2001; Barckley et al. 2006; Guerra et al. 2010). However, because self-compatibility/incompatibility is a gametophytic trait in the targeted crops, knowledge of parental genotypes was sufficient to devise crossing combinations (Fig. 2-6) which essentially ensured all progeny would be self-compatible (Gradziel et al. 2001) without the need for extensive molecular analysis of progeny.

2.5.4 Polygenic Traits

As genetic control for a given trait becomes more complex, Mendelian segregation ratios becomes less discernible against the environmental background and the trait is analyzed instead in terms of the probabilities of its expression using appropriate statistical analysis (Mather and Jinks 1977; Falconer and Mackay 1996; Bos and Caligari 2008). This can occur with genetic control by as few as three genes for low heritability traits, and for four or more genes even for traits showing moderate heritabilities. The statistical or quantitative methods employed are typically reactive in their analysis, (i.e., previously established, segregating populations are prerequisite to predicting future progeny performance). With recurrent selection strategies such quantitative analysis becomes increasingly accurate as each new generation informs and improves upon the overall genetic model. Although quantitative methods have been developed to distinguish additive from dominance effects, the statistical approaches currently used largely precludes a reliable analysis of actual intra- and inter-locus interactions (Luby and Shaw 2001).

Efficient quantitative methods are similarly not available for manipulating genome-genome and associated epigenetic interactions (see Comai 2000; Steimer et al. 2004; Grant-Downton and Dickenson 2006). Part of the reason is that these interactions remain poorly understood and also are not readily captured and manipulated by traditional breeding methods developed for seed propagated crops. Cloning, however, can capture even highly complex and poorly understood genetic interactions making it arguably the most efficient breeding technique for combining, in true-breeding cultivars, the fullest range of desirable genetic, epigenetic and genomic interactions. This capacity makes cloning particularly promising for the characterization and eventual manipulation of these largely underutilized interactions. Towards this goal, however, molecular-based approaches may have to move beyond the current emphasis on DNA-based markers. For example, clone analysis offers unique opportunities for the study of epigenetic interactions since different and often heritable phenotypes (juvenility, imprinting, gene-silencing, etc.) of the same clone (genotype) in the same environment would be the expression of epigenetic rather than genetic factors. For example, Noninfectious Bud-failure in almond appears to be an epigenetic-like clonal aging condition where the genetic (DNA) composition of affected cultivars remains unchanged but where gene activity is altered in a heritable manner (Kester et al. 2004). Although it is a major production problem in almond, it appears a poor candidate for MAS since the DNA sequence appears identical in both affected and unaffected genotypes (see Dangl et al. 2009). Similarly, genome-genome interactions which appear to play important roles in enhancing vegetative vigor as characterized by interspecies hybrid rootstocks (Armstrong 1957; Rom and Carlson 1987; Kester et al. 2001), or fruit quality/productivity in various interspecies plum hybrids (Blazek 1996; Okie and Weinberger 1996) appear be the result of both genetic as well as genomic differences between the parents, possibly including differences in chromosome orientations, scaffold structure, histone composition, methylation patterns, synteny differences, etc. (see Grant-Downton and Dickenson 2006). Although providing valuable tools for a more thorough dissection/characterization of these crop improvement opportunities, molecular-genetic analysis, as currently employed, may ultimately hinder breeder utilization of these germplasm resources because of its very specialized and so inherently reductionistic focus.

2.5.5 Cultivar Development

The definitive aim of plant breeding is the development of successful cultivars. A successful cultivar can be conveniently defined as providing a net improvement over the cultivar to be replaced. That is, it must be at least as good as the cultivar it is to replace in the areas of horticultural, fruit quality, post-harvest, market, etc., yet possess improvements valuable enough to result in sizable commercial plantings. As demonstrated in accompanying chapters, powerful genetic improvement strategies are becoming available to reliably develop superior genotypes. The major barrier to successful stone fruit cultivar development, however, is not the process of genetic improvement but rather the process of simultaneously maintaining commercial quality for the wide range of other essential traits. This is the reason bud-sport mutations have been an important source of new cultivars (Fig. 2-5, Tables 2-4, 2-5 and 2-6) since they can confer a distinct improvement to an otherwise genetically unaltered, commercially proven cultivar. A well established dogma of stone fruit breeding is that the success of a new cultivar is determined not by its exceptional performance in specific areas but rather a uniformly superior performance across a broad range of characteristics or traits. Consequently, it is the absence of serious deficiencies which will ultimately determine its commercial success. This is particularly relevant in tree fruit crops were orchards are expected to be productive for 10 years or more (after a 2 to 5 year vegetative stage) in order to be commercially viable, and where failed cultivars cannot be readily plowed under and replanted as with cereal and vegetable crops. Jared Diamond (1998) has termed this decisive vulnerability to a broad spectrum of potential deficiencies the *Anna Karenina* effect based on Leo Tolstoy's classic opening sentence in his novel of that name: *"All happy families are alike; each unhappy family is unhappy in its own way."* The importance of the *Anna Karenina* effect is shown in results from a recent survey where stone fruit breeders were asked to rank the relative importance of a series of diverse fruit traits in order to identify potential targets of MAS (Table 2-7). While traits differed among crop species (for example, self-fertility was only ranked in the normally self-incompatible sweet cherry), all traits pertinent to any given species were consistently ranked in the highest-priority categories (Yue et al. 2011). In addition to fruit quality, similarly high rankings would be expected for the numerous traits in a broad range of essential categories, including tree structure, productivity and longevity, disease and insect resistances, harvest time, uniformity and ease-of-harvest, precocity, freedom from alternate bearing, post-harvest performance, rootstock compatibility, market type,

Table 2-7 Ratings of selected stone fruit traits as prioritized by applied cultivar breeding programs in the United States as part of a study to identify promising targets for MAB (Yue et al. 2011).

Trait Name	Sweet Cherry	Tart Cherry	Peach	Apricot	Plum
Fruit firmness	5	5	4.88	5	4.67
Skin color	5	5			
Fruit size	5		4.63		4.67
Flavor	4		4.6	5	4.67
Fruit shape		5	4.71		
Flesh color		5			
Sweetness			4.63	5	5
Soluble solids(Brix)			4.5	5	5
Productivity			4.57		5
Production consistency		5	4.63		4.67
Extended harvest season	5			5	4.67
Fruit uniformity		5	4.75		
Pre-harvest dropping				5	4.5
Fruit juiciness	4				
Pit shape and size		5			
Pit splitting and fragments		5			
Machine harvest ability		5			
Titratable acidity				5	
pH				5	
Aromatics/volatiles				5	
Heat tolerance				5	
Storage disorders					4.5
Heat tolerance			4.57		
Resistance to frost injury	5				
Powdery mildew	5				
Other disease-viral	5				
Self fertility	5				
Graft compatibility		5			

consumer preference, etc. Thus, while genetic improvement may benefit from a focused, reductionistic approach to trait improvement, successful cultivar development requires the simultaneous, holistic manipulation of a large number of essential traits. As demonstrated in Fig. 2-3, a traditional additive-gene based MAS approach would quickly become overwhelmed by the number of required markers. This incongruity, while complicating cultivar development may also be undermining future progress. Genetic improvement strategies, including MAS, are becoming increasingly efficient at the partitioning and so manipulating the principal additive genetic interactions affecting the target trait, but because they are resource intensive, these inherently reductionist approaches may lead to reduced effectiveness of successful stone fruit cultivar development if not fully complemented with the equally essential holistic cultivar development approaches.

Tree crops are also unique in that successful cultivars need to demonstrate this superior fitness over a much broader range of environmental variation (position on tree, planting site, age of tree, age of clone, varying disease and insect pressures, changes in climate and weather pattern, etc.). Superior fitness over a range of environments is ultimately more important to final cultivar productivity than exceptional performance within a narrow environmental niche (Janick 2006). Tree crop cultivars thus resemble the clonal colony or 'genet' of ecology such as the "Pando" clone of Quaking Aspen (*Populus tremuloides*) in the Wasatch Mountains of Utah, USA, which cover 43 hectares and so is often considered the world's largest organism by mass (Mitton and Grant 1996). Since the estimated 47,000 individual clonal trees ('ramets' in ecological terms) which constitute the *Pando* clone have developed under a wide range of differing ecological niches, its competitive advantage appears to result from a broad adaptability rather than being highly adapted to a specific niche. In comparison, the *Nonpareil* almond clonal cultivar is planted on over 116,000 hectares (at over 48 trees/ha) in California alone (Almond Board of California 2010), with additional plantings in Europe, Asia, North Africa, South America and Australia. Although there are other cultivars which out-yield or have higher market value than Nonpareil in certain production areas and periods (Micke et al 1997; Socias i Company et al. 2008), Nonpareil continues to dominate this crop because of a superior overall fitness (i.e., economic returns over the typical orchard life of 20+ years) with over 5.6 million trees currently planted over an ecologically diverse 700 km stretch of central California ranging from Redding, CA near the Oregon border to Bakersfield, north of Los Angeles.

Improved environmental buffering has been shown to be associated with the higher genetic heterozygosity typical of most stone fruit cultivars (Mojayad and Planchon 1994; Haussmann et al. 2004; Duvick 2009). Even with predominantly inbreeding species such as peach, with their higher levels of homozygosity, recombination from hybridizations would increase the opportunity for beneficial inter-locus rearrangements. Though relatively rare, such desirable rearrangements once selected would be largely fixed by linkage disequilibrium (Falconer and Mackay 1996; Luby and Shaw 2001) leading to the equivalence of heterosis over extended selection periods (Gore et al. 2009; McMullen et al. 2009). Extended periods of selection for broad environmental adaptability (for many stone fruits, selection has been occurring for hundreds to thousands of years) would thus identify rare, elite selections where the maximum potential of additive, dominance, epistatic, genomic and epigenetic interactions were combined (see Janick 2006). Clonal propagation allows the capture of these rare elite genotypes for future plantings as well as future genetic improvements through bud-sport mutation or further, albeit rare, favorable recombinations. Inbreeding would be deleterious to such buffered fecundity, which could help explain the preponderance of hybrids versus self-pollinations in successful stone fruit cultivars (Fig. 2-5 and Tables 2-4, 2-5, and 2-6). MAS when applied to multiple traits is inherently inbreeding.

Bearing in stone fruits is on shoots and spurs. Spurs production, with its highly efficient fruit-to-vegetative ratio (harvest index) often dominates crop production in mature, highly productive orchards. (Peach would be a notable exception where the cultural practice of severe annual pruning requires strong vegetative as well as reproductive vigor). In stone fruits, these spur or shoot based production-units are also considerably autonomous and competitive, that is carbohydrate flow is primarily from nearby leaves to the more competitive local sink, whether individual developing fruit or vegetative apices (Grossman and DeJong 1995; Lang et al. 2004). While the clone (genet) or individual tree (ramet) may be the target of natural selection of wild clonal colonies, the fruiting-spur/fruiting-shoot (*compet*) may prove as important or more important in the more synthetic and often more intensive breeding selection. Recent research in almond has shown that the most productive orchards are those where the quantity (number), quality (fecundity) and longevity of spurs are optimized during the multi-year, peak production phase of the planting (Gradziel and Lampinen 2011).

Productivity remains the most important attribute in most new cultivars but because of its complexity and all-inclusive nature, is often managed as a nebulous quantitative trait which frustrates a more thorough analysis and manipulation by both traditional as well as biotech approaches. Molecular-approaches such as association mapping, offer unprecedented opportunities to more fully characterize important components of yield as a basis for future genetic and cultural manipulation but require a more detailed understanding of the biological basis. It is informative how biotech progress over the last three decades has advanced to the point where sequencing individual stone fruit breeding lines can now be readily achieved, yet our understanding of the physiological and developmental components of a trait as critical as yield has made only rudimentary progress over the same time period. This precarious biological knowledge-base, along with the traditionally insular nature of molecular genetic analysis remains a major impediment to more efficient cultivar breeding in the stone fruits (Luby and Shaw 2001; Gradziel 2009). The inherent capacity of clone-based cultivars to capture the fullest range of beneficial genetic, epigenetic and genomic interactions for applied crop improvement provides both a prerequisite and unique opportunity for researchers to evolve beyond the current reductionistic additive-gene approach, but would require (perhaps stimulate) significant parallel progress in our understanding of the basic underlying developmental and inheritance mechanisms at the epigenetic and genomic as well as genetic level. An even greater challenge/opportunity would be the progression from the present focus on single trait genetic improvement to an emphasis on the concurrent management/advancement of the multitude of traits required for commercial success, i.e., cultivar breeding.

References

Abbott RJ (1992) Plant invasions, interspecific hybridization and the evolution of new plant taxa. Tree 7: 401–405.

Almond Board of California (2010) Almond industry position report, 2009–2010 crop year. Available at http://www.almondboard.com/files/

Anderson E (1949) Introgressive Hybridization. John Wiley & Sons, New York.

Anonymous (1997) The Brooks and Olmo Register of New Fruit and Nut Varieties. ASHS Press, Alexandria, Virginia.

Armstrong DL (1957) Cytogenetic study of some derivatives of the F_1 hybrid *Prunus amygdalus* × *P. persica*. PhD thesis, University of California, Davis.

Bailey CH, Hough LF (1975) Apricots. In: Janick J, Moore JN (eds) Advances in Fruit Breeding. Purdue University Press, West Lafayette, Indiana, pp 367–384.

Bailey JS, French AP (1942) The inheritance of blossom type and blossom size in the peach. Proc Am Soc Hort Sci 40: 248–250.

Bailey JS, French AP (1949) The inheritance of certain fruit and foliage characters in the peach. MA Agri Exp Sta Bull 452.

Barber HN (1970) Hybridization and the evolution of plants. Taxon 19: 154–160.

Barckley KK, Uratsu SL, Gradziel TM, Dandekar AM (2006) Multidimensional analysis of S-alleles from cross-incompatible groups of California almond cultivars. J Am Soc Hort Sci 131: 632-636.

Bassi D, Rizzo M (2000) Peach breeding for growth habit. Acta Hort 538: 411–414.

Beckman TG, Sherman WB (2003) Probable qualitative inheritance of full red skin color in peach. Hort Science 38: 1184–1185.

Beckman TG, Rodriguez Alcazar J, Sherman WB, Werner DJ (2005) Evidence for qualitative suppression of red skin color in peach. Hort Science 40: 523–524.

Bielenberg DG, Wang Y, Fan S, Reighard GL, Scorza R, Abbott AG (2004) A deletion affecting several gene candidates is present in the evergrowing peach mutant. J Hered 95: 436–444.

Blake MA (1937) Progress in peach breeding. Proc Am Soc Hort Sci 35: 49–53.

Blake MA, Connors CH (1936) Early results of peach breeding in New Jersey. NJ Agri Exp Sta Bull 599.

Blazek J (1996) A survey of genetic resources used in plum breeding. Acta Hort 734: 31–45.

Bos I, Caligari P (2008) Selection Methods in Plant Breeding, 2nd edn. Springer, New York, 461 p.

Brown SK, Iezzoni AF, Fogle HW (1996) Cherries. In: Janick J, Moore JN (eds) Fruit Breeding. Vol 1: Tree and Tropical Fruits. Wiley, New York, pp 213–255.

Burbank L, Hall W (1927) The Harvest of the Years. Houghton Mifflin, Boston, MA, 258 p.

Callahan AM, Scorza RM, Bassett C, Nickerson M, Abeles FB (2004) Deletions in an endopolygalacuronase gene cluster correlate with non-melting flesh texture in peach. Funct Plant Biol 31: 159–168.

CDFA (2011) California Department of Fruit and Agriculture. http://www.cdfa.ca.gov/statistics/

Chaparro JX, Werner DJ, O'Malley D, Sederoff RR (1994) Targeted mapping and linkage analysis of morphological, isozyme, and RAPD markers in peach. Theor Appl Genet 87: 805–815.

Chaparro JX, Werner DJ, Whetten RW, O'Malley DM (1995) Inheritance, genetic interaction, and biochemical characterization of anthocyanin phenotypes in peach. J Hered 86: 32–38.

Comai L (2000) Genetic and epigenetic interactions in allopolyploid plants. Plant Mol Biol 43: 387–399.
Connors CH (1920) Some notes on the inheritance of unit characters in the peach. Proc Am Soc Hort Sci 16: 24–36.
Connors CH (1921) Inheritance of foliar glands of the peach. Proc Am Soc Hort Sci 18: 20–26.
Couranjou J (1995) Genetic studies of 11 quantitative characters in apricot. Sci Hort 61: 61–75.
Dangl, Gerald S, Yang J, Gradziel T, Golino DA (2009) A practical method for cultivar identification and parental analysis using simple sequence repeat markers. Euphytica 168: 41–48.
Diamond J (1998) Guns, Germs, and Steel: the Fates of Human Societies. W.W. Norton, New York.
Dirlewanger E, Cosson P, Boudehri K, Renaud C, Capdeville G, Tauzin Y, Laigret F, Moing A (2006) Development of a second generation genetic linkage map for peach [*Prunus persica* (L.) Batsch.] and characterization of morphological traits affecting flower and fruit. Tree Genet Genomes 1: 1–13.
Dobzhansky T (1952) Nature and origin of heterosis. In: Gowen JW (ed) Heterosis. Iowa State College Press, Ames, IA, pp 218–223.
Duvick DN (2005) The contribution of breeding to yield advances in maize (*Zea mays* L.). Adv Agron 86: 83–145.
Duvick DN (2009) Selection Methods: Hybrid Breeding. In: Ceccarelli S, Guimaraes EP, Weltzien E (eds) Plant Breeding and Farmer Participation. FAO, Rome, pp 229–258.
Falconer DS Mackay TFC (1996) Introduction to Quantitative Genetics. Burgess Press, New York.
Gore MA, Chia J-M, Elshire RJ, Sun Q, Ersoz ES, Hurwitz BL, Peiffer JA, McMullen MD, Grills GS, Ross-Ibarra J, Ware DH, Buckler ES (2009) A first-generation haplotype map of maize. Science 326: 1115–1117.
Gradziel TM, Lampinen BL (2011) Defining the limits of almond productivity to facilitate marker assisted selection and orchard management. Acta Hort 912: 47–52.
Gradziel TM (2011) Almond origin and domestication. In: Janick J (ed) Horticultural Reviews 38: 23–82.
Gradziel TM, McCaa JP (2008) Processing Peach Cultivar Development In: Layne DR, Bassi D (eds) The Peach: Botany, Production and Uses. CAB International, New York, pp 175–192.
Gradziel TM (2009) Tree nut genomics: In danger of missing the forest for the leaves. In: Plant Anim Genomes XVII Conf: http://www.intl-pag.org/17/abstracts/W25_PAGXVII_209.html.
Gradziel TM, Beres W (1993) Semidwarf growth habit in clingstone peach with desirable tree and fruit qualities. Hort Science 28(10): 1045–1047.
Gradziel TM, Martinez-Gomez P, Dandekar AM (2001) The use of S-allele specific PCR analysis to improve breeding efficiency for self-fertility in almond. Hort Science 36: 440.
Grant-Downton RT, Dickinson HG (2006) Epigenetics and its implications for plant biology 2. The epigenetic epiphany': Epigenetics, evolution and beyond. Ann Bot 97: 11–27.
Grossman YL, DeJong TM (1995) Maximum fruit growth potential following resource limitation during peach growth. Ann Bot 75: 561–567.
Grossman YL, DeJong TM (1998) Training and pruning system effects on vegetative growth potential, light interception, and cropping efficiency in peach trees. J Am Soc Hort Sci 123: 1058–1064.
Guerra ME, Rodrigo J, Wünsch A, López-Corrales M (2010) Self-incompatibility in Japanese plum–S-allele genotyping of cultivars. Acta Hort (ISHS) 874: 169–174
Hancock JF (2003) Plant Evolution and the Origin of Crop Species. CABI, London
Hansche PE (1988) Two genes that induce brachytic dwarfism in peach. Hort Science 23: 604–606.

Hansche PE, Hesse CO, Beres V (1972) Estimate of genetic and environmental effects on several traits in peach. J Am Soc Hort Sci 97: 9–12.
Hansche PE, Boynton B (1986a) Heritability of juvenility in peach. Hort Science 21: 1195–1197.
Hansche PE, Boynton B (1986b) Heritability of enzymatic browning in peaches. Hort Science 21: 1197–1198.
Haussmann BIG, Obilana AB, Ayeicho PO, Blum A, Shipprack W, Gieger HH (2000) Yield and yield stability of four population types of grain sorghum in a semi-arid area of Kenya. Crop Sci 40: 319–329.
Hesse CO (1975) Peaches. In: Janick J, Moore JN (eds) Advances in Fruit Breeding. Purdue University Press, West Lafayette, IN, pp 285–335.
Janick J (2002) Ancient Egyptian agriculture and the origins of horticulture. ActaHort 582: 23–39.
Janick J (2006) Origins of fruit culture and plant breeding. In: Lamkey KR, Lee M (eds) Plant Breeding: The Arnel R. Hallauer International Symposium. Blackwell Publ, Victoria, Australia, pp 269–282.
Kester DE, Shackel KA, Micke WC, Viveros M, Gradziel TM (2004) Noninfectious bud failure in 'Carmel' almond: I. Pattern of development in vegetative progeny trees. J Am Soc Hort Sci 129(2): 244–249.
Kester DE, Asay RN, and Gradziel TM (2001) 'Nickels' almond x peach hybrid clonal rootstock. Hort Science 37(2): 415–417.
Kislev ME, Hartmann A and Bar-Yosef O (2006) Early domesticated fig in the Jordan valley. Science 1372–1374.
Lammerts WE (1945) The breeding of ornamental edible peaches for mild climates. I. Inheritance of tree and flower characters. Am J Bot 32: 53–61.
Lang GA, Olmstead JW, Whiting MD (2004) Sweet cherry leaf distributions and leaf populations: Modeling canopy dynamics and management strategies. Acta Hort (ISHS) 636: 591–599.
Ledbetter CA (1996) Characterization and inheritance of the crinkle-leaf trait in peach and peach–almond hybrids. J Genet Breed 50: 5–60.
Lesley JW (1940) A genetic study of saucer fruit shape and other characters in the peach. Proc Am Soc Hort Sci 37: 218–222.
Lownsberry BF, Thomson IJ (1959) Progress in nematology related to horticulture. Proc Am Soc Hort Sci 74: 730–746.
Lu ZX, Reighard GL, Nyczepir AP, Beckman TG, Ramming DW (2000) Inheritance of resistance to root-knot nematodes (*Meloidogyne* sp.) in *Prunus* rootstocks. Hort Science 35: 1344–1346.
Luby JJ, Shaw DV (2001) Does marker assisted selection make dollars and sense in a fruit breeding program? Hort Science 36: 872–879.
Marcotrigiano M, Gradziel TM (1997) Genetic mosaics and plant improvement. In: Janick J (ed) Plant Breeding Reviews 15: 43–84.
Mather K, Jinks JL (1977) Introduction to Biometrical Genetics. Cornell University Press, Ithaca, NY.
McMullen MD, Kresovich S, Villeda HS, Bradbury PJ, Li H, Sun Q, Flint-Garcia S, Thornsberry J, Acharya C, Bottoms C, Brown P, Browne C, Eller M, Guill K, Harjes C, Kroon D, Lepak N, Mitchell SE, Peterson B, Pressoir G, Romero S, Oropeza Rosas M, Salvo S, Yates H, Hanson M, Jones E, Smith S, Glaubitz JC, Goodman M, Ware D, Holland JB, Buckler ES (2009) Genetic properties of the maize nested association mapping population. Science 325: 737–740.
Mehlenbacher SA, Scorza R (1986) Inheritance of growth habit in progenies of 'Com-Pact Redhaven' peach. Hort Science 21: 124–126.
Micke WC, Kester DE, Gradziel TM, Yeager JT, Thorpe MA, Connell JH, Verdegaal PS, Viveros M (1997) Almond cultivar evaluation using regional trials. Acta Hort 470: 91–94.

Mitton JB, Grant MC (1996) Genetic variation and the natural history of Quaking Aspen. Bioscience 25: 1–31.
Mojayad F, Planchon C (1994) Stomatal and photosynthetic adjustment to water deficit as the expression of heterosis in sunflower. Crop Sci 34: 103–107.
Monet R, Bassi D (2008) Botany and Taxonomy. In: Layne DR, Bassi D (eds) The Peach: Botany, Production and Uses. CABI, London
Monet R (1967) Contribution à l'étude génétique du pêcher. Annal l'Amélior Plant 17: 5–11.
Monet R, Bastard Y (1982) Une anomalie du fonctionnement de l'apex à hérédité mendélienne chez le pêcher (*Prunus persica* (L.) Batsch). Agronomie 2: 103–106.
Monet R, Massonié G (1994) Déterminisme génétique de la résistance au puceron vert (Myzus persicae) chez le pêcher. Résultats complémentaires. Agronomie 2: 177–182.
Monet R, Salesses G (1975) Un nouveau mutant de nanisme chez le pecher. Annal Amélior Plant 25: 353–359.
Nicolosi E, Deng ZN, Gentile A, Malfa SL, Continella G, Tribulato E (2000) Citrus phylogeny and genetic origin of important species as investigated by molecular markers. Theor Appl Genet 100: 1155–1166.
Okie VR, Weinberger JH (1996) Plums: In: Janick J, Moore JN (eds) Fruit Breeding. Vol I: Tree and Tropical Fruits. John Willey, New York, pp 559–607.
Okie WR (1988b) Preliminary descriptions of five new peach genes. Acta Hort 465: 107–110.
Okie WR, Prince VE (1982) Surface features of a novel peach × nectarine hybrid. Hort Science 17: 66–67.
Ortiz R, Dochez C, Asiedu R, Moonan F (2006) Breeding vegetatively propagated crops. In: Lamkey KR, Lee M (eds) Plant Breeding: The Arnel R. Hallauer International Symposium. Blackwell Publ, Victoria, Australia, pp 269–282.
Parfitt DE, Arulsekar S, Ramming DW (1985) Identification of plum × peach hybrids by isoenzyme analysis. Hort Science 20: 246–248.
Peace CP, Crisosto CH, Gradziel TM (2005) Endopolygalacturonase: a candidate gene for freestone and melting flesh in peach. Mol Breed 16: 21–31.
Pedrosa A, Schweizer D, Guerra M (2000) Cytological heterozygosity and the hybrid origin of sweet orange [*Citrus sinensis* (L.) Oseck]. Theor Appl Genet 100: 361–367.
Ramming DW (1991) Genetic control of a slow-ripening fruit trait in nectarine. Can J Plant Sci 71: 601–603.
Rodriguez AJ, Sherman WB, Scorza R, Wisniewski M (1994) 'Evergreen' peach, its inheritance and dormant behavior. J Am Soc Hort Sci 119: 789–792.
Rom RC, Carlson RF (1987) Rootstocks for Fruit Crops. John Wiley, New York.
Salesses G, Al-Kai N (1985) Simply inherited grafting incompatibility in peach. Acta Hort 173: 57–62.
Scorza R, Pooler M (1993) Development and testing of F1 hybrid peaches as an alternative peach production strategy. Hort Science 28: 455.
Scorza R, Sherman WB (1996) Peaches. In: Janick J, Moore JN (eds) Fruit Breeding. Vol 1: Tree and Tropical Fruits. Wiley, New York, pp 325–440.
Scorza R, Bassi D, Liverani A (2002) Genetic interaction of pillar (columnar), compact and dwarf peach tree genotypes. J Am Soc Hort Sci 127: 254–261.
Scorza R, Lightner GW, Liverani A (1989) The Pillar peach tree and growth habit analysis of compact × Pillar progeny. J Am Soc Hort Sci 114: 991–995.
Scott D, Cullinan F (1942) The inheritance of wavy leaf character in the peach. J Hered 33: 293–295.
Scott DH, Weinberger JH (1944) Inheritance of pollen sterility in some peach varieties. Proc Am Soc Hort Sci 45: 229–232.
Sharpe RH, Hesse CO, Lownsberry BF, Perry VG, Hansen CJ (1970) Breeding peaches for root-knot nematode resistance. J Am Soc Hort Sci 94: 209–212.
Simmonds NW, Smartt J (1999) Principles of Crop Improvement. Blackwell Science, London
Socias i Company R, Kodad O, Alonso JM, Gradziel JTM (2008) Almond quality: A breeding perspective. In: Janick J (ed) Horticultural Reviews 34: 197–238.

Soost RK, Roose ML (1996) Citrus. In: Janick J, Moore JN (eds) Fruit Breeding. Vol 1: Tree and Tropical Fruits. Wiley, New York, pp 257–322.

Steimer A, Schob H, Grossniklaus U (2004) Epigenetic control of plant development: new layers of complexity. Curr Opin Plant Biol 7: 11–19.

Weinberger JH, Math PC, Scott DH (1943) Inheritance study of root-knot nematode resistance in certain peach varieties. Proc Am Soc Hort Sci 42: 321–325.

Werner DJ, Chaparro JX (2005) Genetic interaction of pillar and weeping peach genotypes. Hort Science 40: 18–20.

Werner DJ, Creller MA (1997) Inheritance of sweet kernel and male sterility. J Am Soc Hort Sci 122: 215–217.

Werner DJ, Creller MA, Chaparro JX (1998) Inheritance of the blood-flesh trait in peach. Hort Science 33: 1243–1246.

Yamane H, Tao R, Sugiura A, Hauck NR, Iezzoni AF (2001) Identification and characterization of S-RNases in tetraploid sour cherry (*Prunus cerasus* L.). Hort Science 33: 1243–1246.

Yamazaki K, Okabe M, Takahashi E (1987) Inheritance of some characteristics and breeding of new hybrids in flowering peaches. Bull Kanagawa Hort Exp Sta 34: 46–53.

Yue C, Gallardo K, Liu L, Yang N, McCracken V, Jussaume R, Taylor M, Luby J (2011) Using socio-economic knowledge of stakeholder values and consumer preferences to inform breeding; http://www.rosbreed.org/socio-economics.

3

Diversity Analysis

María José Aranzana,[1,*] *Teresa Barreneche*[2] and *Pere Arús*[1]

ABSTRACT

Stone fruit trees belong to the *Prunus* genus, originally from Asia, and spread widely to the other continents, where they adapted and diversified. Biodiversity, particularly agrodiversity, is important in agriculture, helping to ensure sustainability and productivity, becoming an important resource in plant breeding. The characterization and evaluation of genetic diversity among cultivars and available germplasm is essential for its preservation. Traditionally, variability has been described visually, however, over the past few years molecular markers have been used. In this chapter we review the current bibliography to describe morphological and molecular diversity of stone fruit trees.

Keywords: agrodiversity, genetic diversity, molecular markers, phenotypic diversity, stone fruit trees

3.1 Introduction

Stone fruits (peach, apricot, European and Japanese plums and sweet and sour cherries) belong to the *Prunus* genus. According to Redher (1947) and Krüsmann (1978) this genus comprises respectively 77 or 140 wild and cultivated species, some economically important as sources of fruits, nuts, oil, rootstock, timber, and ornamentals. Redher divided *Prunus* into five subgenera. Among them, the subgenus *Prunophora* contains plums (section *Euprunus*) and apricots (section *Armeniaca*), the subgenus *Amygdalus* includes peaches and almonds and the subgenus *Cerasus*, cherries. Of Asian origin,

[1]IRTA, Centre de Recerca en Agrigenòmica CSIC-IRTA-UAB-UB. Campus UAB, Bellaterra (Cerdanyola del Vallès), 08193 Barcelona, Spain.
[2]INRA, UR419, Unité de Recherches sur les Espèces Fruitières, BP 81, F-33140 Villenave d'Ornon, France.
*Corresponding author: *mariajose.aranzana@irta.cat*

the centers of diversity of stone fruits are Eastern China (apricot, peach and *P. salicina*) and Central and Western Asia (almond, cherries, *P. insititia, P. domestica*and *P. cerasifera*) from where they spread to Eastern Europe (Moore and Ballington 1991) (Table 3-1).

Stone fruit cultivation is reported to be quite old. Wang (1985) suggests that peach has been grown in China for at least 4,000 years and plums for over 2,000 years (Ramming and Cociu 1991). As stone fruits spread to other areas they adapted and diversified to form landraces. These local populations, resulting from centuries of natural and human selection, represent a valuable source of germplasm (e.g., almond cultivars from Morocco, "vineyard" peaches from different European countries, Mirabelle from Eastern France). Additionally, stone fruit improvement programs have released cultivars which should also be preserved.

Unlike major food crops such as rice, wheat, corn and potato, there is no international center for the preservation of stone fruits, so a world

Table 3-1 Cultivated *Prunus*, closely-related species and origin area (afterRedher 1947; Moore and Ballington 1991).

Subgenus	Section	Species	Common name	Origin
Prunophora	*Euprunus*	*P. spinosa* L.		SW Asia
		P. insititia L.	Damson	W Asia, Europe
		P. domestica L.	Plum	W Asia, Europe
		P. cerasifera Ehrh.	Cherry plum	W Asia
		P. salicina Lindl.	Japaneseplum	China
	Armeniaca	*P. mandshurica* (Maxim.) Koehne		E Manchuria
		P. sibirica L.		Manchuria, N Korea, NE China
		P. armeniaca L.	Apricot	N and NE China
		P. mume (Sieb) Sieb. et Zucc.		Central China, S Korea, SE Japan
		P. X *dasycarpa*		
Amygdalus	*Euamygdalus*	*P. amygdalus* Batsch	Almond	SW Asia
		P. mira		W China-Himalayas
		P. persica (L.) Batsch	Peach	China
		P. davidiana (Carr.) Batsch		N China
		P. tangutica		N China
		P. triloba Lindl		N E China
		P. tenella		SE Europe, W Asia
		P. ferganensis (KostRjab) Kov1 Kost*		NE China
		P. kansuensis Rehd*		NW China
Cerasus	*Eucerasus*	P. *avium* L.	Sweet cherry	W Asia, Europe
		P. *cerasus* L.	Sour cherry	W Asia, SE Europe
		P. *fructicosa* Pall.		

*not mentioned by Rheder

collection of their genetic resources has never been defined. Stone fruit genetic resources are maintained in multiple sites internationally, so an exhaustive analysis of the phenotypic variability available is difficult.

The main countries carrying out stone fruit breeding are involved in the preservation of their genetic resources, including mainly collections of improved materials (Moore and Ballington 1991). In Europe, the preservation of these genetic resources is managed within the framework of the European Cooperative Programme for Crop Genetic Resources (ECPGR) under the auspices of Bioversity International (Gass et al. 1996). In the ECPGR *Prunus* Working Group, 20 countries cooperate on common objectives. These include the survey of national collections, the elaboration of descriptors for their documentation and the development of the European *Prunus* database (EPDB), completed some years ago (Zanetto et al. 2002). The ECPGR *Prunus* Working Group now focuses on the identification of accessions to be included in EURISCO (web-based catalog that provides information on *ex situ* plant collections maintained in Europe), the harmonization of the fingerprinting of stone fruit varieties and the improvement of the EPDB.

The EPDB is hosted and managed at INRA, Bordeaux. It contains data on European collections of all *Prunus* species, cultivated stone fruits and their related species, representing12,756 accessions from 59 institutes in 19 countries (Table 3-2). The online version of the EPDB (http://cbi.labri.fr/outils/EPDB/index.html) currently only displays data for cherry, but, in the future, data will be accessible on five distinct species (almond, apricot, cherry, peach and plum) and inter specific hybrids, providing passport characterization and evaluation data (Maggioni et al. 2010).

In the USA, the preservation of stone fruits is coordinated by the National Plant Germplasm System (NPGS) (http://www.ars-grin.gov/npgs/index.html). In NPGS, different repositories are in charge of stone fruits and related species preservation. Among them, the National Germplasm Repository in Davis, California, maintained the largest collection of stone fruits and related species, with 1,644 accessions of 93 species from 61 different countries (Table 3-3). Thanks to the GRIN (Germplasm Resources Information Network), stone fruit accessions and their passport and evaluation data can be obtained (http://www.ars-grin.gov/npgs/searchgrin.html). In Japan, the National Institute of Agrobiological Sciences (NIAS;http://www.gene.affrc.go.jp/about-plant_en.php) is the central coordinating institute of a network of institutes involved in conservation of genetic resources. The NIAS plant database (http://www.gene.affrc.go.jp/databases-plant_search_en.php) provides passport (8 passport descriptors) and evaluation data (29 descriptors) of stone fruits and related species, with peach as the largest collection (Table 3-4).

Aware of the risk of loss of stone fruit genetic resources, there have been national and international initiatives, more particularly in European

Table 3-2 Number of accessions to be included in the EPDB (European *Prunus* Data Base) (Maggioni et al. 2010).

Country	Almond	Apricot	Cherry	Peach	Plum	Hybrids	Total per country
Belgium	1	2	485		153	35	676
Czech Republic		81	328		128		537
France	50	70	183	337	70		710
Germany			338		188		526
Greece	39	5	30	26			100
Hungary	106	408	209	201	421		1345
Israel	40	169	49	245	102	13	618
Italy	94	129	1063		87		1373
Norway		3			21		24
Poland			123		127	1	251
Portugal			24				24
Romania	49	56	103	34	292	14	548
Slovakia	14	15	43	8	49		129
Slovenia	46						46
Spain	214		121	281	93	21	730
Swizerland					371		371
Turkey	253	236	203	85	276		1053
Ukraine	97	752	1321	97	890		3157
United Kingdom	1		464	20	32	21	538
Total (19 countries)	1004	1926	5087	1334	3300	105	12756

Table 3-3 *Prunus* genetic resources maintained in the different repositories of the NPGS (http://www.ars-grin.gov/npgs/searchgrin.htm).

Site	Accessions	Countries	Species
National Germplasm Repository- Davis California (DAV)	1644	61	93
National Germplasm Repository-Geneva New York (GEN)	130	13	11
National Arboretum Maryland (NA)	84	7	23
North Central Regional PI Station Iowa (NC7)	1	1	1
NationalCenter for Genetic Resources Preservation Colorado (NCGRP)	10	2	5
Holdings at Plant Germplasm Quarantine Program Maryland (PGQO)	498	30	27
Western Regional PI Station Washington (W6)	15	1	4
Total	2382		

Table 3-4 Stone fruits and related species maintained in the NIAS (http://www.gene.affrc.go.jp/about-plant_en.php).

	Species	Accessions
Apricot	*Prunus armeniaca*	12
Cherry	*Prunus avium*	56
Plum	*Prunus cerasifera*	11
Cherry	*Prunus cerasus* hybrid	1
Cherry	*Prunus cerasus*	1
Peach	*Prunus davidiana*	1
Plum	*Prunus domestica*	10
Plum	*Prunus maritima*	1
Japanesea pricot	*Prunus mume*	81
Peach	*Prunus persica*	224
Plum	*Prunus salicina*	45
Peach	*Prunus triloba*	2
Plum	*Prunus spinosa*	1
Total		446

countries, to collect and preserve stone fruit landraces (Leterme 1995; Iezzoni 2005; Nikolić et al. 2010). For instance, the recovery and evaluation of local endangered varieties of almond from the traditional productive areas of the Mediterranean basin has been carried out recently within the SAFENUT project (http://safenut.casaccia.enea.it/). Due to the gradual decrease of diversity within stone fruits and their wild relatives in the countries of origin, caused mainly by extension of improved cultivars, urbanization and the pressure on trees for fuel, there has been a major effort to draw up an inventory, collect, preserve and manage stone fruits genetic resources, wild and cultivated. The countries concerned, e.g., Turkey, Azerbaijan and China, received support from the FAO (State of Plant Genetic Resources for Food and Agriculture in China 2008; Tan A 2009; National report on the state of Plant Genetic Resources for Food and Agriculture in Azerbaijan 2006).

An inventory, the description and the assessment of genetic diversity of stone fruits are essential to optimize the preservation and management of their genetic resources and to provide information and resources useful for breeders. For a long time, despite the great diversity existing and reported in stone fruits and related species, exploration of phenotypic variability was often restricted to a limited number of genotypes. Broad phenotypic diversity studies of large and diverse germplasm collections of stone fruits are still quite rare. In the last fifteen years, with the development of molecular markers, there has been an explosion of genotype-based diversity studies involving large collections as well as the study of relationships between closely related species.

3.2 Molecular Markers for Variability Analysis

Historically, genetic variability in plants was described visually. Despite the considerable quantity of morphological descriptors, the large amount of stone fruit germplasm resources and the prolific development of new cultivars demand additional and more direct indicators of genetic variability. Its study in germplasm became feasible with the development of molecular markers; due to their abundance and importance in plant genomes, they have been widely used to, for example, analyze relationships between wild and cultivated individuals and even between closely related species, to describe germplasm variability and for cultivar identification.

The first molecular marker technique used in plant breeding was based on isozymes (Tanksley and Orton 1983). These are structural variants of an enzyme that have the same catalytic activity but differ from the original in molecular weight and mobility in an electric field. Mobility differences may be the result of polymorphism (point mutations or insertions-deletions) in the coding sequence of a gene. Isozymes were first described in 1957 (Hunter and Markert), but they were first used as molecular markers in plants in 1966 (Brewbacker 1966), and in stone fruit trees in 1985 (Parfitt et al. 1985). Isozymes are codominant markers, robust and reproducible, but the limited number of isoforms may be a drawback in low polymorphic species. In the past, isozymes were amongst the most widely used molecular markers in variability studies. As representative examples of the use of this technique in stone fruit trees we cite the following studies: in apricot (*Prunus armeniaca*) by Badenes et al. (1996), Battistini and Sansavini (1991), Manganaris et al. (1999) and Milatovic et al. (2009); in sweet and sour cherry populations (*Prunus avium* L and *P.*cerasus, respectively) by Beaver et al. (1995), Mariette et al. (1997), Pashkoulov et al. (2000) and Gomory and Paule (2001); and in peach (*Prunus persica* L. Batsch) by Durham et al. (1987), Messeguer et al. (1987), Ibanez et al. (1993) and Monet et al. (1994).

Isozymes were soon superseded by more informative DNA markers with greater rates of polymorphism, including restriction fragment length polymorphism (RFLP) (de Vicente et al. 1998; Quarta et al. 2000), random amplified polymorphic DNA (RAPD) (Warburton and Bliss 1996; Ortiz et al. 1997), amplified fragment length polymorphism (AFLP) (Hagen et al. 2002; Aranzana et al. 2003a), but they are still a cheap and quick method to identify low levels of genetic variability.

However, since the late 1990s simple sequence repeats (SSRs) have been the most popular markers for genetic variability studies. These markers are based on PCR amplification of genetic regions containing highly polymorphic tandem repeats of di- to tetra-nucleotide sequence motifs, which are distributed in the genome. These markers are highly polymorphic even in low variable species (Morgante and Olivieri 1993).

The development of SSR primers in stone fruits was first published by Ciprinai et al. (1999), followed by Sosinski et al. (2000), Testolin et al. (2000), Aranzana et al. (2002) and Dirlewanger et al. (2002). These markers were initially developed from peach genomic sequences, and soon after from other *Prunus* species such a *savium* (Cantini et al. 2001; Clarke and Tobutt 2003), *cerasus* (Downey and Iezzoni 2000; Cantini et al. 2001), *salicina* (Mnejja et al. 2004), *domestica* (Decroocq et al. 2004) and *armeniaca* (Lopes et al. 2002; Hagen et al. 2004; Messina et al. 2004; Vilanova et al. 2006). SSRs have since been broadly used for cultivar identification and variability studies in stone fruit cultivated germplasm as well as for studying relationships between cultivated and wild relatives. Due to their high transferability between closely related species, these markers have been used to study *Prunus* inter specific variability (see Section 3.4 of this chapter).

The major drawback initially was the need for sequence information, then SSRs were discovered by either development of SSR libraries or expressed sequence tag (EST) sequencing. Recent advances in sequencing technologies and the constant reduction in cost have resulted in the massive generation of sequences, allowing the rapid discovery of SSRs. For example, the whole peach genome sequence has recently been released (http://www.rosaceae.org/gb/gbrowse/prunus_persica/), and online tools designed for identifying tandem repeat patterns, such as the Tandem Repeats Finder (TRF, Benson 1999), allow massive development of the markers.

Moreover, massive sequencing technologies (known as next generation sequencing) have resulted in the identification of an enormous amount of single nucleotide polymorphisms (SNPs). These polymorphisms are distributed throughout the genome and are reported as fundamental in genetic variation. Nowadays, SNP markers are widely used in animal and human genome analysis, but their use in plants, and particularly in stone fruit trees, is in the early stages. So far, the recent availability of EST data sets has made possible the discovery of putative SNPs *in silico*, using bioinformatics tools that permit the selection of polymorphisms, which occur multiple times within aligned sequences available in databases. Databases, such as ESTree (http://www.itb.cnr.it/estree/) for peach and the genomic database for Rosaceae (GDR, http://www.rosaceae.org/), now contain thousands of SNPs useful for diversity analysis and for mapping. Recently, the release of the peach genome sequence has triggered the design of peach re-sequencing strategies allowing the discovery of thousands of SNPs (Blas et al. in preparation). In the framework of an international initiative, a set of 9,000 SNPs have been selected for inclusion in a genotyping chip (Verde et al. 2012) for peach variability, linkage mapping and association mapping analysis as part of two major international projects: RosBREED (http://www.rosbreed.org/) and FruitBreedomics (http://fruitbreedomics.com/).

Peach is a model Rosaceae species, so results are expected to be transferable to other closely-related stone fruit trees.

3.3 Phenotypic and Genotypic Variability

As referred in Mehlenbacher et al. (1991), in 1969 Kostina reported the first, and one of the most exhaustive, phenotypic analyses of stone fruit genetic resources (apricot germplasm). Based on 700 genotypes of *Prunus armeniaca* L., the author identified four eco-geographical groups of the cultivated apricot: the Dzhungar-Zailij group, the Central Asian groups, the Irano-Caucasian group and the European group. This classification is still used by breeders.

Apricot is believed to have originated in China. Wang (1998) reported 2,000 varieties or types preserved today in China, corresponding to two-thirds of the world varieties. Central Asia, Afghanistan, Kashmir, Iran, Turkey and Trans-Caucasia are areas where there is diversification of the cultivated apricot and it can still be found associated to wild forms. In order to assess the value for further breeding of an apricot germplasm collection from the European group, Badenes et al. (1998) evaluated 50 cultivars from the main growing areas in the Mediterranean and some American cultivars over three years for 18 quantitative traits (tree habit, phenology and fruit quality). The variability observed was smaller than expected; the authors reported high correlations between phenological traits and identified good candidates for their breeding purposes, in particular concerning local adaptability, fruit size and fruit quality. In another study, dealing with the Irano-Caucasian group, 128 apricot cultivars from different cultivation sites in Anatolia (Turkey) were evaluated for 15 characteristics (tree size, phenology and fruit quality). In this study, wide variation was found in important traits such as harvest season and fruit yield but the authors did not find correlations for phenological traits. High correlations between fruit characteristics were pointed out (Asma and Ozturk 2005).

Apart from pomological and phenotypic descriptors, molecular markers have been used to assess germplasm variability as well as to determine the genetic relationships among genotypes from different eco-geographical groups. Isozymes were the first markers used to assign cultivars to the groups of origin (Byrne and Littleton 1989; Battistini and Sansavini 1991). Isozymes revealed low variability, reaching different levels depending on geographical origin. Milatovic et al. (2009), studying 50 apricot cultivars of diverse geographical origin with 18 enzyme systems, found that variability was highest in the cultivar groups originating from Central Asia and former USSR countries, while European and North American cultivars showed a medium level of variability, and those from Serbia the least polymorphic. A similar east-west gradient of diversity was reported by Hagen et al. (2002) after studying 47 apricot cultivars with AFLP markers.

De Vicente et al. (1998) analyzed a collection of 52 North-American and European apricot cultivars with RFLP markers. These authors identified synonymies and characterized the variability in different growing areas, finding that the Spanish cultivars presented lower variability levels, probably a result of major erosion of variation in the Spanish gene pool due to a random drift having occurred in the past. To overcome the limitations of this reduced pool of variability in Spanish varieties, the authors propose the search of new sources of variability in southern European regions, with similar environmental conditions to Spain. Indeed, a low plasticity has been reported in apricot adaptability to different edaphoclimatic conditions (Layne et al. 1996), which conditions the material to be used in breeding programs. SSR markers have also been used to study apricot genetic variability. For example, Romero et al. (2003) analyzed 40 cultivars of different origin to determine the genetic relationships among their genotypes, detecting that western European and North American subgroups clustered together, in agreement with their common origins from ancient European cultivars.

As apricot, peaches (*Prunus persica* L.) and their wild relatives are native to China, where a broad variability occurs. The term peach refers to the domesticated *Prunus persica*, with the species *davidiana, kasuensis, mira* and *ferganensis* its wild relatives. *P. persica* is the only species grown for fruit besides the local use of *P. mira* and *P. ferganensis* in China (Wang 2005) with yellow and white flesh seedlings of considerably good quality. Crosses between these closely related wild relatives and between wild and domesticated peaches have given good quality rootstocks. Foulongne et al. (2004) studied genetic diversity between cultivated peaches and their wild relatives. A cluster analysis revealed the distinction between the peaches and related wild species according to their geographical origin. Variability analysis indicates the use of wild material as a good source of variability in peach breeding programs.

Wang (1985) reported more than 1,000 peach cultivars in China, but, for a long time, information concerning Chinese germplasm was rare. Surveys of diversity of large germplasm collections have recently started and results will be available soon. Phenotypic variability studies of peach germplasm dealt rather with landraces that evolved in other countries. Pérez et al. (1993) evaluated 52 accessions from Mexico, South American countries and some improved US cultivars. Twenty eight morphological and phenological traits were recorded, and the authors reported high degrees of variability in characters related to local adaptation, such as bud break and harvest season. Two clusters were identified separating Mexican and other Latin American populations from Brazilian and improved US cultivars. In another study, 30 peach accessions selected in a collection of 945 accessions from Serbia were studied, using 21 different traits (tree habit, phenology and fruit). Major correlations were found between the traits representing fruit size and between those representing fruit quality. Hierarchical cluster analysis and principal component analysis revealed a large phenotypical variability in the vineyard peach germplasm. Six clusters were identified, some of them grouping genotypes according to their potential use (fresh consumption versus suitability as rootstocks) (Nikolić et al. 2010).

Although some commercial varieties have their origin in locally adapted landraces, most of the worldwide peach industry now relies on a few, good quality varieties such as 'Chinese Cling', which were extensively used in both Asian and Occidental commercial breeding programs. These programs began intensively 50 to 100 years ago and have been very dynamic in the development of new varieties, focused on the improvement of fresh and canned peaches, nectarines, flat peaches and flat nectarines according to local market preferences.

The high levels of morphologic variability observed are not in agreement with genome-wide variability. Genetic studies have reported low variability levels in peach commercial varieties in both Chinese (Xie et al. 2010) and Occidental markets (Aranzana et al. 2003a, 2003b, 2010), which is a consequence of selfing and the use of limited parental materials in breeding programs. Today, efforts are addressed to explore and detect new sources of variability for introduction in breeding programs, analyzing and comparing variability between germplasm collections from different origins. Warburton et al. (1996) studied, using RAPD markers, the relationship among US, Latin American, Indian, Pakistani, Russian, Japanese and Chinese cultivars. These authors found limited levels of variability and suggested the use of Asian material to expand genetic variability, given that, as the center of origin, peach variability should be higher there.

Several diversity studies in Chinese germplasm collections have recently been tackled. Although variability in Chinese commercial varieties is low, it increases considerably when analyzing landraces, making them a valuable source of new variability (Cheng et al. 2009). Li et al. (2008) selected a primary core collection of *P. persica* containing 51 samples, classified according to fruit characteristics into six variety groups, and analyzed the musing SSRs. Despite the low number of cultivars in each subgroup, they were ranked according to their variability levels as follows: sweet peach, crisp peach, flat peach, nectarine, honey peach and yellow fleshed. The joint analysis of Asian and Occidental commercial peaches with the same set of close to 50 SSRs has revealed that, despite the limited levels of variability found in both collections, they are complementary and the Asian commercial varieties could be used as a source of variability in Occidental programs and vice versa (Li et al. in preparation).

Cherries belong to more than 50 species of the *Prunus* genus. The most important cultivated cherries are sweet cherry (*Prunus avium* L.) and sour cherry (*Prunus cerasus* L). A wide diversity has been reported in European local cultivars and landraces of these species. *P aviumis* a diploid species and includes sweet cherry trees cultivated for human consumption and wild cherry trees used for their timber. Sweet cherry is native to Asia, especially northern Iran, Ukraine and countries south of the Caucasus, while the ancestors of modern cultivated sweet cherries are thought to have originated around the Caspian and Black Seas (Dirlewanger et al. 2009). Sweet cherry cultivation extended through Europe during the time of the Roman Empire, but they were not widely grown until the 16th century (Watkins 1976). Thanks to large levels of genetic variability in the germplasm, sweet cherry adapted to locally diverse climate regions, and is now cultivated in temperate, Mediterranean, and even subtropical areas.

Studies of ancient, or those directly derived from ancient, sweet cherry cultivars with SSR markers clearly divided them according to their geographical origin, separating those from southern and northern Europe (Wünsch and Hormaza 2002). Sweet cherries were not commercially exploited until a century ago, and new varieties were released from breeding programs. Comparison of chloroplast DNA of wild and cultivated cherry collections reveals a reduction of genetic variability in cultivated collections as a consequence of a bottleneck (Panda et al. 2003). Mariette et al. (2010) studied population structure and genetic bottleneck in sweet cherry, analyzing 141 landraces and 66 modern cultivars with 26 SSR markers. Their results suggest that sweet cherry domestication could have occurred through a complex process involving intense gene flow and secondary domestications instead of a single domestication event. On the other hand, the considerable reduction of variability detected by Mariette et al. (2010), with a loss of as much as 40% of alleles in modern cultivars, indicates that the use of few founders in modern breeding programs was the reason for the bottleneck.

Cherries are self-incompatible, which has major implications in the development of new varieties through breeding programs. In natural populations, self-incompatibility is also an important factor with a significant effect on gene flow and genetic diversity. Variability studies in different population have revealed that different selection processes, including balancing selection, have shaped allele frequencies at the *S*-locus as well as high levels of gene flow between populations (Schueler et al.2006). The study of diversity in self-incompatibility alleles by Mariette at al. (2010) quantified the domestication and breeding bottleneck as 20% and 30% respectively, with a total bottleneck of 50% at the *S*-locus.

P. cerasus (sour cherry) is an allotetraploid species and genomic studies have revealed that it is the result of a natural cross between wild *P. avium* (producing unreduced gametes) and *P. fruticosa*, a tetraploid wild species used as rootstock for other *Prunus* species. It is widespread over the major part of central Europe, Siberia and Northern Asia (Hedrick 1915), showing a wide range of variability for both plant habit and fruit characteristics, exhibiting all grades of color, shape, taste and fruit size (Krahl et al. 1991; Turkoglu et al. 2010). Several studies have pursued the characterization of sour cherry variability. Rakonjac et al. (2010) collected 41 genotypes from the most important growing regions in Serbia and analyzed natural variation in 29 morphological traits (tree habit, phenology, yield and fruit quality). The highest degree of variability was observed in traits related to number and composition of fruiting twigs, fruit set and yield. Negative correlations between bloom time and fruit weight were reported. All variables related to fruit size correlated with one another, similar results to those obtained in a study involving more than 300 French and European landraces of sweet cherry over a period of two years (E. Balsemin, personal communication). Three groups of related genotypes were separated and the authors identified good candidates for breeding purposesin the collection: extension of the blooming season, low vigor, large fruit size, good fruit quality and high yield. In contrast to sweet cherry, natural sour cherry selections include both self-incompatible (SI) and self-compatible (SC) types, and several publications have been oriented to find self-incompatibility loss of function alleles in sour cherry (Tsukamoto et al. 2006, 2008, 2010).

P. x gondouinii, duke cherry, is an allotetraploid species resulting from crosses between *cerasus* and *avium*. Duke cherries are less cultivated than sweet and sour cherry, and have fruit and tree characteristics in-between both progenitors. Morphological traits of cherry species vary with geographical distribution. For example, in Western Europe *P. cerasus* trees look like *P. avium*, whereas in Eastern Europe *P. cerasus* trees look like *P. fructicosa* (Krahl et al. 1991). In consequence, the visual distinction between *P. avium*, *P. cerasus* and the hybrid *P. x gondouinii* may not be easy and requires the use of molecular markers. Tavaud et al. (2004) used AFLP markers to study the genetic relationship between these three species, identifying specific markers for the F genome present only in the tetraploid species, so allowing differentiation between the diploid and tetraploid cherries.

Commercial plums are generally classified in two groups: the European plum, *Prunus domestica* L. (hexaploid) and related species, and the Japanese plum *Prunus salicina* Lindl. (diploid). *P. domestica* was initially considered to be derived from a cross between *Prunus espinosa* and *Prunus cerasifera* (Crane et al. 1934). More recent studies discard this hypothesis and propose that the origin of *P. domestica*is in polyploids arising from *P. cerasifera*, at least in its maternal lineage (Horvath et al. 2011), probably followed by secondary hybridization with other species (Eryomine 1990; Zohary and Hopf 1993; Reales et al. 2010; Horvath et al. 2011).

The fruits of plums are quite diverse; several authors have listed 1,000–2,000 plum cultivars (Ramming and Cociu 1991). Genetic diversity of plum has been studied using a wide range of morphological and molecular markers. In order to evaluate phenotypic variability in the French national collection of plums, 80 varieties were studied using 19 quantitative traits related with morphology of flowers, leaves, fruit size and fruit quality, and stone size. The authors reported that fruit characteristics such as weight, length or width, differed among morphological groups of plums traditionally known, such as Agen and mirabelle plums, greengages and damsons. Multivariate analysis did not allow clustering but some plum genotypes belonging to the same morphological group were close together (Horvath et al. 2011). Hend et al. (2009) evaluated 27 accessions (25 collected in northern Tunisia and two introduced varieties) with 25 morphological and phenological traits as well as with RAPD markers. The authors found a wide molecular and morphological variability distributed in two major clusters, independent of geographic origin. RAPD markers were also used by Shimada et al. (1999) to investigate gene flow between European and Japanese plums. Later, to compare diploid and hexaploid plums, Goulao et al. (2001) used AFLP and inter-simple sequence repeat (ISSR) markers to characterize 24 *P. salicina* and four *P. domestica* cultivars, emphasizing a clear separation between diploid and hexaploid plums. Ayanoglu et al. (2007) studied the diversity of economically important Turkish green plum (*P. cerasifera*) accessions, while Ilgin et al. (2009) classified 14 plum cultivars into seven major clusters using four pairs of AFLP markers. SSR markers have also been successfully used to identify plum varieties and rootstocks as well as for parentage studies (Ahmad et al. 2004; Bianchi et al. 2004; Decroocq et al. 2004; Hovarth 2011).

3.4 Variability Comparison between Crop Species of the *Prunus* Genus

The measure of comparative variability between species was only possible after the development of markers based on proteins or DNA. A review of results in a broad range of plant species using isozymes was provided by Gottlieb (1981). Isozymes, particularly those detected with specific substrates, are suitable for variability comparisons as they are based on homologous loci and most isozyme genes have a codominant pattern of inheritance that allows complete genotyping. Gottlieb (1981) found major differences in the level and pattern of distribution between self-pollinating and out crossing species. In the latter, higher levels of variability existed within each population, measured by higher values of polymorphism (P=proportion of polymorphic loci over the total studied) and heterozygosity (H= proportion of heterozygous loci in the average individual). Alleles with high frequency were usually common to all populations of out crossing species whereas populations usually differed due to the presence of rare alleles. In self-pollinating species, the overall level of variability was lower, intra-population variation was very low, and most of the variability could be found between populations.

Considering that most *Prunus* species have a self-incompatibility (SI) system that prevents self-pollination, they would be expected to behave as out crossing species, with a similar variability distribution. Population analysis in *Prunus* crops is particular, however, because they are vegetatively propagated on the commercial scale, and it is not generally possible to work with conventional populations, with data for genetic analysis usually obtained from a collection of commercial cultivars or breeding lines that are treated as a single population. Using this approach, there are several publications of isozyme variation in *Prunus*. One of them (Arulsekar et al. 1986) showed that almond variability for 12 isozyme genes was very high (P=0.75) when compared to that of peach (P=0.08). A more comprehensive study (Byrne 1990) compared the variability of peach, almond, Japanese plum and apricot based on a set of 12 isozyme genes in all species but almond, where nine genes were studied. The results showed that almond and Japanese plum were highly variable, apricot intermediate, and confirmed that peach variability was low. These results are in agreement with the degree of presence of self-compatible genotypes in these species. Almond and Japanese plum, with robust SI systems, behaved essentially as out crossers; peach, which is self-compatible, had variability levels similar to those of autogamous species, and apricot, that includes many self-compatible genotypes within the cultivated varieties, has a reduced level of variability compared to the out crossing species. Cherry was not part of this study, but results from Granger et al. (1993), in a collection of cherry cultivars using a set of 12 isozymes most in common with Byrne (1990), showed that its level of variability was similar to that of apricot. The extensive use of a self-compatible cultivar ('Stella') in the breeding of modern cultivars may explain, in part, these results.

The markers that were developed later did not allow for species variability comparisons as with isozymes. RFLPs required the use of the same set of probes in different species, and for sequence-unspecific markers such as RAPDs or AFLPs, comparisons were inadequate due to their dominant inheritance and homoplasy. SSRs were again an opportunity, but comparisons of the results of different authors were difficult as different sets of SSRs in different species were often used and SSRs are often species-specific, sometimes making it hard to find common sets of SSRs for different species, even within the same genus (Varshney et al. 2005). In *Prunus*, Dirlewanger et al. (2002) showed that transferability across species was high, allowing the use of many common SSRs, the majority from peach, in variability studies. Later, Mnejja et al. (2010) reported the analysis of a large set of SSRs (145) from different *Prunus* in a set of cultivars of nine species, including six *Prunus* (peach, almond, apricot, cherry, Japanese plum, and European plum) and three other rosaceous crops (apple, pear and strawberry). The data established that 63.9% of the SSRs used amplified and were polymorphic in the *Prunus* species studied, and allowed comparison of the variability between different species using a large set of markers and a common number of genotypes (eight per species).

As in the case of isozymes, SSR polymorphism was much higher in species with a strong SI system (almond, Japanese plum), intermediate in apricot and cherry and low in the self-compatible peach (Table 3-5). The ranking of variability in this group of species was exactly the same using either marker type, and far more differences in the amount of variability were detected by SSRs in comparison to isozymes. Therefore, these data confirm the relative levels of molecular variability among *Prunus* species and show the high efficiency of SSRs for detecting polymorphism in any species. This is especially true in the case of species with low variability such as peach, where identification of genotypes is difficult with isozymes (Messeguer et al. 1987) while a few SSRs can separate individually even closely related cultivars (Aranzana et al. 2010).

Given the fact that hybridization between different *Prunus* crops, other than cherry, is often possible, these data provide information on where to search for new alleles to enrich the gene pool of less variable species, such as peach. Due to the close relatedness, almond appears as a potential donor of genes for peach and Japanese plum for apricot, although other sources of variability may be sought within the wild *Prunus* species, including *P. cerasifera, P. mira, P. davidiana, P. kansuensis, P. mume,* which may supply genes for resistance to pests and diseases or genes involved in climate adaptation or fruit quality lacking in the crop species.

Table 3-5 Comparison of variability parameters for isozymes and SSRs of five diplod *Prunus* crops.

	Isozymes				SSRs			
Species	P	A	Ho	Reference	P	A	Ho	Reference
Almond	78	2.29	0.28	Byrne (1990)	74	5.8	0.62	Mnejja et al. (2010)
Japanese Plum	67	2.75	0.25	Byrne (1990)	73	4.5	0.62	Mnejja et al. (2010)
Apricot	58	2.29	0.13	Byrne (1990)	67	3.2	0.55	Mnejja et al. (2010)
Cherry	56	-	-	Granger et al. (1993)	41	3.1	0.51	Mnejja et al. (2010)
Peach	8	2.00	0.02	Byrne (1990)	43	2.6	0.41	Mnejja et al. (2010)

[1]P= % polymorphic loci; A = average number of alleles per polymorphic locus; Ho= observed heterozygosity of polymorphic SSRs

3.5 Conclusions

Nikolai I. Vavilov (1887–1942), a Russian botanist and plant breeder, demonstrated the existence of 'centres of origin' of cultivated plants, in which can be found the highest level of genetic variability of a species. Biodiversity, and particularly agrodiversity, is important in agriculture, helping to ensure sustainability and productivity, making it an important resource in plant breeding. Thus, the need to characterize and preserve the genetic variability is crucial for future breeding projects.

Stone fruit trees (peach, apricot, European and Japanese plums, and sweet and sour cherries) belong to the *Prunus* genus. *Prunus* species are natives of Northern China and from there spread through Asia to Europe following trade routes, and later to the American continent. During this process, stone fruit trees adapted to the local edaphoclimatic conditions, but, genetic variability was eroded. There have been many publications on the genetic diversity of locally adapted genotypes and their relationship with either wild relatives or those recently arisen from breeding programs. In general, these studies reveal the effect of an intense bottleneck in cultivated varieties, translated in a reduction of genetic variability, as consequences of domestication and breeding. The considerably higher levels of variability found in primary and secondary centers of *Prunus* origin suggest the use of these genotypes to broaden the genetic base in breeding programs.

References

Ahmad R, Potter D, Southwick SM (2004) Identification and characterization of plum and pluot cultivars by microsatellite markers. J Hort Sci Biotechnol 79: 164–169.

Aranzana MJ, Abbassi EK, Howad W, Arús P (2010) Genetic variation, population structure and linkage disequilibrium in peach commercial varieties. BMC Genetics 11: 11.

Aranzana MJ, Carbo J, Arús P (2003a) Using amplified fragment-length polymorphisms (AFLPs) to identify peach cultivars. J Am Soc Hort Sci 128: 672–677.

Aranzana MJ, Carbo J, Arús P (2003b) Microsatellite variability in peach [*Prunus persica* (L.) Batsch]: cultivar identification, marker mutation, pedigree inferences and population structure. Theor Appl Genet 106: 1341–1352.

Aranzana MJ, Garcia-Mas J, Carbo J, Arús P (2002) Development and variability analysis of microsatellite markers in peach. Plant Breed 121: 87–92.

Arulsekar S, Parfitt DE, Kester DE (1986) Comparison of isozyme variability in peach and almond cultivars. J Hered 77: 272–274.

Asma BM, Ozturk K(2005) Analysis of morphological, pomological and yield characteristics of some apricot germplasm in Turkey. Genet Resour Crop Evol 52: 305–313.

Ayanoglu H, Bayazit S, Inan G, Bakir M, Akpinar AE, Kazan K, Ergul A (2007) AFLP analysis of genetic diversity in Turkish green plum accessions (*Prunus cerasifera* L.) adapted to the Mediterranean region. Sci Hort 114: 263–267.

Badenes ML, Asins MJ, Carbonell EA, Glacer G (1996) Genetic diversity in apricot, *Prunus armeniaca*, aimed at improving resistance to plum pox virus. Plant Breed 115: 133–139.

Badenes ML, Martinez-Calvo J, Llacer G (1998) Analysis of apricot germplasm from the European ecogeographical group. Euphytica 102: 93–99.
Battistini S, Sansavini S (1991) Electrophoretic analysis of isozyme variability in appricot cultivars. J Genet Breed 45: 117–122.
Beaver JA, Iezzoni AF, Ramm CW (1995) Isozyme diversity in sour, sweet and ground cherry. Theor Appl Genet 90: 847–852.
Benson G (1999) Tandem repeats finder: a program to analyze DNA sequences. Nucl Acids Res 27: 573–580.
Bianchi VJ, Sansavini S, Fachinello JC (2004) Microsatellite markers for identification of *Prunus* spp. rootstocks. Sci Agri 61: 303–306.
Brewbacker JL (1996) Enzyme fingerprints for the plant detective. HI Bot Soc Newsl 5: 1–3.
Byrne DH (1990) Isozyme variability in four diploid stone fruits compared with other woody perennial plants. J Hered 81: 68–71.
Byrne DH, Littleton TG (1989) Characterization of isozyme variability in apricots. J Am Soc Hort Sci 114: 674–678.
Cantini C, Iezzoni AF, Lamboy WF, Boritzki M, Struss D (2001) DNA fingerprinting of tetraploid cherry germplasm using simple sequence repeats. J Am Soc Hort Sci 126: 205–209.
Cheng ZP, Huang HW (2009) SSR fingerprinting Chinese peach cultivars and landraces (*Prunus persica*) and analysis of their genetic relationships. Sci Hort 120: 188–193.
Cipriani G, Lot G, Huang WG, Marrazzo MT, Peterlunger E, Testolin R (1999) AC/GT and AG/CT microsatellite repeats in peach *Prunus persica* (L.) Batsch: isolation, characterisation and cross-species amplification in *Prunus*. Theor Appl Genet 99: 65–72.
Clarke JB, Tobutt KR (2003) Development and characterization of polymorphic microsatellites from *Prunus avium* 'Napoleon'. Mol Ecol Notes 3: 578–580.
Crane MB, Lawrence WJC, Daniel H (1934) The Genetics of Garden Plants. Macmillan, London, 236 p.
de Vicente MC, Truco MJ, Egea J, Burgos L, Arús P (1998) RFLP variability in apricot (*Prunus armeniaca* L.). Plant Breed 117: 153–158.
Decroocq V, Hagen LS, Fave MG, Eyquard JP, Pierronnet A (2004) Microsatellite markers in the hexaploid *Prunus domestica* species and parentage lineage of three European plum cultivars using nuclear and chloroplast simple-sequence repeats. Mol Breed 13: 135–142.
Dirlewanger E, Claverie J, Iezzoni AF, Wünsch A, Folta KM, Gardiner SE (2009) Sweet and sour cherries: linkage maps, QTL detection and marker assisted selection. In: Jorgensen RA (ed) Genetics and Genomics of Rosaceae. Springer, New York, pp 291–313.
Dirlewanger E, Cosson P, Tavaud M, Aranzana MJ, Poizat C, Zanetto A, Arús P, Laigret F (2002) Development of microsatellite markers in peach *Prunus persica* (L.) Batsch and their use in genetic diversity analysis in peach and sweet cherry (*Prunus avium* L.). Theor Appl Genet 105: 127–138.
Downey SL, Iezzoni AF (2000) Polymorphic DNA markers in black cherry (*Prunus serotina*) are identified using sequences from sweet cherry, peach, and sour cherry. J Am Soc Hort Sci 125: 76–80.
Durham RE, Moore GA, Sherman WB (1987) Isozyme banding-patterns and their usefulness as genetic markers in peach. J Am Soc Hort Sci 112: 1013–1018.
Eryomine GV (1990) New data on origin of *Prunus domestica* L. In: Bernhard R (ed) IVth International Symposium on Plum and Prune Genetics, Breeding and Pomology. International Society for Horticultural Science, Wageningen, Netherlands, pp 27–29.
Foulongne M, Pascal T, Pfeiffer F, Kervella J (2004) Re-examination of an introgression procedure of a polygenic resistance to powdery mildew from wild *Prunus davidiana* into peach in the light of molecular data and Monte-Carlo simulations. In: Laurens F, Evans K (eds) Proceedings of the XIth Eucarpia Symposium on Fruit Breeding and Genetics, Vols 1 and 2. International Society for Horticultural Science, Leuven 1, pp 849–854.
Gass T, Tobutt KR, Zanetto A (1996) Report of the Working Group on *Prunus*. Vth Meeting, Menemen-Izmir, Turkey.

Gomory D, Paule L (2001) Spatial structure and mating system in wild cherry (*Prunus avium*) population. Biologia 56: 117–123.
Gottlieb LD (1981) Electrophoretic evidence and plant populations. Phytochemistry 7: 1–45.
Goulao L, Monte-Corvo L, Oliveira CM (2001) Phenetic characterization of plum cultivars by high multiplex ratio markers: Amplified fragment length polymorphisms and inter-simple sequence repeats. J Am Soc Hort Sci 126: 72–77.
Granger AR, Clarke GR, Jackson JF (1993) Sweet cherry cultivar identification by leaf isozyme polymorphism. Theor Appl Genet 86: 458–464.
Hagen LS, Chaib J, Fady B, Decroocq V, Bouchet JP, Lambert P, Audergon JM (2004) Genomic and cDNA microsatellites from apricot (*Prunus armeniaca* L.). Mol Ecol Notes 4: 742–745.
Hagen LS, Khadari B, Lambert P, Audergon JM (2002) Genetic diversity in apricot revealed by AFLP markers: species and cultivar comparisons. Theor Appl Genet 105: 298–305.
Hedrick UP (1915) The history of cultivated cherries. In: Hedrick UP, Howe GH, Taylor OM, Tubergen CB, Wellington R (eds) The Cherries of New York. JB Lyon Company, Albany, NY, pp 39–64.
Hend BT, Ghada B, Sana BM, Mohamed M, Mokhtar T, Amel SH (2009) Genetic relatedness among Tunisian plum cultivars by random amplified polymorphic DNA analysis and evaluation of phenotypic characters. Sci Hort 121: 440–446.
Horvath A, Balsemin E, Barbot JC, Christmann H, Manzano G, Reynet P, Laigret F, Mariette S (2011) Phenotypic variability and genetic structure in plum (*Prunus domestica* L.), cherry plum (*P. cerasifera* Ehrh.) and sloe (*P. spinosa* L.). Sci Hort 129: 283–293.
Hunter RL, Markert CL (1957) Histochemical demonstration of enzymes separated by zone electrophoresis in starch gels. Science 125: 1294–1295.
Ibanez MA, Direnzo MA, Poverene MM (1993) Isozyme diversity among and within peach groups-freestone, clingstone and nectarines. Sci Hort 53: 281–288.
Iezzoni A (2005) Acquiring Cherry Germplasm from Central and Eastern Europe. HortScience 40: 304–308.
Ilgin M, Kafkas S, Ercisli S (2009) Molecular characterization of plum cultivars by AFLP markers. Biotechnol Biotechnol Equip 23: 1189–1193.
Kostina KF (1969) The use of varietal resources of apricots for breeding (in Russian). Trudy Nikit. Bot Sada 40: 45–63.
Krahl KH, Lansari A, Iezzoni AF (1991) Morphological variation within a sour cherry collection. Euphytica 52: 47–55.
Krüssmann G (1978) Manual of Cultivated Broadleaved Trees and Shrubs. Vol III PRU-Z. B. T. Batsford. London.
Layne REC, Bailey CH, Hough LF (1996) Apricots. In: Janick J, Moore JN (eds) Fruit Breeding. John Wiley, New York, pp 79–111.
Leterme E (1995) Les Fruits Retrouvés. Editions du Rouergue. Rodez. France, 287 p.
Li TH, Li YX, Li C, Zhang HL, Qi YW, Wang T (2008) Simple sequence repeat analysis of genetic diversity in primary core collection of peach (*Prunus persica*). J Integr Plant Biol 50: 102–110.
Lopes MS, Sefc KM, Laimer M, Machado AD (2002) Identification of microsatellite loci in apricot. Mol Ecol Notes 2: 24–26.
Maggioni L, Lateur E, Balsemin E, Lipam E (2010) Report of a Working Group on *Prunus*. European Cooperative Programme for Plant Genetic Resources. Bioversity 7–9 Sept 2010, Forlì, Italy.
Manganaris AG, Karayiannis L, Nianious E (1999) Polymorphism and genetic studies of isozymes in appricots. ActaHort 488: 303–307.
Mariette S, Lefranc M, Legrand P, Taneyhill D, FrascariaLacoste N, Machon N (1997) Genetic variability in wild cherry populations in France. Effects of colonizing processes. Theor Appl Genet 94: 904–908.

Mariette S, Tavaud M, Arunyawat U, Capdeville G, Millan M, Salin F (2010) Population structure and genetic bottleneck in sweet cherry estimated with SSRs and the gametophytic self-incompatibility locus. BMC Genetics 11: 77.
Mehlenbacher SA, Cociu V, Hough LF (1991) Apricots. In: Moore JN, Ballington JR (coord) Genetic Resources of Temperate Fruit and Nut Crops. International Society for Horticultural Science, Wageningen, Netherlands, pp 63–107.
Messeguer R, Arús P, Carrera M (1987) Identification of peach cultivars with pollen isozymes. Sci Hort 31: 107–117.
Messina R, Lain O, Marrazzo MT, Cipriani G, Testolin R (2004) New set of microsatellite loci isolated in apricot. Mol Ecol Notes 4: 432–434.
Milatovic D, Nikolic D, Durovic D, Milivojevic J (2009) Isoenzyme polymorphism in apricot cultivars. J Am Pomol Soc 63: 14–23.
Mnejja M, Garcia-Mas J, Audergon JM, Arús P (2010) *Prunus* microsatellite marker transferability across rosaceous crops. Tree Genet Genomes 6: 689–700.
Mnejja M, Garcia-Mas M, Howad W, Badenes ML, Arús P (2004) Simple-sequence repeat (SSR) markers of Japanese plum (*Prunus salicina* Lindl.) are highly polymorphic and transferable to peach and almond. Mol Ecol Notes 4: 163–166.
Monet R, Guye A, Roy M (1994) Genetic study of alcohol-dehydrogenase polymorphism in peach (*Prunus persica* L. Batsch). Agronomie 14: 463–466.
Moore JN, Ballington JR (1994) Genetic Resources of Temperate Fruit and Nut Crops. International Society for Horticultural Science, Wageningen, Netherlands.
Morgante M, Olivieri AM (1993) PCR-amplified micro satellites as markers in plant genetics. Theor Appl Genet 3: 175–182.
National Report on the State of Plant Genetic Resources for Food and Agriculture in Azerbaijan(2006) Country Report on the State of Plant Genetic resources for Food and Agriculture. FAO, Rome, 58 p.
NikolićD, Rakonjac V, Milatović D, Fotirić M (2010) Multivariate analysis of vineyard peach [*Prunus persica* (L.) Batsch] germplasm collection. Euphytica 171: 227–234.
Ortiz A, Renaud R, Calzada I, Ritter E (1997) Analysis of plum cultivars with RAPD markers. JHortSci 72: 1–9.
Panda S, Martin JP, Aguinagalde I, Mohanty A (2003) Chloroplast DNA variation in cultivated and wild *Prunus avium* L: a comparative study. Plant Breed 122: 92–94.
Parffitt DE, Arulsekar S, Ramming DW (1985) Identification of plum x peach hybrids by isozyme analysis. Hort Science 20: 246–248.
Pashkoulov DT, Tobutt KR, Boskovic R (2000) Comparison of isoenzymes in *Prunus avium* separated by two different electrophoretic techniques. Plant Breed 119: 153–156.
Pérez S, Montes S, Mejia C (1993) Analysios of Peach Germplasm in Mexico. J Am Soc Hort Sci 118 (4): 519–524.
Quarta R, Dettori MT, Verde I, Marchesi U, Palomi A (2000) Characterization and evaluation of genetic diversity in peach germplasm using RAPD and RFLP. ActaHort 546: 489–496.
Rakonjac V, Akšić MF, Nikolić D, Milatović D Čolić S (2010) Morphological characterization of 'Oblačinska" sour cherry by multivariate analysis. Sci Hort 125: 679–684.
Ramming DW, Cociu V (1991) Plums. In: Moore JN, Ballington JR (coord) Genetic Resources of Temperate Fruit and Nut Crops. International Society for Horticultural Science, Wageningen, Netherlands, pp 235–287.
Reales A, Sargent DJ, Tobutt KR, Rivera D (2010) Phylogenetics of Eurasian plums, *Prunus* L. section Prunus (Rosaceae), according to coding and non-coding chloroplast DNA sequences. Tree Genet Genomes 6: 37–45.
Redher A (1947) Manual of cultivated trees and shrubs. 2nd edn McMillan, New York, pp 452–481.
Romero C, Pedryc A, Muñoz V, Llácer G, Badenes ML (2003) Genetic diversity of different apricot geographical groups determined by SSR markers. Genome 46: 244–252.

Schueler S, Tusch A, Scholz F (2006) Comparative analysis of the within-population genetic structure in wild cherry (*Prunus avium* L.) at the self-incompatibility locus and nuclear microsatellites. MolEcol 15: 3231–3243.

Shimada T, Hayama H, Haji T, Yamaguchi M, Yoshida M (1999) Genetic diversity of plums characterized by random amplified polymorphic DNA (RAPD) analysis. Euphytica 109: 143–147.

Sosinski B, Gannavarapu M, Hager LD, Beck LE, King GJ, Ryder CD, Rajapakse S, Baird WV, Ballard RE, Abbott AG (2000) Characterization of microsatellite markers in peach, *Prunus persica* (L.) Batsch. Theor Appl Genet 101: 421–428.

Tan A (2009) Second National Report of Turkey on Conservation and Sustainable Utilisation of Plant Genetic resources for Food and Agriculture. FAO, Rome, Italy.

Tanksley SD, Orton TJ (1983) Isozyme in Plant Genetics and Breeding. Elsevier, Amsterdam, Netherlands.

Tavaud M, Zanetto A, David JL, Laigret F, Dirlewanger E (2004) Genetic relationships between diploid and allotetraploid cherry species (*Prunus avium, Prunusxgondouinii* and *Prunus cerasus*). Heredity 93: 631–638.

Testolin R, Marrazzo T, Cipriani G, Quarta R, Verde I, Dettori MT, Pancaldi M, Sansavini S (2000) Microsatellite DNA in peach (*Prunus persica* L. Batsch) and its use in fingerprinting and testing the genetic origin of cultivars. Genome 43: 512–520.

The Ministry of Agriculture People's Republic of China (2008) State of Plant Genetic Resources for Food and Agriculture in China-1996–2007. FAO, Rome, 133 p.

Tsukamoto T, Hauck N, Tao R, Jiang N, Iezzoni A (2006) Molecular characterization of three non-functional *S*-haplotypes in sour cherry (*Prunus cerasus*). Plant Mol Biol 62: 371–383.

Tsukamoto T, Hauck NR, Tao R, Jiang N, Iezzoni AF (2010) Molecular and genetic analyses of four nonfunctional S haplotype variants derived from a common ancestral S-haplotype identified in sour cherry (*Prunus cerasus* L.). Genetics 184: 411–427.

Tsukamoto T, Potter D, Tao R, Vieira CP, Vieira J, Iezzoni AF (2008) Genetic and molecular characterization of three novel S-haplotypes in sour cherry (*Prunus cerasus*L.). J Exp Bot 59: 3169–3185.

Turkoglu Z, Bilgener S, Ercisli S, Bakir M, Koc A, Akbulut M, Gercekcioglu R, Gunes M, Esitken A (2010) Simple sequence repeat-based assessment of genetic relationships among *Prunus* rootstocks. Genet Mol Res 9: 2156–2165.

Varshney RK, Graner A, Sorrels M (2005) Genic microsatellite markers in plants: features and applications. Trends Biotechnol 23: 48–55.

Verde I, Bassil N, Scalabrin S, Gilmore B, Lawley CT, Gasic K, Micheletti D, Rosyara UR, Cattonaro F, Vendramin E, Main D, Aramini V, Blas AL, Mockler TC, Bryant DW, Wilhelm L, Troggio M, Sosinski B, Aranzana MJ, Arús P, Iezzoni A, Morgante M, Peace C (2012) Development and Evaluation of a 9K SNP Array for Peach by Internationally Coordinated SNP Detection and Validation in Breeding Germplasm. PLoS ONE 7: e35668

Vilanova S, Soriano JM, Lalli DA, Romero C, Abbott AG, Llacer G, Badenes ML (2006) Development of SSR markers located in the G1 linkage group of apricot (*Prunus armeniaca* L.) using a bacterial artificial chromosome library. Mol Ecol Notes 6: 789–791.

Wang L (2005) Status Report on Genetic Resources of Peach in East Asia. IPGRI Office for East Asia, Beijing, 44 p.

Wang Y (1985) Peach growing and germplasm in China. Acta Hort 173: 51–55.

Wang YZ (1998) Recommendation of apricot cultivars for commercial growing in China. HortSci (Praha) 25: 121–124.

Warburton ML, Bliss FA (1996) Genetic diversity in peach (*Prunus persica* L. Batsch) revealed by randomly amplified polymorphic DNA (RAPD) markers and compared to inbreeding coefficients. J Am Soc Hort Sci 121: 1012–1019.

Watkins R (1976) Cherry, plum, peach, apricot and almond. In: Simmonds NW (ed) Evolution of Crop Plants. Longman, London, New York, pp 242–247.

Wünsch A, Hormaza JI (2002) Molecular characterisation of sweet cherry (*Prunus avium* L.) genotypes using peach *Prunus persica* (L.) Batsch SSR sequences. Heredity 89: 56–63.

Xie RJ, Li XW, Chai ML, Song LJ, Jia HJ, Wu DJ, Chen MK, Chen KM, Aranzana MJ, Gao ZS (2010) Evaluation of the genetic diversity of Asian peach accessions using a selected set of SSR markers. Sci Hort 125: 622–629.
Zanetto A, Maggioni L, Tobutt KR, Dosba F (2002) *Prunus* genetic resources in Europe: Achievement and perspective of a networking activity. Genet Resour Crop Evol 49: 331–337.
Zohary D, Hopf M (1993) Domestication of Plants in the Old World: The Origin and Spread of Cultivated Plants in West Asia, Europe, and the Nile Valley. 2nd ed. Claredon Press, Oxford, 278 p.

4

Molecular Linkage Maps: Strategies, Resources and Achievements

Elisabeth Dirlewanger,[1,*] *Eudald Illa*[2] and *Werner Howad*[2]

ABSTRACT

In this chapter, we give an overview on the molecular genetic linkage maps available in *Prunus* and summarize the progress made during the last 20 years in the construction of these maps through a brief history. The evolution of the markers, being easier to use, more reliable and more numerous contributed to the development of more and more saturated maps. The different mapping populations used will be examined with respect to their advantages and disadvantages and the restrictions linked to the fruit trees, especially the juvenile phase and the self-incompatibility barrier. We also point out the different strategies that were developed in order to increase the efficiency of mapping, like the 'bin' or 'selective' mapping strategy, which permits rapid mapping of a large numbers of genetic markers to a mapping framework with a low degree of precision, by using only a subset of highly informative progeny individuals. Finally, we will give an overview of the use of all these linkage maps for theoretical analyses as genome comparison or practical purpose, like marker-assisted selection.

Keywords: mapping, markers, bin mapping, comparative mapping

[1]INRA, UR419, Unité de Recherches sur les Espèces Fruitières (UREF), B.P. 81, F-33140 Villenave d'Ornon, France.
[2]IRTA, Centre de Recerca en Agrigenòmica CSIC-IRTA-UAB-UB, Campus UAB, Bellaterra (Cerdanyola del Vallès), 08193 Barcelona, Spain.
*Corresponding author: *dirlewan@bordeaux.inra.fr*

4.1 Introduction

Molecular genetic linkage maps have and continue to be a major tool in genetics, genomics and breeding in plant and animal species. Linkage maps provide opportunities to analyze the complex genetics of quantitatively inherited traits through the localization of quantitative trait loci (QTL), identification and positional cloning of genes, development of genome-wide physical maps using anchor markers. They can also serve as a repository of markers useful in marker-assisted selection. Since the publication of the first genetic maps of *Prunus,* nearly twenty years ago, to those currently constructed, a significant change has occurred increasing their accuracy. They are based on the analyses of larger segregating progenies and include more and more markers which are more informative, some of them corresponding to genes.

4.2 Brief History of Mapping Efforts

The first published maps in *Prunus* were constructed on peach in 1994 (Chaparro et al. 1994; Dirlewanger et al. 1994). They were constructed with random amplified polymorphic DNA (RAPD) markers and isozymes on two F_2 populations segregating for several agronomical characters—the weeping shape for the first one, and several morphological traits for the second. These two maps were only partial maps covering less than 400 cM total map distance. As most of markers were dominant, linkage could be detected only for markers in coupling phase. One year later, the first almond linkage map was constructed using codominant markers—restriction fragment length polymorphism (RFLP) and izoenzymes—using an F_1 progeny between 'Ferragnes' and 'Tuono' (Viruel et al. 1995). In 1996, the first sweet cherry (*Prunus avium*) map was constructed using microspore derived callus from 'Emperor Francis' (Stockinger et al. 1996). For sour cherry (*Prunus cerasus*), the first map was constructed using an F_1 family based on the cross 'Rheinische Schattenmorelle' x 'Erdi Botermo' (Wang et al. 1998). As sour cherry is tetraploid, informative single dose restriction fragment (SDFR, Wu et al. 1992) markers were used in order to identify quantitative trait loci (QTL) for flower and fruit traits (Wang et al. 1998). In Apricot, the first linkage map was built later on, in 2002, in order to identify QTL for sharka resistance using a cross between 'Goldrich' x 'Valenciano' (Hurtado et al. 2002).

Due to the low variability found in intraspecific populations, especially in peach populations, interspecific populations have also been used successfully. Foolad et al. (1995) were the first to publish one such map, using an almond/peach F_2 population. The map constructed with the F_2 of the cross between 'Texas' almond and 'Earlygold' peach has been adopted by the stone fruit community as the reference for the genus, abbreviated TxE. This map, originally constructed by a European consortium group (Joobeur et al. 1998), was obtained only with codominant markers (235 RFLPs and 11 isozymes). Additional markers were added later to the TxE map to produce the current map with 562 markers including 361 RFLPs, 185 simple sequence repeats (SSRs), 11 isoenzymes, and five sequence tagged sites (STSs) arranged in the expected eight linkage groups (G1–G8), covering a total distance of 519 cM, with an average marker density of 0.92 cM (Aranzana et al. 2003; Dirlewanger et al. 2004a; Howad et al. 2005). A common terminology for linkage group numbers has been adopted based on this TxE map, linkage group 1 (G1) being the longest and corresponding to the chromosome 1 identified also as the longest (Saless and Mouras 1977). This map has been a useful resource for *Prunus* community: the makers were used for the construction of other *Prunus* maps, allowing map comparison and construction of framework maps. Its high polymorphism allows mapping of markers that are monomorphic in other populations and markers from this map can be used to saturate specific regions of interest in other maps. The bin mapping strategy, described later on in this chapter, was developed for the first time in Rosaceae using this reference map (Howad et al. 2005). Moreover, RFLPs located in this map have been used as starting points for the construction of the *Prunus* physical map (Zhebentyayeva et al. 2008, http://www.rosaceae.org/gdr/WebAGCoL/peach/WebChrom/).

Various interspecific populations were then constructed, due to their high efficiency for mapping markers but also because they were segregating for interesting agronomical traits. Interspecific populations between almond and peach were constructed to map genes involved in various characters (Jauregui et al. 2001; Bliss et al. 2002). A three-way cross between myrobalan plum (P.2175) and almond x peach rootstock 'Felinem' was used to map two nematode resistance genes, *Mi* from the peach 'Nemared' and *Ma* from the myrobalan plum P.2175 (Dirlewanger et al. 2004b). This myrobalan P.2175 map is the unique one involving a plum species. Interspecific population between peach and related species *P. ferganensis* and *P. davidiana* were analyzed to map QTLs for evergrowing (Quarta et al. 1998, 2000; Verde et al. 2005) and powdery mildew resistance (Dirlewanger et al. 1996; Viruel et al. 1998; Foulongne et al. 2003b), quality (Quilot et al. 2004) and peach tree short life syndrome (Blenda et al. 2007). On cherry, maps were constructed using interspecific population between sweet cherry and *P. incisa* (Boskovic and Tobutt 1998) or *P. nipponica* (Boskovic and Tobutt 1998; Clarke et al. 2009).

In total, 23 mapping populations involving only species of the *Amygdalus* subgenus (10 intraspecific peach, five intraspecific almond, three almond x peach, one peach x *P. ferganensis*,three *P. davidiana* x peach and one from a three way cross) were used to construct genetic linkage maps, their characteristics being summarized in Table 4-1. From the same mapping population, several maps were published, including more markers and often more individuals. This is the case of the peach JxF map, the first one based on 63 individuals (Dirlewanger et al. 1998) and the last one, on 207 individuals and including more codominant markers (Dirlewanger et al. 2006). All the maps involving *P. davidiana* used P1908 clone for its disease resistance property, in several crosses: F_1, F_2, and backcross (BC2).

For Prunophora subgenus, linkage map construction started later, including six intraspecific apricot populations and only one population including myrobalan plum (Table 4-2). For Cerasus subgenus, six mapping populations were used for map construction: three were intraspecific *P. avium* populations, one was an intraspecific *P. cerasus*, and the two others were interspecific populations (Table 4-2). They were all F_1 populations except, the microspore derived callus that were used for the first sweet cherry map.

4.3 Evolution of Marker Types: RFLPs to SNPs

As mentioned earlier, the first *Prunus* maps were constructed with RAPD and isoenzyme markers (Chaparro et al. 1994; Dirlewanger et al. 1994, 1998). RAPDs were very easy to use, with no previous knowledge needed before analyses. However, being dominant markers, they were not very informative and not usable for map comparison: very often several markers were obtained with one primer combination but the homology with amplified fragments with the same primer combination in another population was unknown. Conversely, isoenzymes are very informative but their number remains very low. They were mainly used in cherry maps and the maximum number (67) was used in the interspecific *P. avium* x *P. nipponica* map (Clarke et al. 2009). Amplified fragment length polymorpohism (AFLP) markers were also used in numerous linkage maps because it was possible to obtain rapidly a high number of markers (Abbott et al. 1998; Etienne et al. 2002). Similar to RAPDs they were not very informative, being dominant and not usable for comparative mapping.

From 1995, RFLP markers were then extensively used for mapping in peach (Rajapakse et al. 1995), almond x peach (Foolad et al. 1995) and almond (Viruel et al. 1995). A large set of probes used for RFLP analyses were developed in a European Union project, corresponding to genomic or cDNA sequences from several species (almond, peach, *P. ferganensis*, cherry and plum) (Arùs et al. 1994). The probe nomenclature is as followed: AG (almond genomic), AC (almond cDNA), PC (peach cDNA), FG (*P. ferganensis* genomic), PLG (plum genomic). In the first reference TxE *Prunus* map published, 235 RFLPs were mapped (Joobeur et al. 1998). Additional RFLPs

Table 4-1 Prunus linkage maps constructed with species of the Amygdalus.

Crosses	Name of pop.	Type of pop. (size)	Nb and type of markers	Size of the map	*Location of genes and QTLs*	References
			Peach (*P. persica*)			
Weeping clone (1161:12x2678:47)1:55 x 'Early Sungrand'		F_2 (270)	52 RAPDs	350 cM Partial map	*Pl* (weeping shape)	Dirlewanger and Bodo 1994
Peach clones used to generate the F1: 'Daviell', 'Georgia Belle', 'Honey Glo', 'Marsun', 'Sweet Melody'		F_2 (96)	83 RAPDs, 2 isoenzymes	396 cM + 198 cM, Partials maps	*Br, Dl, Dw3, G, Gr, Ps2, W, Wa, We, Y*	Chaparro et al. 1994
'Ferjalou Jalousia' x 'Fantasia'	JxF	F_2 (63)	50 RFLPs, 92 RAPDs, 8 ISSRs, 115 AFLPs, 1 isoenzyme, SSRs	712 cM Partial map	*D, G, S, Ps, QTLs for fruit quality*	Dirlewanger et al. 1998, 1999; Etienne et al. 2002
		F_2 (207)	82 SSRs, 37 RFLPs, 61 AFLPs, 1 isoenzyme	621 cM	*D, G, S, Af, Ps, F*	Dirlewanger et al. 2006
'New Jersey Pillar' x 'KV 77119'	WV	F_2 (71)	46 RFLPs, 12 RAPDs	465 cM Partial map	*Dl, Pi, Y, F, tree architecture*	Rajapakse et al. 1995; Abbott et al. 1998; Sosinski et al. 1998
'Suncrest' x 'Bailey'		F_2 (48)	51 RFLPs, 12 RAPDs, 82 AFLPs	926 cM Partial map	*QTLs for fruit quality*	Abbott et al. 1998
'Lovell' x 'Nemared'		F_2 (55)	153 AFLPs	1297 cM Partial map	*Mi, Mij, Sh, F, Gr*	Abbott et al. 1998; Lu et al. 1998
'Akame' x 'Juseitou'	AxJ	F_2 (126)	35 AFLPs, 31 RAPDs, 11 SSRs, 5 ISSRs, 1 PCR-RFLP	960 cM Partial map	*F, Cs, Fc, Sc, Gr, Dw, Nl, Mi, Mj*	Yamamoto et al. 2001

'Harrow Blood' x 'Okinawa'	PMP2	F_2 (56)	76 RFLPs		*Mi*	Gillen & Bliss 2005
Dr. Davis' x 'Georgia Belle'	Pop-DG	F_1 (152)	11 ripening candidate genes, 34 cold-response genes, 58 SSRs, 40 RAFs, 23 SRAPs, 14 IMA, 28 CGs	818 cM	*freestone, melting flesh, white/yellow flesh color, showy flower petals*	Ogundiwin et al. 2009
'Contender' x 'Fla.92-2C'		F_2 (378)	96 SSRs, 30 AFLPs	535 cM	*Sh, QTLs blooming date, chilling and heat requirement*	Fan et al. 2010
			Almond (*P. dulcis*)			
'Ferragnès' x 'Tuono'	FxT	F_1 (60)	7 isozymes, 120 RFLPs	393 cM Ferragnès		Viruel et al. 1995
				394 cM Tuono		
			7 isozymes, 120 RFLPs,	415 cM Ferragnès		Joobeur et al. 2000
			54 RAPDs, 6 SSRs	416 cM Tuono		
'Felisia' x 'Bertina'	F1xB	F_1 (134)	81 RFLPs, 5 RAPD-BSA	386 cM Felisia	*Lb(late blooming)*	Ballester 1998; Ballester et al. 2001
				349 cM Bertina		
'R1000' x 'Desmayo Largueta'	RxD	F_1 (167)	48 SSRs (R), 45 SSRs (D)	395 cM	*self-compatibility, blooming date & density, ripening time*	Sanchez-Pérez et al. 2007
Nonpareil' x 'Lauranne'	NxL	F_1 (93)	95 SSRs, 35 ISSRs, 18 SNPs/INDELS, and 12 RAPDs	604 cM		Tavassolian et al. 2010

Table 4-1 contd....

Table 4-1 contd....

Crosses	Name of pop.	Type of pop. (size)	Nb and type of markers	Size of the map	*Location of genes and QTLs*	References
Vivot' x 'Blanquerna'	VxB	F_1 (77)	47 SSRs (V), 43 SSRs (B)	404 cM Vivot 379 cM Blanquerna		Fernández i Martí et al. 2011
			Almond (*P. dulcis*) x Peach (*P. persica*)			
'Padre' x '54P455'	PMP1	F_2 (64)	143 RFLPs, 8 isoenzymes, 2 SSRs, 1 CAP, 1RAPD	1144 cM	*Y, Pcp, D, Sk, St, Dw, Ps*	Foolad et al. 1995; Warburton et al. 1996; Bliss et al. 2002
'Texas' x 'Earlygold'	TxE	F_2 (82)	11 isoenzymes, 5 STS, 361 RFLPs, 185 SSRs	519 cM Saturated map		Joobeur et al. 1998; Aranzana et al. 2003b; Dirlewanger et al. 2004a
		F_2 (75)	10 candidate genes for blooming date		*late blooming (Lb)*	Silva et al. 2005
		F2 (6)	253 SSRs, 961 Candidate genes & RosCOS	Bin map	*fruit quality*	Howad et al. 2005; Chen et al. 2008; Ogundiwin et al. 2008, 2009; Cabrera et al. 2009; Sargent et al. 2009; Illa et al. 2011
'Garfi' x 'Nemared'	GxN	Complex F_2 (78)	46 RFLPs, 5 isozymes		*Mi*	Jàuregui et al. 2001
			P. persica* x *P. ferganensis			
(IF7310828 x *P. ferganensis*) x IF7310828	PxF	BC_1 (297)	74 RFLPs, 17 SSRs, 16 RAPDs	521 cM Partial map	*F, E, QTLs for powdery mildew resistance, fruit quality, blooming and ripening date*	Quarta et al. 1998, 2000; Dettori et al. 2001; Verde et al. 2002
	PxF	BC_1(70)	78 RFLPs, 57 SSRs, 16 RAPDs, 63 AFLPs	665 cM	*evergrowing (evg)*	Verde et al. 2005

			P. davidiana* x *P. persica			
P. davidiana clone P1908 x *P. persica* 'Summergrand'	SD	F_1 (77)	15 RAPDs, 84 RAPDs	83 cM 536 cM	*QTLs for powdery mildew resistance*	Dirlewanger et al. 1996; Viruel et al. 1998
	SD40[2]	F_2 (99)	66 RFLPs, 24 SSRs, 103 AFLPs	874 cM	*QTLs for powdery mildew resistance*	Foulongne et al. 2003a, b
P. davidiana clone P1908 x 'Summergrand' = SD 40 x 'Summergrand' = BC1 x Zéphir	BC_2	BC_2 (162)	41 RFLPs, 29 AFLPs, 59 SSRs	575,1 cM	*QTls for qulity traits*	Quilot et al. 2004; Illa et al. 2009
P. persica 'Guardian' 3-17-7 x Nemaguard (*P. persica* x *P. davidiana*)		F_2 (100)	140AFLPs, 18 SSRs	737 cM	*Peach tree short life syndrome*	Blenda et al. 2007
		Myrobolan plum (*P. cerasifera*) x [almond (*P. dulcis*) x peach (*P. persica*)]				
P.2175 x ('Garfi' x 'Nemared') 22	P.2175 x GN22	F_1 (83)	165 SSRs (GN 22) 91 SSRs (P.2175)	742 cM 525 cM	*Mi, Gr*	Dirlewanger et al. 2004b

Table 4-2 Prunus linkage maps of the Prunophora and Cerasus subgenera.

Crosses	Name of pop.	Type of pop.(size)	Nb and type of markers	Size of the map	*Location of genes and QTLs*	References
			Apricot (*P. armeniaca*)			
'Goldrich' x 'Valenciano'	GxV	F_1 (81)	33 RAPDs, 82 AFLPs, 4 RFLPs (G)	511 cM (G)	*sharka*	Hurtado et al . 2002
			19 RAPDs, 48 AFLPs, 5 RFLPs (V)	467 cM (V)		
'Lito'	LxL	F_2 (76)	180 AFLPs, 29 SSRs	602 cM	*sharka, self incompatibility*	Vilanova et al. 2003
'Polonais' x 'Stark Early Orange'	P x SEO	F_1 (142)	88 AFLPs, 27 SSRs, 38 RFLPs (P)	538 cM (P)		Lambert et al. 2004
			54 AFLPs, 30 SSRs, 45 RFLPs (SEO)	699 cM (SEO)		
'Lito' x 'BO 81604311'	LxB	F_1 (125)	144 SSRs (L) 144 SSRs (B)	504 cM (L) 620 cM (B)		Dondini et al. 2007
Goldrich' x 'Moniquí'	GxM	F_1 (126)	85 SSRs (G) 63 SSRs (M)	468 cM (G) 707 cM (M)		Illa et al. 2009
A.1740 x 'Perfection'		F_1 (100)	592 (SSRs & AFLPs) (A.1740)	455 cM	*Chilling requirement*	Olukolu et al. 2009
			655 (SSRs & AFLPs) ('Perfection')	550 cM		
			Sweet cherry (*P. avium*)			
'Emperor Francis'	EF	56 microspore -derived callus	87 RAPDs, 2 allozymes	503 cM		Stockinger et al. 1996
'Emperor Francis' x 'New York 54'	EFxNY	F_1 (190)	84 SSR, 41 AFLP, 4 SRAP, 15 gene derived (EF)	711 cM (EF)		Olmstead et al. 2008
			51 SSR, 23 AFLP, 3 SRAP, 14 gene derived (NY)	566 cM (NY)		

'Regina' x 'Lapins'	RxL	F_1 (120)	68 SSRs (R) 54 SSRs (L)	639 (R) 495 (L)		Dirlewanger et al. 2004
			Sour cherry (*P. cerasus*)			
'Rheinische Schattenmorelle' x 'Erdi Botermo'	RS x EB	F_1 (86)	126 SDRF (RS), 95 SDRF (EB)	461 cM (RS), 279 cM (EB)	*QTL for flower and fruit traits*	Wang et al. 1998, 2000
			P. avium* x *P. incisa			
'Napoleon' x *P. incisa* E621		F_1 (63)	19 isoenzymes			Boskovic & Tobutt 1998
			P. avium* x *P. nipponica			
'Napoleon' x P. nipponica F1292		F_1 (47)	19 isoenzymes			Boskovic & Tobutt 1998
	PA x PN	F_1 (94)	158 SSRs, 6 gene-specific markers	680 cM	*S(incompatibility locus)*	Clarke et al. 2009
		F_1 (47)	67 isoenzymes, 166 STS			
		Myrobolan plum (*P. cerasifera*) x [almond (*P. dulcis*) x peach (*P. persica*)]				
P.2175 x ('Garfi' x 'Nemared')22	P.2175 x GN22	F_1 (83)	98 SSRs (*P. cerasifera*)	525 cM (*P. cerasifera*)	*Ma*	Dirlewanger et al. 2004b

were added and the TxE map includes now 361 RFLPs (Dirlewanger et al. 2004a). As those markers, like isozymes can be easily transferred to other populations of the different *Prunus* species, they were included in nearly all linkage maps constructed until 2004. RFLPs included in the TxE map were finally used to anchor the genetic linkage map with the physical map. However, RFLPs were very time consuming and could not be automated. That is why, as soon as simple sequence repeats (SSRs) were available, they were used extensively.

The first *Prunus* SSRs were published in 1999 (Cipriani et al. 1999) on peach and correspond to the UDP set. During the years 2000 to 2010, hundreds of SSRs were developed in various *Prunus* species (Table 4-3). They were developed either from enriched genomic libraries, more frequently enriched in CT repeats (Aranzana et al. 2002; Dirlewanger et al. 2002; Lopes et al. 2002; Mnejja et al. 2004), or from expressed sequence tag (EST) collections (Yamamoto et al. 2002, 2005; Howad et al. 2005; Vendramin et al. 2007; Li et al. 2010). The excellent properties of SSRs, especially their high level of polymorphism, have made SSRs the markers of choice for many uses in *Prunus* genetics and breeding. Rapidly, they were used in most linkage maps, some of them including only this type of markers, for example the sweet cherry RxL map (Dirlewanger et al. 2004), the almond RxD map (Sanchez-Perez et al. 2007) and the apricot LxB map (Dondini et al. 2007).

In the last five years, with the development of the candidate gene approach, many EST libraries were constructed and sequenced, providing a valuable resource of molecular markers based on single nucleotide polymorphisms (SNPs) (Grimplet et al. 2005; Horn et al. 2005; Ogundiwin et al. 2008; Le Dantec et al. 2010). Because of their abundance in all regions of the genome, SNPs can be developed readily from ESTs using bioinformatics tools. Then, candidate genes for flowering time were mapped in the TxE map (Silva et al. 2005) and candidate genes involved in fruit quality traits were mapped in peach (Ogundiwin et al. 2009; Le Dantec et al. 2010). A large set of 273 candidate genes involved in metabolic pathways affecting fruit growth and maturity, texture, sugar and organic acid content, aroma and color, as well as putative allergen genes were bin mapped in TxE (Chen et al. 2008; Illa et al. 2011a). Gene-anchored SNPs were also mapped in the almond linkage map constructed by using 'Nonpareil' x 'Lauranne' F_1 progeny (Tavassolian et al. 2010).

Once large amount of Rosaceae ESTs were publicly available, Rosaceae Conserved Orthologous Set (RosCOS) were developed (Cabrera et al. 2009). After comparing Rosaceae EST (412,827, available in 2007), to *Arabidopsis* single copy genes, 30,801 putative orthologs were identified. Only unigenes that were represented by at least two ESTs (3.818) were kept in the following steps of selection to optimize the chance of successful amplification of

Table 4-3 *Prunus* microsatellites.

SSR names	*Prunus* species	Repeats	Origins	References
AMPA	*P. armeniaca*	TC, TG, AAG	genomic and cDNA libraries from 'Polonais'	Hagen et al. 2004
aprigms1	*P. armeniaca*	TA, GA	G1 specific BAC clones from 'Goldrich'	Vilanova et al. 2006
BPPCT	*P. persica*	CT	Enriched genomic library from 'O'Henry'	Dirlewanger et al. 2002
CGPPB	*P. persica*	Various	EST collection	Le Dantec et al. 2010
CPDCT	*P. dulcis*	CT	Enriched genomic library from 'Texas'	Mnejja et al. 2005
CPPCT	*P. persica*	CT	Enriched genomic library from 'O'Henry'	Aranzana et al. 2002
CPSCT	*P. salicina*	CT	Enriched genomic library from 'Santa Rosa'	Mnejja et al. 2004
EPDCU	*P. dulcis*	Various	EST collection	Genome Database for Rosaceae (GDR), Howad et al. 2005
EPPCU	*P. persica*	Various	EST collection	Genome Database for Rosaceae (GDR), Howad et al. 2005
EPPB	*P. persica*	Various	EST collection	Howad et al. 2005
EPPISF	*P. persica*	Various	mesocarp cDNA library from 'Yumyeong'	Vendramin et al. 2007
EMPA	*P. avium*	AC, AG, ACC, CCG, CTG, AAT	Enriched genomic library from 'Napoleon'	Clarke and Tobutt 2003
EMPaS	*P. avium*	CT, GA	Enriched genomic library from 'Charger'	Vaughan and Russell 2004
M	*P. persica*	CT, GA	cDNA library from 'Akatsuki'	Yamamoto et al. 2002
MA	*P. persica*	GA, tri- and hexa motives	enriched genomic libraries from 'Akatsuki'	Yamamoto et al. 2002; Nishitani et al. 2007
MD	*P. persica*	Various	EST collection	GeneBank, Yamamoto et al. 2005
ES	*P. mume*	Various	Fruit and flower cDNA library from 'Xiyeqing'	GeneBank, Li et al. 2010
PAC	*P. armeniaca*	CA, GA	leaf cDNA enriched library from 'Stark Early Orange'	Decroocq et al. 2003

Table 4-3 contd....

Table 4-3 contd....

SSR names	*Prunus species*	Repeats	Origins	References
PaCITA	*P. armeniaca*	CT	Genomic library from 'Ungarische Beste'	Lopes et al. 2002
PceGA	*P. cerasus*	GA	Genomic library from 'Erdi Botermo'	Cantini et al. 2001; Downey and Iezzoni 2000
pchcms	*P. persica*	CA, GA, TA	genomic library from 'Bicentennial'	Sosinski et al. 2000
pchgms	*P. persica*	CT, CA	BAC clones from 'Nemared' leaf library, cDNA library from 'Suncrest'	Georgi et al. 2003, 2005; Wang et al. 2002; Sosinski et al. 2000
Pgra	*P. grayana*	$(TC)_6$ (AC)n, $(AC)_6$ (AG)n, $(AC)_6$ (TC)n	DNA library	Sawa et al. 2010
PMS	*P. avium*	CT, CA, GA	Genomic library from 'Valerij Tschakhalov'	Cantini et al. 2001
PS	*P. avium*	GA, GT, GTT	Enriched genomic library from 'Napoleon'	Sosinski et al. 2000; Joobeur et al. 2000; Cantini et al. 2001; Downey and Iezzoni 2000
PTCR	*P. pseudocerasus*	GA, CAA, CTG, CAG, CAT, ACT, GAC	Enriched genomic library from 'Laiyang Short cherry'	Ai et al. 2008
Pver	*P. verecunda*	(AC)6(TC)n, (AG)6(AC)n	DNA library	Sawa et al. 2010
UCD-CH	*P. avium*	CA, CT	genomic library from 'Valerij Tschkalov'	Struss et al. 2003
UDA	*P. dulcis*	AC, AG	Two enriched genomic library from 'Ferragnes'	Testolin et al. 2004
UDP	*P. persica*	CT, GT	Two enriched genomic libraries from 'Redhaven'	Cipriani et al. 1999; Testolin et al. 2000
UDPp	*P. armeniaca*	AG	Enriched genomic library from 'Portici'	Messina et al. 2004
TPScp	*P. salicina*	T, G, TAAAA, ATAGAT, TCAAA	cpDNA	Ohta et al. 2005

Prunus DNA with designed primers. A final set of 1,039 RosCOS was then established. For 857 RosCOS, primers flanking introns were designed and used to amplify the corresponding region from the TxE peach parent 'Earlygold', the F_1 hybrid, and the six selected F_2 bin mapping individuals. For 784 couple of primers (91%) DNA amplification was obtained and the percentage of RosCOS that exhibited only one SNP was 18%, whereas 43% harbored at least two SNPs. In total, 607 RosCOS were bin mapped in TxE. From these, 126 were located to defined mapping bins on the *Fragaria* FV x FN map (Sargent et al. 2009) and 545 showed a high level of DNA sequence conservation with apple, allowing them to be located to precise positions on the *Malus* chromosomes. They were used to reveal collinearity, translocation and inversion events between *Prunus, Fragaria* and *Malus* genomes (Illa et al. 2011b). COS markers, that are conserved gene-based molecular markers appear extremely useful for the analysis of genome evolution among closely and distantly related species (Illa et al. 2011b). However, as they are defined in conserved regions their polymorphism is lower than the polymorphism obtained for SNPs identified in intergenic regions.

SNP-based genotyping were rapidly adopted and SNPs emerged as the markers of choice for trait genetic mapping. The main factors responsible for the success of SNP technique are the ubiquity of this type of genetic variation, easy standardization of results thanks to the inherently biallelic nature of SNPs, and the possibility of increasing throughput by multiplexing. In addition to the desired properties of the SNP markers, the recent high capacity of sequencing offered by the Next Generation Sequencing platforms such as Applied Biosystems SOLiD sequencer, the Roche(454) sequencer and the Illumina Genome Analyzer, allowed the extensive development of SNPs. The need to define a universal core set of SNPs for genetic studies in a given species has been recognized and SNP arrays were recently developed for peach and cherry in a collaborative initiative, part of two projects, RosBreed (http://www.rosbreed.org/) and the European Fruit Breedomics (KBBE-2010-4). For the construction of these arrays, including respectively 9,000 and 6,000 SNPs for the peach and cherry array, numerous analyses were performed to optimize their efficiency. SNP assays were restricted to conserved genomic region around each target SNP. The SNP transferability—probability that a nucleotide position known to be heterozygous in one accession is heterozygogous in a second accession—was evaluated. SNPs were chosen in order to cover the complete genome but with a higher density in specific regions involved in interesting agronomical traits. Selected SNPs were located in genomic or intergenomic regions and some were located in candidate gene for specific traits. Then we can expect that the following linkage maps for peach and cherry will include large amount of SNP markers.

4.4 Mapping Populations Used

4.4.1 Population Types

Several different population types may be utilized for mapping, with each type possessing advantages and disadvantages (Kole and Abbott 2008). In fruit trees the choice of the population type is, apart from breeding purpose considerations, mainly influenced by two factors, namely the juvenile phase and the self-incompatibility barrier. The juvenile phase of several years hampers the creation of populations that need various generations to be obtained such as recombinant inbred lines (RILs), whereas the self-incompatibility character prevents the creation of populations that include self-fertilisation steps like F_2 type, RILs and collections of near isogenic lines (NILs). The majority of *Prunus* species exhibit varying levels of gametophytic self-incompatibility that makes it usually impossible to produce F_2 type populations. As a consequence the most common population type found in *Prunus* is of F_1 type except for peach, which is predominantly a self-fertilizing species. Thus F_2 type population is the population type of choice in this species because of its highly informative character, although an additional generation is needed if F_1 hybrids are not already available. However, the self-incompatibility barrier has also been overcome by a few cultivars in other *Prunus* species like almond (Channuntapipat et al. 2003; Fernández i Martí et al. 2011), apricot (Halász 2007), plum (Dragan and Dragan 2010) as well as sweet and sour cherry (Hauck et al. 2002; Sonneveld et al. 2005). For example in apricot an F_2 type population has been generated based on the self-compatible cultivar 'Lito' (Vilanova et al. 2003).

The F_2 type populations are also common in interspecific crosses of the subgenus *Amygdalus* although the corresponding maps contain a distortion towards the self-incompatibility locus due to the absence of almond alleles in homozygosis in this region. In the subgenera *Prunophora* and *Cerasus* interspecific crosses are possible, but fertile hybrids are rare (Scorza and Sherman 1996). The most prominent example of an almond x peach interspecific F_2 population is T×E , which was obtained by crossing the almond cultivar 'Texas' with the peach cultivar 'Earlygold'. This population is regarded to be the reference population of the genus *Prunus* as well as related genera (Joobeur et al. 1998; Aranzana et al. 2003; Dirlewanger et al. 2004a). The corresponding map contains 562 markers, all of them transferable (RFLP, SSR and STS) that identify eight linkage groups with a total distance of 519 cM (average density 0.92 cM/marker and largest gap of 7 cM) (Dirlewanger et al. 2004a). This population is highly polymorphic and suitable for linkage studies, gives a set of markers that can be used as anchors for map construction in other progenies as well as to anchor on it bacterial artificial chromosome (BAC) contigs of the peach physical map

(Zhebentyayeva et al. 2008) or scaffolds for peach genome sequencing. Due to common markers it is possible to locate on the T×E *Prunus* reference map major genes and QTLs that segregate in different populations (Abbott et al. 2007). In total, 28 loci controlling some simple-inherited characteristics of the flower, fruit and plant vegetative structure were assigned to specific positions on the T×E map (Dirlewanger et al. 2004a). For polygenic traits 35 QTLs for bloom and maturity time, fruit quality characters, tree architecture, disease resistance and chilling requirement were also placed on the map (Abbott et al. 1998; Viruel et al. 1998; Dirlewanger et al. 1999; Etienne et al. 2002; Verde et al. 2002; Foulongne et al. 2003b; Ogundiwin et al. 2008; Olukolu et al. 2009; Fan et al. 2010).

A subset of this population that represents the bin mapping population has been used to apply for the first time the bin or selective mapping approach (Howad et al. 2005). Six individuals were selected by inspecting visually numbers and positions of recombination breakpoints to obtain the maximum number of total bins equally distributed all over the genome and, at the same time, decrease to a minimum the number of duplicated bins (bins with the same bin genotype) as well as the size of the largest bin. Each recombination breakpoint in any of these selected individuals creates a bin limit, and, thus a bin can be defined as an interval between one recombination breakpoint or telomere and the next one within the selected bin mapping individuals. The joined individual genotype of the selected bin mapping individuals represents the bin genotype. At the moment the T×E bin map contains 67 different bin genotypes plus two duplicated ones, which adds to a total of 69 bins with an average bin length of 7.5 cM per bin. The bin terminology consists of two figures separated by a colon, where the figure before the colon indicates the linkage group number and the figure after the colon the position of the last marker within that bin (e.g., bin 1:14 => bin mapped marker is located in G1 where the last marker that defines the lower bin limit is located at 14 cM). A display of the T×E bin map is shown in GDR web site (http://www.rosaceae.org/cgi-bin/gdr/cmap/viewer?ref_map_set_acc=30&ref_map_accs=-1) representing the minimum bin length, in which the markers closest to the recombination breakpoints define the bin limits, causing gaps between two neighboring bins due to the absence of markers in these regions. To place any kind of marker onto this T×E bin map the six selected bin mapping individuals plus two parents, 'Earlygold' and the F_1 hybrid MB 1–73, need to be genotyped instead of the whole population. Using these eight individuals (bin set) 253 SSR markers have been added onto the T×E reference map (Howad et al. 2005). The bin mapping approach has been proven to be particularly useful for mapping candidate genes (CGs) in order to find gene QTL co-localizations as well as the RosCOS for synteney analysis. So far 348 CGs and 613 RosCOS markers have been placed onto the T×E reference map (Chen et al. 2008; Ogundiwin et al. 2008, 2009; Cabrera et al. 2009; Sargent et al. 2009; Illa et al. 2011a). Bin sets have also been developed in other species such as apple (Han et al. 2009; van Dyke et al. 2007), strawberry (Sargent et al. 2009) and melon (Fernandez-Silva et al. 2008).

Backcross populations have been used to map major traits that are of agronomic interest but absent in commercial varieties. In *Prunus*, an interspecific BC_2 progeny was generated by crossing *P. davidiana* with the peach cultivar 'Summergrand'. One of the F_1 hybrids (SD40) was selected to generate a backcross population (BC_1) by using 'Summergrand' as the recurrent parent. A mixture of pollen of the resulting BC_1 individuals was used to pollinate another peach cultivar 'Zephyr®' to generate the BC_2 population. This population was analyzed to demonstrate the potential of *Prunus davidiana* to introgress characters of interest into peach (Foulongne et al. 2003a; Quilot et al. 2004). *P. davidiana* is described as resistant to several peach pests and diseases, including powdery mildew (Smykov et al. 1982), green aphid (Massonié et al. 1982; Sauge et al. 1998), plum pox virus (Pascal et al. 1998) and leaf curl (Hesse 1975).

4.4.2 Population Size

In general, the generation of large populations is desirable for an increased mapping resolution. However, in fruit trees the generation of large populations is largely limited because of extended orchard maintenance and space over several years (Hansche and Beres 1980). As a consequence the average population size found in fruit trees for general mapping purpose is rather small with around 110 individuals ranging in the majority between 50 and 200 individuals. It has been shown that small populations can be useful to map characters like fruit quality, tree architecture and pest resistance as for the F_2 peach population (55 individuals) 'Suncrest'×'Bailey' or for the F_2 peach population (48 individuals) 'Lovell' × 'Nemared' (Abbott et al. 1998; Lu et al. 1998). Only few populations exceed considerably the population size of 200 individuals as for the F_2 peach population (270 individuals) Weeping clone (1161:12x2678:47)1:55 × 'Early Sungrand' that is segregating for morphological characters such as aphid resistance, nectarine/peach and weeping shape (Dirlewanger and Bodo 1994) or the F_2 peach population (378 individuals) 'Contender' × 'Fla.92-2C' used for QTL mapping of blooming date as well as chilling and heat requirement (Fan et al. 2010).

However, to do high-resolution mapping or positional cloning much bigger populations are needed. One example in peach is the *D* locus (see Chapter 8) that was described as dominant and controlling the low-acid character of the fruit (Yoshida 1970; Monet 1979). In 1998, Dirlewanger et al. (1998) analyzed a rather small F_2 population of 63 individuals generated by crossing the two peach cultivars 'Ferjalou Jalousia' and 'Fantasia' to map the *D* locus on the proximal end of G5 that co-localized with major QTLs controlling fruit pH, titrable acidity and organic acid concentration (Dirlewanger et al. 1999; Etienne et al. 2002). In 2006, a population of 208 individuals of this cross was used to map the *D* locus with more precision (Dirlewanger et al. 2006). Recently, 1,510 additional individuals summing up to a total of 1,718 individuals were screened to locate the *D* locus to a genetic interval of 0.4 cM (~100 Kb) (Boudehri et al. 2009).

An example for high-resolution mapping and subsequent positional cloning in fruit trees is the *Ma* gene from Myrobalan plum (*Prunus cerasifera*). This gene confers a complete-spectrum dominant resistance to root-knot nematode *Meloidogyne* species (Esmenjaud et al. 1996; Lecouls et al. 1997). A total of 288 progenies from the Myrobalan plum accession P.2175 (carrying the *Ma* gene in heterozygosis) crossed with three susceptible parents were used to localize the *Ma* gene on G7 close to SSR marker pchgms6 (Claverie et al. 2004a). Later on up to 1,332 progenies were genotyped to map the *Ma* gene within an interval of 0.8 cM. Flanking markers were used to isolate a single 280 kb BAC clone that carries the *Ma* gene (Claverie et al. 2004b). By using more than 3,000 progenies from Myrobalan plum accession P.2175 the *Ma* locus containing interval could be reduced to a 32-kb cluster of three Toll/Interleukin1 Receptor-nucleotide binding site-leucine-rich repeat (LRR) genes (TNL1–TNL3) from which TNL1 was the only non-truncated gene conferring the complete-spectrum resistance (Claverie et al. 2011).

4.4.3 Future Prospects

Several populations, mainly of F_1 and F_2 type, have been generated in order to map traits of interest within the *Prunus* genome. Among them are economically important traits like resistance genes, self-incompatibility or fruit quality characters such as acidity, flesh-texture and -color. For these traits, MAS can already be applied using flanking markers or, if known, genes itself to speed-up breeding programs. Nevertheless applying MAS for complex traits remain to be a challenge. In case of blooming time in peach at least three genes of known position are involved (Howad and Arús 2007; Fan et al. 2010). For other economically important traits like sugar content or aroma the inherence seem to be even more complex. To dissect such complexity and gene interactions the available mapping resources in *Prunus* are limited. However, to analyze the effect of a single QTL, a collection of NILs is produced as a powerful tool in horticultural species such as tomato (Maxon Smith and Ritchie 1983). NILs are identical to each other except for a single introgression in homozygosis of a donor line within the genetic background of a selected cultivar. Therefore a collection of NILs represent the whole genome of the donor line piece by piece within the background of a selected cultivar, and any phenotypic differences respect to this cultivar must be caused by the corresponding introgression. As many crosses are needed to be done to obtain such a collection of NILs, it has not been attempted in species that contain a juvenile phase of several years like in fruit trees. However, a project in this direction based on the T×E population has been started in IRTA.

4.5 Comparative Mapping

The comparison of genomes of species phylogenetically separated allows an evolutionary view of the genome by identification of conserved fragments followed by comparative mapping. The obtained information might also be useful as a starting point for future analysis in less characterized species.

Numerous of genetic maps have been produced in *Prunus* (Table 4-2 and 3) that are essentially collinear (Table 4-4; Dirlewanger et al. 2004a; Arús et al. 2005) as for example between peach and apricot (Lambert et al. 2004; Dondini et al. 2007) or between peach and sweet cherry (Olmstead et al. 2008; Clarke et al. 2009). However, occasionally small discrepancies between the positions of some markers can be observed, mainly attributed to the existence of duplicate loci. Apart from these minor rearrangements two major chromosomal reorganizations have been found in *Prunus*. The first one, a reciprocal translocation between G6 and G8 was identified in a F_2 population derived from a cross between 'Garfi' (almond) × 'Nemared' (peach rootstock, Jauregui et al. 2001). The same translocation could also be detected in the cross between 'Akame' × 'Juseitou' (Yamamoto et al. 2005). Another reciprocal translocation was observed in a Myrobalan plum (*P. ceresifera*) genotype affecting G3 and G5 (Lambert et al. 2004).

Although most individuals within the genus *Prunus* have the same general genome configuration few major chromosomal reorganizations occur and apparently remain stable during evolution.

4.6 Conclusion

Even with the complete genome sequence available for *Prunus* species, molecular genetic linkage maps will continue to be a major tool in genetics, genomics and breeding. Firstly constructed on peach, they were rapidly developed on other *Prunus* species, apricot, almond, cherry, myrobalan plum. Thanks to the development of new marker types, easier to use, more reliable, more informative, the construction of linkage maps is becoming less and less time consuming. The production of mapping populations and phenotyping them over years is now becoming the bottleneck for linkage map construction and subsequent QTL analysis, especially for species with long juvenile phase. Consequently, the choice of the parents used

Table 4-4 Prunus linkage maps compared with the TxE reference map.

Crosses	Name of pop.	Type of pop. (size)	Parental maps	Markers in common with T × E	Anchor marker% in same group as T × E[a]	Non-colinear markers[b]	% T × E coverage[c]	% Common distance[d]	Paired t-test comparison with T × E[e]	Mapping software used	References
Peach (*P. persica*)											
'Ferjalou Jalousia' x Fantasia'	J x F	F2 (63)	-	49	95	2	57	142	4.21**	MAPMAKER	Dirlewanger et al. 1998, 1999; Etienne et al. 2002
		F_2 (207)	-	51	88	3	57	140	3.27*	MAPMAKER	Dirlewanger et al. 2006
		F_2 (208)	-	48	96	1	61	141	2.95*	MAPMAKER	Illa et al. 2009
'Akame' x 'Juseitou'	A x J	F_2 (126)	-	45	98	2	52	109	0.79	MAPMAKER	Yamamoto et al. 2001
Dr. Davis' x ' Georgia Belle'	Pop-DG	F_1 (152)	-	40	98	4	54	178	3.42*	JOINMAP	Ogundiwin et al. 2009
'Contender' x 'Fla.92-2C'		F_2 (378)	-	36	100	1	49	128	2.88*	JOINMAP	Fan et al. 2010
Almond (*P. dulcis*)											
'Ferragnès' x 'Tuono'	F x T	F_1 (60)	Ferragnès	56	96	2	65	96	0.53	MAPMAKER	Viruel et al. 1995
			Tuono	41	93	0	48	107	0.27	MAPMAKER	
			Ferragnès	53	100	3	64	100	0.01	MAPMAKER	Joobeur et al. 2000
			Tuono	41	100	1	48	111	0.76	MAPMAKER	
'Felisia' x 'Bertina'	F x B	F_1 (134)	Felisia	32	100	0	57	111	0.80	MAPMAKER	Ballester 1998; Ballester et al. 2001
			Bertina	28	100	2	57	112	1.23	MAPMAKER	
'R1000' x 'Desmayo Largueta'	R x D	F_1 (167)	R1000	42	100	1	48	133	4.21**	MAPMAKER	Sanchez-Pérez et al. 2007
			Desmayo Largueta	39	100	2	53	130	3.08*	MAPMAKER	
Nonpareil' x 'Lauranne'	N x L	F_1 (93)	combined map	59	98	11	62	139	3.15*	JOINMAP	Tavassolian et al. 2010

Table 4-4 contd....

Table 4-4 contd....

Crosses	Name of pop.	Type of pop. (size)	Parental maps	Markers in common with T × E	Anchor marker% in same group as T × E[a]	Non-colinear markers[b]	% T × E coverage[c]	% Common distance[d]	Paired t-test comparison with T × E[e]	Mapping software used	References
Vivot' x 'Blanquerna'	V x B	F_1 (77)	Vivot	37	100	0	62	117	1.22	MAPMAKER	Fernández i Martí et al. 2011
			Blanquerna	36	100	0	58	114	1.63		
Almond (*P. dulcis*) x Peach (*P. persica*)											
'Texas' x 'Earlygold'	T x E	F_2 (82)	-	562	100	0	100	100	-	MAPMAKER	Joobeur et al. 1998; Aranzana et al. 2003b; Dirlewanger et al. 2004a
'Garfi' x 'Nemared'	G x N	Complex F_2 (78)	-	51	100	0	78	117	3.57**	MAPMAKER	Jàuregui et al. 2001
P. persica* x *P. ferganensis											
	P x F	BC_1 (70)	-	63	98	3	66	134	3.11*	JOINMAP	Verde et al. 2005
P. davidiana* x *P. persica											
	SD40[2]	F_2 (99)	-	57	100	0	70	196	4.62**	MAPMAKER	Foulongne et al. 2003a, b
P. davidiana clone P1908 x 'Summergrand'= SD 40 x 'Summergrand'	BC2	BC_2 (162)	-	40	100	0	58	144	1.90	MAPMAKER	Quilot et al. 2004
		BC_2 (162)	-	82	100	4	85	117	0.96	MAPMAKER	Illa et al. 2009
Myrobolan plum (*P. cerasifera*) x [almond (*P. dulcis*) x peach (*P. persica*)]											
P.2175 x ('Garfi' x 'Nemared')22	P.2175 x GN22	F_1 (83)	P.2175 GN22	47 92	94 95	93 89	47 70	203 168	6.87** 4,70**	MAPMAKER	Dirlewanger et al. 2004b

					Apricot (*P. armeniaca*)						
'Polonais' x 'Stark Early Orange'	P x SEO	F_1 (142)	Polonais	49 (81)	93	2	63	115	2.19	MAPMAKER	Lambert et al. al. 2004
			SEO	61	95	4	79	152	3.59**	MAPMAKER	
'Lito' x 'BO 81604311'	L x B	F_1 (125)	Lito	27	100	1	58	113	0.66	MAPMAKER	Dondini et al. 2007
			BO81604311	31	97	0	57	176	2.16	MAPMAKER	
					Sweet cherry (*P. avium*)						
'Regina' x 'Lapins'	R x L	F_1 (120)	Regina	30	97	1	34	233	3.34**	MAPMAKER	Dirlewanger et al. 2004
			Lapins	28	96	1	21	323	3.19**	MAPMAKER	
					Myrobolan plum (*P. cerasifera*) x [almond (*P. dulcis*) x peach (*P. persica*)]						
P.2175 x ('Garfi' x 'Nemared')22	P.2175 x GN22	F_1 (83)	P.2175	47	94	93	47	203	6.87**	MAPMAKER	Dirlewanger et al. 2004b
			GN22	92	95	89	70	168	4,70**	MAPMAKER	

[a]Percentage of anchor markers located on the same linkage group as T x E.
[b]Number of markers placed on the same linkage group as T x E, but in a different order (generally pairs of markers in inverted order; only one marker is considered).
[c]Percentage of the T x E map covered by the other map calculated as: distance of T x E covered x100/total T x E distance.
[d]Relative size of the common maps: common distance covered in a map x 100/total distance covered in T x E.
[e]Paired t-test of the comparison between the distances of the two furthest-apart common markers in each linkage group.
*P ≤ 0.05; **P ≤ 0.01.

to produce a mapping population as to be highly examined in order to optimize the polymorphism for the makers and for the studied characters. Moreover, the use of transferable markers, like COS markers, useful for comparative mapping, will allow to take benefit of results obtained on a mapping population to an other. We can expect that in a near future the effort conducted on the construction on linkage maps will lead to the integration of MAS for numerous characters in *Prunus* species breeding programs.

References

Abbott AG, Arús P, Scorza R (2007) Peach. In: Kole C (ed) Genome Mapping and Molecular Breeding in Plants. Vol 4: Fruits and Nuts. Springer, Berlin, Heidelberg, pp 137–156.

Abbott AG, Rajapakse S, Sosinski B, Lu ZX, Sossey-Alaoui K, Gannavarapu M, Reighard G, Ballard RE, Baird WV, Scorza R, Callahan A (1998) Construction of saturated linkage maps of peach crosses segregating for characters controlling fruit quality, tree architecture and pest resistance. Acta Hort 465: 41–49.

Ai CX, Yu XM, Dahlenburg A, Zhao HJ, Zhao Y, Liu QZ (2008) Development and characterization of SSR markers in Chinese cherry (*Prunus pseudocerasus* Lindl.). Eur J Hort Sci 73(3): 104–110.

Aranzana MJ, Garcia-Mas J, Carbo J, Arús P (2002) Development and variability analysis of microsatellite markers in peach. Plant Breed 121: 87–92

Aranzana MJ, Pineda A, Cosson P, Ascasibar J, Dirlewanger E, Ascasibar J, Cipriani G, Ryder CD, Testolin R, Abbott A, King GJ, Iezzoni AF, Arùs P (2003) A set of simple-sequence repeat (SSR) markers covering the *Prunus* genome. Theor Appl Genet 106: 819–825.

Arús P, Messeguer R, Viruel M, Tobutt K, Dirlewanger E, Santi F, Quarta R, Ritter E (1994) The European *Prunus* mapping project—progress in the almond linkage map. Euphytica 77: 97–100.

Arús P, Yamamoto T, Dirlewanger E, Abbott AG (2005) Synteny in the *Rosaceae*. In: Janick J (ed) Plant Breeding Reviews. Vol 27. John Wiley, London, UK, pp 175–211.

Ballester J, Company RSI, Arús P, de Vicente MC (2001) Genetic mapping of a major gene delaying blooming time in almond. Plant Breed 120: 268–270.

Blenda A, Verde I, Georgi L, Reighard G, Forrest S, Muñoz-Torres M, Baird W, Abbott A (2007) Construction of a genetic linkage map and identification of molecular markers in peach rootstocks for response to peach tree short life syndrome. Tree Genet Genomes 3: 341–350

Bliss FA, Arulsekar S, Foolad MR, Becerra V, Gillen AM, Warburton ML, Dandekar AM, Kocsisne GM, Mydin KK (2002) An expanded genetic linkage map of *Prunus* based on an interspecific cross between almond and peach. Genome 45: 520–529.

Bošković R, Tobutt KR (1998) Inheritance and linkage relationships of isoenzymes in two interspecific cherry progenies. Euphytica 103: 273–286.

Boudehri K, Bendahmane A, Cardinet G, Troadec C, Moing A, Dirlewanger E (2009) Phenotypic and fine genetic characterization of the *D* locus controlling fruit acidity in peach. BMC Plant Biol 9: 59.

Cabrera A, Kozik A, Howad W, Arús P, Iezzoni AF, van der Knaap E (2009) Development and bin mapping of a Rosaceae Conserved Ortholog Set (COS) of markers. BMC Genomics 29: 562.

Campoy J, Martínez-Gómez P, Ruiz D, Rees J, Celton J (2010) Developing microsatellite multiplex and megaplex PCR systems for high-throughput characterization of breeding progenies and linkage maps spanning the apricot (*Prunus armeciaca* L.) genome. Plant Mol Biol Rep 28(4): 560–568.

Cantini C, Iezzoni AF, Lamboy WF, Bortizki M, Struss D (2001) DNA fingerprinting of tetraploid cherry germplasm using simple sequence repeats. J Am Soc Hort Sci 126: 205–209.
Channuntapipat C, Wirthensohn M, Ramesh SA, Battlle I, ARÚS P, Sedgley M, Collins G (2003) Identification of incompatibility genotypes in almond (*Prunus dulcis* Mill.) using specific primers based on the introns of the S-alleles. Plant Breed 122: 164–168.
Chaparro JX, Werner DJ, O'Malley D, Sederoff RR (1994) Targeted mapping and linkage analysis of morphological, isozyme, and RAPD markers in peach. Theor Appl Genet 87: 805–815.
Chen L, Zhang SM, Illa E, Song LJ, Wu SD, Howad W, Arús P, van de Weg E, Chen KS, Gao ZS (2008) Genomic characterization of putative allergen genes in peach/almond and their synteny with apple. BMC Genomics 9: 543.
Cipriani G, Lot G, Huang WG, Marrazzo MT, Peterlunger E, Testolin R (1999) AC/GT and AG/CT microsatellite repeats in peach [*Prunus persica* (L.) Batsch]: isolation, characterization and cross-species amplification in *Prunus*. Theor Appl Genet 99: 65–72.
Clarke J, Sargent D, Bošković R, Belaj A, Tobutt K (2009) A cherry map from the inter-specific cross *Prunus avium* 'Napoleon' × *P. nipponica* based on microsatellite, gene-specific and isoenzyme markers. Tree Genet Genomes 5: 41–51.
Clarke JB, Tobutt KR (2003) Development and characterization of polymorphic microsatellites from *Prunus avium* 'Napoleon'. Mol Ecol Notes 3: 578–580.
Claverie M, Bosselut N, Lecouls AC, Voisin R, Poizat C, Dirlewanger E, Kleinhentz M, Lafargue B, Laigret F, Esmenjaud D (2004a) Location of independent root-knot nematode resistance genes in plum and peach. Theor Appl Genet 108: 765–773.
Claverie M, Dirlewanger E, Cosson P, Bosselut N, Lecouls AC, Voisin R, Kleinhentz M, Lafargue B, Caboche M, Chalhoub B, Esmenjaud D (2004b) High-resolution mapping and chromosome landing at the root-knot nematode resistance locus *Ma* from Myrobalan plum using a large-insert BAC DNA library. Theor Appl Genet 109: 1318–1327.
Claverie M, Dirlewanger E, Bosselut N, Van Ghelder C, Voisin R, Kleinhentz M, Lafargue B, Abad P, Rosso MN, Chalhoub B, Esmenjaud D (2011) The *Ma* gene for complete-spectrum resistance to *Meloidogyne* species in *Prunus* is a TNL with a huge repeated C-terminal post-LRR region. Plant Physiol 156: 779–792.
Decroocq V, Fave MG, Hagen L, Bordenave L, Decroocq S (2003) Development and transferability of apricot and grape EST microsatellite markers across taxa. Theor Appl Genet 106: 912–922.
Dettori MT, Quarta R, Verde I (2001) A peach linkage map integrating RFLPs, SSRs, RAPDs and morphological markers. Genome 44: 783–790.
Dirlewanger E, Bodo C (1994) Molecular genetic mapping of peach. Euphytica 77: 101–103.
Dirlewanger E, Cosson P, Tavaud M, Aranzana MJ, Poizat C, Zanetto A, Arús P, Laigret F (2002) Development of microsatellite markers in peach [*Prunus persica* (L.) Batsch] and their use in genetic diversity analysis in peach and sweet cherry (*Prunus avium* L.). Theor Appl Genet 105: 127–138.
Dirlewanger E, Cosson P, Howad W, Capdevill G, Bosselu N, Claverie M, Voisin R, Poizat C, Lafargue B, Baron O, Laigret F, Kleinhentz M, Arús P, Esmenjaud D (2004b) Microsatellite Genetic linkage maps of Myrobalan Plum and an Almond-Peach hybrid—Location of root-knot nematode resistance genes. Theor Appl Genet 109: 827–838.
Dirlewanger E, Cosson P, Boudehri K, Renaud C, Capdeville G, Tauzin Y, Laigret F, Moing A (2006) Development of a second-generation genetic linkage map for peach [*Prunus persica* (L.) Batsch] and characterization of morphological traits affecting flower and fruit. Tree Genet Genomes 3: 1–13.
Dirlewanger E, Graziano E, Joobeur T, Garriga-Calderé F, Cosson P, Howad W, Arús P (2004a) Comparative mapping and marker assisted selection in Rosaceae fruit crops. Proc Natl Acad Sci USA 101: 9891–9896.
Dirlewanger E, Moing A, Rothan C, Svanella L, Pronier V, Guye A, Plomion C, Monet R (1999) Mapping QTLs controlling fruit quality in peach (*Prunus persica* (L.) Batsch). Theor Appl Genet 98: 18–31.

Dirlewanger E, Pascal T, Zuger C, Kervella J (1996) Analysis of molecular markers associated with powdery mildew resistance genes in peach (*Prunus persica* (L.) Batsch) x *Prunus davidiana* hybrids. Theor Appl Genet 93: 909–919.

Dirlewanger E, Pronier V, Parvery C, Rothan C, Guy A, Monet R (1998) Genetic linkage map of peach (*Prunus persica* (L.) Batsch) using morphological and molecular markers. Theor Appl Genet 97: 888–895.

Dondini L, Lain O, Geuna F, Banfi R, Gaiotti F, Tartarini S, Bassi D, Testolin R (2007) Development of a new SSR-based linkage map in apricot and analysis of synteny with existing *Prunus* maps. Tree Genet Genomes 3: 239–249.

Downey SL, Iezzoni AF (2000) Polymorphic DNA markers in black cherry (*Prunus serotina*) are identified using sequences from sweet cherry, peach, and sour cherry. J Am Soc Hort Sci. 125: 76–80.

Dragan N, Dragan M (2010) Examining self-compatibility in plum (*Prunus domestica* L.) by fluorescence microscopy. Genetika 42: 387–396.

Esmenjaud D, Minot JC, Voisin R (1996) Effect of durable inoculum pressure and high temperature on root galling, nematode numbers and survival of Myrobalan plum genotypes (*Prunus cerasifera* Ehr.) highly resistant to *Meloidogyne* spp. Theor Appl Genet 92: 873–879.

Etienne C, Rothan C, Moing A, Plomion C, Bodénès C, Svanella-Dumas L, Cosson P, Pronier V, Monet R, Dirlewanger E (2002) Candidate genes and QTLs for sugar and organic acid content in peach [*Prunus persica* (L.) Batsch]. Theor Appl Genet 105: 145–159.

Fan S, Bielenberg DG, Zhebentyayeva TN, Reighard GL, Okie WR, Holland D, Abbott AG (2010) Mapping quantitative trait loci associated with chilling requirement, heat requirement and bloom date in peach (*Prunus persica*). New Phytol 185: 917–930.

Fernández i Martí A, Howad W, Tao R, Alonso Segura JM, Arús P, Socias i Company R (2011) Identification of quantitative trait loci associated with self-compatibility in a *Prunus* species. Tree Genet Genomes 7: 629–639.

Fernandez-Silva I, Eduardo I, Blanca J, Esteras C, Picó B, Nuez F, Arús P, Garcia-Mas J, Monforte A J (2008) Bin mapping of genomic and EST-derived SSRs in melon (*Cucumis melo* L.). Theor Appl Genet 118: 139–150.

Foolad MR, Arulsekar S, Becerra V, Bliss FA (1995) A genetic map of *Prunus* based on an interspecific cross between peach and almond. Theor Appl Genet 91: 262–269.

Foulongne M, Pascal T, Arús P, Kervella J (2003a) The potential of *Prunus davidiana* for introgression into peach [*Prunus persica* (L.) Batsch] assessed by comparative mapping. Theor Appl Genet 107: 227–238.

Foulongne M, Pascal T, Pfeiffer F, Kervella J (2003b) QTLs for powdery mildew resistance in peach x *Prunus davidiana* crosses: consistency across generations and environments. Mol Breed 12: 33–50.

Georgi LL, Wang Y, Yvergniaux D, Ormsbee T, Iñigo M, Reighard G, Abbott AG (2002) Construction of a BAC library and its application to the identification of simple sequence repeats in peach [*Prunus persica* (L.) Batsch]. Theor Appl Genet 105: 1151–1158.

Gillen AM and Bliss FA (2005) Identification and mapping of markers linked to the Mi gene for root-knot nematode resistance in peach. J Am Soc Hort Sci 130: 24–33.

Grimplet J, Romieu C, Audergon JM, Marty I, Albagnac G, Lambert P, Bouchet JP, Terrier N (2005) Transcriptomic study of apricot fruit (*Prunus armeniaca*) ripening among expressed sequence tags. Physiol Plant 125: 281–292.

Halász J (2007) Review of self-(in)compatibility in Apricot (*Prunus armeniaca* L.) World J Agri Sci 3: 670–677.

Hagen LS, Chaib J, Fady B, Decroocq V, Bouchet JP, Lambert P, Audergon JM (2004) Genomic and cDNA microsatellites from apricot (*Prunus armeniaca* L.). Mol Ecol Notes 4: 24–26.

Han Y, Chagné D, Gasic K, Rikkerink E H A, Beever J E, Gardiner S E, Korban S S (2009) BAC-end sequence-based SNPs and Bin mapping for rapid integration of physical and genetic maps in apple. Genomics 93: 282–288.

Hansche PE (1983) Response to selection. In: Moore JN, Janick J (eds) Methods in Fruit Breeding. Purdue Univ Press, West Lafayette, IN, pp 154–171.
Hansche PE, Beres W (1980) Genetic remodeling of fruit and nut trees to facilitate cultivar development. HortScience 15: 710–715.
Hauck NR, Yamane H, Tao R, Iezzoni AF (2002) Self-compatibility and incompatibility in tetraploid sour cherry (*Prunus cerasus* L.). Sex Plant Reprod 15: 39–46.
Hesse CO (1975) Peaches. In: Janick J, Moore JN (eds) Advances in fruit breeding Temperate fruits. Purdue Univ Press, West Lafayette, IN, pp 285–335.
Horn R, Lecouls AC, Callahan A, Dandekar A, Garay L, McCord P, Howad W, Chan H, Verde I, Main D, et al. (2005) Candidate gene database and transcript map for peach, a model species for fruit trees. Theor Appl Genet 110: 1419–1428.
Howad W, Yamamoto T, Dirlewanger E, Testolin R, Cosson P, Cipriani G, Monforte AJ, Georgi L, Abbott AG, Arús P (2005) Mapping with a few plants: Using selective mapping for microsatellite saturation of the *Prunus* reference map. Genetics 171: 1305–1309.
Howad W, Arús P (2007) QTL Analysis of blooming time in peach. In: Plant Anim Genome XV Conf, San Diego, CA, P476.
Hurtado MA, Romero C, Vilanova S, Abbott AG, Llácer G, Badenes ML (2002) Genetic linkage maps of two apricot cultivars (*Prunus armaniaca* L.), and mapping of PPV (sharka) resistance. Theor Appl Genet 105: 182–191.
Illa E, Eduardo I, Audergon JM, Barale F, Dirlewanger E, Li X, Moing A, Lambert P, Le Dantec L, Gao Z, Poëssel JL, Pozzi C, Rossini L, Vecchietti A, Arus P, Howad W (2011a) Saturating the *Prunus* (stone fruits) genome with candidate genes for fruit quality. Mol Breed: DOI 10.1007/s11032-010-9518-x.
Illa E, Sargent DJ, Lopez Girona E, Bushakra J, Cestaro A, Crowhurst R, Pindo M, Cabrera A, van der Knaap E, Iezzoni A, Gardiner S, Velasco R, Arús P, Chagné D, Troggio M (2011b) Comparative analysis of rosaceous genomes and the reconstruction of a putative ancestral genome for the family. BMC Evol Biol 11: 9.
Jáuregui B, de Vicente MC, Messeguer R, Felipe A, Bonnet A, Salesses G, Arús P (2001) A reciprocal translocation between 'Garfi' almond and 'Nemared' peach. Theor Appl Genet 102: 1169–1176.
Joobeur T, Periam N, de Vicente MC, King GJ, Arús P (2000) Development of a second generation linkage map for almond using RAPD and SSR markers. Genome 43: 649–655.
Joobeur T, Viruel MA, de Vicente MC, Jáuregui B, Ballester J, Dettori MT, Verde I, Truco MJ, Messeguer R, Batlle I, Quarta R, Dirlewanger E, Arús P (1998) Construction of a saturated linkage map for *Prunus* using an almond x peach F_2 progeny. Theor Appl Genet 97: 1034–1041.
Jung S, Staton M, Lee T, Blenda A, Svancara R, Abbott A, Main D (2008) GDR (Genome Database for Rosaceae): integrated web-database for Rosaceae genomics and genetics data. Nucl Acids Res 36: D1034–D1040.
Kole C, Abbott AG (2008) Fundamentals of plant genome mapping. In: Kole C, Abbott AG (eds) Principles and Practices of Plant Genomics. Vol 1: Genome Mapping. Science Publishers, Enfield, NH, pp 1–22.
Lalli D, Decroocq V, Blenda A, Schurdi-Levraud V, Garay L, Le Gall O, Damsteegt V, Reighard G, Abbott A (2005) Identification and mapping of resistance gene analogs (RGAs) in *Prunus*: a resistance map for *Prunus*. Theor Appl Genet 11: 1504–1513.
Lambert P, Hagen LS, Arus P, Audergon JM (2004) Genetic linkage maps of two apricot cultivars (*Prunus armeniaca* L.) compared with the almond Texas x peach Earlygold reference map for *Prunus*. Theor Appl Genet 108: 1120–1130.
Le Dantec L, Cardinet G, Bonet J, Fouché M, Boudehri K, Monfort A, Poëssel JL, Moing A, Dirlewanger E (2010) Development and mapping of peach candidate genes involved in fruit quality and their transferability and potential use in other Rosaceae species. Tree Genet Genomes 6: 995–1012.

Lecouls AC, Salesses G, Minot JC, Voisin R, Bonnet A, Esmenjaud D (1997) Spectrum of the *Ma* genes for resistance to *Meloidogyne* spp. in Myrobalan plum. Theor Appl Genet 85: 1325–1334.

Li X, Shangguan L, Song C, Wang C, Gao Z, Yu H, Fang J (2010) Analysis of expressed sequence tags from *Prunus* mume flower and fruit and development of simple sequence repeat markers. BMC Genetics 2010 11: 66.

Lopes MS, Sefc KM, Laimer M, Da Camara Machado A (2002) Identification of microsatellite loci in apricot. Mol Ecol Notes 2: 24–26.

Lu ZX, Sosinski B, Reighard GL, Baird WV, Abbott AG (1998) Construction of a genetic linkage map and identification of AFLP markers for resistance to root-knot nematodes in peach rootstocks. Genome 41: 199–207.

Massonié G, Monet R, Bastard Y, Grasselly C (1982) Résistance au puceron vert du pÞcher, Myzus persicae Sulzer (Homoptera aphididae) chez *Prunus persica* (L.) Batsch et d'autres espèces de *Prunus*. Agronomie 2: 63–70.

Maxon Smith JW, Ritchie DB (1983) A collection of near-isogenic lines of tomato: Research tool of the future? Plant Mol Biol Rep 1: 41–45.

Messina R, Lain O, Marrazzo MT, Cipriani G, Testolin R (2004) New set of microsatellite loci isolated in apricot. Mol Ecol Notes 4: 432–434.

Monet R (1979) Genetic transmission of the 'fruit sweetness' character. Incidence on selection for quality. In: INRA (ed) Proceedings of Eucarpia Fruit Section Symp. Tree Fruit Breed Angers, France, pp 273–276.

Mnejja M, Garia-Mas J, Howad W, Arús P (2005) Development and transportability across *Prunus* species of 42 polymorphic almond microsatellites. Mol Ecol Notes 5: 531–535.

Mnejja M, Garcia-Mas J, Howad W, Badenes ML, Arús P (2004) Simple-sequence repeat (SSR) markers of Japanese plum (*Prunus salicina* Lindl.) are highly polymorphic and transferable to peach and almond. Mol Ecol Notes 4: 163–166.

Nishitani C, Kimura T, Ueda E, Howad W, Arús P, Yamamoto T (2007) Tri-/hexanucleotide microsatellite markers in peach derived from enriched genomic libraries and their application in rosaceae. Breed Sci 57: 289–296.

Ogundiwin EA, Peace CP, Gradziel TM, Parfitt DE, Bliss FA, Crisosto CH (2009) A fruit quality gene map of *Prunus*. BMC Genomics 10: 587.

Ogundiwin EA, Martı´C, Forment J, Pons C, Granell A, Gradziel TM, Peace CP, Crisosto CH (2008) Development of ChillPeach genomic tools and identification of cold-responsive genes in peach fruit. Plant Mol Biol 68: 379–397.

Olmstead JW, Sebolt AM, Cabrera A, Sooriyapathirana SS, Hammar S, Iriarte G, Wang D, Chen CY, van der Knaap E, Iezzoni AF (2008) Construction of an intra-specific sweet cherry (*Prunus avium* L.) genetic linkage map and synteny analysis with the *Prunus* reference map. Tree Genet Genomes 4: 897–910.

Olukolu BA, Trainin T, Fan S, Kole C, Bielenberg DG, Reighard GR, Abbott AG, Holland D (2009) Genetic linkage mapping for molecular dissection of chilling requirement and budbreak in apricot (*Prunus armeniaca* L.). Genome 52: 819–828.

Pascal T, Kervella J, Pfeiffer F, Sauge MH, Esmanjaud D (1998) Evaluation of the interspecific progeny *Prunus persica* cv Summergrand x *Prunus davidiana* for disease resistance and some agronomic features. Acta Hort 465: 185–191.

Paterson AH (1996) Mapping genes responsible for differences in phenotype. In: Paterson AH (ed) Genome Mapping in Plants. R G Landes Company, San Diego, CA; Academic Press, Austin, TX, pp 41–54.

Quarta R, Dettori MT, Verde I, Gentile A, Broda Z (1998) Genetic analysis of agronomic traits and genetics linkage mapping in a BC1 peach population using RFLPs and RAPDs. Acta Hort 465: 51–59.

Quarta R, Dettori MT, Sartori A, Verde I (2000) Genetic linkage map and QTL analysis in peach. Acta Hort 521: 233–241.

Quilot B, Wu BH, Kervella J, Génard M, Foulongne M, Moreau K (2004) QTL analysis of quality traits in an advanced backcross between *Prunus persica* cultivars and the wild relative species *P. davidiana*. Theor Appl Genet 109: 884–897.

Rajapakse S, Belthoff LE, He G, Estager AE, Scorza R, Verde I, Ballard RE, Baird WV, Callahan A, Monet R, Abbott AG (1995) Genetic linkage mapping in peach using morphological, RFLP and RAPD markers. Theor Appl Genet 91: 964–971.

Salesses G and Mouras A (1977) Tentative d'utilisation des protoplastes pour l'etude des chromosomes chez les *prunus*. Ann Amel Plant 27(3): 363–368.

Sánchez-Pérez R, Howad W, Dicenta F, Arús P, Martínez-Gómez P (2007) Mapping major genes and quantitative trait loci controlling agronomic traits in almond. Plant Breed 126: 310–318.

Sauge MH, Pascal T, Lacroze JP, Pfeiffer F, Kervella J (2001) Mapping of a genetic factor of partial resistance to *Myzus persicae* in the wild peach *Prunus davidiana*, that impedes phloem sap ingestion by the aphid. Aphids in a new Millenium, 6th Int Symp on Aphids, Rennes, France, pp 3–7.

Sargent DJ, Marchese A, Simpson DW, Howad W, Fernández-Fernández F, Monfort A, Arús P, Evans KM, Tobutt KR (2009) Development of 'universal' gene-specific markers from *Malus pumila* cDNA sequences, their mapping and use in synteny studies within Rosaceae. Tree Genet Genomes 5: 133–145.

Sawa A, Kaneko S, Isagi Y, Mariko S, Masaki T (2010) Development and characterization of microsatellite markers for *Prunus verecunda* and *Prunus grayana* (Rosaceae). Conserv Genet 11: 1167–1169.

Scorza R, Melnicenco L, Dang P, Abbott AG (2002) Testing a microsatellite marker for selection of columnar growth habit in peach [*Prunus persica* (L.) Batsch]. Acta Hort 592: 285–289.

Scorza R, Sherman WB (1996) Peaches. In: Janick J, Moore JN (ed) Fruit Breeding, vol I. John Wiley, New York, pp 325–440.

Silva C, Garcia-Mas J, Sanchez AM, Arús P, Oliveira M (2005) Looking into flowering time in almond (*Prunus dulcis* (Mill) DA Webb): the candidate gene approach. Theor Appl Genet 110: 959–968.

Sonneveld T, Tobutt KR, Vaughan SP, Robbins TP (2005) Loss of Pollen-S function in two self-compatible selections of *Prunus avium* is associated with deletion/mutation of an S haplotype-specific F-box gene. Plant Cell 17: 37–51.

Sosinski B, Gannavarapu M, Hager LD, Beck LE, King GJ, Ryder CD, Rajapakse S, Baird WV, Ballard RE, Abbott AG (2000) Characterization of microsatellite markers in peach [*Prunus persica* (L.) Batsch]. Theor Appl Genet 97: 1034–1041.

Smykov VK, Ovcharenko GV, Perfilyeva ZN, Shoferistov EP (1982) Estimation of the peach hybrid resources by its mildew resistance against the infection background. Byull Gos Nikitsh Bot Sada 88: 74–80.

Stockinger EJ, Mulinix CA, Long CM, Brettin TS, Iezzoni AF (1996) A linkage map of sweet cherry based on RAPD analysis of a microspore-derived callus culture population. J Hered 87 (3): 214–218.

Struss D, Ahmad R, Southwick SM, Boritzki M (2003) Analysis of sweet cherry (*Prunus avium* L.) cultivars using SSR and AFLP markers. J Am Soc Hort Sci 128: 904–909.

Tavassolian I, Rabiei G, Gregory D, Mnejja M, Wirthensohn MG, Hunt PW, Gibson JP, Ford CM, Sedgley M, Wu S-H (2010) Construction of an almond linkage map in an Australian population Nonpareil × Lauranne. BMC Genomics 11: 551.

Testolin R, Marrazzo T, Cipriani G, Quarta R, Verde I, Dettori MT, Pancaldi M, Sansavini S (2000) Microsatellite DNA in peach (*Prunus persica* L. Batch) and its use in fingerprinting and testing the genetic origin of cultivars. Genome 43: 512–520.

Testolin R, Messina R, Lain O, Marrazzo MT, Huang WG, Cipriani G (2004) Microsatellites isolated in almond from an AC-repeat enriched library. Mol Ecol Notes 4: 459–461.

Van Dyk MM, Baison J, Hove P, Marondedze C, Klein A, Soeker K, Rees DJG (2007) BIN mapping of EST-SSRs and EST-SNPs in apple (*Malus* x *domestica Borkh.*). Eucarpia XII Fruit Section Symp, Sept 16–20, 2007, Zaragoza, Spain, pp 57–58.

Vaughan SP, Russell K (2004) Characterization of novel microsatellites and development of multiplex PCR for large-scale population studies in wild cherry, *Prunus avium*. Mol Ecol Notes 4: 429–431.

Vendramin E, Dettori MT, Giovinazzi J, Micali S, Quarta R, Verde I (2007) A set of EST-SSRs isolated from peach fruit transcriptome and their transportability across *Prunus* species. Mol Ecol Notes 7: 307–310.

Verde I, Quarta R, Cedrola C, Dettori MT (2002) QTL analysis of agronomic traits in a BC1 peach population. Acta Hort 592: 291–297.

Verde I, Lauria M, Dettori MT, Vendramin E, Balconi C, Micali S, Wang Y, Marrazzo MT, Cipriani G, Hartings H, Testolin R, Abbott AG, Motto M, Quarta R (2005) Microsatellite and AFLP markers in the *Prunus persica* [L. (Batsch)]·*P. ferganensis* BC1linkage map: saturation and coverage improvement Theor Appl Genet 111: 1013–1021.

Vilanova S, Romero C, Abbott AG, Llácer G, Badenes ML (2003) An apricot (*Prunus armeniaca* L.) F_2 progeny linkage map based on SSR and AFLP markers, mapping plum pox virus resistance and self-incompatibility traits. Theor Appl Genet 107: 239–247.

Vilanova S, Soriano JM, Lalli DA, Romero C, Abbott AG, Llacer G, Badenes ML (2006) Development of SSR markers located in the G1 linkage group of apricot (*Prunus armeniaca* L.) using a bacterial artificial chromosome library. Mol Ecol Notes 6: 789–791.

Viruel MA, Madur D, Dirlewanger E, Pascal T, Kervella J (1998). Mapping quantitative trait loci controlling peach leaf curl resistance. Acta Hort 465: 79–88.

Viruel MA, Messeguer R, de Vicente MC, Garcia-Mas J, Puigdomènech P, Vargas F, Arús P (1995) A linkage map with RFLP and isozyme markers for almond. Theor Appl Genet 91: 964–971.

Vision TJ, Brown DG, Shmoys DB, Durrett RT, Tanksley SD (2000) Selective mapping: A strategy for optimizing the construction of high-density linkage maps. Genetics 155: 407–420.

Wang D, Karle R, Brettin TS, Iezzoni AF (1998) Genetic linkage map in sour cherry using RFLP markers. Theor Appl Genet 97: 1217–1224.

Wang D, Karle R, Iezzoni AF (2000) QTL analysis of flower and fruit traits in sour cherry. Theor Appl Genet 100: 535–544.

Wang Y, Georgi LL, Zhebentyayeva TN, Reighard GL, Scorza R, Abbott AG (2002) High-throughput targeted SSR marker development in peach (*Prunus persica*). Genome 45: 319–328.

Warburton ML, Becerra-Velasquez VL, Goffreda JC, Bliss FA (1996) Utility of RAPD markers in identifying genetic linkages to genes of economic interest in peach. Theor Appl Genet 93: 920–925.

Wu S-B, Wirthensohn MG, Hunt P, Gibson JP, Sedgley M (2008) High resolution melting analysis of almond SNPs derived from ESTs. Theor Appl Genet 118: 1–14.

Wu S-B, Franks TK, Hunt P, Wirthensohn MG, Gibson JP, Sedgley M (2010) Discrimination of SNP genotypes associated with complex haplotypes by high resolution melting analysis in almond: implications for improved marker efficiencies. Mol Breed 25: 351–357.

Yamamoto T, Mochida K, Imai T, Shi Z, Ogiwara I, Hayashi T (2002) Microsatellite markers in peach [*Prunus persica* (L.) Batsch] derived from an enriched genomic and cDNA libraries. Mol Ecol Notes 2: 298–301.

Yamamoto T, Shimada T, Imai T, Yaegaki H, Haji T, Matsuta N, Yamaguchi M, Hayashi T (2001) Characterization of morphological traits based on a genetic linkage map in peach. Breed Sci 51: 271–278.

Yamamoto T, Yamaguchi M, Hayashi T (2005) An integrated genetic linkage map of peach by SSR, STS, AFLP and RAPD. J Jpn Soc Hort Sci 74(3): 204–213.

Yoshida M (1970) Genetical studies on the fruit quality of peach varieties. Bull Hortic Res Stn A9 Jpn 1–15.

Zhebentyayeva TN, Swire-Clark G, Georgi LL, Garay L, Jung S, Forrest S, Blenda AV, Blackmon B, Mook J, Horn R, Howad W, Arús P, Main D, Tomkins JP, Sosinski B, Baird WV, Reighard GL, Abbott AG (2008) A framework physical map for peach, a model Rosaceae species. Tree Genet Genomes 4: 745–756.

5

Mapping and Tagging of Simply Inherited Traits

Jose Miguel Soriano[1] and *Maria Luisa Badenes*[1,*]

ABSTRACT

The genus *Prunus* comprises more domesticated species of temperate fruits than any other genera of the Rosaceaes family. The identification of traits of interest, determination of their inheritance and establishing efficient methods of selection are the keys for progress in *Prunus* breeding. Selection assisted by molecular markers (MAS) is possible if previous identification of molecular markers tightly linked to the target trait is carried out. In this chapter, single inheritance traits identified in *Prunus* species and results from two methods for searching trait-linked markers: bulked segregant analysis (BSA) and mapping by association analysis between traits of interest and molecular markers are reviewed.

Keywords: linkage maps, bulked segregant análisis, marker-assisted breeding

5.1 Importance of the Target Traits

The genus *Prunus* comprises more domesticated species of temperate fruits than any other genera, e.g., *Malus, Pyrus, Sorbus, Cydonia, Rubus,* and *Fragaria,* of the family Rosaceae. Many species and cultivars are grown for their edible fruits, while others are planted for their ornamental values. Stone fruits are appreciated worldwide either for fresh consumption, or in the processing industry: drying, distillation, canning and production of jams, syrups and fruit juices (Druart and Gruselle 1986). The range of

[1]Instituto Valenciano de Investigaciones Agrarias, Apartado Oficial 46113 Moncada, Valencia, Spain.
*Corresponding author: *badenes_mlu@gva.es*

species and their different uses resulted in a wide range of traits that need to be improved in order to meet the demands of the industry, growers or consumers.

In most of the stone fruit species, the breeding goals are focused at those characteristics that improve the orchard management and production, for instance improvement in tree architecture for avoiding pruning, increase in yield which is strongly related to reproductive biology and self-compatibility, resistance to biotic and abiotic stresses which allowed a better adaptability to the environment and reduced the use of pesticides and finally, fruit quality adapted to the consumers' demand. Recently, other concerns as the increased awareness of the health benefits of fruit consumption and practices friendly with the environment became new goals towards the demand for healthier food.

These vast amount of target traits are usually improved through conventional breeding. However, improvement of fruit trees through traditional breeding methods is a long-term effort because of their long juvenile period, complex inheritance of most of the target traits and high degree of heterozygosity.

The identification of traits of interest, determination of their inheritance and to establish efficient methods of selection are the keys for progress in *Prunus* breeding programs. All together development of high quality cultivars makes the scope possible. The new cultivars should meet the demand of the growers, industry and consumers. In this context, the most time consuming step in the procedure of breeding fruit tree species is the screening of the traits of interest. Improvement of methods for selection of seedlings resulted in a direct increase of the breeding efficiency. Thus, new approaches are needed to achieve selection of the traits of interest in a reasonable time frame. Biotechnology has broadened these approaches by adding *in vitro* techniques, such as embryo rescue (Laibach 1929), that allows recovering of inmature seeds; regeneration and protoplast fusion (Bhojwani et al. 1977) that allows overcoming intercrossability barriers; development of genetic maps with molecular markers, enabling marker-assisted selection (MAS), comparative genomics (Chahal and Gosal 2002; Schmidt 2002) that allow identification of homology of the target genes in the species of interest; and genetic modification (GM) that allows introgression of genes without undesired genetic background. Among these biotechnological techniques, the identification of the genes involved in a trait and/or molecular markers tightly linked to them, make marker assisted selection (MAS) possible, which results in a faster and more powerful selection of the progenies. Setting methods of selection based on molecular markers provide highly efficient procedures for characterization of varieties and progenies under selection enabling molecular-assisted breeding (MAB).

A characteristic of the *Prunus* species is the high heterozygousity of genotypes. This phenomenon in annual species is a handicap for genetic studies because that makes it difficult to obtain specific types of populations such as recombinant inbred lines (RILs), near isogenic lines (NILs) or inbreed homozygous lines. However in trees with long juvenile period it becomes an advantage because it allows studies on inheritance from single crosses. Many of the simply inherited traits were studied in single crosses or F_1. In peach, where the breeding activity goes back to the beginning of the 20th century, many simply inherited traits were identified by means of observation of the hybrids segregating from crosses into breeding programs, mainly the traits identified were related to morphological characteristics, easy to be noted by breeders.

Some examples of simply inherited traits in peach are related to (1) tree arquitecture: standard versus weeping (Monet et al. 1988), standard versus dwarf (Hansche 1988), and standard versus pillar (Scorza et al. 1987; Chaparro et al. 1994) and presence of anthocianines (Monet 1967); (2) leaves: red colour versus green (Blake 1937), shape of glands in the leaves, reniform versus globose; (3) flowers: non showy versus showy (Weinberger 1944), white color versus pink, single versus double, pink versus red (Lammersts 1945); and (4) fruits: white versus yellow flesh, pubescent skin versus glabrous (Blake 1932), freestone versus clingstone, melting flesh versus non-melting (Bailey and French 1949), slow ripening versus standard (Ramming 1991).

In apricot the breeding programs are more recent and only observations made on crosses between Central Asian and European groups were referenced. These were related mostly to segregation observed in fruit characteristics such as size, acidity, color and firmness but none of them was identifyed as single traits (Layne et al. 1996). Recent genetic studies in the framework of the active breeding programs in apricot indicated the quantitative nature of the traits related to fruit quality and phenology (Zebentyayeva et al. 2011). A few traits inherited in a discrete manner, suggesting an oligogenic inheritance pattern, are seed bitterness (Gómez et al. 1998), male sterility (Burgos and Ledbetter 1994) and resistance to sharka virus (Karayiannis et al. 2008).

In cherries, similar to apricot, the main target traits are related to fruit characteristics and harvest season, all being of quantitative nature (Kappel et al. 2011). A similar case occurrs in plum (Topp et al. 2011).

An important simple trait related to productivity is self-fertility. It has been a target trait in the self-incompatibible species such as almond (Socias and Company 1990), cherry (Lansari and Iezzoni 1990) and apricot (Burgos et al. 1997). Molecular studies initiated in almond and followed in cherry and apricot resulted in establishing markers for selecting this important trait.

In this chapter, results from two methods for searching trait-linked markers: bulked segregant analysis (BSA) and mapping by association analysis between traits of interest and molecular markers are reviewed.

5.2 Employment of BSA and Genes Tagged

Bulked segregant analysis (BSA) is a widely used method for rapidly identifying molecular markers linked to a trait of interest. It involves genotyping of two pools (bulks) of DNA samples from individual plants originating from a cross. Plants are assigned to bulks that differ with regard to the trait of interest. The bulks are screened with a large number of markers to identify those that distinguish both bulks and, by inference, must be genetically linked to the trait (Michelmore et al. 1991).

The most frequent types of molecular markers used for carrying out a BSA approach are those based on the PCR (polymerase chain reaction; Saiki et al. 1988) such as random amplified polymorphic DNA (RAPD, Williams et al. 1990) and amplified fragment length polymorphism (AFLP, Vos et al. 1995). These markers do not require to have previous genetic information of the species and the trait, they are randomly distributed in the genome and in the case of AFLPs they generate a high number of markers in a single reaction. The main problems of AFLP and RAPD markers is their dominance and in the case of RAPDs its low reproducibility. Both of them require their conversion to codominant markers after the confirmation of their association with the trait of interest. The most frequent conversion into a codominant marker is the sequence characterized amplified region (SCAR, Paran and Michelmore 1993), which consists on the sequence of the fragment linked to the trait and designing of specific primers from the ends. Other marker technique used in BSA is restriction fragment length polymorphism (RFLP, Botstein et al. 1980). It was developed earlier and used until the markers based on the PCR became available, mainly due to the complexity and low polymorphism. Recently, microsatellites or simple sequence repeats (SSRs) (Hamada et al. 1982) have been used for BSA analysis (Cachi and Wünsch 2011; Claverie et al. 2011). These markers need to develop enriched libraries for identification of the genome tandem repeats and determination of the sequences flanking the microsatellites. SSRs became available for *Prunus* species much later. Table 5-1 summarizes the SSR markers developed in stone fruit species. They have been used mainly in mapping and in a few cases for BSA.

Peach is the most studied species at genetic level in the *Prunus* genus and many simple traits were mapped in this crop. Chaparro et al. (1994) were the first to map simple traits including leaf color, skin hairiness and pollen fertility. Later on Warburton et al. (1996) constructed a map based

Table 5-1 SSR markers developed in *Prunus* species.

Acronyme	Species	Reference
aprigms	*P. armeniaca*	Vilanova et al. (2006); Lalli et al. (2008)
AMPA	*P. armeniaca*	Hagen et al. (2004)
BPPCT	*P. persica*	Dirlewanger et al. (2002)
Cd	*P. persica*	Sicard et al. (2008)
CPDCT	*P. dulcis*	Mnejja et al. (2005)
CPPCT	*P. persica*	Aranzana et al. (2002)
CPSCT	*P. salicina*	Mnejja et al. (2004)
EpDCu	*P. amigdalus*	Aranzana et al. (2003)
EppCu	*P. persica*	Howad et al. (2005)
M/MA	*P. persica*	Yamamoto et al. (2002)
PacA/PacD	*P. armeniaca*	Decroocq et al. (2003)
PceGA	*P. cerasus*	Cantini et al. (2001)
pchcms/pchgms	*P. persica*	Sosinski et al. (2000); Verde et al. (2004); Verde et al. (2005); Wang et al. (2002b)
PGS	*P. persica*	Soriano et al. (2011)
ssrPaCITA	*P. armeniaca*	Lopes et al. (2002)
UDA	*P. dulcis*	Testolin et al. (2004)
UDAp	*P. armeniaca*	Messina et al. (2004)
UDP	*P. persica*	Testolin et al. (2000); Cipriani et al. (1999)

on RAPDs and mapped flesh color, adhesion to the pit, firmness and pollen fertility. Traits related to pest resistance were mapped by Blis et al. (2002), Claverie et al. (2004a) and Blenda et al. (2006). The most recent traits mapped were related to fruit quality: acidity by Wu et al. (2007) and slow ripening by Tataranni et al. (2010).

In almond species, the first map available was based on RFLP markers in the framework of a revision of the progress made in *Prunus* mapping (Arus et al. 1998). This map included anther color, self-compatibility, shell hardness, kernel bitterness and leaf color. Blooming time was determined by Ballester et al. (2001) but using RAPD markers employing BSA approach.

In apricot, the BSA employing RAPD markers was used to map two simple traits related to fertility: male sterility and self-compatibility (Badenes et al. 2000). AFLP markers were used by Salava et al. (2002) and Lalli et al. (2008) for mapping resistance to plum pox virus (PPV).

In plum, the only simple trait mapped was the resistance to root-knot nematode. It was first mapped using RAPD markers by Lecouls (1999) and later, as SSRs became available, it was mapped by Claverie et al. (2004a) using them.

Table 5-2 shows the traits mapped through BSA in *Prunus*.

Table 5-2 Target traits in *Prunus* mapped through Bulked Segregant Analysis.

Species	Trait	Marker type	Map References
Almond	Anther color	RFLP	Arús et al. (1998)
	Self-incompatibility	RFLP	Arús et al. (1998)
	Kernel bitterness	RFLP	Arús et al. (1998)
	Shell hardness	RFLP	Arús et al. (1998)
	Leaf color	RFLP	Arús et al. (1998)
	Blooming time	RAPD	Ballester et al. (2001)
Apricot	Male sterility and self-compatibility	RAPD	Badenes et al. (2000)
	PPV resistance	AFLP	Salava et al. (2002)
	PPV resistance	AFLP	Lalli et al. (2008)
Cherry	Self-compatibility	SSR	Cachi and Wünsch (2011)
Peach	Leaf color	RAPD	Chaparro et al. (1994)
	Skin hairiness	RAPD	Chaparro et al. (1994)
	Pollen fertility	RAPD	Chaparro et al. (1994); Warburton et al. (1996)
	Flesh color	RAPD	Warburton et al. (1996)
	Adhesion	RAPD	Warburton et al. (1996)
	Texture	RAPD	Warburton et al. (1996)
	Plant stature	RAPD	Warburton et al. (1996)
	Evergrowing	RAPD	Wang et al. (2002a)
	Root-knot nematode resistance	RAPD	Bliss et al. (2002)
	Polycarpel	RAPD	Bliss et al. (2002)
	Root-knot nematode resistance	SSR	Claverie et al. (2004a)
	Short life syndrome	AFLP	Blenda et al. (2006)
	Non-acid/acid fruit	AFLP	Wu et al. (2007)
	Slow ripening	AFLP	Tataranni et al. (2010)
Plum	Root-knot nematode resistance	RAPD	Lecouls et al. (1999)
	Root-knot nematode resistance	SSR	Claverie et al. (2004a)

5.3 Genes Mapped

Although in *Prunus* species a high number of linkage maps have been developed from intra- and interspecific crosses, the interspecific cross of the almond 'Texas' with the peach 'Earlygold' (Joobeur et al. 1998; Aranzana et al. 2003; Dirlewanger et al. 2004) is the most saturated map and considered as the reference map for *Prunus*.

Numerous simply inherited traits have been positioned on the available maps (Table 5-3), which make them a target for the implementation of marker-assisted breeding (MAB) by searching tightly linked markers in the genomic region or to use the information for positional gene cloning.

Table 5-3 Description of major genes for single traits located in genetic maps of *Prunus*.

LG	Trait	Symbol	Species	Population	Position[1]	Closest marker(s)	Reference
1	Fruit flesh color	Y	Peach	New Jersey Pillar x KV7719	0.0	B7F6I	Rajapakse et al. (1995)
			Almond x Peach	Padre x 54P455	72.4	RGA400C1	Bliss et al. (2002)
	Senesced leaf color	LFCLR	Almond x Peach	Padre x 54P455	72.4	RGA400C1	Bliss et al. (2002)
	PPV resistance	PPVres	Apricot	Goldrich x Currot	54.0	UDP96-018	Hurtado et al. (2002)
			Apricot	Lito x Lito	75.0	EAG-MCAC (1)	Vilanova et al. (2003)
			Apricot	LE-3246 x Vestar	38.3	EAA-MCAG8b	Lalli et al. (2008)
			Apricot	Various progenies	23.4	PGS1.21, PGS1.23, PGS1.24	Soriano et al. (2011)
	Evergrowing	Evg	Peach Almond x Peach	Empress op op dwarf x P1442380 Texas x Earlygold	58.0 82.0	EAT/MCAC PLG26A	Wang et al. (2002) Silva et al. (2005)
	Green peach aphid resistance	Rm2	Peach	Pamirskij 5 x Rubira	96.6	UDAp-467	Lambert and Pascal (2011)
2	Root-knot nematode resistance	Mij R_{Mia} Mj Mi	Peach	Lovell x Nemared Garfi x Nemared Akame x Juseitou Akame x Juseitou	77.7 7.0 63.5 68.3	EAA/MCAT10 OPS14a ISSR834-1/0.4 EAA/MCAC135	Lu et al. (1998) Claverie et al. (2004a) Yamamoto et al. (2001) Yamamoto et al. (2001)
	Growth habit	Br	Peach	New Jersey Pillar x KV7719	17.6	Double flowers	Rajapakse et al. (1995)
			Peach	Various progenies	34.0	pchgms1	Scorza et al. (2002)
	Double flowers	Dl	Peach	NC174RL x P1 New Jersey Pillar x KV7719	0.0 0.0	OPG-13/0.78	Chaparro et al. (1994) Rajapakse et al. (1995)
3	Flesh colour around stone	Cs	Peach	Akame x Juseitou	50.8	OPO2/0.6	Yamamoto et al. (2001)

Table 5-3 contd....

Table 5-3 contd....

LG	Trait	Symbol	Species	Population	Position[1]	Closest marker(s)	Reference
	Anther color	Ag	Almond x Peach	Texas x Earlygold	25.0	AG57A	Joobeur et al. (1998)
	Flower color	Fc	Peach	Akame x Juseitou	77.9	EACA/MCTG220	Yamamoto et al. (2001)
4	Blooming time	Lb	Almond	D.3.5 x Bertina	20.0	AG6	Ballester et al. (2001)
	Flesh adhesion	F	Peach	*P. ferganensis* x IF7310828 Akame x Juseitou Ferjalou Jalousia x Fantasia	44.0 93.8 68.0	AG12b EAC/MCTC180 UDAp431	Dettori et al. (2001) Yamamoto et al. (2001) Dirlewanger et al. (2006)
5	Non acid fruit	D	Peach	Ferjalou Jalousia x Fantasia Ferjalou Jalousia x Fantasia Ferjalou Jalousia x Fantasia	6.0 22.5 5.9	A09-1.3 EAC/MCAA2 pGT-TTG	Dirlewanger et al. (1998) Etienne et al. (2002) Dirlewanger et al. (2006)
	Kernel taste	Sk	Almond x Peach Almond	Padre x 54P455 R1000 x Desmayo Largueta R1000 x Desmayo Largueta	89.9 28.0 35.7	AG19-3 PceGA025 EPPCU2584	Bliss et al. (2002) Sánchez-Pérez et al. (2007) Sánchez-Pérez et al. (2010)
	Skin hairiness	G	Almond x Peach Peach	Padre x 54P455 Ferjalou Jalousia x Fantasia Ferjalou Jalousia x Fantasia	89.9 98.0 81.4	AG19-3 EAC/MCAA1 eAC/CAA	Bliss et al. (2002) Dirlewanger et al. (1998) Dirlewanger et al. (2006)
6	Leaf shape	Nl	Peach	Akame x Juseitou	184.2	EAC/MCAC180	Yamamoto et al. (2001)
	Dwarf/normal	Dw	Peach	Akame x Juseitou	184.2	EAC/MCAC180	Yamamoto et al. (2001)
	Fruit skin color[2]	Sc	Peach	Akame x Juseitou	119	UDP96-015	Yamamoto et al. (2001)
	Leaf color[2]	Gr	Peach	Akame x Juseitou	119	UDP96-015	Yamamoto et al. (2001)
	Male sterility	Ps	Peach	Ferjalou Jalousia x Fantasia Ferjalou Jalousia x Fantasia	11.0 0.0	EAA/MCAT4 FG40	Dirlewanger et al. (1998) Dirlewanger et al. (2006)
	Nonaborting fruit	Af	Peach	Ferjalou Jalousia x Fantasia	72.6	MA040a	Dirlewanger et al. (2006)
	Fruit shape	S*	Peach	Ferjalou Jalousia x Fantasia Ferjalou Jalousia x Fantasia	90.0 72.6	EACA/MCAT4 MA040a	Dirlewanger et al. (1998) Dirlewanger et al. (2006)

	Self-incompatibility	S	Almond x Peach	Padre x 54P455	123.5	CPM8	Bliss et al. (2002)
			Almond	D.3.5 x Bertina	52.0	Pgl-1	Ballester et al. (1998)
				R1000 x Desmayo Largueta	56.0	CPPCT030	Sánchez-Pérez et al. (2007)
			Apricot	Lito x Lito	57.0	EAC/MCAT10	Vilanova et al. (2003)
			Sweet Cherry	Emperor Francis x New York 54	72.0	EAC/MCTA350	Olmstead et al. (2008)
				Napoleon × *Prunus nipponica*	86.1	CPPCT030	Clarke et al. (2009)
7	Leaf gland shape	E	Peach	*P. ferganensis* x IF310828	23.0	AG104	Dettori et al. (2001)
	Root-knot nematode resistance	Ma	Myrabolan plum	Different progenies	14.7	SCAL19	Claverie et al. (2004a)
				Ferjalou Jalousia x Fantasia	12.4	$SCAFLP2_{202}$	Dirlewanger et al. (2004)
				Different progenies	18.0[3]	OPAL19	Lecouls et al. (1999)
		Rjap	Japanese plum	J.222 x J.13	16.0	CPPCT022	Claverie et al. (2004a)
		Rmja	Almond	P.2175 x Alnem 1	18.6[3]	CPPCT022	Van Ghelder et al. (2010)

[1]Map position in the original maps (cM).
[2]Genes located close to the translocation breakpont between LG6 and LG8.
[3]Position inferred on the basis of the SSR location in the reference map.

The most representative maps that included simple traits in chronological order are discused below.

Rajapakse et al. (1995) used a peach F_2 progeny derived from the self-fertilization of four F_1 individuals from the cross 'New Jersey Pillar x KV7719'. The progeny segregated for the tree architecture pillar versus standard (*Br*). The map was aimed at mapping this interesting trait but the authors mapped other segregating traits such as fruit flesh color (*Y*) on LG1 and double flowers (*Dl*) on LG2 (Fig. 5-1).

Joobeur et al. (1998) using the interspecific F_2 population derived from the cross between almond ('Texas') and peach ('Earlygold'), located the anther color (*Ag*) trait on LG3. This mapping population became later the resource for the *Prunus* reference map (Fig. 5-1).

Ballester et al. (2001) in the framework of their studies on determination of the genetic basis of late blooming time (*Lb*) in almond located the trait in the LG4 using an intraspecific family derived from a cross between the varieties 'Felisia' and 'Bertina'. (Fig. 5-1).

Dettori et al. (2001) used a family obtained from the cross between a peach selection 'IF7310828' and one accession of *P. ferganensis* carrying a source of resistance to powdery mildew. These authors located the flesh adhesion (*F*) trait on LG4 and the leaf gland shape (*E*) on LG7 (Fig. 5-1).

Yamamoto et al. (2001) mapped both Mendelian and quantitative traits using the peach F_2 progeny 'Akame x Juseitou'. Among the different traits mapped, two were resistance genes against the root-knot nematode and were located in LG2 (*Mj* and *Mi*); other traits mapped were flesh color around the stone (*Cs*) and flower color (*Fc*) both mapped in LG3; the flesh adhesion (*F*) was located on LG4 in agreement with results of Dettori et al. (2001); leaf shape (*Nl*) and plant height (*Dw*) mapped in the lower part of LG6; and finally two traits were mapped on a translocation breakpoint between LG6 and LG8, the fruit skin color (*Sc*) and the leaf color (*Gr*) (Fig. 5-1).

Due to the low varability in the peach species an option that allows finding more heterozygous markers was to use families from interspecific crosses between peach and almond. Bliss et al. (2002) using an interspecific F_2 population derived from the cross between an almond cultivar 'Padre' and a dwarf peach selection '54P455' mapped five trait loci—the fruit flesh color (*Y*), which agreed with the results obtained previously by Rajapakse et al. (1995) and other traits as the senesced leaf color (*LFCLR*) cosegregating with the flesh color on LG1; the kernel bitterness (*Sk*) and the skin hairiness (*G*) on LG5; other reproductive traits as the self-incompatibility (*S*) and the male sterility (*Ps*), were mapped on LG6 as well as nonaborting fruit (*Af*) and the fruit shape (*S**) (Fig. 5-1).

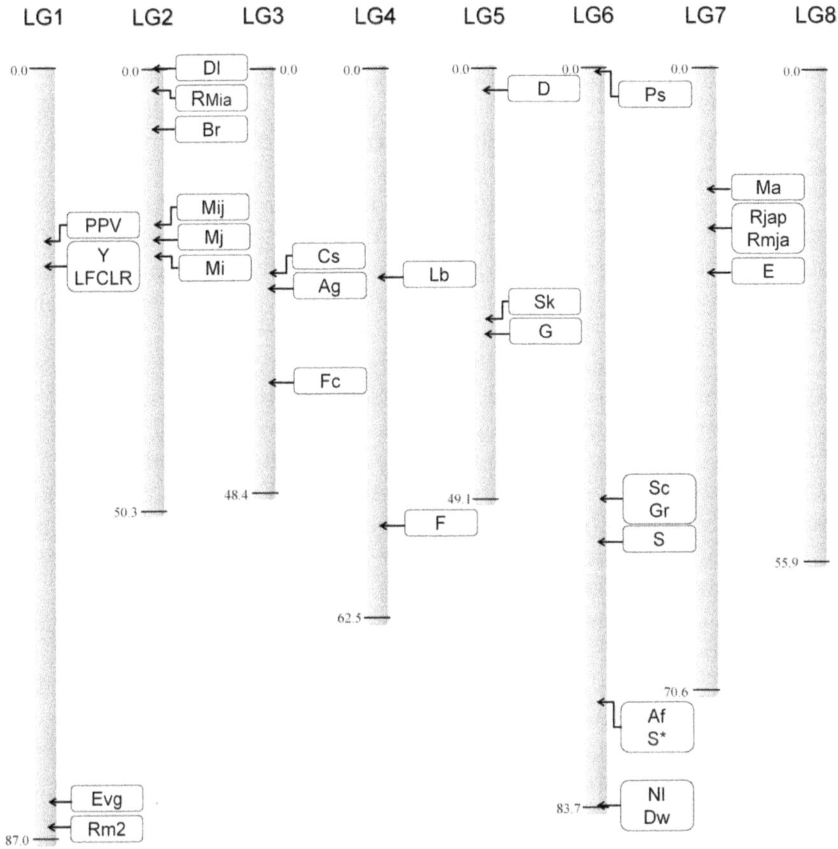

Figure 5-1 Map position of simple inherited traits in Prunus. It is shown the relative position according to the *Prunus* CMap available at http: //www.rosaceae.org. PPV: Plum Pox Virus Resistance (Soriano et al. 2011); Y: Fruit Flesh Colour (Bliss et al. 2002); LFCLR: Senesced Leaf Colour (Bliss et al. 2002); Evg: Evergrowing (Wang et al. 2002); Rm2: Green Peach Aphid Resistance (Lambert and Pascal 2011); Dl: Double Flowers (Rajapakse et al. 1995); RMia, Mij, Mj and Mi: Root-knot nematode resistance (Claverie et al. 2004a; Lu et al. 1998; Yamamoto et al. 2001; Yamamoto et al. 2001); Br: Growth habit (Rajapakse et al. 1995); Cs: Flesh colour around stone (Yamamoto et al. 2001); Ag: Anther Colour (Joobeur et al. 1998); Fc: Flower Colour (Yamamoto et al. 2001); Lb: Blooming time (Ballester et al. 2001); F: Flesh adhesión (Dirlewanger et al. 2006); D: Non AcidFruit (Dirlewanger et al. 2006); Sk: Kernel Taste (Bliss et al. 2002); G: Skin Hairiness (Dirlewanger et al. 2006); Ps: MaleSterility (Dirlewanger et al. 2006); Sc: Fruit Skin Colour (Yamamoto et al. 2001); Gr: LeafColour (Yamamoto et al. 2001); S: Self-Incompatibility (Vilanova et al. 2003); Af: Non-AbortingFruit (Dirlewanger et al. 2006); S*: Fruit Shape (Dirlewanger et al. 2006); Nl: LeafShape (Yamamoto et al. 2001) ; Dw: Dwarf/ Normal (Yamamoto et al. 2001); Ma, Rjap, Rmja: Root-Knot Nematode Resistance (Claverie et al. 2004a; Claverie et al. 2004a; Van Ghelder et al. 2010); E: Leaf Gland Shape (Dettori et al. 2001).

A mutated genotype for dormancy identified as evergreen (named *Evg* gene) was first found in peach trees in Mexico and later identified as a simply inherited trait by Rodriguez et al. (1994). The trait was located on the LG1 of the peach population 'Empress op op dwarf x P1442380' 'Empress op op dwarf' is a deciduous peach cultivar whereas 'P1442380' is an evergrowing peach (Wang et al. 2002a) (Fig. 5-1).

Claverie et al. (2004a) mapped two independent root-knot nemadote resistance genes in peach and plum. The resistance gene of peach (R_{Mja}) was located on LG2 using the 'Garfi x Nemared' progeny. Orthologous resistant genes *Ma* and *Rjap* were mapped in Myrabolan plum and Japanese plum, respectively on LG7 (Fig. 5-1).

Dirlewanger et al. (2006) developed a second generation linkage map from the intraspecific peach F_2 population 'Ferjalou Jalousia x Fantasia' and characterized different morphological traits affecting flower and fruit (Fig. 1). Among these traits, the authors located the flesh adhesion (*F*) on LG4 in agreement with the previous results of Dettori et al. (2001) and Yamamoto et al. (2001); the nonacid fruit trait (*D*) and the skin hairiness (*G*) in a similar position on LG5 as it was mapped previously by the same research group (Dirlewanger et al. 1998).

The most recent maps included genes related to resistance to pathogenes. Lambert and Pascal (2011) located on the LG1 of the peach population 'Pamirskij 5 x Rubira' the green peach aphid resistance (*Rm2*) (Fig. 5-1).

The resistance to plum pox virus (PPV), the most important disease affecting *Prunus* species have been mapped in several apricot progenies. The first map including the PPV resistance locus (*PPVres*) was mapped by Hurtado et al. (2002). It was a low-density map composed mainly of dominant markers. Development of SSR markers in *Prunus* species supported establishment of new maps based on codominant markers. Maps of a new generation were saturated with codominant SSR markers, allowing better localization of the *PPVres* locus in progenies with different ancestral backgrounds (Lambert et al. 2007; Lalli et al. 2008; Soriano et al. 2008; Marandel et al. 2009; Pilarova et al. 2010; Dondini et al. 2011). The most saturated map of the LG1 containing the *PPVres* locus constructed by Vera-Ruiz et al. (2011) included several apricot populations, an F_2 from self-pollination of the resistant cultivar 'Lito' and the progenies from intraspecific crosses with the varieties 'Goldrich x Currot' and 'Goldrich x Canino'. These authors identified a 2.1 Mb region containing the resistance locus. A bacterial artificial chromosome (BAC) contig spanning the identified region allowed the implementation of MAS by identifying tightly linked SSR markers (Soriano et al. 2011) (Fig. 5-2). These results allowed implementation of MAB by means of identification of markers tightly linked to the PPV resistance.

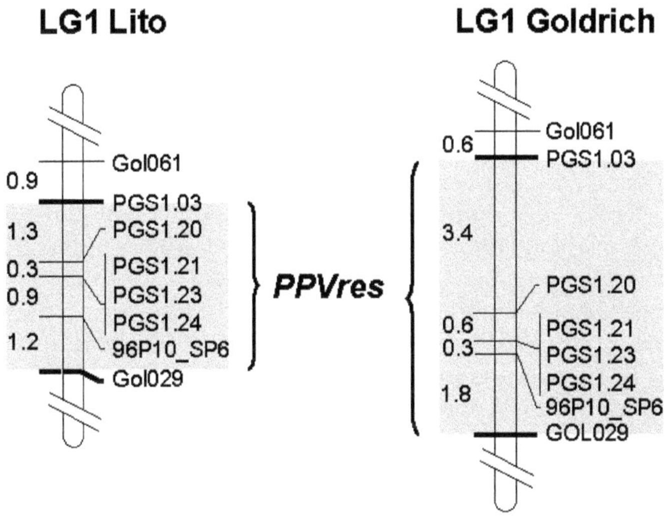

Figure 5-2 Local linkage maps of the *PPVres* locus of the resistant cultivars 'Lito' and 'Goldrich'. From Soriano et al. (2011).

Construction of linkage maps is the first step for positional cloning of a gene of interest. A useful approach to locate a trait of interest once a previous map has been developed is the genome scan approach (GSA) developed by Patocchi et al. (2005) in apple. In this approach, few markers from each linkage group (LG) are used to look for marker trait association, and then when the trait is positioned the region is saturated with new molecular markers. The iterative procedure ends with the identification of markers tightly linked to the trait. This is the case of the *Ma* gene of Myrabolan plum conferring resistance to the root-knot nematode. This gene has been physical-mapped in a BAC contig spanning the resistance locus (Claverie et al. 2004b) and cloned (Claverie et al. 2011).

In addition to the Mendelian traits located in the available *Prunus* maps, in the last years a number of candidate genes (CGs) have been mapped on the basis of homologous sequences from related species or using degenerate sequences from non-related species. Using this approach, Silva et al. (2005) mapped several CGs related to flowering time in the 'Texas x Early Gold' population, Olmstead et al. (2008) mapped four gene-derived markers representing orthologs of a tomato vacuolar invertase and a fruit size gene in an intraspecific sweet cherry population: 'Emperor Francis x New York 54'. In peach, the inheritance of fruit firmness and softening traits has been widely studied. The melting flesh trait (*M*) segregated in a Mendelian fashion along with the adherence of the flesh to the pit (*F*), both traits being controlled by a single locus (Bailey and French 1949). Several linkage analyses in peach found both *F* and *M* loci in a common region in linkage group 4 (Abbott et al. 1998; Quarta et al. 1998; Dettori et al. 2001; Dirlewanger et al. 2006; Ogundiwin et al. 2009). It has been suggested that multiple alleles of a gene coding for the cell wall-metabolizing enzyme endoPG could control these two traits (Peace et al. 2005). Lester et al. (1994) identified an endoPG-like gene from peach fruit and designated as *PRF5*, the expression of which was associated with texture differences. Based on single nucleotide polymorphism (SNP) analysis, Leida et al. (2011) identified an ortholog in apricot named *PaPG* that mapped in syntenic position to the *PRF5* locus from peach.

Another Mendelian trait mapped based on CG approach that allowed implementation of MAB was self-incompatibility (SI) in *Prunus*, mechanism of which in Rosaceae is gametophytic (GSI). It affects mainly almond, sweet and sour cherry, apricots and Japanese plums. Since the discovery of the SI determinants in the Rosaceae, molecular genotyping systems have been developed based on the protein structure and DNA sequence polymorphism. The development of molecular markers for *S* genotypes has led to MAS of self-compatibility (SC). Tao et al. (1997) were the first to conduct a molecular study to determine the pistil component for GSI in almond. They then sequenced the N-terminus of almond S-RNases. This was the first sequence information for *Prunus* S-RNases and opened the vistas for subsequent cloning of cDNAs for *Prunus* S-RNases. Following these results, SI determinants were cloned in almond (Ushijima et al. 2001), sweet cherry (Tao et al. 1999) and molecular typing systems were set (Yamane et al. 2001, 2003). In Japanese apricot, Yaegaki et al. (2001) determined specific genotypes that allowed MAS. In apricot, specific alleles were determined by Romero et al. (2004) and Vilanova et al. (2005). Three *S*-haplotypes, *S1, S2*, and *S2m*, have been found in the peach germplasm (Tao et al. 2007). In European and Japanese plums the trait has similar mechanism and alleles have been identified for screening the trait (Beppu et al. 2005, 2010).

Using new approaches, such us high-resolution melting (HRM) analysis (Liew et al. 2004), improvement on the detection of SNPs proved to be an efficient and cost-effective approach. Wu et al. (2009) located in peach and almond maps ortologous genes corresponding to known proteins described in other species. *Prunus* unigenes were searched and compared with known proteins in the public databases and SNP markers were developed.

Expressed sequence tag (EST) databases and peach genome sequence allowed the location of a high number of candidate genes in different genetic maps (Ogundiwin et al. 2009; Le Dantec et al. 2010; Illa et al. 2011).

References

Abbott AG, Rajapakse S, Sosinski B, Lu ZX, Sossey-Alaoui K, Gannavarapu M, Reighard G, Ballard RE, Baird WV, Scorza R, Callahan A (1998) Construction of saturated linkage maps of peach crosses segregating for characters controlling fruit quality, tree architecture and pest resistance. Acta Hort 465: 41–50.

Aranzana MJ, Carbo J, Arus P (2002) Microsatellite variability in peach [*Prunus persica* (L.) Batsch]: cultivar identification, marker mutation, pedigree inferences and population structure. Theor Appl Genet 106: 1341–1352.

Aranzana MJ, Pineda A, Cosson P, Dirlewanger E, Ascasibar J, Cipriani G, Ryder CD, Testolin R, Abbott A, King GJ, Iezzoni AF, Arús P (2003) A set of simple-sequence repeat (SSR) covering the *Prunus* genome. Theor Appl Genet 106: 819–825.

Arus P, Ballester J, Jauregui B, Joobeur T, Truco M, Vicente MC (1998) The European *Prunus* mapping project: update on marker development in almond. Acta Hort 484: 331–338.

Badenes ML, Hurtado MA, Sanz F, Archelos MD, Burgos L, Egea J, Llácer G (2000) Searching for molecular markers linked to male sterility and self-compatibility in apricot. Plant Breed 119: 157–160.

Bailey JS, French AP (1949) The inheritance of certain fruit and foliage characters in the peach. MA Agri Exp Sta Bull 452.

Ballester J, Boskovic R, Batlle I, Arus P, Vargas F, de Vicente MC (1998) Location of the selfincompatibility gene on the almond linkage map. Plant Breed 117: 69–72.

Ballester J, Socias i Company R, Arus P, de Vicente MC (2001) Genetic mapping of a major gene delaying blooming time in almond. Plant Breed 120: 268–270.

Beppu K, Komatsu N, Yamane H, Yaegaki H, Yamaguchi M, Tao R and Kataoka I (2005) Se-haplotype confers self-compatibility in Japanese plum (*Prunus salicina* Lindl.). J Hort Sci Biotechnol 80: 760–764.

Beppu K, Syogase K, Yamane H, Tao R, Kataoka I (2010) Inheritance of self-compatibility conferred by the Se-haplotype of Japanese plum and development of Se-RNase gene-specific PCR primers. J Hort Sci Biotechnol 85: 215–218.

Bhojwani SS, Powar JB, Cocking EL (1977) Isolation, culture and division of protoplast. Plant Sci Lett 8: 85–89.

Blake MA (1932) The JH Hale as a parent in peach crosses. Proc Am Soc Hort Sci 29: 131–136.

Blake MA (1937) Progress in peach breeding. Proc Am Soc Hort Sci 35: 49–53.

Blenda AV, Wechter WP, Reighard GL, Baird WV, Abbott AG (2006) Development and characterisation of diagnostic AFLP markers in *Prunus persica* for its response to peach tree short life syndrome. J Hort Sci Biotechnol 81: 281–288.

Bliss FA, Arulsekar S, Foolad MR, Becerra V, Gillen AM, Warburton ML, Dandekar AM, Kocsisne GM, Mydin KK (2002) An expanded genetic linkage map of *Prunus* based on an interspecific cross between almond and peach. Genome 45: 520–529.

Botstein D, White RL, Skolnick M, Davis RW (1980) Construction of a genetic map in man using restriction fragment length polymorphism. Am J Hum Genet 32: 314–330.

Burgos L, Ledbetter CA (1994) Observations on inheritance of male sterility in apricot. Hort Science 29: 127.

Burgos L, Ledbetter CA, Pérez-Tornero O, Ortín-Párraga F, Egea J (1997) Inheritance of sexual incompatibility in apricot. Plant Breed 116: 383–386.

Cachi AM, Wünsch A (2011) Characterization and mapping of non-S gametophytic self-compatibility in sweet cherry (*Prunus avium* L.). J Exp Bot 62: 1847–1856.

Cantini C, Iezzoni AF, Lamboy WF, Boritzki M, Struss D (2001) DNA fingerprinting of tetraploid cherry germplasm using simple sequence repeats. J Am Soc Hort Sci 126: 205–209.

Chahal GS, Gosal SS (2002) Principles and Procedures of Plant Biotechnological and Conventional Approaches. Alpha Science International, Pangbourne, UK.

Chaparro JX, Werner DJ, O'Malley DO, Sederoff RR (1994) Targeted mapping and linkage analysis of morphological isozyme, and RAPD markers in peach. Theor Appl Genet 87: 805–815.

Cipriani G, Lot G, Huang WG, Marrazzo M-T, Peterlunger E, Testolin R (1999) AC/GT and AG/CT microsatellite repeats in peach [*Prunus persica* (L.) Batsch]: isolation, characterisation and cross species amplification in *Prunus*. Theor Appl Genet 99: 65–72.

Clarke JB, Sargent DJ, Boskovic RI, Belaj A, Tobutt KR (2009) A cherry map from the inter-specific cross Prunus avium 'Napoleon' x P. nipponica based on microsatellite, gene-specific and isoenzyme markers. Tree Genet Genomes 5: 41–51.

Claverie M, Bosselut N, Lecouls AC, Voisin R, Lafargue B, Poizat C, Kleinhentz M, Laigret F, Dirlewanger E, Esmenjaud D (2004a) Location of independent root-knot nematode resistance genes in plum and peach. Theor Appl Genet 108: 765–773.

Claverie M, Dirlewanger E, Cosson P, Bosselut N, Lecouls AC, Voisin R, Kleinhentz M, Lafargue B, Caboche M, Chalhoub B, Esmenjaud D (2004b) High-resolution mapping and chromosome landing at the root-knot nematode resistance locus *Ma* from Myrobalan plum using a large-insert BAC DNA library. Theor Appl Genet 109: 1318–1327.

Claverie M, Dirlewanger E, Bosselut N, Van Ghelder C, Voisin R, Kleinhentz M, Lafargue B, Abad P, Rosso MN, Chalhoub B, Esmenjaud D (2011) The Ma gene for complete-spectrum resistance to *Meloidogyne* species in *Prunus* is a TNL with a huge repeated C-terminal post-LRR region. Plant Physiol 156: 779–792.

Decroocq V, Fave MG, Hagen L, Bordenave L, Decroocq S (2003) Development and transferability of apricot and grape EST microsatellite markers across taxa. Theor Appl Genet 106: 912–922.

Dettori MT, Quarta R, Verde I (2001) A peach linkage map integrating RFLPs, SSRs, RAPDs, and morphological markers. Genome 44: 783–790.

Dirlewanger E, Pronier V, Parvery C, Rothan C, Guye A, Monet R (1998) A genetic linkage map of peach (*Prunus persica* L. Batsch) using morphological, RFLP, isoenzyme, RAPD and AFLP markers. Theor Appl Genet 97: 888–895.

Dirlewanger E, Cosson P, Tavaud M, Aranzana J, Poizat C, Zanetto A, Arus P and Laigret F (2002) Development of microsatellite markers in peach [*Prunus persica* (L.) Batsch] and their use in genetic diversity analysis in peach and sweet cherry (*Prunus avium* L.). Theor Appl Genet 105: 127–138.

Dirlewanger E, Cosson P, Howad W, Capdeville G, Bosselut N, Claverie M, Voisin R, Poizat C, Lafargue B, Baron O, Laigret F, Kleinhentz M, Arus P, Esmenjaud D (2004) Microsatellite genetic linkage maps of myrobalan plum and an almond-peach hybrid-location of root-knot nematode resistance genes. Theor Appl Genet 109: 827–838.

Dirlewanger E, Cosson P, Boudehri K, Renaud C, Capdeville G, Tauzin Y, Laigret F, Moing A (2006) Development of a second-generation genetic linkage map for peach [*Prunus persica* (L.) Batsch] and characterization of morphological traits affecting flower and fruit. Tree Genet Genomes 3: 1–13.

Dondini L, Lain O, Vendramin V, Rizzo M, Vivoli D, Adami M, Guidarelli M, Gaiotti F, Palmisano F, Bazzoni A, et al. (2011) Identification of QTL for resistance to plum pox virus strains M and D in Lito and Harcot apricot cultivars. Mol Breed 27: 289–299.

Druart P, Gruselle R (1986) Plum (*Prunus domestica*). In: Bajaj YPS (ed) Biotechnology in Agriculture and Forestry Trees I. Vol 1. Springer, Berlin, Heidelberg, pp 130–156.

Etienne C, Rothan C, Moing A, Plomion C, Bodnes C, Svanella-Dumas L, Cosson P, Pronier V, Monet R, Dirlewanger E (2002) Candidate genes and QTLs for sugar and organic acid content in peach [*Prunus persica* (L.) Batsch]. Theor Appl Genet 105: 145–159.

Gómez E, Burgos L, Soriano C, Marín J (1998) Amygdalin content in the seeds of several apricot cultivars. J Sci Food Agri 77: 184–186.

Hagen LS, Chaib J, Fady B, Decroocq V, Bouchet JP, Lambert P, Audergon JM (2004) Genomic and cDNA microsatellites from apricot (*Prunus armeniaca* L.). Mol Ecol Notes 4: 742–745.

Hamada H, Petrino MG, Kakunaga T (1982) A novel repeated element with Z-DNA-forming potential is widely found in evolutionary diverse eukariotic genomes. Proc Natl Acad Sci USA 79: 6465–6469.

Hansche PE (1988) Two genes induce brachytic dwarfism in peach. Hort Science 23: 604–606.

Howad W, Yamamoto T, Dirlewanger E, Testolin R, Cosson P, Cipriani G, Monforte AJ, Georgi L, Abbott AG, Arús P (2005) Mapping with a few plants: using selective mapping for microsatellite saturation of the *Prunus* reference map. Genetics 171: 1305–1309.

Hurtado MA, Romero C, Vilanova S, Abbott AG, Llácer G, Badenes ML (2002) Genetic linkage maps of two apricot cultivars (*Prunus armeniaca* L.) and mapping of PPV (sharka) resistance. Theor Appl Genet 105: 182–191.

Illa E, Eduardo I, Audergon JM, Barale F, Dirlewanger E, Li X, Moing A, Lambert P, Le Dantec L, Gao Z, Poëssel JL, Pozzi C, Rossini L, Vecchietti A, Arús P, Howad W (2011) Saturating the *Prunus* (stone fruti) genome with candidate genes for fruit quality. Mol Breed: DOI 10.1007/s11032-010-9518-x.

Joobeur T, Viruel MA, de Vicente MC, Jauregui MB, Ballester J, Dettori MT, Verde I, Truco MJ, Messeguer R, Batlle I, Quarta R, Dirlewanger E, Arus P (1998) Construction of a saturated linkage map for *Prunus* using an almond × peach F2 progeny. Theor Appl Genet 7: 1034– 1041.

Kappel F, Granger A, Hrotkó K, Schuster M (2011) Cherry. In: Byrne D, Badenes ML (eds) Fruit Breeding. Horticultural Series Vol 6. Springer, Berlin, Heidelberg: ISBN 978-1-4419-0762-2 (in press).

Karayiannis I, Thomidis T, Tsaftaris A (2008) Inheritance of resistance to *Plum pox virus* in apricot (*Prunus armeniaca* L.). Tree Genet Genomes 4: 143–148.

Laibach F (1929) Ectogenesis in plants—Methods and genetic possibilities of propagating embryos otherwise dying in the seed. J Hered 20: 201–208.

Lalli DA, Abbott AG, Zhebentyayeva TN, Badenes ML, Damsteegt V, Polák J, Krška B, Salava J (2008) A genetic linkage map for an apricot (*Prunus armeniaca* L.) BC1 population mapping plum pox virus resistance. Tree Genet Genomes 4: 481–493.

Lambert P, Pascal T (2011) Mapping *Rm2* gene conferring resistance to the green peach aphid (*Myzus persicae* Sulzer) in the peach cultivar "Rubira®". Tree Genet Genomes: DOI 10.1007/s11295-011-0394-2.

Lambert P, Dicenta F, Rubio M, Audergon JM (2007) QTL analysis of resistance to sharka disease in the apricot (*Prunus armeniaca* L.) 'Polonais' × 'Stark Early Orange' F1 progeny. Tree Genet Genomes 3: 299–309.

Lammerts WE (1945) The breeding of ornamental edible peaches for mild climates, 1: inheritance of tree and flower characters. Am J Bot 32: 53–61.

Lansari A, Iezzoni A (1990) A preliminary analysis of self-incompatibility in sour cherry. Hort Science 25: 1636–1638.

Layne REC, Bailey CH, Hough LF (1996) Apricots. In: Janick J, Moore JN (eds) Fruit Breeding. Vol 1: Tree and Tropical Fruits. John Wiley, New York, pp 79–111.

Le Dantec L, Carinet G, Bonet J, Fouché M, Boudehri K, Monfort A, Poëssel JL, Moing A, Dirlewanger E (2010) Development and mapping of peach candidate genes involved in fruit quality and their transferability and potential use in other Rosaceae species. Tree Genet Genomes 6: 995–1012.

Lecouls AC, Rubio-Cabetas MJ, Minot JC, Voisin R, Bonnet A, Salesses G, Dirlewanger E, Esmenjaud D (1999) RAPD and SCAR markers linked to the Ma1 root-knot nematode resistance gene in Myrabolan plum (*Prunus cerasifera* Ehr.). Theor Appl Genet 99: 328–335.

Leida C, Rios G, Soriano JM, Pérez B, Llácer G, Crisosto CH, Badenes ML (2011) Identification and genetic characterization of an ethylene-dependent polygalacturonase from apricot fruit. Postharvest Biol Technol 62: 26–34.

Lester DR, Speirs J, Orr G, Brady CJ (1994) Peach (*Prunus persica*) endopolygalacturonase cDNA isolation and mRNA analysis in melting and nonmelting peach cultivars. Plant Physiol 105: 225–231.

Liew M, Pryor R, Palais R, Meadows C, Erali M, Lyon E, Wittwer C (2004) Genotyping of single-nucleotide polymorphisms by highresolution melting of small amplicons. Clin Chem 50: 1156–1164.

Lopes MS, Sefc KM, Laimer M, da Camara Machado A (2002) Identification of microsatellite loci in apricot. Mol Ecol Notes 2: 24–26.

Lu ZX, Sosisnski A, Reighard GL, Baird WV, Abbott AG (1998) Construction of a genetic linkage map and identification of AFLP markers for resistance to root-knot nematodes in peach rootstocks. Genome 41: 199–207.

Marandel G, Salava J, Abbott A, Candresse T, Decroocq V (2009) Quantitative trait loci meta-analysis of Plum pox virus resistance in apricot (*Prunus armeniaca* L.): new insights on the organization and the identification of genomic resistance factors. Mol Plant Pathol 10: 347–360.

Messina R, Lain O, Marrazzo MT, Cipriani G, Testolin R (2004) New set of microsatellite loci isolated in apricot. Mol Ecol Notes 4: 432–434.

Mnejja M, García-Mas J, Howad W, Badenes L and Arus P (2004) Simple-sequence repeat (SSR) markers of Japanese plum (*Prunus salicina* Lindl.) are highly polymorphic and transferable to peach and almond. Mol Ecol Notes 4: 163–166.

Mnejja M, Garcia-Mas J, Howad W and Arús P (2005) Development and transportability across Prunus species of forty-two polymorphic almond microsatellites. Mol Ecol Notes 5: 531–535.

Michelmore RW, Paran I, Kesseli RV (1991) Identification of markers linked to disease-resistance genes by bulked segregant analysis: a rapid method to detect markers in specific genomic regions by using segregating populations. Proc Natl Acad Sci USA 88: 9828–9832.

Monet R (1967) A contribution to the genetics of peaches. Ann Amel Plant 17: 5–11.

Monet R, Bastard Y, Gibault B (1988) Genetic study of the weeping habit in peach. Agronomie 8: 127–132.

Ogundiwin EA, Peace CP, Gradziel TM, Parfitt DE, Bliss FA, Crisosto CH (2009) A fruit quality gene map of *Prunus*. BMC Genomics 10: 587.

Olmstead JW, Sebolt AM, Cabrera A, Sooriyapathirana SH, Iriarte G, Wang D, Chen CY, van der Knaap E, Iezzoni AF (2008) Construction of an intra-specific sweet cherry (*Prunus avium* L.) genetic linkage map and synteny analysis with the *Prunus* reference map. Tree Genet Genomes 4: 897–910.

Paran I, Michelmore RW (1993) Development of reliable PCR-based markers linked to downwy mildew resistance genes in lettuce. Theor Appl Genet 85: 985–993.

Patocchi A, Walser M, Tartarini S, Broggini GAL, Gennari F, Sansavini S, Gessler C (2005) Identification by genome scanning approach (GSA) of a microsatellite tightly associated with the apple scab resistance gene *Vm*. Genome 48: 630–636.

Peace CP, Crisosto CH, Gradziel TM (2005) Endopolygalacturonase: a candidate gene for *Freestone* and *Melting Flesh* in peach. Mol Breed 16: 21–31.

Pilarova P, Marandel G, Decroocq V, Salava J, Krka B, Abbott AG (2010) Quantitative trait analysis of resistance to plum pox virus in the apricot F_1 progeny "Harlayne" x "Vestar". Tree Genet Genomes 6: 467–475.

Quarta R, Dettori MT, Verde I, Gentile A, Broda Z (1998) Genetic analysis of agronomic traits and genetic linkage mapping in a BC1 peach population using RFLPs and RAPDs. Acta Hort 465: 51–60.

Rajapakse S, Belthoff LE, He G, Estager AE, Scorza R, Verde I, Ballard RE, Baird WV, Callahan A, Monet R, Abbott AG (1995) Genetic Linkage mapping in peach using morphological, RFLP and RAPD markers. Theor Appl Genet 90: 503–510.

Ramming DW (1991) Genetic control of a slow-ripening fruit trait in nectarine. Can J Plant Sci 71: 601–603.

Rodriguez J, Sherman WB, Scorza R, Wisniewski M, Okie WR (1994) Evergreen peach, its inheritance and dormant behavior. J Am Soc Hort Sci 119: 789–792.

Romero C, Vilanova S, Burgos L, Martínez-Calvo J, Vicente M, Llácer G, Badenes ML (2004) Analysis of the *S*-locus structure in *Prunus armeniaca* L. Identification of *S* haplotype specific *S-RNase* and F-box genes. Plant Mol Biol 56: 145–157.

Saiki R, Gelfand D, Stoffel S, Scharf S, Higuchi R, Horn G, Mullis K, Erlich H (1988) Primer-directed enzymatic amplification of DNA with a thermostable DNA polymerase. Science 239: 487–491.

Salava J, Wang Y, Krska B, Polak J, Kominek P, Miller W, Dowler W, Reighard GL, Abbott AG (2002) Identification of molecular markers linked to resistance of apricot (*Prunus armeniaca* L.) to plum pox virus. J Plant Dis Protec 9: 64–67.

Sánchez-Pérez R, Howad W, Dicenta F, Arús P, Martínez-Gómez P (2007) Mapping major genes and quantitative trait loci controlling agronomic traits in almond. Plant Breed 126: 310–318.

Sánchez-Pérez R, Howad W, García-Mas J, Arús P, Martínez-Gómez P, Dicenta F (2010) Molecular markers for kernel bitterness in almond. Tree Genet Genomes 6: 237–245.

Scorza R, Liverani A, Lightner GW (1987) Growth habit analysis of seedlings of columnar peach trees. Hort Science 22: 1075.

Scorza R, Melnicenco L, Dang P, Abbott AG (2002) Testing a microsatellite marker for selection of columnar growth habit in peach [*Prunus persica* (L.) Batsch]. Acta Hort 592: 285–289.

Schmidt R (2002) Plant genome evolution: lessons from comparative genomics at the DNA level. Plant Mol Biol 48: 21–37.

Sicard O, Marandel G, Soriano JM, Lalli DA, Lambert P, Salava J, Badenes ML, Abbott A and Decroocq V (2008) Flanking the major Plum pox virus resistance locus in apricot with co-dominant markers (SSRs) derived from candidate resistance genes. Tree Genet Genomes 4: 359–365.

Silva C, García-Mas J, Sánchez AM, Arús P, Oliveira MM (2005) Looking into flowering time in almond (*Prunus dulcis* (Mill) DA Webb): the candidate gene approach. Theor Appl Genet 110: 959–968.

Socias and Company R (1990) Breeding self-compatible almonds. Plant Breed Rev 8: 313–338

Soriano JM, Domingo ML, Zuriaga E, Romero C, Llácer G, Badenes ML (2011) Genetic tools for selecting resistance to sharka disease in apricot. XV International Symposium on Apricot Breeding and Culture, Yereban (Armenia). Acta Hort: in press.

Soriano JM, Vera-Ruiz EM, Vilanova S, Martinez-Calvo J, Llacer G, Badenes ML, Romero C (2008) Identification and mapping of a locus conferring plum pox virus resistance in two apricot improved linkage maps. Tree Genet Genomes 4: 391–402.

Sosinski B, Gannavarapu M, Hager LD, Beck LE, King GJ, Ryder CD, Rajapakse S, Baird WV, Ballard RE, Abbott AG (2000) Characterization of microsatellite markers in peach (*Prunus persica* L. Batsch). Theor Appl Genet 101: 421–4.

Tao R, Yamane H, Sassa H, Mori H, Gradziel TM, Dandekar AM, Sugiura A (1997) Identification of stylar RNases associated with gametophytic self-incompatibility in almond (*Prunus dulcis*). Plant Cell Physiol 38: 304–311.

Tao R, Yamane H, Sugiura A, Murayama H, Sassa H, Mori H (1999) Molecular typing of S-alleles through identification, characterization and cDNA cloning for S-RNases in sweet cherry. J Am Soc Hort Sci 124: 224–233.

Tao R, Watari A, Hanada T, Habu T, Yaegaki H, Yamaguchi M, Yamane H (2007) Self-compatible peach (*Prunus persica*) has mutant versions of the *S* haplotypes found in self-incompatible *Prunus* species. Plant Mol Biol 63: 109–123.

Tataranni G, Spada A, Pozzi C, Bassi D (2010) AFLP-based bulk segregant analysis for tagging the slow-ripening trait in peach [*Prunus persica* (L.) Batsch]. J Hort Sci Biotechnol 85: 78–82.

Testolin R, Marrazzo MT, Cipriani G, Quarta R, Verde I, Dettori MT, Pancaldi M, Sansavini S (2000) Microsatellite DNA in peach (*Prunus persica* L. Batsch) and its use in fingerprinting and testing the genetic origin of cultivars. Genome 43: 512–520.

Testolin R, Messina R, Lain O, Marrazzo MT, Huang WG, Cipriani G (2004) Microsatellites isolated in almond from an AC-repeat enriched library. Mol Ecol Notes 4: 459–461.

Topp B, Russell DM, Neumüller M, Dalbó MA, Liu W (2011) Plum. In: Byrne D, Badenes ML (eds) Fruit Breeding. Horticultural series Vol 6. Springer, Berlin, Heidelberg: ISBN 978-1-4419-0762-2 (in press).

Ushijima K, Sassa H, Dandekar AM, Gradziel TM, Tao R, Hirano H (2003) Structural and transcriptional analysis of the self-incompatibility locus of almond (*Prunus dulcis*): identification of a pollen-expressed F-box gene with haplotype-specific polymorphism. Plant Cell 15: 771–781.

Van Ghelder C, Lafargue B, Dirlewanger E, Ouassa A, Voisin R, Polidori J, Kleinhentz M, Esmenjaud D (2010) Characterization of the *RMja* gene for resistance to root-knot nematodes in almond: spectrum, location, and interest for *Prunus* breeding. Tree Genet Genomes 6: 503–511.

Vera Ruiz EM, Soriano JM, Romero C, Zhebentyayeva T, Terol J, Zuriaga E, Llacer G, Abbott AG, Badenes ML (2011) Narrowing down the apricot *Plum pox virus* resistance locus and comparative analysis with the peach genome syntenic region. Mol Plant Pathol 12: 535–547.

Verde I, Vendramin E, Dettori MT, Quarta R, Wang Y, Lecouls AC, Abbott AG (2004) Target SSR development in peach and SSR mapping in a peach BC1 progeny. Acta Hort 663: 63–68.

Verde I, Lauria M, Dettori MT, Vendramin E, Balconi C, Micali S, Wang Y, Marrazzo MT, Cipriani G, Hartings H, Testolin R, Abbott AG, Motto M, Quarta R (2005) Microsatellite and AFLP markers in the *Prunus persica* [L. (Batsch)] × *P. ferganensis* BC1 linkage map: saturation and coverage improvement. Theor Appl Genet 111: 1013–1021.

Vilanova S, Romero C, Abbott AG, Llacer G, Badenes ML (2003) An apricot (*Prunus armeniaca* L.) F_2 progeny linkage map based on SSR and AFLP markers, mapping plum pox virus resistance and self-incompatibility traits. Theor Appl Genet 107: 239–247.

Vilanova S, Romero C, Burgos L, Llácer G, Badenes ML (2005) Identification of Self (in)compatibility alleles in apricot by PCR sequence analysis. J Am Soc Hort Sci 130: 893–898.

Vilanova S, Soriano JM, Lalli DA, Romero C, Abbott AG, Llácer G, Badenes ML (2006) Development of SSR markers located in the G1 linkage group of apricot (*Prunus armeniaca*) using a bacterial artificial chromosome library. Mol Ecol Notes 6: 789–791.

Vos P, Hogers R, Bleeker M, Reijans M, van de Lee T, Hornes M, Frijters A, Pot J, Peleman J, Kuiper M, Zabeau M (1995) AFLP: a new technique for DNA fingerprinting. Nucl Acids Res 23: 4407–4414.

Wang Y, Georgi LL, Reighard GL, Scorza R, Abbott AG (2002a) Genetic mapping of the evergrowing gene in peach [*Prunus persica* (L.) Batsch]. J Hered 93: 352–358.

Wang Y, Georgi LL, Zhebentyayeva TN, Reighard GL, Abbott AG (2002b) High-throughput targeted SSR marker development in peach (*Prunus persica.*). Genome 45: 319–348.

Warburton ML, Becerra-Velásquez VL, Goffreda JC, Bliss FA (1996) Utility of RAPD markers in identifying linkages to genes of economic interest in peach. Theor Appl Genet 93: 920–925.

Weinberger JH (1944) Characteristics of the progeny of certain peach varieties. Proc Am Soc Hort Sci 45: 233–238.

Williams JG, Kubelik AR, Livak KJ, Rafalski JA, Tingey SV (1990) DNA polymorphisms amplified by arbitrary primers are useful as genetic markers. Nucl Acids Res 18: 6531–6535.

Wu J, Shu HR, Zhang KC, Jiang LJ (2007) Development of molecular markers linked to the traits of non-acid/acid fruit of peach. Acta Hort 763: 131–137.

Wu SB, Tavassolian I, Rabiei G, Hunt P, Wirthensohn M, Gibson JP, Ford CM, Sedgley M (2009) Mapping SNP-anchored genes using high-resolution melting analysis in almond. Mol Genet Genom 282: 273–281.

Yaegaki H, Shimada T, Moriguchi T, Hayama H, Haji T, Yamaguchi M (2001) Molecular characterization of S-RNase genes and S-genotypes in the Japanese apricot (*Prunus mume* Sieb. et Zucc.). Sex Plant Reprod 13: 251–257.

Yamamoto T, Shimada T, Imai T, Yaegaki H, Haji T, Matsuta N, Yamaguchi M, Hayashi T (2001) Characterization of morphological traits based on a genetic linkage map in peach. Breed Sci 51: 271–278.

Yamamoto T, Mochida K, Imai T, Shi YZ, Ogiwara I, Hayashi T (2002) Microsatellite markers in peach [*Prunus persica* (L.) Batsch] derived from an enriched genomic and cDNA libraries. Mol Ecol Notes 23: 298–301.

Yamane H, Tao R, Sugiura A, Hauck NR, Iezzoni AF (2001) Identification and characterization of S-RNases intetraploid sour cherry (*Prunus cerasus*). J Am Soc Hort Sci 126: 661–667.

Yamane H, Ikeda K, Ushijima K, Sassa H, Tao R (2003) A pollen-expressed gene for a novel protein with an F-box motif that is very tightly linked to a gene for S-RNase in two species of cherry, *Prunus cerasus* and *P. avium*. Plant Cell Physiol 44: 764–769.

Zhebentyayeva T, Ledbetter C, Burgos L, Llacer G (2011) Apricot. In: Byrne D, Badenes ML (eds) Fruit Breeding. Horticultural series Vol 6. Springer Berlin, Heidelberg: ISBN 978-1-4419-0762-2 (in press).

6

Molecular Mapping of Complex Traits

Bode A. Olukolu[1,*] and *Chittaranjan Kole*[2]

ABSTRACT

This chapter reviews on the understanding of the genetics underlying complex traits of stone fruits. While stable and major QTLs across multiple environments and mapping populations are desirable, knowledge of population-specific and minor QTLs (particularly in traits controlled by only minor QTLs) are crucial for efficient and precise breeding. Sources of desirable traits lacking in cultivars require exploration in wild species as demonstrated in interspecific crosses in breeding programs and generation of multiple mapping populations. Utility of these findings for crop improvement are impeded by genetic-by-environment interactions, linked QTLs and interacting QTLs that have a tendency to alter complex traits in a manner that is difficult to predict. Besides the use of QTL linkage analysis, a LD-based association mapping strategy that exploits numerous ancient recombination events in diverse germplasms is now gaining importance for fine mapping QTLs as well as detecting polymorphisms lacking in bi-parental mapping populations. The availability of a *Prunus* reference map and the high genome colinearity among stone fruit species now present an unprecedented ease for comparative mapping, hence, the capability to validate conserved QTLs across several stone fruit species. While genetic improvement based on MAB approach is plausible for complex traits with one or two major QTLs, traits with several minor QTLs still remain intractable for breeding purposes using MAB. The decreasing cost of

[1]Department of Plant Pathology, North Carolina State University, 112 Derieux Place, Raleigh, NC 27607, USA.
[2]Department of Genetics and Biochemistry, Clemson University, 109 Jordan Hall, Clemson, SC 29634, USA.
*Corresponding author: *baolukol@ncsu.edu*

sophisticated and high-throughput techniques required for genomics-based breeding now offer promise for the genetic improvement of complex traits controlled by minor QTLs.

Keywords: linkage analysis, genome-wide association mapping (GWAS),genotyping-by-sequencing (GBS), marker-assisted breeding (MAB), genomics-based breeding

6.1 Introduction

Complex traits are usually controlled by several loci and population-specific alleles that necessitate the use of multiple mapping populations. Consequently, a favorable allele in one genetic background can turn out to be an unfavorable allele in another genetic background. In many cases, each of these quantitative trait loci (QTLs) explains small proportions of total phenotypic variance of the trait. QTLs controlling complex traits in stone fruits have been mapped in the three major subgenera (Rehder 1940) that comprise species of horticultural importance. These include *Amygdalus* (peaches and almonds), *Prunophora* or *Prunus* (apricots and plums) and *Cerasus* (sweet and sour cherries). The high level of genome colinearity, development of transferable simple sequence repeat (SSR) markers and availability of a *Prunus* reference map (Joobeur et al. 1998; Aranzana et al. 2003) have highly facilitated these efforts. However, in stone fruits, most of the successes at mapping QTLs have been reported in the more tractable diploid species that comprise peach, apricot, almond and sweet cherry, while the polyploid species including plums and sour cherries have proven to be more challenging. Nevertheless, studies have identified QTLs within syntenic genomic regions underlying specific complex traits across stone fruits, thus lending diploid species as models for understanding some complex traits in polyploid *Prunus* species.

Diversity studies have revealed remarkable variations in the extent of genetic diversity across stone fruit species ranging from a narrow genetic base in peach; intermediate for apricot, sweet cherry and sour cherry; and high genetic variability in almond and plum (Byrne and Bacon 1990; Granger et al. 1993). Utilization of wild relatives and available germplasm for breeding of complex traits is vital to identifying alleles lacking in commercial cultivars. Although horticulturally inferior wild relatives can provide favorable alleles for introgression, reduction of the wild parental genome while retaining the regions harboring the favorable alleles undermines the utility of genetic resources in wild species (Bernacchi et al. 1998a, b, c; Moncada et al. 2001; Foulongne et al. 2003a; Pillen et al. 2003; Quilot et al. 2004). QTLs detected from bi-parental crosses or association studies involving wild species now provide molecular markers, which can be used in marker-assisted selection (MAS) in a targeted manner to exploit alleles not present in the existing cultivars (Quilot et al. 2004; Decroocq et al. 2005; Marandel et al. 2009; Rubio et al. 2010; see for review Potter 2011).

Slow breeding to improve complex traits in stone fruits due to long generation time, large plant size limiting the number of progenies that can be screened in breeding programs and the high level of within-species heterozygosity therefore necessitate the use of powerful genomic and statistical tools to catalog QTLs, their interactions inter se and with environment for precision breeding.

6.2 Target Complex Traits

Breeding goals in stone fruits vary with the fruit species (Table 6-1); nevertheless, a few traits such as high fruit quality, climatic adaptability

Table 6-1 Traits relevant to the breeding objectives of horticulturally important stone fruit species.

Trait	Peach	Almond	Apricot	Plum	Sweet cherry	Sour cherry
Fruit quality						
Skin color	+	–	+	+	–	–
Flesh color	+	–	+	+	–	–
Flesh color around stone	+	–	+	+	–	–
Flesh adhesion	+	+	+	+	+	+
Fruit shape	+	–	+	+	+	+
pH	+	–	+	+	+	+
Sugar content	+	–	+	+	+	+
Ripening	+	+	+	+	+	+
Fruit texture	+	–	+	+	–	–
Skin cracking	–	–	+	–	+	+
Shell hardness and kernel quality						
Shell hardness	–	+	+	–	–	–
Kernel bitterness	–	+	+	–	–	–
Wide ranges of production time and climatic adaptation						
Blooming date	+	+	+	+	+	+
Flowering time	+	+	+	+	+	+
Chilling requirement	+	+	+	+	+	+
Cold tolerance/hardiness	+	+	+	+	+	+
Freeze tolerance	+	+	+	+	+	+
Disease resistance						
Plum-pox virus	+	+	+	+	+	+
Powdery Mildew	+	–	–	–	+	+
Short life syndrome	+	–	–	–	–	–

Note: + indicates trait is relevant to crop, while – indicates trait not very important to crop genetic improvement.

for cultivation in wide geographical range and disease resistance remain cornerstone to their genetic improvement. The description of fruit quality also varies widely from one species to another or even within the same species for varied consumer preference. While most stone fruit species are primarily grown for their flesh, almonds stand out since they are primarily bred for the seeds within their stone; hence, the focus is on the shell and kernel quality rather than the fruit itself. Other traits common to stone fruits include fruit cracking in cherries and apricots.

This chapter does not attempt to report all the trait loci mapped in stone fruits, rather, it focuses only on complex or polygenic traits reported to be controlled by two or more QTLs (usually many QTLs). Depending on genetic background, QTLs can be unique to specific mapping populations and environments. Stable QTLs across several mapping populations and environments are highly desirable since they are more predictive for breeding. The ability to detect QTLs is also affected by the population size, which affects the degree of recombination events and the map resolution, a function of the number of markers used for map construction and genome size. Accurate proportions of contribution to phenotypic variance require accounting for QTL interactions since values are usually over-estimated when QTL interactions are not accounted for.

6.3 Framework Maps and Markers

A recent study, which revealed that 63.9% of a set of 145 SSR markers produced polymorphic amplicons among six *Prunus* species (peach, almond, apricot, cherry, Japanese plum and European plum), now provides further validation for marker transferability among *Prunus* species (Mnejja et al. 2010). The SSR marker transferability among stone fruit species and high levels of genome colinearity are probably the most important resource for understanding complex traits in *Prunus* species. Thirteen *Prunus* maps constructed with a subset of 562 markers of the reference map (Joobeur et al. 1998) also reveal high conservation of synteny among seven diploid ($2n = 2x = 16$) species (almond, peach, apricot, cherry, *Prunusfer ganensis*, *P. davidiana* and *P. cerasifera*). Consequently, tightly linked markers for 28 major genes have been detected in these maps (Dirlewanger et al. 2004). The use of comparative genomics has facilitated the validation of QTLs within similar genomic regions. The TxE *Prunus* reference map, peach x almond, provides a reference genome that other *Prunus* species can be compared to. The genome sizes of the diploid species, ranging from 280 Mb in peach to 338 Mb in sweet cherry, are fairly equivalent. Conversely, the tetraploid sour cherry (599 Mb) and the hexaploid plum (883 Mb) genomes limit the ability of studies to identify QTLs.

The conservation of synteny among *Prunus* species have been exploited for mapping complex traits segregating in interspecific crosses especially between peach and other *Prunus* species (Dirlewanger et al. 1996; Viruel et al. 1998; Foulongne et al. 2003b; Quilot et al. 2004; Silva et al. 2005; Blenda et al. 2007; Ogundiwin et al. 2009; Illa et al. 2011). Several QTL studies in *Prunus* species are based on genetic linkage maps that have all chromosomes (basic chromosome number, $n = 8$) represented on contiguous linkage groups and with a good coverage of the genome except in sour cherry.

The QTL analysis in the tetraploid sour cherry is all based on a partial map of 19 linkage groups covering 461 cM and 16 linkage groups covering 279 cM for the Reinische Schattenmorelle (RS) x Erdi Botermo (EB) parental maps, respectively. With a limited number of shared markers with the reference map, only the *Prunus* linkage groups 2, 4, 6 and 7 were successfully identified as the putative homologous linkage groups (Wang et al. 1998). The commercial hexaploid plum lacks a genetic map till date, although genetic maps based on interspecific crosses involving Myrobolan plum (*P. cerasifera*) have been achieved (Dirlewanger et al. 2004).

6.4 Fruit Quality Traits Mapped

Breeding in stone fruits like many other crops initially focused on yield, which has increased steadily due to progress in genetic improvement and cropping practices (Calderini and Slafer 1998; Cassman 1999, 2001). Following the leveling off of yield increases, focus is now shifting to fruit quality traits such as flesh adhesion, flesh color around stone, skin color, flesh color, shape, sugar content, kernel taste and shell hardness. These traits are the major driving force in the market for sale of stone fruits. Despite increasing fruit yield, the consumption of raw fruits have not increased because of the low quality of fruits, which is not necessarily due to the genetics of the cultivars; rather, to the fact that fruits are usually harvested for storage and shipping before they ripen (Etienne et al. 2002).

Consequently, breeding objectives are now targeted towards finding the optimal balance between fruit quality and immaturity (Genard and Bruchou 1992; Fideghelli et al. 1998), which is determined by the sugar-acid ratio or also referred to as the quality index (Robertson et al. 1989; Bassi et al. 1990). The components that affect fruit taste include the right balance of soluble sugars (such as sucrose, fructose, glucose and sorbitol), organic acids (such as malic acid, citric acid, quinic acid and shikimic acid) and aromatic compounds (Pangborn 1963; Yoshida 1970; Meredith et al. 1989; Moing et al. 1998a, b; Bassi et al. 1990; Etienne et al. 2002). The kernel quality is the most important fruit quality trait in almond since the kernel is the edible part. There is also an interest in consuming apricot kernels, thus prompting the need to breed against bitterness that is conferred by amygdalin, a cyanogenic glucoside. Enzymatic cleavage of amygdalin during crushing of seeds leads to the release of toxic hydrogen cyanide in these seeds (Cervellati et al. 2011).

Breeding for large fruit size and desirable fruit quality traits in stone fruits along with other horticultural traits has been limited because the genetic control of fruit quality is poorly understood (Tucker 1993). For example, in sweet cherries, breeders are hesitant to use small-fruited lines to improve traits such as disease resistance, since several more generations are required to recover the desirable fruit size and qualities (Zhang et al. 2010). Like many complex traits, the difficulty of understanding the genetics underlying fruit quality is compounded by the fact that the fruit is the most complex plant organ comprised of diverse tissues and contains an array of plant metabolites. Consequently, the overlapping chain of physiological processes from bud formation to seed/fruit development is controlled by several interacting genes that in turn interact with the environment. Reports from other well studied fruits implicate over 100 genes that each usually explains very low proportion of the phenotypic variance (Causse et al. 2004, 2007; Bermudez et al. 2008). The critical consideration for mapping QTLs for fruit quality borders on combining genetic resources for fruit quality while maintaining yield. The challenge of these interrelated processes can pose a problem for breeding fruit quality and yield simultaneously, since enhancing a trait such as water flux into fruits increases fruit size but leads to an unfavorable reduction in the concentration of sugars (Génard and Lescourret 2004). Accordingly, a model for fruit quality must take into account several traits, underlying processes and their interactions (Lescourret and Genard 2005).

Similar to other studies conducted in apples (Baumgärtner et al. 1984; Seem et al. 1986), grapes (Gutierrez et al. 1985), kiwifruit (Buwalda 1991), olives (Abdelrazik 1989) and tomatoes (Heuvelink and Bertin 1994), carbon relationships such as photosynthetic-driven models are now been adapted to peach (Grossman and Dejong 1994; Lescourret and Genard 2005) as a model for stone fruit species. Throughout the final stages of fruit development, ongoing models simultaneously simulate changes in several fruit quality traits such as fruit size, the most important trait for growers; proportion of total mass containing fruit flesh; dry matter content of flesh and sugar concentrations in the flesh (Lescourret and Genard 2005). In consolidation with QTL analysis of specific complex fruit quality traits, the model can offer an exceptional tool (Struik et al. 2005) for understanding complex pathways that comprise gene-by-gene and genetic-by-environment interactions (Fig. 6-1). Although, studies are underway to model fruit quality, the definitions and standards for fruit quality are subject to what the consumers desire (Barritt 2001; Elia 2002; Kupferman 2002), which tend to be variable with individual preferences. The breeder is beset with a challenge to

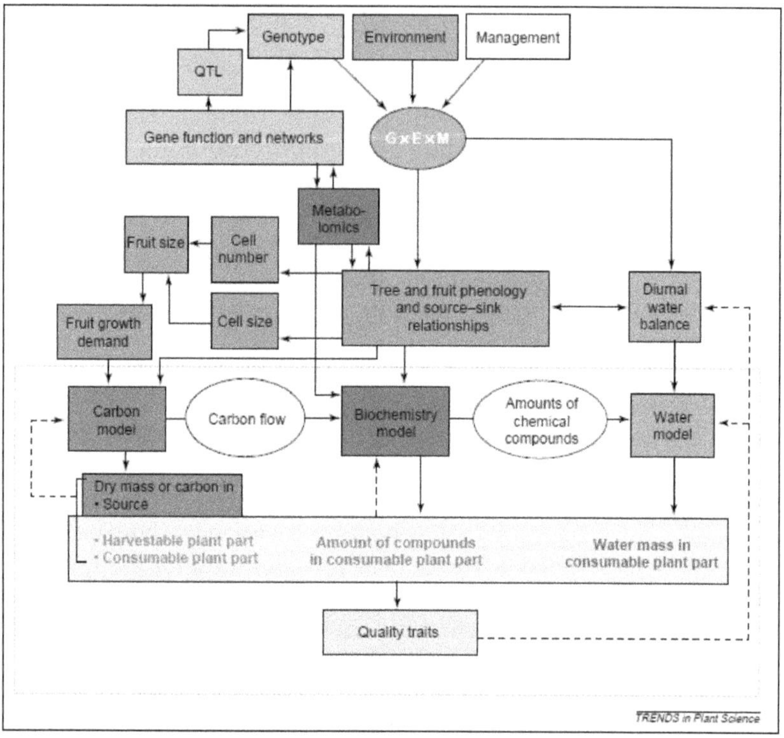

Figure 6-1 A schematic diagram for the virtual peach fruit model. Genomics and metabolomics information can also be integrated to understand complex traits using a computational systems biology approach for predicting genotype-by-environment-by-management interactions on fruit quality characteristics (Struik et al. 2005; reproduced with permission).

Color image of this figure appears in the color plate section at the end of the book.

understand how consumers perceive taste in order to combine alleles in an appropriate manner during marker-assisted or genomics-based breeding.

Several candidate genes mined from metabolic pathways for fruit quality have been mapped and shown to co-localize with previously reported QTLs (Ogundiwin et al. 2009; Illa et al. 2011). The comprehensive integrated fruit quality gene maps contain 133 putative genes involved in fruit texture, pigmentation, flavor and chilling injury (CI) resistance (Ogundiwin et al. 2009) and 273 expressed sequence tags (ESTs) corresponding to candidate genes involved in fruit growth and maturity, texture, content of sugar and organic acid, aroma and color (Illa et al. 2011).

6.4.1 Skin and Flesh Coloration

QTL studies on fruit skin speckle, skin coloration and flesh coloration were detected in a peach BC_2: *P. davidiana* clone P1908 x *P. persica* cv. Summergrand (S) x S x Zéphyr population (Quilot et al. 2004). The study identified four QTLs for fruit skin speckle (Fig. 6-2). Two of the QTLs were localized on linkage group (LG) 6 and one on LG3. A fruit coloration QTL was mapped on LG5, while two fruit flesh QTLs were mapped on LG1 and LG3. The fruit flesh coloration QTLs co-localized to the same genomic region controlling fruit skin speckle on LG3. A QTL detected for each of fruit coloration and fruit flesh coloration were mapped to LG9 that could not be linked to any of the eight LGs that represent the eight basic chromosome number in *Prunus*. Co-localization of some of these coloration and speckle QTLs suggest that the traits share common biochemical pathways for production of the underlying pigmentations. In a sweet cherry F_1: Emperor Francis x New York 54 mapping population, skin and flesh color QTLs correspond to the same genomic region as the peach skin speckle and flesh color QTLs on LG3 and LG6.

6.4.2 Fruit Flesh and Seed Taste

The QTLs controlling fruit juiciness and sweetness were also detected in the same peach mapping population as skin and flesh coloration (Quilot et al. 2004), while the seed taste QTLs were detected in an apricot F_1: Lito x BO81604311 population (Cervellati et al. 2011). Two fruit flesh juiciness QTLs were detected on LG4 and LG7, while three fruit flesh sweetness QTLs were localized on LG1, LG3 and LG5 (Fig. 6-2). A total of three QTLs were identified for seed bitterness on LG1, LG4 and LG6 (Fig. 6-2).

6.4.3 Content of Soluble Sugars and Organic Acids

The QTL underlying content of soluble sugars and organic acids were all studied in peach (Dirlewanger et al. 1999; Etienne et al. 2002; Quilot et al. 2004). These traits include contents of sucrose, glucose, fructose, sorbitol, soluble-solids, total sugar, malic acid, citric acid, quinic acid and shikimic acid. The QTLs underlying total acid, pH and titratable acidity (TA) were also detected in peach. Several overlapping QTLs were found to be underlying some of these traits (Fig. 6-2), thus suggesting shared biochemical pathways.

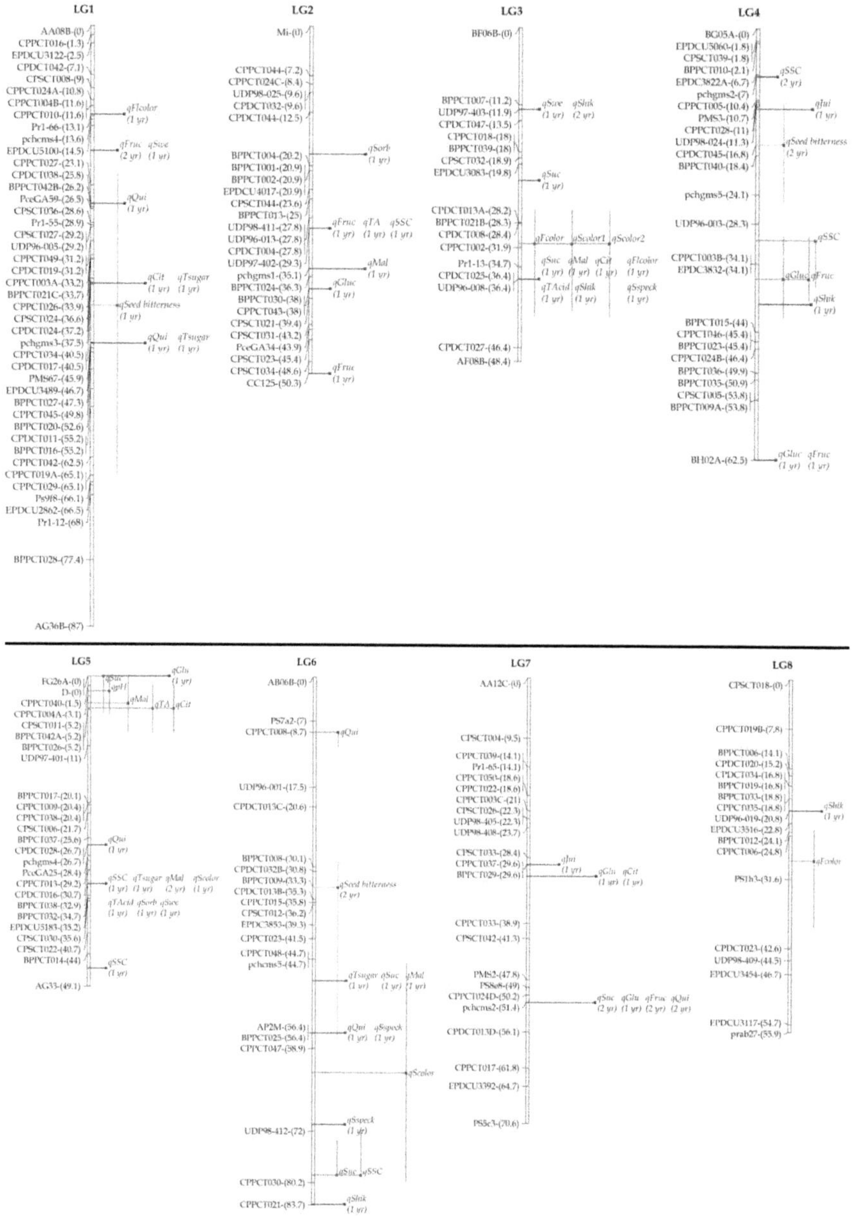

Figure 6-2 contd....

6.5 Fruit Yield and Yield-Component Traits

The various yield parameters (Table 6-2) that are phenotyped for QTL detection in stone fruits include dry flesh weight (peach); fruit weight (peach, sweet cherry and sour cherry); stone weight (peach); fruit length and diameter (sweet cherry); fruit Polar diameter; fruit cheek diameter; fruit suture diameter; mesocarp length and diameter (sweet cherry); pit length and diameter (sweet cherry); productivity (peach and almond); in-shell weight (almond) and kernel weight (almond). QTLs from three studies shown on the *Prunus* reference map (Fig. 6-3) include a peach F_1: FerjalouJalousiax Fantasia population (Etienne et al. 2002); an almond F_1: R1000 x Desmayo Largueta population (Sánchez-Pérez et al. 2007) and a sweet cherry F_1: Emperor Francis x New York 54 population (Zhang et al. 2010). QTLs underlying fruit weight, mesocarp length and diameter, pit length and diameter were all mapped to the same genomic region on LG2 and LG6. Of the almond kernel yield components, only the in-shell weight co-localized with some fruit yield components on LG2 (Fig. 6-3). Likewise, fruit weight QTLs in sour cherry (Wang et al. 2000) were detected on LG2 and LG4 but the lack of transferable markers limited comparative QTL analysis.

In the peach BC_2: (*P. davidiana* x *P. persica*cv. Summergrand (S)) x S x Zéphyr population phenotyped for dry flesh weight trait, two QTLs each were detected on LG3 and LG5 (Quilot et al. 2004). The dry flesh weight QTL on LG5 was co-localized to the same region for fruit weight phenotyped on the same population (Quilot et al. 2004). The fruit weight QTLs includes two QTLs on LG4 and one QTL each on LG1, LG2, LG5 and LG7 (Table 6-2). QTL analysis for fruit polar diameter, fruit cheek diameter and fruit suture diameter in peach reveal common QTLs were underlying these

---

Figure 6-2 contd....

Figure 6-2 Fruit quality QTLs, soluble solid content (SSC), sucrose (Suc), glucose (Gluc), fructose (Fruc), sorbitol (Sorb), total sugar (T sugar) concentrations, malic acid (Mal), citric acid (Cit), quinic acid (Qui), shikimique acid (Shik), total acid (TAcid) concentrations, skin coloration (S color), sweet cherry skin color on cheeks of fruit (Scolor1) and on suture of fruits (Scolor2), skin speckle (S speck), flesh coloration (Fl color), juiciness (Jui) and sweetness (Swe) and seed bitterness, mapped in various mapping populations and studies shown localized on the framework of the *Prunus* reference map (Joobeur et al. 1998). Approximate QTL positions and confidence/support intervals on the reference map are inferred based on common transferable SSR markers at and that flank the QTL peak. QTLs detected over multiple years (yrs) indicated in parenthesis. QTLs from different studies indicated by colored lines include a peach F_1: Ferjalou Jalousia x Fantasia population (red line; Etienne et al. 2002); a peach BC_2: *P. davidiana* clone P1908 x *P. persica* cv. Summergrand (S) x S x Zéphyr population (black line; Quilot et al. 2004); a sweet cherry F_1: Emperor Francis x New York 54 (blue line; Sooriyapathirana et al. 2010); and an apricot F_1: Lito x BO81604311 population (green line; Cervellati et al. 2011).

Color image of this figure appears in the color plate section at the end of the book.

Table 6-2 A list of complex traits mapped in stone fruit species.

Trait	Cross	Reference
Fruit quality and yield		
Skin color	BC_2: (*P. davidiana* x *P. persica* cv. Summergrand (S)) x S x Zéphyr	Quilot et al. 2004
	F_1: Emperor Francis x New York 54 (sweet cherry)	Sooriyapathirana et al. 2010
Flesh color	BC_2: (*P. davidiana* x *P. persica* cv. Summergrand (S)) x S x Zéphyr	Quilot et al. 2004
	F_1: Emperor Francis x New York 54 (sweet cherry)	Sooriyapathirana et al. 2010
Skin speckle	BC_2: (*P. davidiana* x *P. persica* cv. Summergrand (S)) x S x Zéphyr	Quilotet al. 2004
Juiciness		
Sweetness		
Fruit Polar diameter		
Fruit cheek diameter		
Fruit suture diameter		
Stone weight		
Dry flesh weight		
Seed bitterness	F_1: Lito x BO81604311 (apricot)	Cervellati et al. 2011
Fruit weight	F_1: ReinischeSchattenmorelle x ErdiBotermo (sour cherry)	Wang et al. 2000
	F_2: Akame x Juseitou (peach)	Yamamoto et al. 2001
	BC_2: (*P. davidiana* x *P. persica* cv. Summergrand (S)) x S x Zéphyr	Quilot et al. 2004
	F_1: Emperor Francis x New York 54 (sweet cherry)	Zhang et al. 2010
	F_1: FerjalouJalousia x Fantasia (peach)	Etienne et al. 2002
Fruit length	F_1: Emperor Francis x New York 54 (sweet cherry)	Zhang et al. 2010
Fruit diameter		
Pit length		
Pit diameter		
Mesocarp length		
Mesocarp diameter		

Productivity	F_1: FerjalouJalousia x Fantasia (peach)	Dirlewanger et al. 1999
In-shell weight	F_1: R1000 x DesmayoLargueta (almond)	Sánchez-Pérez et al. 2007
Kernel weight		
Sucrose	F_1: FerjalouJalousia x Fantasia (peach)	Dirlewanger et al. 1999
		Etienne et al. 2002
	BC_2: (*P. davidiana* x *P. persica* cv. Summergrand (S)) x S x Zéphyr	Quilot et al. 2004
Glucose	F_1: FerjalouJalousia x Fantasia (peach)	Dirlewanger et al. 1999
		Etienne et al. 2002
	BC_2: (*P. davidiana* x *P. persica* cv. Summergrand (S)) x S x Zéphyr	Quilot et al. 2004
Fructose	F_1: FerjalouJalousia x Fantasia (peach)	Dirlewanger et al. 1999
		Etienne et al. 2002
	BC_2: (*P. davidiana* x *P. persica* cv. Summergrand (S)) x S x Zéphyr	Quilotet al. 2004
Sorbitol	F_1: FerjalouJalousia x Fantasia (peach)	Dirlewanger et al. 1999
	BC_2: (*P. davidiana* x *P. persica* cv. Summergrand (S)) x S x Zéphyr	Quilot et al. 2004
Total sugar content	BC_2: (*P. davidiana* x *P. persica* cv. Summergrand (S)) x S x Zéphyr	Quilot et al. 2004
Soluble-solids content	F_1: FerjalouJalousia x Fantasia (peach)	Dirlewanger et al. 1999
	F_1: FerjalouJalousia x Fantasia (peach)	Etienne et al. 2002
	BC_2: (*P. davidiana* x *P. persica* cv. Summergrand (S)) x S x Zéphyr	Quilot et al. 2004
Malic acid	F_1: FerjalouJalousia x Fantasia (peach)	Dirlewanger et al. 1999
	F_1: FerjalouJalousia x Fantasia (peach)	Etienne et al. 2002
	BC_2: (*P. davidiana* x *P. persica* cv. Summergrand (S)) x S x Zéphyr	Quilot et al. 2004
Citric acid	F_1: FerjalouJalousia x Fantasia (peach)	Dirlewanger et al. 1999
	F_1: FerjalouJalousia x Fantasia (peach)	Etienne et al. 2002
	BC_2: (*P. davidiana* x *P. persica* cv. Summergrand (S)) x S x Zéphyr	Quilot et al. 2004

Table 6-2 contd....

Table 6-2 contd....

Trait	Cross	Reference
Quinic acid	F_1: Ferjalou Jalousia x Fantasia (peach)	Dirlewanger et al. 1999
	F_1: Ferjalou Jalousia x Fantasia (peach)	Etienne et al. 2002
	BC_2: (*P. davidiana* x *P. persica* cv. Summergrand (S)) x S x Zéphyr	Quilot et al. 2004
Shikimic acid	BC_2: (*P. davidiana* x *P. persica* cv. Summergrand (S)) x S x Zéphyr	Quilot et al. 2004
Total acid		Quilot et al. 2004
pH	F_1: FerjalouJalousia x Fantasia (peach)	Dirlewanger et al. 1999
		Etienne et al. 2002
Titratable acidity		Dirlewanger et al. 1999
		Etienne et al. 2002
Wide ranges of production time and climatic adaptation		
Blooming time	F_2: Akame x Juseitou (peach)	Yamamoto et al. 2001
	F_1: FerjalouJalousia x Fantasia (peach)	Dirlewanger et al. 1999
	F1: Schattenmorelle (RS) x ErdiBotermo (EB) (sour cherry)	Wang et al. 2000
	BC_2: *P. davidiana* x *P. persica* cv. Summergrand (S)) x S x Zéphyr	Quilot et al. 2004
	F_1: R1000 x Desmayo Largueta (almond)	Sánchez-Pérez et al. 2007
	F_2: Contender x Fla.92-2C (peach)	Fan et al. 2010
Trait	Cross	Reference
Maturation time	F_1: FerjalouJalousia x Fantasia (peach)	Dirlewanger et al. 1999
		Etienne et al. 2002
	F_2: Akame x Juseitou (peach)	Yamamoto et al. 2001
	BC_2: (*P. davidiana* x *P. persica* cv. Summergrand (S)) x S x Zéphyr	Quilot et al. 2004
Ripening	F1: Schattenmorelle x ErdiBotermo (sour cherry)	Wang et al. 2000
	F_1: R1000 x DesmayoLargueta (almond)	Sánchez-Pérez et al. 2007
Fruit dropping time	F_2: Akame x Juseitou (peach)	Yamamoto et al. 2001

Chilling requirement	F_1: Perfection x A.1740 (apricot)	Olukolu et al. 2009
	F_2: Contender x Fla. 92-2C (peach)	Fan et al. 2010
Heat Requirement	F_2: Contender x Fla. 92-2C (peach)	Fan et al. 2010
Disease resistance		
Plum-pox virus	F_1:Goldrich x Valenciano(apricot)	Hurtado et al. 2002
	F_1, F_2: *P. persica x P. davidiana* (peach)	Decroocq et al. 2005
	F_1: Polaris x Stark Early Orange (apricot)	Lambert et al. 2007
	F_1: Goldrich x Currot (apricot)	Soriano et al. 2008
	F_2: Lito x Lito (apricot)	Soriano et al. 2008
	F_2: Summergrand x *P. davidiana* (peach)	Marandel et al. 2009
	F_1: *P. persica* cv Rubira x *P. davidiana* (peach)	Rubio et al. 2010
Short life syndrome	F_2: Nemaguard x GuardianBY520-9 (peach)	Blenda et al. 2007
Powdery mildew	F_1, F_2, BC_2: *P. persica* var. Summergrand x *P. davidiana* (peach)	Dirlewanger et al. 1996
		Foulongne et al. 2003a

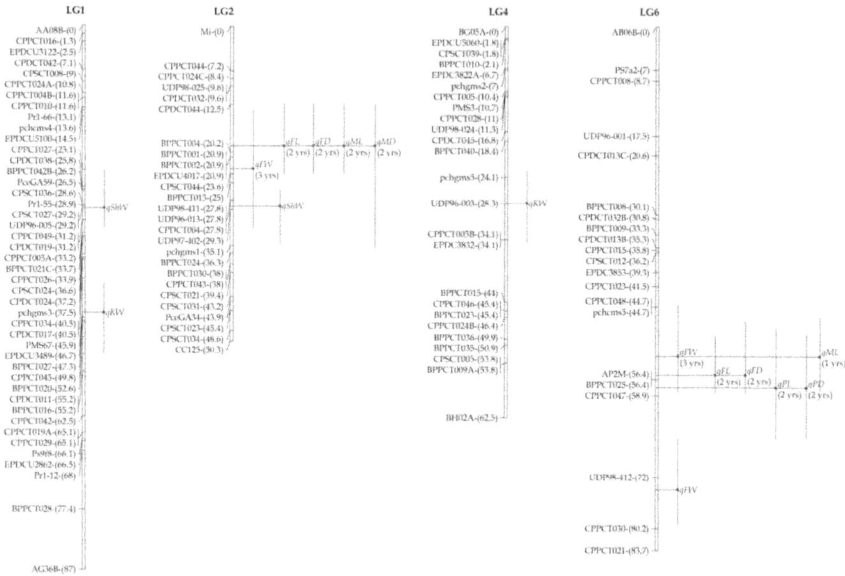

Figure 6-3 Fruit yield and yield component QTLs, fruit weight (FW), fruit length (FL), fruit diameter (FD), pit length (PL), pit diameter (PD), mesocarp length (ML), mesocarp diameter (MD), in-shell weight (ShW) and kernel weight (KW), mapped in various mapping populations and studies shown localized on the framework of the *Prunus* reference map (Joobeur et al. 1998). Approximate QTL positions and confidence/support intervals on the reference map are inferred based on common transferable SSR markers at and that flank the QTL peak. QTLs detected over multiple years (yrs) indicated in parenthesis. QTLs from different studies indicated by colored lines include a peach F_1: Ferjalou Jalousia x Fantasia population (red line; Etienne et al. 2002); an almond F_1: R1000 x Desmayo Largueta population (blue line; Sánchez-Pérez et al. 2007) and a sweet cherry F_1: Emperor Francis x New York 54 population (purple line; Zhang et al. 2010).

Color image of this figure appears in the color plate section at the end of the book.

yield component traits (Quilot et al. 2004). Fruit cheek and fruit suture diameter have five common QTLs underlying them, with two QTLs on LG1 and one QTL each on LG2, LG4 and LG5. A common QTL on LG5 also control fruit polar and fruit suture diameter.

6.6 Climatic Adaptation and Environmental Stress Tolerance

The most important limiting factor affecting cultivation and production range of all stone fruits are complex traits such as chilling requirement (CR), blooming time, cold hardiness and freezing tolerance that improve the plants' survivability in extreme cold regions. Other complex traits of importance include heat requirement (HR), maturation time and fruit dropping time (Table 6-2). Commercial production of stone fruits is highly influenced by climatic conditions. The synchronization of the development and cyclic climatic conditions ensures survival and reproduction, which is crucial for fruit production. Late winter or early spring frosts can lead to yield losses in temperate regions due to unfulfilled CR. The need to expand production to regions with a warmer climate is dependent on the ability to breed cultivars with low CR. The ability of cultivars to bloom late is essential to avoid frost damage of fruits, while the ability to produce both early and late blooming cultivars provides the basis for extending production time within the growing season.

Blooming time is dependent on weather conditions required for induction and release of endodormancy in floral and vegetative buds (Lang et al. 1987). Variations in blooming time are highly related to observed differences among genotype CR (Tabuenca 1972) and HR (Tabuenca et al. 1972). Several studies reveal that blooming time, CR and HR are all quantitatively inherited traits in stone fruit species (Kester 1965; Grasselly and Gall 1967; Olukolu et al. 2009; Fan et al. 2010). Similarly, maturation and fruit dropping time are also quantitatively inherited. Despite the importance of both cold hardiness and freezing tolerance to production in very cold climates, QTL studies are currently not available for these traits. Nevertheless, cultivars with high CR are expected to possess a high measure of cold hardiness and freezing tolerance since high CR cultivars are mostly adapted to very cold climates where their ability to bud break and reproduce is higher. On the other hand, low chill cultivars are adapted to warmer winter climates where there is no requirement to be cold hardy or freezing tolerant.

6.6.1 Chilling and Heat Requirement

Only two QTL studies have been performed recently for CR QTLs in stone fruit species. These studies in apricot (Olukolu et al. 2009) and peach (Fan et al. 2010) have identified several loci, some of which are remarkably conserved between both the species (Fig. 6-4). The apricot study identified a total of 12 QTLs with three QTLs on LG2; two QTLs each on LG1, LG5, LG6 and LG7; and one QTL on LG8. Following phenotyping over two years, the peach study localized a total of eight QTLs with two QTLs each on LG1 and LG4; and one QTL each on LG5, LG6, LG7 and LG8. QTLs on the LG1, LG4, LG5 and LG7 were reproducible between the two trial years in peach. Both studies in peach and apricot have five QTLs consistently detected and conserved within the same genomic regions on LG1, LG5, LG6 andLG7 based on overlapping QTL intervals. The QTL on the distal end of LG1 is particularly interesting because it explains the highest contribution to phenotypic variance in peach and co-localizes with the dormancy-associated ever growing (*EVG*) region (Bielenberg et al. 2004).

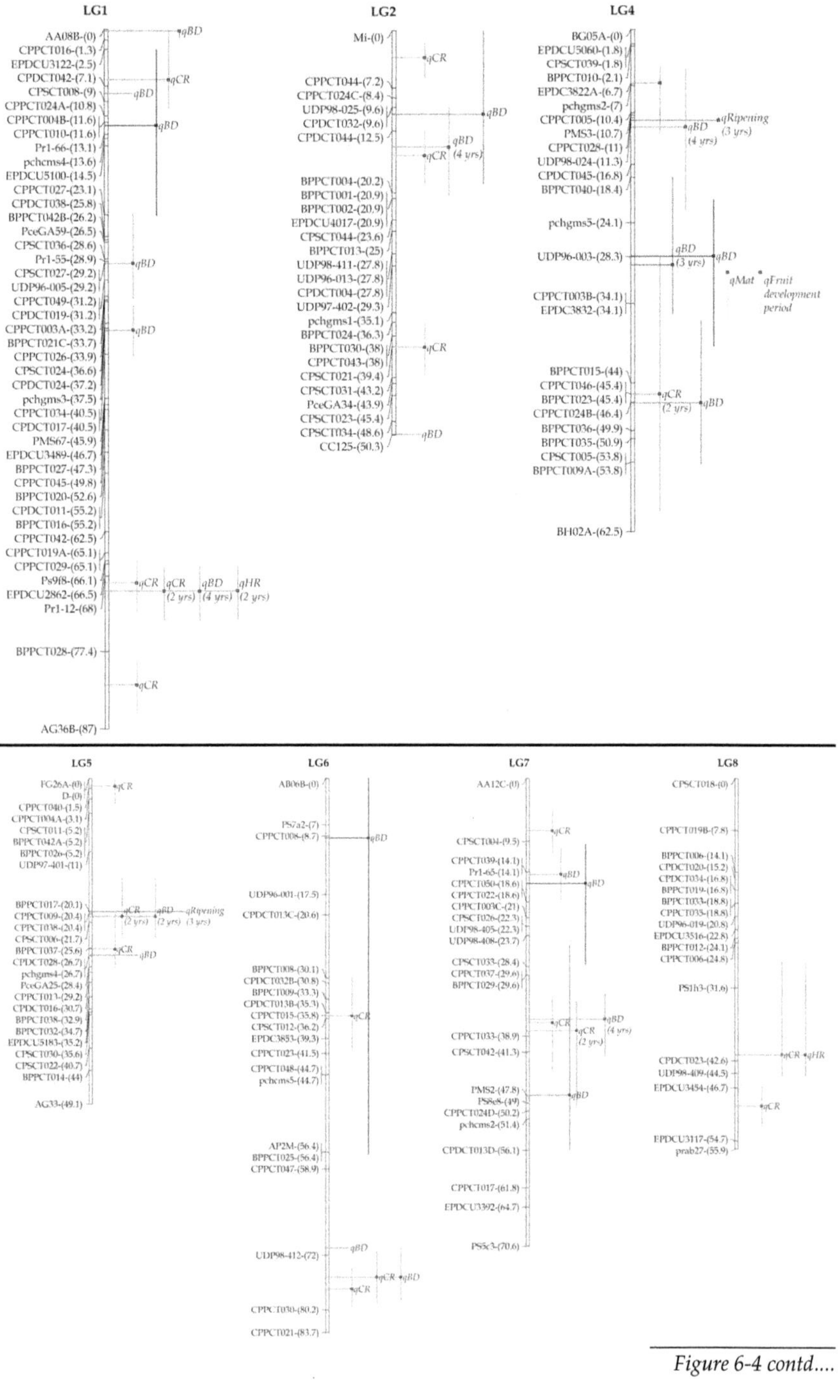

Figure 6-4 contd....

In the peach study, the QTLs explaining the higher contribution to phenotypic variance were localized towards the distal end of LG1 and on LG7. These two QTLs were also detected in the apricot study but the contribution to phenotypic variance in the apricot study was not as high as that estimated in the peach study. The peach study also identified two HR QTLs on LG1 and LG8, both of which co-localized with the CR QTLs.

6.6.2 Blooming Date, Maturation Time, Ripening and Fruit Dropping Time

QTL mapping for blooming time have been conducted in several peach populations and one population each in almond and sour cherry (Fig. 6-4). Several blooming date (BD) QTLs detected in all these studies are co-localized with most of the CR QTLs in the apricot and peach study. BD QTLs are also co-localized with QTLs for ripening date and maturation time. In the peach study, BD QTLs on LG1, LG2, LG4, LG5 and LG7 were consistently detected across multiple environments. Some of the BD QTLs were also detected consistently in independent studies based on overlapping QTL intervals.

Only a few QTLs for maturation time, ripening and fruit dropping time are shown on the *Prunus* reference map in this chapter due to lack of transferable SSR markers. All QTL studies for maturation time were performed in peach (Dirlewanger et al. 1999; Yamamoto et al. 2001; Etienne et al. 2002) and in a cross between the wild relative *P. davidiana* species and a peach cultivar (Quilot et al. 2004). Two studies that utilized the F_1: Ferjalou Jalousia x Fantasia population both detected a maturation time QTL localized to LG4 (Dirlewanger et al. 1999; Etienne et al. 2002), while the former study detected an additional QTL on LG3 (Dirlewanger et al. 1999).

Figure 6-4 contd....

Figure 6-4 Climatic adaptation QTLs, chilling requirement (CR), heat requirement (HR), blooming date (BD), maturation date (Mat), ripening and fruit development period, mapped in various mapping populations and studies shown localized on the framework of the *Prunus* reference map (Joobeur et al. 1998). Approximate QTL positions and confidence/support intervals on the reference map are inferred based on common transferable SSR markers at and that flank the QTL peak. QTLs detected over multiple years (yrs) indicated in parenthesis. QTLs from different studies indicated by colored lines include a peach F_1: Ferjalou Jalousia x Fantasia population (blue line; Dirlewanger et al. 1999); a peach F_1: FerjalouJalousia x Fantasia population (yellow line; Etienne et al. 2002); a peach BC_1: *Prunuspersica* x *P. ferganensis* population (red line; Verde et al. 2002); a peach BC_2: *P. davidiana* clone P1908 x *P. persica* cv. Summergrand (S)) x S x Zéphyr population (brown line; Quilot et al. 2004); an almond F_1: R1000 x DesmayoLargueta population (purple line; Sánchez-Pérez et al. 2007); an apricot F_1: Perfection x A.1740 population (green line; Olukolu et al. 2009) and a peach F_2: Contender x Fla. 92-2C population (orange line; Fan et al. 2010).

Color image of this figure appears in the color plate section at the end of the book.

Another study based on a peach F_2: Akame x Juseitou population identified QTLs on LG3, LG6 and LG7 (Yamamoto et al. 2001). The peach BC_2: *P. davidiana* x *P. persica* cv. Summergrand (S) x S x Zéphyrpopulation detected three other unique QTLs on LG2, LG3 and LG4 (Quilot et al. 2004).

Ripening QTLs in sour cherry based on a F_1: Schattenmorelle x Erdi Botermo population were identified on LG4 and LG6 (Wang et al. 2000), while almond ripening QTLs (Fig. 6-4) were detected on LG4 and LG5 (Sánchez-Pérez et al. 2007). The fruit dropping time was detected on LG3, LG6 and LG7 in Yamamoto et al. (2001) and fruit development period on LG4 (Fig. 6-4) by Etienne et al. (2002).

6.7 Disease Resistance

Although, various stone fruit species are bred for resistance to different diseases that are of major economic importance to them, the plum pox virus (PPV) is the most studied (in peaches and apricots) and devastating among diseases affecting all stone fruit tree species including peaches, almonds, apricots, plums, sweet cherries and sour cherries. QTLs for powdery mildew disease and peach short life (PTSL) are also exclusively studied in peach, although they adversely impact production in some other stone fruit species such as sour cherry and apricot.

PPV, also known as sharka disease, is a serious viral disease caused by a *Potyvirus* (family *Potyviridae*) and transmitted by aphids. Over the past 20 to 30 years, the disease has spread rapidly worldwide through transfer of plant material (Atanasof 1932). Once trees are infected, the only way to eradicate the disease is by destroying the trees, hence, resulting in significant long-term economic losses. Infection leads to as much as 90–100% yield losses and ruins the marketability of the fruits because it causes acidity and deformities of fruits (Kegler et al. 1998; Water worth and Hadidi 1998; Hadidi and Candresse 1999, 2001). The disease has been reported to lead to worldwide economic losses estimated at over 10 billion Euros (Cambra et al. 2006).

Of the six strains (Candresse and Cambra 2006), PPV-C (cherry), PPV-D (Dideron), PPV-EA (El Amar), PPV-M (Marcus), PPV-Rec (a natural recombinant between PPV-D and PPV-M), PPV-W (Winona), PPV-M is the most aggressive in peach and the only strain known to be transmitted through seeds. The PPV-M and PPV-D strains are virulent in peaches, apricots and plums, while the PPV-C strain is the only strain known to cause disease in sweet and sour cherry. Studying the genetic basis of resistance has been difficult due to the available methods for evaluation, which are expensive, tedious and time-consuming. Also, the discrete assignment of phenotypes to susceptibility/resistance classes is not reliable (Dondini et al. 2011).

Studies on PPV resistance have proposed a monogenic (Dicenta et al. 2000), digenic (Moustafa et al. 2001) and oligenic (Guillet-Bellanger and Audergon 2001) mode of inheritance but QTL studies with sources of resistance from apricot cultivars (Lambert et al. 2007) and the Chinese wild peach species, *P. davidiana* (Decroocq et al. 2005; Marandel et al. 2009; Rubio et al. 2010) suggest a polygenic mode of inheritance. Despite several studies conducted, the genetic basis underlying resistance to PPV still remains unclear.

The powdery mildew disease, caused by an obligate pathogenic fungus, has been studied in several crop species. It is a disease with an apparent symptom that displays circular white powdery spots on leaves and stems that may induce necrosis and cause premature fall that leads to serious yield losses. In stone fruits, this disease is mostly common on cherries and peaches, although more serious on sour cherries than on sweet cherries. It is caused by *Podosphaeraclandestina* in cherry and *Sphaerothecapannosa* (Wallr.) var. *persicae* in peach. It results in uneven ripening of fruits in sour cherry, thus, inability to employ mechanized harvesting. It can also infect fruits and petioles of sweet cherry, leading to poor fruit quality.

Resistance to powdery mildew in several species has described complete monogenic or digenic mode of inheritance (Knight and Alston 1968; Mccreight et al. 1987; Lundqvist et al. 1991; Lohnes and Bernard 1992; Reddy et al. 1994). Nevertheless, polygenic partial resistance has also been reported in barley (Saghai-Maroof et al. 1984; Heun 1992) and rye (Miedaner et al. 1993). In peach (Saunier 1973; Tsukanova et al. 1982; Dabov 1983), attempts have been made to employ a simple Mendelian mode of inheritance for breeding resistance to powdery mildew but the narrow genetic base necessitates searching for resistance from wild related species that are highly resistant (Smykov et al. 1982). Consequently, studies based on interspecific cross with *P. davidiana* now suggest a polygenic mode of inheritance for resistance to powdery mildew (Dirlewanger et al. 1996; Foulongne et al. 2003b).

The PTSL syndrome results in a sudden spring collapse and death of young peach trees at ages between three and seven years. The limitations around studying the genetic factors underlying the disease can be attributed to the fact that the tree survival or death is not obvious until three to five years (Blenda et al. 2007). The symptoms associated with the disease include wilting during or shortly after bloom; shriveling of one year-old wood; death and souring of the bark; and tree dying back to the bud union or soil surface (Ritchie and Clayton 1981; Sherman and Andersen 1994). Suckers are usually produced in late spring to summer, while the primary root system persists (Sherman and Andersen 1994).

The primary biotic factor that initiates the disease is the ectoparasitic feeding habit of the ring worm, *Criconemoidesxenopla* x Raski synonymous

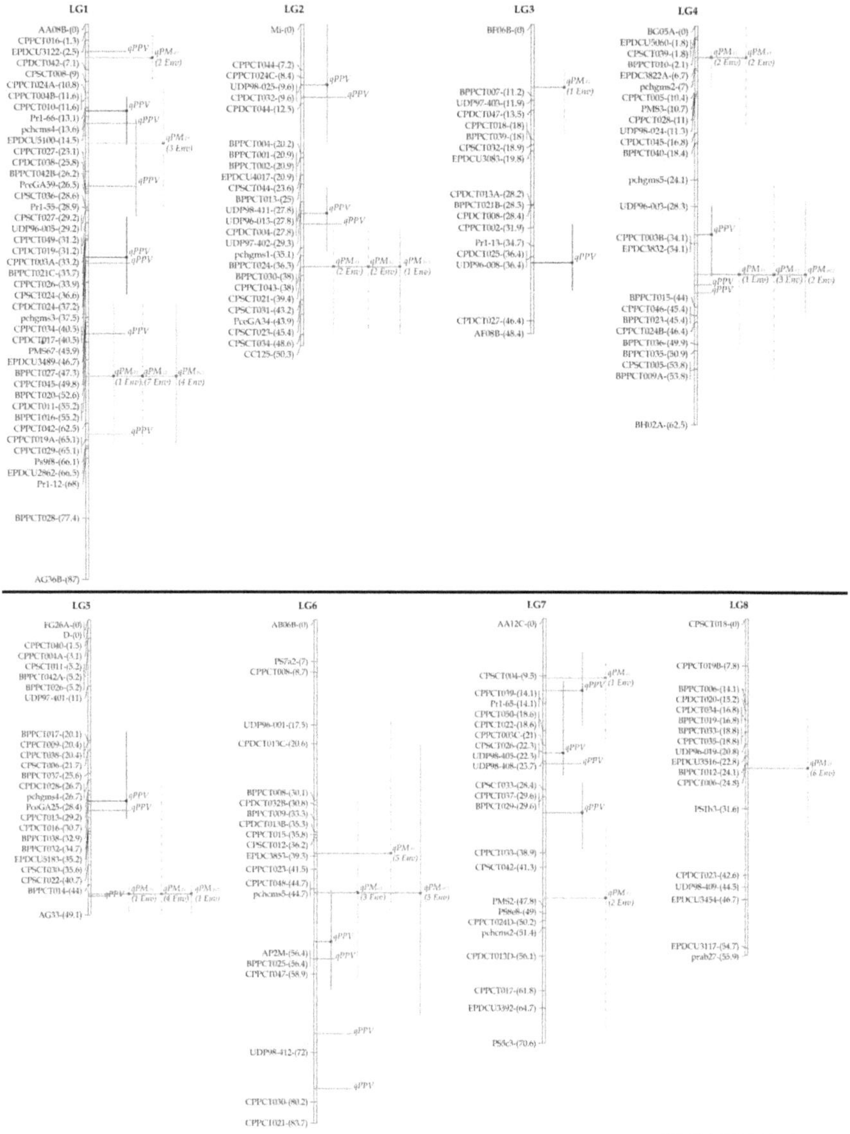

Figure 6-5 contd....

Figure 6-5 Disease resistance QTLs, plum pox virus (PPV) and powdery mildew (PM), mapped in various mapping populations and studies shown localized on the framework of the *Prunus* reference map (Joobeur et al. 1998). Approximate QTL positions and confidence/support intervals on the reference map are inferred based on common transferable SSR markers at and that flank the QTL peak. QTLs detected in multiple environments (Env) indicated in parenthesis. QTLs from 7 different studies indicated by colored lines include an apricot F_1: Goldrich x Valenciano population (brown line; Hurtado et al. 2002); 3 peach (F_1, F_2, BC_2: *P. davidiana*clone P1908 × *P. persica*var. Summergrand® clone S3971) populations (even dashed,

to *Mesocriconemaxenoplax* (Raski) Loof and de Grisse (Nyczepir and Wood 1988), that predisposes the plant to infection by a bacterial canker caused by *Pseudomonas syringae* pv. *syringae* and/or cold injury (CI) (Dowler and Petersen 1966; Sherman and Andersen 1994). The fungal infection of *Cytospora* canker, *Cytosporaleucostoma* or *C. cincta*, follows and ensures the death of the trees. The disease severity is affected by a complex of these biotic factors as well as abiotic factors that include rootstocks that enhance or reduce incidence of PTSL (Lownsbery et al. 1977; Brittain and Miller 1978; Dozier et al. 1984; Yadava and Doud 1989), time of pruning (Carter 1978), cultural practices (Devay et al. 1967; Daniell 1973; Nesmith and Dowler 1976) and rapid fluctuation in late/early spring temperatures.

Genetically improving cultivars for resistance to the PTSL syndrome was unknown until a rootstock, Guardian®'BY520-9', was shown to provide tolerance. Scion cultivars have also been shown to influence the survival of trees in a comparable way to the rootstock, thus, indicating that resistance to PTSL can also be strongly conferred by the scion (Dozier et al. 1984; Beckman et al. 1994; Sherman and Andersen 1994).

6.7.1 Plum-Pox Virus

QTL studies conducted to understand the genetic basis of PPV resistance are mostly based on crosses using apricot cultivars (Lambert et al. 2007; Soriano et al. 2008) and *P. persica* x *P. davidiana* (Decroocq et al. 2005; Marandel et al. 2009; Rubio et al. 2010) segregating for variable disease response (Table 6-2). Although, most of these studies suggest a polygenic mode of inheritance, some other studies suggest a monogenic (Hurtado et al. 2002; Lalli et al. 2008; Pilarova et al. 2010; Dondini et al. 2011) or digenic (Soriano et al. 2008) mode of inheritance with a major QTL on the top half of LG1. The proposed major QTL on LG1, mostly from studies in apricot, has prompted interest in saturating the QTL window for map-based cloning of a candidate gene to further characterize the genetic mechanism underlying the trait (Ruiz et al. 2011).

Based on transferable SSR markers close to and flanking QTL peaks, the QTLs from six PPV studies were localized on the *Prunus* reference map (Fig. 6-5). Several QTL regions were consistently mapped in different

Figure 6-5 contd....

uneven dashed and solid orange line, respectively; Foulonge et al. 2003a); an apricot F_2: Lito x Lito population (green line; Vilanova et al. 2003); 2 peach (F_1, F_2: *P. davidiana* clone P1908 × *P. persica*var. Summergrand® clone S3971) population (blue line; Decroocq et al. 2005);an apricot F_1: Polaris x Stark Early Orange population (black line; Lambert et al. 2007); a peach F_2: *P. davidiana*clone P1908 × *P. persica*var. Summergrand® clone S3971population (purple line; Marandel et al 2009) and a peach F_1: *P. persica* cv Rubira x *P. davidiana* P1908 population (red line; Rubio et al. 2010).

Color image of this figure appears in the color plate section at the end of the book.

studies with overlapping QTL intervals on LG1, LG2, LG4, LG5, LG6 and LG7. These overlapping QTLs were observed to be consistent in 2 to 3 independent studies as shown in Fig. 6-5. Most of the QTLs were validated in independent studies.

6.7.2 Powdery Mildew

Based on an interspecific cross between *P. persica* var. Summergrand and *P. davidiana*to generate F_1, F_2 and BC_2 populations, two independent studies were conducted over several years and environments (Dirlewanger et al. 1996; Foulongne et al. 2003b). The more recent study (Foulongne et al. 2003b) included all three populations for the study and comprised several more transferable SSR markers. Common markers among all the three populations (F_1, F_2 and BC_1) facilitated map alignment and comparison of QTL positions. The random amplified polymorphic DNA (RAPD) markers used in the earlier study (Dirlewanger et al. 1996) are not transferable, hence, limiting comparison with the three latter maps (Foulongne et al. 2003b). Additionally, the 10 linkage groups on the earlier RAPD-based map reflect that eight *Prunus* linkages corresponding to the eight chromosomes were not well represented in a contiguous manner, thus further limiting the ability to compare QTLs detected from these independent studies.

Based on overlapping QTL intervals, QTLs on LG1, LG2, LG4, LG5 and LG6 were detected consistently in all three mapping populations (Foulongne et al. 2003b). Several of these QTLs were also detected in multiple environments (Fig. 6-5). Interestingly, some of the powdery mildew QTLs were observed to localize within close proximity of the PPV QTLs (Fig. 6-5), hence, suggesting the possibility of multiple disease resistance (MDR) genes.

6.7.3 Short Life Syndrome

Following a bulked segregant analysis (BSA) approach based on 151 amplified fragment length polymorphism (ALFP) markers and 21 *Prunus* reference map-anchored SSR markers, QTLs underlying the PSTL syndrome were localized to the *Prunus* LG1, LG2, LG4, LG5 and LG6. A total of 38 of the 140 AFLP markers were shown to be associated with PTSL response (Blenda et al. 2007). The *Prunus* reference map-anchored SSRs provide the ability to identify linkage groups, align them and localize the AFLP markers by flanking SSR markers. The study was based on an F_2 population generated from a cross made between the PTSL-tolerant Guardian® 'BY520-9' (selection 3-17-7) and the PTSL-susceptible Nemaguard cultivar. Based on the map resolution that was generated using 100 individuals and clustering of the associated AFLP markers, 16 QTLs could be inferred from this study.

The study also attempted to follow a candidate gene approach to suggest plausible candidate genes that comprised resistance gene analogs (RGA), nucleotide binding site-leucine rich repeat (NBS-LRR) putative genes and a eukaryotic translation initiation factor 4E.

6.8 QTL Mapping Software Used

The strategies used for mapping genomic regions controlling complex traits following the bi-parental linkage analysis include the bulk segregant analysis (BSA), analysis of variance (ANOVA), Kruskal-Wallis test, simple interval mapping (SIM), composite interval mapping (CIM) and multiple interval mapping (MIM). These methods vary in the amount of statistical power to detect QTLs and QTL interactions as well as their ability to define confidence or support intervals and to estimate genetic effects. The software used for implementing these methods in stone fruits species are listed in Table 6-3. The approaches that are based on the maximum likelihood algorithm for software such as MAPMAKER/QTL and MAPMAKER are best suited for data with a normal distribution, while those based on a stepwise multiple regression analysis such as QTL CARTOGRAPHER and PlabQTL are more robust for data with non-normal distribution. Other studies (Decroocq et al. 2005; Lambert et al. 2007; Zhang et al. 2010) have also implemented the non-parametric Kruskal-Wallis (KW) test (Kruglyak and Lander 1995; Asins 2002).

6.9 Conclusions and Perspectives

The utility of QTLs underlying complex trait for marker-assisted breeding (MAB) is limited by map resolution resulting in fairly large genomic regions. This limits the ability to identify tightly linked or functional markers for precise breeding following a marker-assisted selection (MAS) approach. Studies in other crops have performed fine mapping by increasing the number of recombination events following generation of several thousand progenies in a mapping population. This strategy is impracticable in crops with large growth habit, such as the trees of stone fruit species, which require long generation time and large areas of land for planting.

Researchers are now turning to linkage disequilibrium (LD)-based genome-wide association study (GWAS), which exploits numerous ancient recombination events available in a diverse germplasm. The diversity observed in the germplasm of most *Prunus* species makes them a good candidate for this type of genetic analysis. Compared to the high power achievable in linkage analysis following composite and multiple interval mapping, the association mapping strategy lacks as much power, hence, it is only complimentary to linkage analysis where resolution is lacking. The fine

Table 6-3 QTL mapping software and methods implemented for various studies.

Cross	Reference	Method	Software
F_1, F_2, BC_2: *P. persica* var. Summergrand x *P. davidiana* (peach)	Dirlewanger et al. 1996	SIM	MAPMAKER/QTL
F_1: FerjalouJalousia x Fantasia (peach)	Dirlewanger et al. 1999	SIM	MAPMAKER/QTL, QGENE
F_1: Schattenmorelle x ErdiBotermo (sour cherry)	Wang et al. 2000	CIM	QTL CARTOGRAPHER
F_2: Akame x Juseitou (peach)	Yamamoto et al. 2001	SIM	MAPMAKER/QTL
F_1: FerjalouJalousia x Fantasia (peach)	Etienne et al. 2002	SIM, CIM	MAPMAKER/QTL, QTL CARTOGRAPHER
F_1:Goldrich x Valenciano(apricot)	Hurtado et al. 2002	SIM, CIM	QTL CARTOGRAPHER
F_1, F_2, BC_2: *P. persica* var. Summergrand x *P. davidiana* (peach)	Foulongne et al. 2003a	SIM, CIM	QTL CARTOGRAPHER
BC_2: (*P. davidiana* x *P. persica* cv. Summergrand (S)) x S x Zéphyr	Quilot et al. 2004	SIM	S-plus
F_1, F_2: *P. persica* x *P. davidiana* (peach)	Decroocq et al. 2005	KW, CIM	S-plus, SAS, QTL CARTOGRAPHER
F_1: Polaris x Stark Early Orange (apricot)	Lambert et al. 2007	KW, SIM, CIM	MAPQTL, QTL CARTOGRAPHER
F_1: R1000 x DesmayoLargueta (almond)	Sánchez-Pérez et al. 2007	SIM	MAPQTL
F_1: Goldrich x Currot (apricot)	Soriano et al. 2008	KW	MAPQTL
F_2: Lito x Lito (apricot)	Soriano et al. 2008	KW	MAPQTL
F_2: Summergrand x *P. davidiana* (peach)	Marandel et al. 2009	KW, CIM	MAPQTL
F_1: Perfection x A. 1740 (apricot)	Olukolu et al. 2009	CIM, MIM	PlabQTL
F_2: Contender x Fla. 92-2C (peach)	Fan et al. 2010	CIM	PlabQTL
F_1: *P. persica* cv. Rubira x *P. davidiana* (peach)	Rubio et al. 2010	KW, CIM	MAPQTL, QTL CARTOGRAPHER
F_1: Emperor Francis x New York 54 (sweet cherry)	Sooriyapathirana et al. 2010	KW, SIM, MQM	MAPQTL
F_1: Emperor Francis x New York 54 (sweet cherry)	Zhang et al. 2010	KW, SIM, MQM	MAPQTL
F_1: Lito x BO81604311 (apricot)	Cervellati et al. 2011	KW, SIM	MAPQTL

Note: SIM = simple interval mapping; CIM = composite interval mapping; MIM = multiple interval mapping, MQM = multiple QTL mapping

mapping achievable using association analysis offers the ability to directly detect causative genes in genomes with very low LD. The limitations in the ability to generate and saturate the genome with markers have now given way to new technologies such as genotyping by genome-wide sequencing. Additionally, a single well selected association mapping population panel can offer the ability to identify loci underlying multiple traits as well as detecting several alleles that are not available in bi-parental crosses.

Preliminary studies have been conducted in peach to estimate LD (Aranzana et al. 2010) and in apricot for a local LD-based association mapping (Olukolu 2010). As expected, the less diverse peach germplasm revealed LD extending over large genomic regions, thus rendering it a poor choice for the fine-scale association mapping strategy. A preliminary local LD analysis within the *EVG* region and the overlapping CR QTL of apricot revealed low LD as well as detection of markers significantly associated to the CR trait. Based on this result in apricot and the comparable level of variability in other *Prunus* germplasms, stone fruit species can be ideal for GWAS. Subsequently, the utility of MAB for complex traits with several underlying loci will gradually give way to genomics-based breeding for faster and precise breeding following identification of several loci.

Other major advancements now facilitating the study of complex traits in an unprecedented manner include faster and cheaper whole genome sequencing and genotyping-by-sequencing (GBS) technologies, a well assembled peach genome and the development of fast-track breeding strategies. Mapping resolution in stone fruits has always been limited by population size and marker density but with the advent of cheap genotyping technologies such as GBS, any stone fruit genome can be densely saturated with markers. This is in contrast to the more tedious and slow approaches for marker development that requires prior sequence information. The requirement of an assembled genome for LD-based association mapping (fine mapping strategy that exploits numerous recombination events in a diverse germplasm) has now been attained with the sequenced and asembled peach genome serving as a reference for other *Prunus* species. Recent studies based on shot-gun Illumina whole genome sequencing and GBS of several genotypes following a bulked segregant analysis strategy (broomy/pillar trait in peach, Dardick 2012; PPV resistance in apricot, T. Zhebentyayeva 2011, pers. Comm.) validate a high conservation of synteny after mapping assembled contigs spanning regions of the apricot genome to the assembled peach genome.

Furthermore, fast-track breeding in stone fruits such as in plum offers an attractive approach for validating QTLs and candidate genes as well as for developing superior cultivars following short generation cycles, previously a major limiting factor in breeding tree species. At present, the availability of several resources and technologies hold great promise for crop improvement in stone fruits. Nevertheless, understanding QTL interactions in a mechanistic way and how they affect expression of complex traits in a predictive manner will still require ingenuity.

References

Abdelrazik M (1989) A model of the productivity of olive trees under optional water and nutrient Supply in desert conditions. Ecol Modell 45: 179–204.

Aranzana MJ, Abbassi EK, Howad W, Arus P (2010) Genetic variation, population structure and linkage disequilibrium in peach commercial varieties. BMC Genetics 11.

Aranzana MJ, Carbo J, Arus P (2003) Using amplified fragment-length polymorphisms (AFLPs) to identify peach cultivars. J Am Soc Hort Sci 128: 672–677.

Asins MJ (2002) Present and future of quantitative trait locus analysis in plant breeding. Plant Breed 121: 281–291.

Atanasof D (1932) Plum Pox. A New Virus Disease. Yearbook University of Sofia, Faculty of Agriculture and Silviculture.

Barritt BH (2001) Apple quality for consumers. Intl Dwarf Fruit Tree Assoc, pp 54–46.

Bassi D, Muzzi E, Negri P, Selli R (1990) A computer program for record keeping and seedling evaluation in fruit trees. Hort Science 25: 121–121.

Baumgärtner J, Graf B, Zahner P (1984) A stochastic population model to simulate the annual growth pattern of mature Golden Delicious apple tree. Schweizerische Landwirtschaftliche Forschung 23: 489–501.

Beckman TG, Okie WR, Nyczepir AP (1994) Influence of scion cultivar on incidence of peach tree short life. Proc 6th Stone Fruit Tree Decline Workshop on New Insights and Alternative Management Strategies, Oct 26–28, 1992, Fort Valley, Georgia, pp 45–48.

Bermudez L, Urias U, Milstein D, Kamenetzky L, Asis R, Fernie AR, Van Sluys MA, Carrari F, Rossi M (2008) A candidate gene survey of quantitative trait loci affecting chemical composition in tomato fruit. J Exp Bot 59: 2875–2890.

Bernacchi D, Beck-Bunn T, Emmatty D, Eshed Y, Inai S, Lopez J, Petiard V, Sayama H, Uhlig J, Zamir D, Tanksley S (1998a) Advanced back-cross QTL analysis of tomato. II. Evaluation of near-isogenic lines carrying single-donor introgressions for desirable wild QTL-alleles derived from *Lycopersicon hirsutum* and *L. pimpinellifolium*. Theor Appl Genet 97: 1191–1196.

Bernacchi D, Beck-Bunn T, Eshed Y, Inai S, Lopez J, Petiard V, Sayama H, Uhlig J, Zamir D, Tanksley S (1998b) Advanced backcross QTL analysis of tomato. II. Evaluation of near-isogenic lines carrying single-donor introgressions for desirable wild QTL-alleles derived from *Lycopersicon hirsutum* and *L. pimpinellifolium*. Theor Appl Genet 97: 170–180.

Bernacchi D, Beck-Bunn T, Eshed Y, Lopez J, Petiard V, Uhlig J, Zamir D, Tanksley S (1998c) Advanced backcross QTL analysis in tomato. I. Identification of QTLs for traits of agronomic importance from *Lycopersicon hirsutum*. Theor Appl Genet 97: 381–397.

Bielenberg DG, Wang Y, Fan S, Reighard GL, Scorza R, Abbott AG (2004) A deletion affecting several gene candidates is present in the Ever growing peach mutant. J Hered 95: 436–444.

Blenda AV, Verde I, Georgi LL, Reighard GL, Forrest SD, Munoz-Torres M, Baird WV, Abbott AG (2007) Construction of a genetic linkage map and identification of molecular markers in peach rootstocks for response to peach tree short life syndrome. Tree Genet Genomes 3: 341–350.

Brittain JA, Miller RWJ (eds) (1978) Managing Peach Tree Short Life in the Southeast. Coop Agri Ext Serv Circ, Clemson Univ.

Buwalda JG (1991) A mathematical model of carbon acquisition and utilization by kiwi fruit vines. Ecol Modell 57: 43–64.

Byrne DH, Bacon TA (1990) *Prunus* breeders meeting. Fruit Var J 44: 192–193.

Calderini DF, Slafer GA (1998) Changes in yield and yield stability in wheat during the 20th century. Field Crops Res 57: 335–347.
Cambra M, Capote N, Myrta A, Llacer G (2006) *Plum pox virus* and the estimated costs associated with sharka disease. OEPP/EPPO Bull 36: 202–204.
Candresse T, Cambra M (2006) Causal agent of sharka disease: historical perspective and current status of Plum pox virus strains. EPPO Bull 36: 239–246.
Carter GE (1978) Effect of soil fumigation and pruning date on resumption of growth of vascular cambium of peach trees in a short life site. Hort Science 13: 156–158.
Cassman KG (1999) Ecological intensification of cereal production systems: Yield potential, soil quality, and precision agriculture. Proc Natl Acad Sci USA 96: 5952–5959.
Cassman KG (ed) (2001) Crop Science Research to Assure Food Security. CAB International, New York.
Causse M, Chaib J, Lecomte L, Buret M, Hospital F (2007) Both additivity and epistasis control the genetic variation for fruit quality traits in tomato. Theor Appl Genet 115: 429–442.
Causse M, Duffe P, Gomez MC, Buret M, Damidaux R, Zamir D, Gur A, Chevalier C, Lemaire-Chamley M, Rothan C (2004) A genetic map of candidate genes and QTLs involved in tomato fruit size and composition. J Exp Bot 55: 1671–1685.
Cervellati C, Paetz C, Dondini L, Tartarini S, Bassi D, Schneider B, Masia A (2011) A qNMR approach for bitterness phenotyping and QTL identification in an F_1 apricot progeny. J Biotechnoldoi: 10.1016/j.jbiotec.2011.09.004.
Dabov S (1983) Inheritance of powdery mildew resistance in the peach. IV. Data supporting the hypothesis about the main role of 2 loci controlling the reaction to the pathogen. Genet Sel 16: 349–355.
Daniell JW (1973) Effects of time of pruning on growth and longevity of peach trees.J Am Soc Hort Sci 98: 383–386.
Dardick C(2012) Simultaneous Mapping and Candidate Gene Identification in Peach via Sequencing of Pooled Genomes (Pnomes). In: Plant & Animal Genome 20th Conference, San Diego, California, USA.
Decroocq V, Foulongne M, Lambert P, Le Gall O, Mantin C, Pascal T, Schurdi-Levraud V, Kervella J (2005) Analogues of virus resistance genes map to QTLs for resistance to sharka disease in *Prunus davidiana*. Mol Genet Genom 272: 680–689.
Devay JE, Lownsber BF, English WH, Lembrigh H (1967) Activity of soil fumigants in relation to increased growth response and control of decline and bacterial canker in trees of *Prunus Persica*. Phytopathology 57: 809.
Dicenta F, Martinez-Gomez P, Burgos L, Egea J (2000) Inheritance of resistance to plum pox potyvirus (PPV) in apricot, *Prunus armeniaca*. Plant Breed 119: 161–164.
Dirlewanger E, Graziano E, Joobeur T, Garriga-Caldere F, Cosson P, Howad W, Arus P (2004) Comparative mapping and marker-assisted selection in Rosaceae fruit crops. Proc Natl Acad Sci USA 101: 9891–9896.
Dirlewanger E, Moing A, Rothan C, Svanella L, Pronier V, Guye A, Plomion C, Monet R (1999) Mapping QTLs controlling fruit quality in peach (*Prunus persica* (L.) Batsch). Theor Appl Genet 98: 18–31.
Dirlewanger E, Pascal T, Zuger C, Kervella J (1996) Analysis of molecular markers associated with powdery mildew resistance genes in peach (*Prunus persica* (L.) Batsch) x *Prunus davidiana* hybrids. Theor Appl Genet 93: 909–919.
Dondini L, Lain O, Vendramin V, Rizzo M, Vivoli D, Adami M, Guidarelli M, Gaiotti F, Palmisano F, Bazzoni A, Boscia D, Geuna F, Tartarini S, Negri P, Castellano M, Savino V, Bassi D, Testolin R (2011) Identification of QTL for resistance to plum pox virus strains M and D in Lito and Harcot apricot cultivars. Mol Breed 27: 289–299.
Dowler WM, Petersen DH (1966) Induction of bacterial canker of peach in field. Phytopathology 56: 989–990.
Dozier WA, Knowles JW, Carlton CC, Rom RC, Arrington EH, Wehunt EJ, Yadava UL, Doud SL, Ritchie DF, Clayton CN, Zehr EI, Gambrell CE, Britton JA, Lockwood DW (1984) Survival, growth, and yield of peach trees as affected by rootstock. Hort Science 19: 26–30.

Elia L (2002) Producing a profitable peach. Fruit Grower, June 6.
Etienne C, Rothan C, Moing A, Plomion C, Bodenes C, Svanella-Dumas L, Cosson P, Pronier V, Monet R, Dirlewanger E (2002) Candidate genes and QTLs for sugar and organic acid content in peach [*Prunus persica* (L.) Batsch]. Theor Appl Genet 105: 145–159.
Fan S, Bielenberg DG, Zhebentyayeva TN, Reighard GL, Okie WR, Holland D, Abbott AG (2010) Mapping quantitative trait loci associated with chilling requirement, heat requirement and bloom date in peach (*Prunus persica*). New Phytol 185: 917–930.
Fideghelli C, Della Strada G, Grassi F, Morico G (1998) The peach industry in the world: Present situation and trend. Acta Hort 465: 29–40.
Foulongne M, Pascal T, Arus P, Kervella J (2003a) The potential of *Prunus davidiana* for introgression into peach [*Prunus persica* (L.) Batsch] assessed by comparative mapping. Theor Appl Genet 107: 227–238.
Foulongne M, Pascal T, Pfeiffer F, Kervella J (2003b) QTLs for powdery mildew resistance in peach x *Prunus davidiana* crosses: consistency across generations and environments. Mol Breed 12: 33–50.
Genard M, Bruchou C (1992) Multivariate analysis of within-tree factors accounting for the variation of peach fruit quality. Sci Hort 52: 37–51.
Génard M, Lescourret F (eds) (2004) Modeling Fruit Quality: Ecophysiological, Agronomical and Ecological Perspectives. Kluwer Academic Press, Dordrecht, Netherlands.
Granger AR, Clarke GR, Jackson JF (1993) Sweet cherry cultivar identification by leaf isozyme polymorphism. Theor Appl Genet 86: 458–464.
Grasselly C, Gall H (1967) Étude sur la possibilité de combinaison de quelques charactères agronomiques chez l'amandier. Ann Amélior Plant 17: 83–91.
Grossman YL, Dejong TM (1994) Peach—a simulation model of reproductive and vegetative growth in peach trees. Tree Physiol 14: 329–345.
Guillet-Bellanger I, Audergon JM (2001) Inheritance of the stark early orange apricot cultivar resistance to *Plum pox virus*. Acta Hort 550: 111–115.
Gutierrez AP, Williams DW, Kido H (1985) A model of grape growth and development—the mathematical structure and biological considerations. Crop Sci 25: 721–728.
Hadidi A, Candresse T (1999) Data sheet on plum pox virus. Global Crop Prot Compendium, CAB international, Wallingford, UK.
Hadidi A, Candresse T (2001) Plum pox. In: Maloy OC, Murray TD (eds) Encyclopedia of Plant Pathology. John Wiley, New York, pp 788–791.
Heun M (1992) Mapping quantitative powdery mildew resistance of barley using a restriction fragment length polymorphism map. Genome 35: 1019–1025.
Heuvelink E, Bertin N (1994) Dry-matter partitioning in a tomato crop—Comparison of 2 simulation models. J Hort Sci 69: 885–903.
Hurtado MA, Romero C, Vilanova S, Abbott AG, Llacer G, Badenes ML (2002) Genetic linkage maps of two apricot cultivars (*Prunus armeniaca* L.), and mapping of PPV (sharka) resistance. Theor Appl Genet 105: 182–191.
Illa E, Eduardo I, Audergon JM, Barale F, Dirlewanger E, Li X, Moing A, Lambert P, Le Dantec L, Gao Z, Poëssel JL, Pozzi C, Rossini L, Vecchietti A, Arús P, Howad W (2011) Saturating the *Prunus* (stone fruits) genome with candidate genes for fruit quality. Mol Breed 28: 667–682.
Joobeur T, Viruel MA, de Vicente MC, Jauregui B, Ballester J, Dettori MT, Verde I, Truco MJ, Messeguer R, Batlle I, Quarta R, Dirlewanger E, Arus P (1998) Construction of a saturated linkage map for *Prunus* using an almond x peach F_2 progeny. Theor Appl Genet 97: 1034–1041.
Kegler H, Fuchs E, Gruntzig M, Schwarz S (1998) Some results of 50 years of research on the resistance to plum pox virus. Acta Virol 42: 200–215.
Kester DE (1965) Inheritance of time of bloom in certain progenies of almond. J Am Soc Hort Sci 87: 214–221.
Knight RL, Alston FH (1968) Sources of field immunity to mildew (*Podosphaera Leucotricha*) in apple. Can J Genet Cytol 10: 294–298.

Kruglyak L, Lander ES (1995) A nonparametric approach for mapping quantitative trait loci. Genetics 139: 1421–1428.
Kupferman E (2002) Critical aspects of harvest and quality management: http://postharvest.tfrec.wsu.edu/EMK2002A.pdf.
Lalli DA, Abbott AG, Zhebentyayeva TN, Badenes ML, Damsteegt V, Polak J, Krska B, Salava J (2008) A genetic linkage map for an apricot (*Prunus armeniaca* L.) BC(1) population mapping plum pox virus resistance. Tree Genet Genomes 4: 481–493.
Lambert P, Dicenta F, Rubio M, Audergon JM (2007) QTL analysis of resistance to sharka disease in the apricot (*Prunus armeniaca* L.) Polonais' x 'Stark Early Orange' F1 progeny. Tree Genet Genomes 3: 299–309.
Lang GA, Early JD, Martin GC, Darnell RL (1987) Endodormancy, paradormancy, and ecodormancy—Physiological terminology and classification for dormancy research. Hort Science 22: 371–377.
Lescourret F, Genard M (2005) A virtual peach fruit model simulating changes in fruit quality during the final stage of fruit growth. Tree Physiol 25: 1303–1315.
Lohnes DG, Bernard RL (1992) Inheritance of resistance to powdery mildew in soybeans. Plant Dis 76: 964–965.
Lownsbery BF, English H, Noel GR, Schick FJ (1977) Influence of Nemaguard and Lovell rootstocks and *Macroposthonia Xenoplax* on bacterial canker of peach. J Nematol 9: 221–224.
Lundqvist U, Meyer J, Lundqvist A (1991) Mutagen specificity for 71 lines resistant to barley powdery mildew race D1 and isolated in four highbred barley varieties. Hereditas 115: 227–239.
Marandel G, Salava J, Abbott A, Candresse T, Decroocq V (2009) Quantitative trait loci meta-analysis of Plum pox virus resistance in apricot (*Prunus armeniaca* L.): new insights on the organization and the identification of genomic resistance factors. Mol Plant Pathol 10: 347–360.
Mccreight JD, Pitrat M, Thomas CE, Kishaba AN, Bohn GW (1987) Powdery mildew resistance genes in muskmelon.J Am Soc Hort Sci 112: 156–160.
Meredith FI, Robertson JA, Horvat RJ (1989) Changes in physical and chemical parameters associated with quality and postharvest ripening of harvester peaches. J Agri Food Chem 37: 1210–1214.
Miedaner T, Schmidt HK, Geiger HH (1993) Components of variation for quantitative adult-plant resistance to powdery mildew in winter rye. Phytopathology 83: 1071–1075.
Mnejja M, Garcia-Mas J, Audergon JM, Arus P (2010) *Prunus* microsatellite marker transferability across rosaceous crops. Tree Genet Genomes 6: 689–700.
Moing A, Svanella L, Rolin D, Gaudillere M, Gaudillere JP, Monet R (1998a) Compositional changes during the fruit development of two peach cultivars differing in juice acidity. J Am Soc Hort Sci 123: 770–775.
Moing A, Svanella L, Rothan C, Just D, Rolin D, Monet R, Diakou P, Gaudillere JP (1998b) Organic acid metabolism during the fruit development of two peach cultivars. Acta Hort 465: 425–432.
Moncada P, Martinez CP, Borrero J, Chatel M, Gauch H, Guimaraes E, Tohme J, McCouch SR (2001) Quantitative trait loci for yield and yield components in an *Oryza sativa* x *Oryza rufipogon* BC2 F_2 population evaluated in an upland environment. Theor Appl Genet 102: 41–52.
Moustafa TA, Badenes ML, Martinez-Calvo J, Llacer G (2001) Determination of resistance to sharka (plum pox) virus in apricot. Sci Hort 91: 59–70.
Nesmith WC, Dowler WM (1976) Cultural practices affect cold hardiness and peach tree short life.J Am Soc Hort Sci 101: 116–119.
Nyczepir AP, Wood BW (1988) Peach leaf senescence delayed by *Criconemella Xenoplax*. J Nematol 20: 585–589.
Ogundiwin EA, Peace CP, Gradziel TM, Parfitt DE, Bliss FA, Crisosto CH (2009) A fruit quality gene map of *Prunus*. BMC Genomics 10.

Olukolu BA (2010) The genetics of chilling requirement in apricot (*Prunus armeniaca* L.). Dept of Genetics and Biochemistry. Clemson Univ, USA.

Olukolu BA, Trainin T, Fan SH, Kole C, Bielenberg DG, Reighard GL, Abbott AG, Holland D (2009) Genetic linkage mapping for molecular dissection of chilling requirement and bud-break in apricot (*Prunus armeniaca* L.). Genome 52: 819–828.

Pangborn RM (1963) Relative taste intensities of selected sugars and organic acids. J Food Sci 28: 726–733.

Pilarova P, Marandel G, Decroocq V, Salava J, Krska B, Abbott AG (2010) Quantitative trait analysis of resistance to plum pox virus in the apricot F1 progeny "Harlayne" x "Vestar". Tree Genet. Genomes 6: 467–475.

Pillen K, Zacharias A, Leon J (2003) Advanced backcross QTL analysis in barley (*Hordeumvulgare* L.). Theor Appl Genet 107: 340–352.

Quilot B, Wu BH, Kervella J, Genard M, Foulongne M, Moreau K (2004) QTL analysis of quality traits in an advanced backcross between *Prunus persica* cultivars and the wild relative species *P. davidiana*. Theor Appl Genet 109: 884–897.

Reddy KS, Pawar SE, Bhatia CR (1994) Inheritance of powdery mildew (*Erysiphe Polygoni* DC) resistance in mungbean (*Vigna Radiata* L Wilczek). Theor Appl Genet 88: 945–948.

Rehder A (ed) (1940) Manual of Cultivated Trees and Shrubs Hardy in North America. 2nd edn. New York: Macmillan, 996 p.

Ritchie DF, Clayton CN (1981) Peach-tree short life—a complex of interacting factors. Plant Dis 65: 462–469.

Robertson JA, Meredith FI, Russell RB, Scorza R (1989) Physical, chemical and sensory evaluation of high- and low-quality peaches. Acta Hort 254: 155–159.

Rubio M, Pascal T, Bachellez A, Lambert P (2010) Quantitative trait loci analysis of Plum pox virus resistance in *Prunus davidiana* P1908: new insights on the organization of genomic resistance regions. Tree Genet Genomes 6: 291–304.

Ruiz EMV, Soriano JM, Romero C, Zhebentyayeva T, Terol J, Zuriaga E, Llacer G, Abbott AG, Badenes ML (2011) Narrowing down the apricot Plum pox virus resistance locus and comparative analysis with the peach genome syntenic region. Mol Plant Pathol 12: 535–547.

Saghai-Maroof MA, Soliman KM, Jorgensen RA, Allard RW (1984) Ribosomal DNA spacer length polymorphism in barley: Mendelian inheritance, chromosomal location and population dynamics. Proc Natl Acad Sci USA 81: 8014–8018.

Saunier R (1973) Contribution to study of relationships between certain characteristics of simple genetic determination in peach tree and susceptibility of peach cultivars to Oidium, *Sphaerotheca Pannosa* (Wallr) Lev (in French). Ann Amelior Plantes 23: 235–243.

Seem RC, Elfving DC, Oren TR, Eisensmith SP (1986) A carbon balance model for apple tree growth and production. Acta Hort 184: 129–137.

Sherman WB, Andersen PC (1994) Observations on peach tree short life in the university of Florida peach breeding program. Proc Fla State Hort Soc 107: 329–331.

Silva C, Garcia-Mas J, Sanchez AM, Arus P, Oliveira M (2005) Looking into flowering time in almond (*Prunus dulcis* (Mill) DA Webb): the candidate gene approach. Theor Appl Genet 110: 959–968.

Smykov VK, Ovcharenko GV, Perfilyeva ZN, Shoeferistov EP (1982) Estimation of the peach hybrid resources by its mildew resistance against the infection background [in Russian]. Bull Gos Nik Bot Sada, pp 74–80.

Sooriyapathirana SS, Khan A, Sebolt AM, Wang DC, Bushakra JM, Lin-Wang K, Allan AC, Gardiner SE, Chagne D, Iezzoni AF (2010) QTL analysis and candidate gene mapping for skin and flesh color in sweet cherry fruit (*Prunus avium* L.). Tree Genet Genomes 6: 821–832.

Soriano JM, Vera-Ruiz EM, Vilanova S, Martinez-Calvo J, Llacer G, Badenes ML, Romero C (2008) Identification and mapping of a locus conferring plum pox virus resistance in two apricot-improved linkage maps. Tree Genet Genomes 4: 391–402.

Struik PC, Yin XY, de Visser P (2005) Complex quality traits: now time to model. Trends Plant Sci 10: 513–516.
Tsukanova ZG, Sokolova SA, Gatina ES, Smykov VK (1982) Inheritance of mildew resistance by peaches [in Russian]. Bull Gos Nik Bot Sada, pp 72–75.
Tucker GA (1993) Introduction. In: Seymour GB, Taylor JE, Tucker GA (eds) Biochemistry of Fruit Ripening. Chapman and Hall, London.
Viruel MA, Madur D, Dirlewanger E, Pascal T, Kervella J (1998) Mapping quantitative trait loci controlling peach leaf curl resistance. Acta Hort 465: 79–87.
Wang D, Karle R, Brettin TS, Iezzoni AF (1998) Genetic linkage map in sour cherry using RFLP markers. Theor Appl Genet 97: 1217–1224.
Wang D, Karle R, Iezzoni AF (2000) QTL analysis of flower and fruit traits in sour cherry. Theor Appl Genet 100: 535–544.
Waterworth HE, Hadidi A (eds) (1998) Economic Losses Due to Plant Viruses. APS Press, St. Paul, MN.
Yadava UL, Doud SL (1989) Rootstock and scion influence growth, productivity, survival, and short life-related performance of peach trees.J Am Soc Hort Sci 114: 875–880.
Yamamoto T, Shimada T, Imai T, Yaegaki H, Haji T, Matsuta N, Yamaguchi M, Hayashi T (2001) Characterization of morphological traits based on a genetic linkage map in peach. Breed Sci 51: 271–278.
Yoshida M (1970) Genetical studies on the fruit quality of peach varieties. Bull Hort Res Sta A9 Japan, pp1–15.
Zhang GR, Sebolt AM, Sooriyapathirana SS, Wang DC, Bink MCAM, Olmstead JW, Iezzoni AF (2010) Fruit size QTL analysis of an F(1) population derived from a cross between a domesticated sweet cherry cultivar and a wild forest sweet cherry. Tree Genet Genomes 6: 25–36.

7

Molecular Breeding

Daniel Esmenjaud[1,*] and *Chinnathambi Srinivasan*[2]

ABSTRACT

Prunus breeding programs face continual challenges to select for cultivars with durable resistance to pests and diseases, with taste- and nutrition-improved characteristics and adaptation to increasing changes in environmental conditions. Historically various molecular fingerprinting techniques mainly RFLPs, RAPDs, SCARs, AFLPs, and SSRs successively contributed for identification of stone fruit germplasm, characterization of its genetic diversity, and assistance in depicting distinctness, uniformity and stability (DUS) of cultivars. AFLPs and SRRs both generate the highest-resolution patterns to identify cultivars in species with a very low polymorphism such as peach. Nevertheless, SSRs, as the most transportable markers across species, are now a preferred technique for marker-assisted breeding (MAB) in stone fruits, particularly peach, apricot and almond. Breeding programs primarily rely on the evaluation of genetic diversity of collection of material before choosing the parental material that will allow to maximize this diversity for subsequent optimal combination. The online release of the peach genome sequence has opened up the way to a huge step forward by giving a direct access to all genomic data and particularly to multiple new SSR loci as an optimal tool for improved germplasm characterization. In *Prunus* spp. a diversification of markers, which is mainly based on gene sequences (SNPs, RGAs, CGAs) and gene expression (ESTs), is now in progress. Current examples of MAB relate to very diverse features including columnar growth habit, self-incompatibility, environmental adaptation, resistance to biotic stresses

[1]Equipe Interactions Plantes-Nématodes, INRA, Centre de recherche Provence-Alpes-Côte d'Azur, UMR ISA (Institut Sophia Agrobiotech), INRA / CNRS / Université de Nice, 400 route des Chappes, BP 167, 06903 Sophia Antipolis Cedex, France.
[2]USDA-ARS Appalachian Fruit Research Station, 2217 Wiltshire Road, Kearneysville, WV 25430, USA.
*Corresponding author: *daniel.esmenjaud@sophia.inra.fr*

such as green peach aphid (*Myzus persicae*), sharka (*Plum pox virus*-PPV), powdery mildew, peach tree short life (PTSL), and root-knot nematodes (RKN). In this context, one of the predominant goals of MAB is to promote gene pyramiding as a guarantee for resistance durability to biotic stresses. A first example is the ongoing breeding strategy for RKN resistance in *Prunus* rootstocks, which is based on pyramiding of the major genes *Ma* (plum), *RMia* (peach) and *RMja* (almond) in order to confer a high-level, wide-spectrum and heat-stable resistance. Another example is the Fruit Breedomics project that focused notably on several genes or QTLs for resistance to peach green aphid and powdery mildew. In the near future, new MAB challenge will be to emphasize on the selection at the genome level and develop genome-wide association mapping (linkage disequilibrium studies) enabling genome-wide surveys of variability based on markers evolving from SSRs to SNPs. Due to the availability of peach genome sequence, molecular breeding will rely more and more on peach as a model species for stone fruits. Investigations aiming at developing transgenic breeding of stone fruits report *Agrobacterium tumefaciens*-mediated transformation of both mature and juvenile tissues. However successful transformation and regeneration of commercial cultivars are still far from routine and are limited to a few genotypes or to seedlings. Thus transgenic breeding efforts focus on solving major bottlenecks, e.g., the recalcitrance to regenerate adventitious shoots from clonal tissues such as leaf explants and the long (3 to 10 years) juvenile period in obtaining transgenic stone fruits. Nevertheless, PPV-resistant transgenic plum has been produced that is now deregulated for commercial planting in USA.

Keywords: Prunus, RFLP, RAPD, AFLP, SSR, gene pyramiding, transgenic breeding

7.1 Introduction

Prunus crops within the Rosaceae globally form a homogeneous entity: they are geographically restricted to Mediterranean and temperate climates (Ramming and Cociu 1991) and comprise three major subgenera *Prunophora* (plums and apricots), *Amygdalus* (peaches and almonds) and *Cerasus* (sweet and sour cherries) (Rehder 1954). All wild species from which they have been domesticated are characterized by a low basic chromosome number of 8 ($2n = 2x = 16$) where *Amygdalus* and *Cerasus* are diploid while *Prunophora* range from diploid (apricots, Japanese and Myrobalan plums) to hexaploid (the domestic and European plums; $2n = 6x = 48$) with intermediate ploidy levels existing within the wild plum species (e.g., *P. spinosa*; $2n = 4x = 32$). Besides this low basic chromosome number, the size of the haploid *Prunus* genome of 280–320 Mb (Arumuganathen and Earle 1991; confirmed by the peach genome sequence released in April 2010) makes it one of the smallest among plants. This situation and coordinated research efforts favored the publication of a saturated map as earlier as 1998 (Joobeur et al.).

Nevertheless as for other perennials, classical breeding is hampered by the long generation cycles and, in particular, the juvenile period of several years is a real obstacle to studies concerning for example fruit yield or self-incompatibility genotyping. Identification and characterization of *Prunus* genotypes classically relies on morphological and physiological traits based on conventional markers, which are not always easy to analyze and may be influenced by environmental conditions. Molecular breeding for such perennials is clearly of the highest importance as it allows to shorten the duration of selection experiments at each plant cycle to the minimum (Rom and Carlson 1987; Ramming and Cociu 1991; Dosba et al. 1994). Linkage maps are the basic tools for molecular breeding. Generation of different types of such maps (reference *Prunus* maps at the genus level and species or cultivar specific maps) has greatly facilitated breeding together with correlative variability, evolutionary and taxonomic studies. This chapter will first consider the contribution of molecular breeding for *Prunus* germplasm identification and characterization purposes by illustrating the molecular techniques that have successively been used in *Prunus*. It will then report the work conducted until now using molecular approaches for marker-assisted breeding (MAB) (strategies and major current results) in order to complete or fasten breeding schemes for an efficient creation of new and original plant material. MAB will mainly focus on physiological traits and on resistance to biotic stresses. Finally this chapter will deliberate on gene pyramiding with examples from ongoing or planned studies and transgenic breeding.

7.2 Germplasm Characterization

Characterization of germplasm is particularly useful to guarantee the identity of a given material at any stage of its production or for any legal process especially for patented materials. In that objective, tools that meet the demand of all people involved in creation or in management of plants to have the *Prunus* accessions characterized unequivocally, have thoroughly evolved with molecular techniques during the last decades. Description of distinctness, uniformity and stability (DUS) of a new variety is needed to state its unique status in comparison to previously registered plant material. Isoenzymes were the first genetic markers used in *Prunus* identification and characterization. These codominant markers have been very useful in outcrossing species showing a high level of polymorphism such as almond and plum but showed their limits for an autogamous species with a low variability such as in peach (Arulsekar et al. 1986; Durham et al. 1987). Fingerprinting genomes of any living organisms with PCR arbitrary primers has been first proposed in the early 90s (Welsh and McClelland 1990; Williams et al. 1990). In *Prunus*, DNA information has then readily replaced data based on isozyme polymorphism—for example peach, apricot, almond, and plum (Byrne 1990), apricot (Battistini and Sansavini 1991; Byrne Littleton 1992) and sweet cherry (Granger et al. 1993).

DNA markers that have provided passport data for many varieties have been developed roughly on a historical scale in the following order: restriction fragment length polymorphisms (RFLPs), random amplified polymorphic DNAs (RAPDs), sequence characterized amplified regions (SCARs), amplified fragment length polymorphisms (AFLPs) and simple sequence repeats (SSRs, also called as microsatellites). Using these markers, the overall objective is to reach the generation of a single profile that will also be useful to assist with breeders' rights issues. In that objective the development of the reliable characterization of cultivated material shed light on the importance of having reference *Prunus* genetic resources readily available (Dosba et al. 1994; Horn et al. 2005). Whatever their origin, any markers will add new information to the existing structure of *Prunus* database generated, very importantly, for conservation and management of European (EPDB http://cbi.labri.fr/outils/EPDB/index.html) and US germplasm collections. In this context, a major interest of DNA markers also consists in the ability they give to users to correct misidentifications and state homonyms and synonyms among plant accessions.

7.2.1 Overview of the Diversity and History of Fingerprinting Techniques

Several fingerprinting techniques have been used mainly for peach, apricot, almond and cherry (Table 7-1). We report hereafter an overview of the diversity and history of RFLPs, RAPDs, SCARs, AFLPs and SSRs.

7.2.1.1 RFLP

RFLPs (Botstein et al. 1980) were the first DNA markers used in plants (Tanksley et al. 1989). Based on the differential hybridization of DNA fragments obtained from restriction enzyme digestion, they are codominant, resolute (unlimited number of markers) and reproducible but very difficult to use routinely because of the high amount of DNA required. In USA the first RFLP *Prunus* maps were published in 1995 (Foolad et al. 1995; Rajapakse et al. 1995). In Europe, through the collaboration of six research groups, a saturated linkage map for *Prunus* using an interspecific F_2 population (namely the TxE map) between almond (cv. Texas) and peach (cv. Earlygold) was constructed with 246 markers (11 isozymes and 235 RFLPs) (Arus et al. 1994; Joobeur et al. 1998). RFLPs were obtained using 213 probes

Table 7-1 Illustration of techniques used for identification and molecular characterization of *Prunus* germplasm.

Crop species	Marker type	Reference
Peach	RAPD	Lu et al. 1996
		Warburton and Bliss 1996
	AFLP	Aranzana et al. 2003a
	SSR	Aranzana et al. 2003b
		Marchese et al. 2005
Apricot	RFLP	De Vicente et al. 1998
	SSR	Hormaza 2002
		Dondini et al. 2007
		Donoso et al. 2008
		Martinez-Mora et al. 2009
		Campoy et al. 2010
Almond	RAPD	Bartolozzi et al. 1998
		Ryan et al. 2001
	SSR	Kadkhodaei et al. 2010
Cherry	SSR and AFLP	Struss et al. 2003
	PCR-RFLP AFLP	Panda et al. 2003
	SSRs	Zhou et al. 2005
	SSRs, isozymes	Dirlewanger et al. 2008
	SSR	Clarke et al. 2009
		Lacis et al. 2009
Diverse species	RAPD	Casas et al. 1999
	AFLP	Aradhya et al. 2004
	SSR and others	Arus 2006
Plum	RAPD	Ortiz et al. 1997

from the genomic and cDNA libraries of different species (almond, peach, *P. ferganensis*, cherry, plum and apple), including 16 almond probes, which corresponded to known genes. All markers were distributed in eight linkage groups (LGs), the same as the basic chromosome number of the genus, covering a total distance of 491 cM. The average map density was 2.0 cM/marker and only four gaps of 10 cM or more were found. This first high density RFLP was considered as the reference *Prunus* map as RFLP markers could be used as anchor loci for comparative mapping with any other *Prunus* species.

7.2.1.2 RAPD

RAPDs, based on PCR amplification of random locations from the genome (Welsh and McClelland 1990; Williams et al. 1990) were the first markers that have been widely used for identification and mapping mainly because they are easy to develop and are cost- and time-effective. First applications were on peach (Lu et al. 1996; Warburton and Bliss 1996) where they confirmed the low genetic diversity of cultivars at the world scale. In plum (Ortiz et al. 1997) hexaploid and diploid plum cultivars showed a high degree of polymorphism and the 31 tested genotypes could be distinguished using single cultivar-specific primers or combined primers. In almond, genetic diversity revealed by RAPDs was found to be more limited than expected despite its need for obligate outcrossing (Bartolozzi et al. 1998). A Spanish study (Casas et al. 1999) mentioned the first use of RAPD markers to characterize *Prunus* rootstocks from different species, both commercial and selected clones, and state intra- and interspecific variations. This screening used RAPD primers of which seven were selected to produce a combined classification of the whole set of rootstock clones belonging to *P. amygdalus* x *P. persica,* and *P. persica* x *P. davidiana* hybrids, *P. cerasifera, P. domestica,* and *P. insititia.* The classification scheme widely used to characterize *Prunus* clones, mainly based on morphological descriptors, was confirmed by the clusters based on this set of seven primers. Using information from supplemental primers, the authors could logically form a first group with separate clusters for *P. amygdalus x P. persica* and for *P. persica x P. davidiana* hybrids, a second group with Myrobalan (*P. cerasifera*) and Marianna (*P. cerasifera* x *P. munsoniana*) plums. Hexaploid plum (*P. insititia* and *P. domestica*) clones formed another group where clones of both species were mixed, an observation that authors attributed to their genetic interchange during domestication and breeding history.

Nevertheless the lack of repeatability of the amplification of some RAPD fragments needed particular caution and required involvement in the selection of stable marker fragments.

7.2.1.3 SCAR

Difficulties in using RAPDs repeatedly have been overcome by generating sequence-characterized amplified regions (SCARs) (Paran and Michelmore 1993). Sequencing the fragment of interest and thus designing reliable and less temperature-sensitive primers also often allows to obtain user-friendly codominant markers (Lecouls et al. 1999).

7.2.1.4 AFLP

Another highly performing tool is AFLP technique (Vos et al. 1995), which is based on the selective amplification of a subset of genomic restriction fragments. However this highly resolute identification technique needs long preparative steps (DNA restrictions and pre-amplifications) that render it particularly delicate for friendly or routine use. Nevertheless fluorochrome labeling of fragments may offer the possibility of automatic readings when coupled with an automated genetic analyzer in a high throughput strategy. As for RAPDs, AFLP informative fragments are generally sequenced and converted into user-friendly codominant SCAR markers (Lecouls et al. 2004). The high potential of AFLPs for cultivar identification has been first stressed in a peach study by Aranzana et al. (2003a). A sample of 210 cultivars with a wide range of fruit and plant characteristics was studied for variability using combinations of nine primer pairs. Forty-seven AFLPs allowed identification of 196 (93%) different genotypes, 187 of which could be distinguished with only three primer combinations. Eleven cultivars with the same AFLP phenotype corresponded to known somatic mutations (sports), but from the four sports of the 'Springcrest' group, two ('Maycrest' and 'Queencrest') differed at three AFLP fragments from the others ('Starcrest' and 'Early Maycrest'). In sweet cherry, a total of 63 (21%) polymorphic fragments were recorded in 15 cultivars using four AFLP primer combinations (Struss et al. 2003). AFLP markers thus generated unique fingerprints for all sweet cherry cultivars. On the same crop, Zhou et al. (2005) could characterize 67 cultivars and selections with six pre-selected pairs of AFLP primers that generated a total of 118 polymorphic marker fragments (i.e., an average of 20 markers per primer pair) and thus produced a unique binary code for each accession.

Another AFLP study within 113 diverse accessions representing seven cultivated and seven wild species of *Prunus* from the three subgenera and four sections described within the genus *Prunus* evidenced for four well-supported groups corresponding to the four sections *Amygdalus*, *Armeniaca*, *Cerasus* and *Prunophora* (Aradhya et al. 2004). The section *Armeniaca* showed significant differentiation from its sister section *Euprunus* within the subgenus *Prunophora*. Within-species estimates of molecular variation ranged as expected in decreasing order from apricots to hexaploid plums, almonds, cherries, diploid plums and peaches.

7.2.1.5 SSR or Microsatellite Technique

This technique appeared in 1989 (Tautz 1989) and was adapted to *Prunus* in the late 1990 to early 2000s (first publication in 1999 by Cipriani et al. in peach) as a revolutionary tool for efficient molecular identification to complement pomological studies, being highly informative and reproducible. Generating SSR primers was costly and time- and space-consuming as it involved the construction of a genomic library (or cDNA library) and included obtaining sequence data, primer design and synthesis together with testing of the selected fragments. Nevertheless these steps could be reduced in *Prunus* given that the same primers were often used for different species. This has been shown in 2001 by Hormaza by using apricot SSR primer pairs designed for peach by Cipriani et al. (1999) as well as at Clemson University by Abbott (cited in Hormaza 2001). Consequently microsatellites, because of their cross-species transportability and their facility of use in comparison with the time consuming AFLPs or the unstable RAPDs, appeared as the markers of choice for cultivar identification. An interesting example was reported by Struss et al. (2003) who, together with the identification of sweet cherry cultivars from their unique SSR fingerprints, opened up the way to a possible application of the technique for quality control of varietal identity and purity using fruits directly collected in the market place. This work also showed the transportability of the SSR technique across *Prunus* species through cross-species amplification: more than 80% of the sweet cherry primer pairs could be amplified in plum (*P. salicina*), apricot (*P. armeniaca*) and peach (*P. persica*), confirming the high intrageneric relationships within *Prunus* species. The transportability within the Rosaceae family was also studied more in details by Mnejja et al. (2010) from 145 microsatellite primer pairs from *Prunus* DNA sequences in a set of eight cultivars from nine Rosaceous species (almond, peach, apricot, Japanese plum, European plum, cherry, apple, pear, and strawberry), ranging into 25 each of almond genomic, peach genomic, peach expressed sequence tags (EST), and Japanese plum genomic, 22 of almond EST, and 23 of apricot (13 EST and 10 genomic). Transferability, i.e., the proportion of microsatellites that amplified and were polymorphic, was also high within *Prunus* (63.9%). These data confirmed that almond and Japanese plum were the most variable species and peach the least polymorphic, a result that Byrne had firstly pointed out as early as 1990 from isoenzyme studies. Thirty-one microsatellites amplified and were polymorphic in all *Prunus* species studied, 12 of which, covering its whole genome, are proposed as the "universal *Prunus* set" completing the previous approach of Aranzana et al. (2003c). In contrast, only 16.3% were transferable in species of other Rosaceae genera (apple, pear, and strawberry). Polymorphic *Prunus* microsatellites also detected lower levels of variability in the non-congeneric species.

In peach, the high economic value of the new cultivars and the low level of variability of the species rendered urgent an effective fingerprinting method (Arus 2006) and SSRs brought an important but not necessarily sufficient contribution in this objective as reported further in this chapter. Another advantage of these markers is that fluorochrome labeling of SSR fragments offers the possibility of a high throughput usage by automatic readings from an automated genetic analyzer.

7.2.1.6 Misidentifications, Homonyms and Synonyms

DNA markers brought highly powerful tools to reveal misidentifications and assist in depicting DUS of cultivars. One of the first examples to illustrate this has been described for the plum rootstock 'Myrobalan 29C' (Casas et al. 1999). From phylogenetic data based on RAPDs, 'Myrobalan 29C' was shown to be situated among *P. cerasifera* x *P. munsoniana* hybrids.

Prunus munsoniana is an American diploid *Prunus* species that may have generated Mariana hybrid rootstocks when hybridized with *P. cerasifera*. Thus Myrobalan 29C, morphologically close to Mariana (Grasselly, cited in Crossa-Reynaud and Audergon 1987), is not the pure Myrobalan plum stated by its name but an hybrid rootstock. This hybrid status is highlighted by the many common RAPD marker bands shared by Myrobalan 29C and Mariana reference rootstocks.

Another major interest of genomic markers for DUS is the detection of homonyms and synonyms, as a powerful tool to go much further than phenotypic observations. For example a combined approach with RAPD and microsatellite markers have been used to identify and characterize 50 apricot cultivars originating from different Spanish geographic areas among which they differentiated putatively cloned accessions of a single cultivar, that were designed as 'Moniqui' and were homonyms (Hormaza 2001).

7.2.2 Illustration of the Use of Fingerprinting Techniques for Germplasm Characterization in Different Prunus Species

Since their first use by Tautz (1989), microsatellites have become the marker of choice for fingerprinting and genetic diversity studies in a wide range of living organisms. Most studies published in *Prunus* in the last five to ten years relate to SSRs and illustrate the generalization of their use in stone fruit crops and particularly in peach, apricot and almond (Table 7-1).

7.2.2.1 Peach

Cultivar identification, marker mutation, and pedigree data have been reported by Aranzana et al. (2003b) with 16 SSRs applied to a collection of 212 peach and nectarine cultivars covering wide species variation. Benefiting from an average of seven alleles per locus, 87% of the cultivars studied could be individually identified. Cultivars with the same SSR fingerprint included known peach mutants. Five unexpected genotypes, due to a mutation at five SSR loci were found when comparing the SSR fingerprint of 14 known mutant cultivars and putative synonymous cultivars. The pedigree data were not consistent with the observed data in 11 out of 38 cases that could be analyzed. This study was completed with more details in 2010 (Aranzana et al.) on 224 peach cultivars using 50 SSRs evenly distributed along the eight LGs of the *Prunus* reference map. The whole set of 50 SSRs used was able to identify 209 out of the 224 cultivars studied. It is particularly interesting to note that ten groups of cultivars (one group with seven cultivars and nine with two cultivars each), that had resulted as genotypically identical with the 16 SSRs of the previous study, still displayed identical genotypes with 50 SSRs, stating that fingerprinting at a higher genotyping scale did not bring additional discriminating information. These data suggest that the differences existing between these genotypes may not be pointed out with neutral genomic markers such as SSRs but rather at the gene level with point mutations affecting the phenotype. The mean observed heterozygosity of 0.34, and the mean *F*-index of 0.26 indicated a very high level of inbreeding. Three non-melting cultivars 'Cofrentes-6' and 'Auberge Blanc', and the nectarine 'Independence' were even homozygous at all of their SSR loci. The distribution of genotype differences between pairs of cultivars showed that only a very small percentage of pairs of cultivars (0.3%), including all groups of known sports (cultivars originated by somatic mutation), differed at six or less loci. Similarly, molecular characterization of diverse indigenous Sicilian peaches, nectarines, and platicarpa (flat peaches) cultivars (Marchese et al. 2005) showed a wide range of fruit quality and ripening periods. Such diverse assortment of local cultivars was a resource, not yet exploited, for breeding programs. When challenged with 15 primer pairs, the 49 cultivars displayed a mean of five alleles per locus and 36 cultivars showed a unique fingerprint. Eleven of the 79 alleles were found only in the 40 Sicilian cultivars and their mean observed heterozygosity of 0.44, and the mean *F*-index of 0.3 indicated a high level of inbreeding (despite being lower than in the studies by Aranzana et al. (2003b, 2010)). Although the 15 SSR markers were useful for detecting homonyms and synonyms, they could not be distinguished in some cases phenotypically because they were derived from different indigenous accessions. Inbreeding, co-ancestry and mutation are plausible explanations for the co-identity of these genotypes with microsatellites.

7.2.2.2 Apricot

Molecular characterization and similarity relationships among apricot genotypes were first conducted by Hormaza (2002) in Spain in a collection of 48 apricot genotypes, originating from diverse geographic areas, with 37 SSR primer pairs recovered in different species of *Prunus*. Twenty of those primer pairs produced repeatable polymorphic amplification patterns with a total of 82 alleles detected. All the genotypes studied could be unequivocally distinguished with the combination of SSRs used. The results obtained confirmed the cross-species transportability of microsatellite sequences, allowing the discrimination among different genotypes of a given fruit-tree species with sequences developed in other species. UPGMA cluster analysis of the similarity data grouped the genotypes studied according to their geographic origin and/or their pedigree information. Genetic relationships in apricot using SSR markers and their implications for breeding were also investigated in France (Hagen et al. 2004) and Chile (Donoso et al. 2008). Hagen et al. (2004) did not observe any difference in allele numbers in

EST- versus genomic-derived loci when they compared the 24 polymorphic microsatellites (13 originated from genomic libraries and 11 from cDNA libraries). Donoso et al. (2008) evaluated 31 apricot cultivars with 11 primer pairs developed in peach and found that seven successfully amplified primer pairs generated a mean of six polymorphic bands, ranging from two to eight alleles. This observation further confirmed the existence of high degree of homology for the SSR loci between peach and apricot. Diversity studies by these later authors showed that cultivars of the American and European groups share common genetic backgrounds, which confirms that apricot breeding has been mainly based on the hybridization and selection from European genotypes.

7.2.2.3 Almond

In this species, diversity of 250 genotypes from different origins distributed throughout Iran together with some foreign reference cultivars have been evaluated with nine SSR primer pairs cloned from almond (Kadkhodaei et al. 2011). Microsatellites exhibited the high mean of 16 (9 to 20) alleles per locus and five loci permitted to distinguish 98% of the genotypes. Cluster analysis displayed two major groups that were the foreign cultivars and the Iranian genotypes. Some specific markers were recorded among the germplasm which can be efficiently used for rapid and precise identification of the related genotypes and also in breeding programs through MAB. These authors defined a coding method for genotype molecular identification of almond genotypes.

7.2.2.4 Concluding Remarks

Examples reported here clearly illustrate the power of the DNA markers and the position of the SSR technique as the today's tool for germplasm identification (Arus 2006). Across *Prunus* species, cross-transportability of SSR primers allows a generic use of the best markers. Several SSR libraries have been constructed primarily on peach and they have provided a wide range of SSR series in various laboratories in Europe, USA and Japan (Cipriani et al. 1999; Sosinski et al. 2000; Testolin et al. 2000; Dirlewanger et al. 2002; Yamamoto et al. 2002; Aranzana et al. 2003c). At least part of these data has been merged in the reference TxE map (Dirlewanger et al. 2004b): this map is available for the selection of markers that cover the entire genome, which greatly increase the probability to use them successfully for the characterization and discrimination of any cultivars of interest. The release of the first online version of the peach genome sequence (cv. Lovell; http://www.rosaceae.org/) in 2010 has permitted to reach a huge step forward to access all genomic data and particularly to multiple new microsatellite loci. It appears that one of the primary valorization of sequence data will be for germplasm characterization through the use of such new markers. It is very likely that microsatellites will remain the optimal tool for this purpose. Nevertheless neutral markers such as SSRs may not be sufficient to identify certain peach cultivars as difficulties due to the low peach polymorphism are not completely solved. In such cases a come back to AFLPs might be a more resolute and efficient identification method because these fragments reveal marker locations at random in the genome, that are indifferently either neutral or placed in DNA regions submitted to selection pressure. Anyhow, in *Prunus,* a diversification of markers which is mainly based on gene sequences (single-nucleotide polymorphisms, SNPs), resistance gene analogs (RGA), cloned gene analogs (CGAs), within-gene markers (allelic diversity) and gene expression (ESTs) is now available. It is expected that the online availability of the peach genome sequence, by facilitating the access to genomic and cDNA sequences of genes and the subsequent detection of putative point mutations within them, will accelerate the use of these new or complemental identification tools.

7.3 Marker-assisted Gene Introgression

7.3.1 Strategies

Marker-assisted breeding (MAB) first refers to genetic linkage maps that provide the direct linkage between markers and the trait of interest (or breeding criterion). Traits of interest may be monogenic (single major genes) or polygenic (controlled by quantitative trait loci, QTLs). A wide range of maps has been constructed in *Prunus* in the 1990s (Foolad et al. 1995; Warburton et al. 1996; see the specific chapter in this book). Among them the TxE interspecific *Prunus* map (Joobeur et al. 1998) was first mainly constructed with transportable RFLP markers and progressively completed later with SSRs (Aranzana et al. 2003b; Dirlewanger et al. 2004b) and is today a major tool for MAB in *Prunus* crops. These RFLP and SSR transportable markers, considering the high synteny between *Prunus* species, have allowed to locate a large number of traits (Table 7-3) on the same reference map for all *Prunus* species (Dirlewanger et al. 2004b). Although the location of these genes had been established in low-density maps in most cases, their position could be further defined by using the information provided by the network of maps available for *Prunus* and additional markers can be found in the regions of interest and used for marker-assisted selection (MAS) without additional SSR development. Moreover, locating a new SSR marker on the *Prunus* reference map is facilitated by the bin mapping approach based on only a few plants of the corresponding TxE reference progeny (Werner et al. 2005). With the online reference Lovell peach genome sequence now available, molecular mapping can go a step further by permitting to land on the peach sequence and directly define putative markers from the orthologous region in this species. If the character or QTL of interest has been mapped with a high resolution in its specific map, it is possible to span a physical interval encompassing the corresponding putative orthologous candidate genes in peach. The ultimate goal is the identification of these candidate genes either through a candidate gene approach or by positional cloning. For the *Prunus* species in question, corresponding EST libraries from the GDR (http://www.rosaceae.org/node/355) can be screened for those putative genes to provide the corresponding cDNA candidate sequences (Horn et al. 2005). Differential expression of the candidate genes may then allow to select the best final candidate. This aspect is described in Chapter 8 dealing with 'Map-based Cloning'.

Breeding programs often first rely on the evaluation of genetic diversity of collection material before choosing the parental material that will allow to maximize this diversity for subsequent optimal combination. Studies on the genetic diversity sensu stricto have been deliberated in another chapter of this book. In our previous paragraphs on germplasm characterization, genetic diversity has also been discussed indirectly as the molecular marker techniques applied to a wide range of *Prunus* material. The work of Sanchez-Perez et al. (2006) is an example of how one aspect of this diversity, i.e., the heterozygosity of putative breeding material, could be explored and evaluated using SSRs in apricot cultivars. Data from 17 primer pairs showed that local cultivars from Murcia (Spain) and their local descendants had very low genetic heterozygosity (<0.30), that patented cultivars from France and Spain (notably 'Moniqui' and 'Bergeron') and their descendents had intermediate levels (approx. 0.45) while North-American ('Orange Red' and 'Goldrich') and Greek ('Lito') cultivars together with their descendants, displayed the highest values (>0.50). Thus North American cultivars were shown to be the best material to increase the genetic heterozygosity in the Spanish apricot breeding programs to a level higher than in any of the crosses among local Spanish cultivars.

We report hereafter the work done until now on MAB dealing mainly with morphological and physiological features or with biotic stress resistance. In *Prunus* crops, MAB approaches focused on a definite trait have been conducted mainly i) by mapping progenies segregating for this trait and ii) by constructing bulks with the alternate genotypes using the bulked-segregant analysis method (BSA; Michelmore et al. 1991). These techniques have been coupled with various DNA marker types as soon as they appeared. Examples of MAB studies are reported in Table 7-2 to illustrate diverse strategies used in *Prunus*.

Table 7-2 Examples of studies dealing with molecular breeding in *Prunus* crops.

Crop species	Objective	Methodology	Reference
Peach	Enhancing genetic heterozygosity	SSR polymorphism	Sanchez-Perez et al. 2006
	Selection of columnar growth habit	BSA coupled with SSRs	Scorza et al. 2002
	Powdery mildew resistance	QTL mapping	Foulongne et al. 2002, 2003, 2004;
	Root-knot nematode resistance	BSA coupled with AFLPs; Elaboration of SCAR markers	Claverie et al. 2004a
		Mapping with SSRs	Dirlewanger et al. 2004a
Almond	Root-knot nematode resistance	BSA coupled with AFLPs; Elaboration of SCAR markers	Van Ghelder et al. 2010
Apricot	Self-incompatibility genotyping	DNA chromatography (DHPLC)	Raz et al. 2009
	Characterization of self-incompatible genotypes	Sequence polymorphism of genes involved	Yaegaki et al. 2001
	Idem	Idem	Yamane et al. 2009
Plum	Idem	Idem	Zhang et al. 2008
	Root-knot nematode resistance	BSA coupled with RAPDs; Elaboration of SCAR markers	Lecouls et al. 1999; Rubio-Cabetas et al. 2000
		BSA coupled with AFLPs; Elaboration of SCAR markers	Lecouls et al. 2004; Claverie et al. 2004a
		Mapping with SSRs Positional cloning	Dirlewanger et al. 2004a; Claverie et al. 2004b
Cherry	Characterization of self-incompatible genotypes	Sequence polymorphism of genes involved	Zhu et al. 2004

Table 7-3 Major genes affecting morphological or agronomical characters in different *Prunus* crops that can be located on the reference map (reproduced from Dirlewanger et al. (2004b) with permission of PNAS and completed with recent references).

LG*	Characters	Species	Symbols	Populations	References
LG1	Fruit flesh color (white/yellow)	Peach	*Y*	Padre x 54P455	Bliss et al. 2002
	Sharka resistance	Apricot	*sharka*	Lito x Lito; Goldrich x Currot Polonais x Stark Early Orange	Vilanova et al. 2003; Lambert et al. 2007; Soriano et al. 2008 Pilarova et al. 2010
	Evergrowing	Peach	*Evg*	Empress op op dwarf x P1442380	Wang et al. 2002
	Flower color	Almond x peach	*B*	Garfi x Nemared	Jauregui 1998
	Peach aphid resistance	Peach	*Rm2*	Rubira x Pamirskij 5	Pascal et al. 2002; Lambert and Pascal 2011
LG2	Root-knot nematode resistance	Peach	*RMia* (*Mi, Mia*)	P.2175 x Felinem, Akame x Juseitou, Lovell x Nemared, Garfi x Nemared; Harrow Blood x Okinawa	Jauregui 1998; Lu et al. 1998, 1999, 2000; Yamamoto et al. 2001; Claverie et al. 2004a; Dirlewanger et al. 2004a; Gillen and Bliss 2005
	Shell hardness	Almond	*D*	Ferragnes x Tuono	Arus et al 1998
	Broomy (or pillar) growth habit	Peach	*Br*	Various progenies	Scorza et al. 2002
	Double flower	Peach	*Dl*	NC174RL x P1	Chaparro et al. 1994
LG3	Flesh color around the stone	Peach	*Cs*	Akame x Jusetou	Yamamoto et al. 2001
	Anther color (yellow/ anthocyanic)	Almond x peach	*Ag*	Texas x Earlygold	Joobeur 1998
	Polycarpel	Peach	*Pcp*	Padre x 54P455	Bliss et al. 2002
	Flower color	Peach	*Fc*	Akame x Jusetou	Yamamoto et al. 2001
LG4	Blooming time	Almond	*Lb*	D.3.5 x Bertina	Ballester et al. 2001
	Flesh adhesion (clingstone/freestone)	Peach	*F*	(*P. ferganensis* x IF310828)BCI; Akame x Juseitou	Yamamoto et al. 2001 Foulongne et al. 2003
LG5	Non-acid fruit	Peach	*D*	Ferjalou Jalousia x Fantasia	Dirlewanger et al. 1998, 1999; Etienne et al. 2002
	Kernel taste (bitter/sweet)	Almond	*Sk*	Padre x 54P455	Bliss et al. 2002
	Skin hairiness (nectarine/peach)	Peach	*G*	Ferjalou Jalousia x Fantasia; Padre x 54P455	Dirlewanger et al. 1998, 1999; Bliss et al. 2002

LG6	Leaf shape (narrow/wide)	Peach	*Nl*	Akame x Juseitou	Yamamoto et al. 2001
	Plant height (normal/dwarf)	Peach	*Dw*	Akame x Juseitou	Yamamoto et al. 2001
	Male sterility	Peach	*Ps*	Ferjalou Jalousia x Fantasia	Dirlewanger et al. 1998
	Fruit shape (flat/round)	Peach	*S**	Ferjalou Jalousia x Fantasia	Dirlewanger et al. 1998; 1999
	Self-incompatibility	Almond	*S*	Ferragnes x Tuono; D.3.5 x Bertina; Padre x 54P455	Ballester et al. 1998; Bliss et al. 2002;
		Apricot	*S*	Lito x Lito	Vilanova et al. 2003
LG6–LG8	Fruit skin color	Peach	*Sc*	Akame x Juseitou	Yamamoto et al. 2001*
	Leaf color (red/green)	Peach	*Gr*	Garfi x Nemared; P.2175 x Felinem; Akame x Juseitou	Yamamoto et al. 2001*
	Resistance to powdery mildew	Peach	*Vr2*	Rubira x Pamirskij 5	Pascal et al. 2010*
LG7	Root-knot nematode resistance	Myrobalan plum	*Ma*	P.2175 x Felinem	Lecouls et al. 1999; Claverie et al. 2004a;
	Root-knot nematode resistance	Almond	*RMja*	P.2032 x (Alnem x P.2175)	Kochba and Spiegel-Roy 1975, 1976, Van Ghelder et al. 2010
	Resistance to powdery mildew	Peach	*Sf*	(*P. ferganensis* x IF310828)BCI	Dettori et al. 2001
	Leaf gland (reniform/globose)	Peach	*E*	(*P. ferganensis* x IF310828)BCI	Dettori et al. 2001

*LG, linkage group; LG6–LG8 genes located close to the translocation breakpoint between these two linkage groups.

7.3.2 Illustration of MAB for Morphological/Physiological Traits

Morphological and physiological traits used to be the classical method for identification and characterization of *Prunus* material. These conventional markers often cannot be evaluated during the long juvenile period of stone fruit crops (e.g., characteristics of the fruit shape and fruit quality traits) or may be highly variable with environmental conditions (e.g., soil and climate adaptive traits). Using mainly RFLP and SSR anchor markers, the diverse *Prunus* maps involving parental cultivars segregating for various features have been merged into the reference TxE almond-peach (Aranzana et al. 2003b; Dirlewanger et al. 2004a). This reference map has allowed to locate an increasing number of traits on the eight LGs of the *Prunus* genome and these data are shown in Table 7-3. Hereafter we will report studies on the current development of the MAB strategy for some of these traits.

7.3.2.1 Columnar Growth Habit

One of the first examples of MAS for a morphological/physiological trait in *Prunus* has been the selection of columnar growth (broomy or pillar growth) habit in peach (Scorza et al. 2002; Table 7-3), a profitable feature for many objectives including the development of high-density production systems. Using the BSA approach (Michelmore et al. 1991) in a source of columnar growth habit, the showy-flowered peach, this feature was shown to be presumably conferred by a single gene and linked to the microsatellite marker pchgms1 on LG2.

7.3.2.2 Self-incompatibility

Gametophytic self-compatibility (or self-incompatibility) is one of the characters currently selected by molecular methods in breeding programs of almond, apricot, and cherry. The *S/s* locus that controls this character is located on the consensus map in the distal part of LG6 (Ballester et al. 1998), which allows to work to some extent in a complementary mode on any of these three *Prunus* crops as illustrated below. Indeed knowing both the *S*-allele composition and the flowering period of given cultivars makes it possible to predict their cross-compatibility and the best combinations of cultivars to use in an orchard. Cloning and characterization of *S*-alleles has been conducted as an initial step of breeding programs in diverse *Prunus* species for the early identification of *S*-allele combinations using vegetative tissue. Following a report of correlation between the stylar ribonuclease zymograms with incompatibility alleles in sweet cherry (Boskovic and Tobutt 1996), S-RNase gene sequences at the *S*-locus were identified in almond (Tao et al. 1997). This opened up the way to develop species-specific or allele-specific DNA markers in this latter crop. For example, Tamura et al. (2000) developed codominant *S*-specific markers with primers in conserved gene regions flanking introns of variable sizes (and thus amplifying specific sizes in the different cultivars). Markers for PCR-RFLP identification of *S*-alleles were developed by Ma and Oliveira (2001) whereas Channuntapipat et al. (2003) designed *S*-specific PCR primers from the sequences of the introns. These markers, although dominant, were found more reliable than the previous codominant markers of which some alleles appeared difficult to amplify from conserved primers. In sweet cherry, identification of alleles wa proposed first by a PCR-RFLP approach using the enzymes *Hind*III, *Eco*RI, *Dra*I, or *Xba*I, which allowed the determination of a reference nomenclature based on fragment sizes for each *S*-allele/restriction enzyme combination (Hauck et al. 2001). Zhu et al. (2004) have completed these data in cherry with the characterization of the self-compatible S4′ haplotype, derived by X-ray mutagenesis from the self-incompatible S4 haplotype, and have proposed a simple molecular marker for its specific dominant detection based on a primer designed in the deleted region in comparison with the S4′ haplotype. But genetic compatibility, governed by the *S*-locus, also holds, besides the S-RNase gene of pistil, the S-haplotype-specific F-Box (SFB) gene of pollen and alleles of both the S-RNase and SFB genes are distinctive for each S-haplotype. In European cultivars of apricot, the SFB-based S-haplotyping has been performed by Raz et al. (2009) using denaturing high-performance liquid chromatography (DHPLC), a technique that is amenable to high-throughput automation for high-quality plant typing in breeding programs. In Japanese apricot (*Prunus mume*), MAB of the pollen SFB gene has also been considered by Yamane et al. (2009). Cloning and sequence analysis of a self-incompatible allele (SI S-7) together with a self-compatible mutant (SC-S′) revealed an insertion that causes the transcription of a truncated dysfunctional SFB in the S-3′ mutant that resembles the mutation of SFBF, another SC S-haplotype in Japanese apricot. Taking advantage of the similarity of insertions in the two SC S-haplotypes of distinct origin (S-3′ and S-f), a universal PCR marker for SC has been designed and validated for future breeding programs in Japanese apricot.

7.3.2.3 Environmental Adaptation

MAS for physiological adaptation of trees to their edapho-climatic environment is at its beginning. Chilling requirement, together with heat requirement, determines the bloom date, which has a high impact on the climatic distribution of the genotypes. From two peach mapping populations from parents with contrasting chilling requirements, 20 QTLs with additive effects were identified for three traits, including one major

QTL for chilling requirement and two major QTLs for bloom date (Fan et al. 2010). Most QTLs colocalized with QTLs for other trait(s) and in particular, one genomic region of 2 cM, pleiotropic for the three traits, overlapped with the sequenced peach *EVG* (evergreen) region (Bielenberg et al. 2008). This first report on the QTL mapping of floral bud chilling requirement should facilitate MAB for low chilling requirement cultivars and the map-based cloning of genes controlling this physiological trait.

The same type of molecular dissection for chilling requirement and budbreak has been performed for apricot (Olukolu et al. 2009) which is particularly sensitive to cycling between vegetative or floral growth and dormancy. A mapping population of F_1 individuals (using a two-way pseudo-testcross mapping strategy) between the apricot cultivars, Perfection and A.1740, respectively with high and low chilling requirement, allowed to construct high-density male and female maps of respectively, 655 and 592 markers (SSR and AFLP) spanning 550.6 and 454.9 cM (average marker intervals of 0.84 and 0.77 cM). Chilling requirement, evaluated in two seasons on potted trees forced to break buds after variable cold treatments, was shown to be linked to 12 putative QTLs detected on six LGs by composite interval mapping and multiple regression fit. QTL main effects accounted for approx. 58 and 66% of the total phenotypic variance in each parental cultivar, respectively. QTLs corresponding to map positions of differentially expressed transcripts in the two high-density maps suggested candidate genes controlling chilling requirements.

7.3.3 Resistance to Biotic Stresses

The three main applications of MAS in this field have been described on peach, apricot and rootstocks. Major studies concern peach breeding for resistance to the green aphid (*Myzus persicae*), to powdery mildew (*Sphaerotheca pannosa*) and leaf curl (*Taphrina deformans*) fungi, and to the sharka virus (*Plum pox virus*; PPV). Studies in apricot also deal with sharka disease. Rootstock studies are focused on root-knot nematodes *Meloidogyne* spp. and will be reported separately as an illustration of the pyramiding strategy.

7.3.3.1 Resistance to Biotic Stresses in Peach

In this species, the lack of variability is a real concern for further breeding programs and thus very few sources are available (Table 7-4). *Prunus davidiana*, a wild species with poor agronomic performance but closely related to peach, is a putative source of resistance genes for some of the most important peach pests and diseases previously mentioned (Kervella et al. 1998; Pascal et al. 1998; Foulongne et al. 2002, 2003).

Table 7-4 Predominant resistance factors to pests and disease identified and mapped in peach germplasm.

Resistance type	Bio-agressor	Source	Gene or QTL designation and chromosomal location	References
Monogenic	Green peach aphid (*Myzus persicae*)	Weeping flower peach	*Rm1*	Monet et Massonié 1994
		Rubira	*Rm2*; LG1	Pascal et al. 2002 ; Lambert and Pascal 2011
	Mildew (*Podosphaera pannosa*)	Malo Konare	*Vr1*	Dabov 1985
		Pamirskij 5	*Vr2*; LG8	Pascal et al. 2010
	Green peach aphid	*P. davidiana* var. P. 1908	Two QTLs MP.SD-3.1; LG3 MP.SD-5.1; LG5	Sauge et al. 2004
Polygenic	Mildew	Idem	Two QTLs PM.SD-6.1; LG6 PM.SD-8.1; LG8	Foulongne et al. 2003
	Leaf curl	Idem	Several QTLs	Viruel et al. 1998
	Sharka (Plum pox virus)	Idem	Six QTLs ranging into: - Stable QTLs PPV-1b: LG1 PPV-4a: LG4 PPV-6a: LG6 PPV-7a: LG7 - Environment-dependant QTLs: PPV-1a : LG1 PPV-5a : LG5	Decroocq et al. 2005 Marandel et al. 2009 Rubio et al. 2010

7.3.3.1.1 Myzus persicae

This major pest of peach crops in most areas of the world (Pascal et al. 2002) damages both leaves and stems, resulting in leaf curl, shoot stunting and fruit punctures that decrease fruit quality. Consequently deploying resistant peach cultivars would contribute to a better management of the aphid. Resistance has been investigated among peach and related species (Massonie et al. 1982) and the single dominant genes *Rm1* on the weeping flower peach (Monet and Massonié 1994) and *Rm2* in the red peach rootstock cultivar 'Rubira' (Pascal et al. 2002; Lambert and Pascal 2011) have been detected. Another factor of partial resistance to the green peach aphid has been evidenced in *P. davidiana* that impedes phloem sap ingestion by *M. persicae* (Sauge et al. 2004). Studies for resistance conferred by *Rm2* using F_1 and F_2 progenies between Rubira and the susceptible peach cultivar 'Pamirskij 5' showed that the reddish spots on the plant apices, corresponding to necrotic reactions (Massonie et al. 1982) probably due to aphid feeding punctures, are associated with, but not themselves responsible for, *M. persicae* resistance in Rubira. Indeed aphids leave the plants within 2–3 days of the initial infestation, a characteristic behavior of antixenosis resistance but differences in response to aphid puncture, suggest that the underlying resistance genes might not be the same in *Rm1* and *Rm2* (Sauge et al. 1998a, 2002). This resistance induced by *M. persicae* in Rubira was further investigated by comparing the effects of a previous infestation of 48 h in this later cultivar and in the susceptible control cultivar GF305. There were significantly fewer aphids settling on preinfested than on uninfested plants and induced resistance in preinfested plants was increased with the duration of the first infestation, the rejection of Rubira as a host plant being confirmed by electrical penetration graph (EPG) technique (Sauge et al. 2002). Nothing is yet known about the location of *Rm1* but *Rm2* has been recently mapped in the lower part of LG1 (Lambert and Pascal 2011), which will facilitate MAS of this later gene.

The partial resistance factor to *M. persicae* in *P. davidiana* P.1908, that impedes phloem sap ingestion by the aphid (Sauge et al. 2004) was located on LG3 (RAPD marker Q06-350 and RFLP marker AG50A) in the cross (called SD) between the susceptible peach cultivar 'Summergrand', and *P. davidiana*. To increase understanding of the physiological process that alters suitability of P1908 as a host plant for the aphid, the feeding response of the insect with the EPG technique on the SxD peach hybrids was studied and showed that hybrids carrying the QTL markers produced significantly less continuous phloem sap ingestion and more phases of salivary secretion into the sieve element than hybrids lacking them. Thus the resistance factor in P. 1908 might be linked to a phloem seating system of the plant in response to injury of the sieve tubes by the aphid stylet that would prevent aphid from sustained sap uptake.

7.3.3.1.1.1 Powdery mildew

Among peach bio-agressors at the world scale, powdery mildew is the most important fungus parasitizing peach orchards. In order to improve peach resistance to this pathogen, a backcross breeding scheme has been carried on from the F_1 cross SD previously mentioned, involving *P. davidiana* P.1908 which expresses a polygenic resistance to powdery mildew (Foulongne et al. 2002). The resistant F_1 individual SD40, tested under controlled conditions, was used as parent for a backcross (BC1) with 'Summergrand'. Scoring for resistance to powdery mildew also under controlled conditions allowed to select a resistant BC_1 individual as a parent of a pseudo-BC_2 progeny with another susceptible commercial cultivar. Meanwhile the genetic basis of the resistance was investigated in 77 hybrids of the F_1 (SD) progeny and the first quantitative trait alleles (QTAs) were located on the P1908 map (Dirlewanger et al. 1996). Further search for possible homozygous resistance genes of P1908 was done using SD40 selfed progeny (named F_2).

New quantitative trait loci controlling powdery mildew resistance were then detected using the three maps (F_1, F_2 and BC_2) by submitting all three populations to natural exposure of the fungus, in several locations and over several years. Globally thirteen QTLs were detected of which nine favourable alleles were contributed by the resistant parent with five of them consistently detected across the three populations. Nevertheless as the F_1 individual SD40 used to produce F_2 and BC_2 progenies had not inherited the favourable QTL alleles on LG3 and LG8 from P1908, those QTLs were not detected in the corresponding regions in F_2 and BC_2 populations. In two other genomic areas, the favourable alleles were inherited from Summergrand in the F_2 and BC_2 populations. Based on the SD F_1 progeny, comparative analysis of the phenotypic data reported under controlled conditions in Dirlewanger et al. (1996) suggested an important qualitative change, confirmed by QTL analysis, in the distribution of powdery mildew resistance in the tests under field conditions: for example, a dramatic decrease and a substantial increase of the effects of the major QTLs detected, respectively on LG6 (PM.SD-6.1) and LG8 (PM.SD-8.1) were observed (Table 7-4). These data were the first ones in a perennial crop that established the consistency of a QTL detected across generations and environments. This peach breeding program based on *P. davidiana* powdery mildew resistance is pursued through the Fruit Breedomics European program (7th framework programme 2010). Recently an additional resistance factor, which is the major gene *Vr2*, has been detected in the source Pamirskij5 and mapped close to the *Gr* gene of leaf color in a telomeric position on LG8 (Pascal et al. 2010). Another resistance factor (designated *Sf*; Table 7-3) to powdery mildew had been mapped in the wild peach source *P. ferganensis* on LG7 (Dettori et al. 2001) but no data have confirmed this location in *P. davidiana*.

7.3.3.1.1.2 Sharka disease

Plum pox virus (PPV), the causative agent of sharka disease, is one of the most serious problems affecting stone fruit production in Europe and America. In Europe, sharka is the most important viral disease affecting *Prunus* fruit crops. Several hypotheses on the genetic inheritance of sharka resistance in *Prunus* species have been put forward and a quantitative trait was hypothesized in *P. davidiana* clone P1908. An analysis of QTLs was firstly performed on the F_1 interspecific peach population SD (Summergrand x *P. davidiana*) previously described (Decroocq et al. 2005) to decipher the complex pattern of inheritance in this latter species. Six genomic regions involved in PPV resistance explained between 22 and 51% of the phenotypic variance. In another study (Marandel et al. 2009), based on the F_2 population $SD40^2$ (by selfing the individual SD40) also used for powdery mildew resistance, six QTLs were identified and the total phenotypic variation explained ranged from 29 to 58%. The genetic map of the F_2 cross and the *P. davidiana* parent map (from the SxD cross) could be aligned unambiguously which showed consistency of QTL over generations. The four QTLs, PPV-1b, PPV-4a, PPV-6a and PPV-7a, located respectively on LGs 1, 4, 6 and 7 and contributed from the *P. davidiana* parent, were the most stable, being detected with all mapping methods and at the different generations. In $SD40^2$ two QTLs, PPV-1a and PPV-5a, located respectively on LGs 1 and 5, appeared to be temporally specific, reflecting the environmental dependence of PPV-resistance scoring (Table 7-4). Thus two of the QTLs, one stable and one unstable, were mapped onto LG1 where PPV resistance factors were also mapped in the apricot. cvs. 'Stark Early Orange' or cv. 'Goldrich' (Hurtado et al. 2002; Vilanova et al. 2003; Lambert et al. 2007). To circumvent the limiting factors for a precise mapping of QTLs [size and structure (F_1, F_2, BC_1) of the population, ability of reliable phenotyping], Marandel et al. (2009) reinforced a candidate gene approach that had been initiated previously (Decroocq et al. 2005) and looked for colocalization of QTLs with resistance gene analogs (RGA). Candidate gene fragments were amplified by PCR, isolated and mapped on the peach interspecific linkage map detecting three analogs of virus resistance genes co-localizing to two distinct genomic regions linked to PPV resistance in *P. davidiana* of which one was an eukaryotic translation initiation factor *eIF(iso)4E* (Decroocq et al. 2005). Development and mapping of new microsatellite markers linked to candidate genes (Marandel et al. 2009) revealed a striking co-localization of three of the detected QTLs with gene copies coding for eukaryotic translation initiation factors [eIF4E and eIF(iso)4G] and strengthened the idea that components of the eIF complex are correlated with resistance to PPV in *P. davidiana* P1908. A more recent study by Rubio et al. (2010) reporting PPV evaluation of an F_1 cross between the susceptible peach cultivar Rubira and *P. davidiana* P.1908 confirmed the overall results of the previous studies involving *P. davidiana* but did not recover the main QTL from P.1908 in LG6. This result highlights the possible strong interaction between the genetic backgrounds of the susceptible parent with that of the resistance source *P. davidiana* P1908.

7.3.3.1.1.3 Peach tree short life (PTSL)

It is worth mentioning an original study of MAS for peach resistance to PTSL, a complex disease syndrome caused by multiple factors including two bio-agressors, the ectoparasitic nematode *Criconemella xenoplax* and the bacterium *Pseudomonas syringae* (Blenda et al. 2006). This study identified one diagnostic AFLP marker associated to susceptibility to the PTSL syndrome in peach rootstocks (Blenda et al. 2006).

7.3.3.2 Resistance to Biotic Stresses in Apricot

In apricot sharka disease has become a serious threat since the late 1980s. Major natural sources of resistance to PPV have been identified in North American apricot cultivars (Stella', 'Stark Early Orange' and 'Goldrich') and most breeding programs currently use one of these as a source of resistance (Audergon et al. 1994; Egea et al. 1999; Karyiannis and Mainou 1999; Badenes et al. 2003). In this species control of the trait by a single (Dicenta et al. 2000; Martinez-Gomez and Dicenta 2000), two (Vilanova et al. 2003; Lalli et al. 2008; Soriano et al. 2008) or three dominant or recessive genes (Guillet-Bellanger and Audergon 2000; Soriano et al. 2008) have been proposed but sharka resistance is now widely considered as a qualitative trait. Nevertheless, to alleviate the concern of researchers about the number of loci involved, apricot genetic maps have been constructed from progenies derived from diverse resistance sources identified in the US germplasm, mainly 'Stark Early Orange' (Vilanova et al. 2003; Lambert et al. 2004), 'Goldrich' (Hurtado et al. 2002) or 'Harlayne' (Pilarova et al. 2010). To re-visit the location of PPV resistance, interval mapping, composite interval mapping, and the Kruskal-Wallis non-parametric test were performed in the F_1 progeny of the susceptible cv. 'Polonais' and resistant 'Stark Early Orange' cross (Lambert et al. 2007). Four genomic regions, respectively in LG1 (2 regions), LG3 and LG5, were identified as carrying QTLs conferring this resistance. One of these QTLs (PPV1-1s) accounted for 56% of the phenotypic variance and was mapped to the upper region of LG1 of 'Stark Early Orange'. This illustrated the strong effect of this factor in comparison with all others and its putative designation as the *sharka* gene (Table 7-3). Interestingly its location was close to the region previously identified in 'Goldrich' (Hurtado et al. 2002) and *P. davidiana* SD (PPV1-2, Decroocq et al. 2005) considering the variability between the different evaluation methods, the variable sizes of progenies and the diverse incidence of PPV between experiments. Thus it is plausible that all the three previous studies identify the same region on LG1 of the *Prunus* genome and that the PPV resistance factor carried by 'Goldrich' and 'Stark Early Orange' are the same locus or belong to the same gene cluster. In addition, two putative candidate genes [(O-linked N-acetylglucosamine (GlNAc) transferase; Fernandez-Fernandez et al. (2002) and Cd199 protein kinase; Decroocq et al. (2005)] were found to be in the vicinity of AG51a, the closest marker. Fernandez-Fernandez et al. (2002) demonstrated that an O-GlcAcylation reaction occurred on PPV capsid protein while Chen et al. (2005) showed that this protein had a role in the modification of PPV coat protein and participated in PPV infection even though no record existed about its possible involvement in PPV resistance. The protein kinase Cd199 was an analog of resistance gene detected and mapped by Decroocq et al. (2005) in the vicinity of AG51 in the SD40 ('Summergrand' peach x *P. davidiana*) F_2 map. The second QTL on LG1 (PPV-1-2s, Lambert et al. 2007) had a low effect (R2 = 8%) and was mapped to the homologous LG1 region of the F_1 progeny of this latter cross (PPV1-1, Decroocq et al. 2005). The two other putative QTLs on LG3 of 'Polonais' and five of both 'Polonais' and 'Stark Early Orange' (SEO) were detected at logarithm of odds (LOD) scores slightly above the detection threshold, the last QTL being only detected in the early stage of the infection. PPV resistance in apricot is, thus, controlled by a major dominant factor, designated the *sharka* gene, located on LG1 but complementary recessive factors with lower effect are not excluded (Lambert et al. 2007). This location of a QTL with a strong effect or a major dominant locus in the upper part of LG1 is confirmed in maps involving both sources: Goldrich and SEO (Soriano et al. 2008), which reported the SSR ssrPaCITA5 as the best marker for MAS in the 20 cM interval mapped, and in the source Harlayne in a study that refers to symptom ratings of two independent experiments with the PPV-M and D major strains, respectively (Pilarova et al. 2010).

7.4 Gene Pyramiding

An example of gene pyramiding is the ongoing breeding program for the creation of *Prunus* rootstocks that harbor durable resistance to root-knot nematodes (RKN) caused by *Meloidogyne* spp. Another project is the construction of peach cultivars with multiple pest and disease resistance features.

7.4.1 Pyramiding for Resistance to Root-knot Nematodes in Prunus *Rootstocks*

Root-knot nematodes (RKN) caused by *Meloidogyne* spp. can be considered worldwide as the most damaging nematodes (Lamberti 1979; Sasser and Freckman 1987). Among these, the predominant nematodes are the polyphagous and parthenogenetic species namely *M. arenaria*, *M. incognita* and *M. javanica* which are distributed in the Mediterranean, Tropical and Equatorial areas (Sasser 1979). RKN reduce fruit and nut production in several economically important *Prunus* species, including peaches, almonds and plums. Control of RKN was classically based on fumigation or nonvolatiles applied as pre-plant treatments. The ban of nematicides has rendered urgent environment-friendly alternatives and, in this new context, the use of RKN-resistant rootstocks appeared more than ever the most economic and environmentally sound method for managing *Meloidogyne* in *Prunus* species (Layne 1987; Cook and Evans 1987; Greco and Esmenjaud 2003). The source of resistance was first detected in peach in the first half of the 20th century (Tufts 1929; Day and Tufts 1939; Weinberger et al. 1943). Since then other sources of resistance have been found in other *Prunus* species, particularly in plum and almond (Kochba and Spiegel-Roy 1976; Scotto La Massèse et al. 1984; Esmenjaud et al. 1997). In USA, breeding programs have been mainly focused on the creation of peach or peach-almond rootstocks using peach germplasm (such as 'Nemaguard' and 'Okinawa') as the resistance sources. The peach rootstock 'Nemared' (Ramming and Tanner 1983) used 'Nemaguard' as its source, while the interspecific hybrids 'Hansen 2168' and 'Hansen 536' (Kester and Asay 1986) and 'Flordaguard' (Sherman et al. 1991) used 'Okinawa' and *P. davidiana* as RKN resistance source. Resistance in Guardian™ is inherited from 'Nemaguard' and 'S-37' (Okie et al. 1994a). In Europe, resistance evaluations of known or new germplasm sources have been performed to characterize their spectrum of activity (Marull et al. 1994; Pinochet et al. 1996a; Esmenjaud et al. 1997) and their level of resistance under high inoculum pressure or high temperatures (Fernández et al. 1994b; Esmenjaud et al. 1996a, b, 1997). A summary of resistance sources, their spectrum and the genes involved is furnished in Table 7-5. Resistance in tested peach and almond cultivars is not complete and does not control *M. floridensis* (Esmenjaud et al. 1997). In contrast, 'Myrobalan' plum (*P. cerasifera* from the subgenus *Prunophora*) comprises certain clones that also resist *M. floridensis*, the 'peach RKN' (Handoo et al. 2004) and carry a complete spectrum and high level of resistance to other *Meloidogyne* spp. (Esmenjaud et al. 1996a, 1997). Some of these clones also confer favorable agronomic features for peach cultivars such as broad graft compatibility and high adaptation to waterlogged soils (Kester and Grasselly 1987; Layne 1987; Salesses et al. 1992) (Table 7-6).

Table 7-5 Phenotype of main sources for resistance to RKN in *Prunus* spp. (corresponding genes with their spectrum, *Prunus* map location and type of segregating material) and of existing or putative interspecific materials.

<table>
<tr><th rowspan="2">Species or hybrid</th><th rowspan="2">Accession or cross</th><th colspan="4">Resistance (phenotype) and gene spectrum</th><th rowspan="2">Linkage group</th><th rowspan="2">Type of segregating material (references)</th></tr>
<tr><th>M. arenaria</th><th>M. incognita</th><th>M. javanica</th><th>M. floridensis (USA)</th></tr>
<tr><td>Putative parental material
Myrobalan plum</td><td>P.2175
P.2980</td><td colspan="4">Ma</td><td>LG7</td><td>Diallel cross (Esmenjaud et al. 1996; Lecouls et al. 1977); Diallel cross (Rubio-Cabetas et al. 1999)</td></tr>
<tr><td>Japanese plum</td><td>J.222</td><td colspan="4">Rjap</td><td>LG2</td><td>F1 cross (Claverie et al. 2004a)</td></tr>
<tr><td>Peach</td><td>Nemared
Nemaguard</td><td colspan="2">RMia</td><td>R/S</td><td>S</td><td>LG2*</td><td>F2 cross (Lu et al. 1998; 1999; 2000); Interspecific crosses (Claverie et al. 2004a; Esmenjaud et al. 2009)</td></tr>
<tr><td></td><td>Okinawa</td><td>R</td><td>R</td><td>R</td><td>S</td><td>LG2**</td><td>F2 crosses (Gillen and Bliss 2005)</td></tr>
<tr><td></td><td>Shalil (GF.557)
Bockara
Yunnan</td><td colspan="2">RMia</td><td>S</td><td>S</td><td>LG2</td><td>F2 crosses (Weinberger et al. 1943 ; Sharpe et al. 1969) ; Interspecific crosses (Claverie et al. 2004a; Esmenjaud et al, 2009)</td></tr>
<tr><td></td><td>Juseitou</td><td colspan="2">RMia</td><td>R</td><td>S</td><td>LG2***</td><td>F2 cross (Yamamoto at al. 2001; Yamamoto and Hayashi 2002)</td></tr>
<tr><td>Almond</td><td>Alnem 1
Alnem 88
Alnem 201</td><td>RMja ?</td><td>S</td><td>RMja</td><td>S</td><td>LG7</td><td>F2 crosses (Kochba and Spiegel-Roy 1975; 1976)
BC1 crosses (Esmenjaud et al. 2009; Van Ghelder et al. 2009)</td></tr>
<tr><td></td><td>Garfi</td><td>S</td><td>S</td><td>S</td><td>S</td><td></td><td>Interspecific crosses (Claverie et al. 2004a; Esmenjaud et al. 2009)</td></tr>
<tr><td>Existing hybrid material</td><td>P.2175 x
(Garfi x Nemared)22</td><td colspan="4">RMia Ma</td><td></td><td>3-way hybrid (Esmenjaud et al. 2009)</td></tr>
<tr><td>Putative hybrid material</td><td>P.2175 x
(Alnem x Nemared)</td><td colspan="4">RMia Ma RMja</td><td></td><td></td></tr>
</table>

R resistant, *S* susceptible, *R/S* phenotype depending on the isolate; * Different genes and locations on LG2 have been proposed (see text); ** A single R gene *(Mi)* to *M. incognita* (LG2); *** A single R gene to *M. incognita (Mia)* and a single closely linked gene to *M. javanica (Mja)*

Table 7-6 Agronomical and physiological properties of plum, peach and almond parents used for rootstock breeding and expected characteristics of three-way hybrid rootstock material. RKN = root-knot nematodes; 0: none, *: low, **: mean, ***: high, ?: unknown.

	Prunophora	**Amygdalus**			**Three-way hybrids**	
	P.2175 (M) Myrobalan	**Nemared peach (N)**	**Garfi (G) almond**	**Alnem (A) almond**	**M x (G x N)**	**M x (A x N)**
RKN gene	*Ma*	*RMia*	None	*RMja*	*Ma* + *RMia*	*Ma* + *RMia* + *RMja*
RKN spectrum	Complete	Uncomplete	None	Uncomplete	Complete	Complete
Adaptation to drought	?	0	***	***	**	**
Adaptation to chlorosis	?	0	***	***	**	**
Adaptation to waterlogging	***	0	0	0	**	**
Peach graft compatibility	**	***	**	**	**	**
Vigor	**	**	**	**	***	***
Propagation	***	*	**	*	***	**

7.4.1.1 Genetics of RKN Resistance

Because Myrobalan plum is a self-incompatible species, studies of the genetics of its RKN resistance were based on a diallel mating cross (Esmenjaud et al. 1996). Resistance to *M. arenaria, M. incognita, M. javanica,* and *M. floridensis* in 'Myrobalan' plum is conferred by a single major gene *Ma,* which behaves as completely dominant in the accessions P.2175 (allele *Ma1,* heterozygous) (Esmenjaud et al. 1996b; Lecouls et al. 1997) and P.2980 (allele *Ma3,* heterozygous) (Rubio-Cabetas et al. 1998) (Table 7-5). This gene, located on LG7 (Claverie et al. 2004a), also controls *M. mayaguensis* (Rubio-Cabetas et al. 1999), a species of tropical origin that overcomes the resistance conferred by the *Mi* gene from tomato (Rammah and Hischmann 1988; Fargette et al. 1996). Not surprisingly, in the Japanese diploid plum *P. salicina,* a close and interfertile species with Myrobalan plum, but grown for fruit and not as a rootstock, another gene named *Rjap* has been identified from a F_1 segregating cross and shown as lying on LG7 and cosegregating with *Ma* (Claverie et al. 2004a). The cosegregation of both genes confirms the existence of a locus for RKN resistance in this genomic region, presumably with a high allelic variability. Interestingly the *Ma* gene has also been demonstrated to have a protective effect against crown gall, a bacterial disease caused by *Agrobacterium tumefaciens* when this soil bacterium is present in the soil together with RKN (Rubio-Cabetas et al. 2001). This protective effect of *Ma* and presumably of other RKN resistance genes against *Meloidogyne*-transmitted crown gall is an additional argument for their introgression into *Prunus* rootstocks.

Nemaguard, Nemared and Okinawa sources do not control *M. floridensis* and Nemaguard and Nemared only control some isolates of *M. javanica* (Esmenjaud et al. 1996; Esmenjaud et al. 2009). In both later rootstocks, resistance to *M. arenaria* and *M. incognita* is conferred by the *RMia* gene located on LG2 (Claverie et al. 2004a) and thus independent from the *Ma* gene (Claverie et al. 2004a). Yamamoto and Hayashi (2002) also reported two tightly linked genes controlling either *M. incognita* (*Mia*) or *M. javanica* (*Mja*) in the Japanese accession Juseitou (Table 7-5). The SCAR markers designed by these authors for *Mia* were closely linked to *RMia* (Claverie et al. 2004a), which suggests that *RMia* and *Mia* might be allelic. They might also lay on the same locus as the so-called *Mi* (for *M. incognita*) and *Mij* genes (for both *M. incognita* and *M. javanica*) reported by Lu et al. (1998, 1999, 2000) in 'Nemared', even though they were found distantly linked to *RMia* by Claverie et al. (2004a).

In bitter almond, resistance to *M. javanica* has been characterized in Alnem series by Kochba and Spiegel-Roy (1976). The incomplete spectrum of this almond material was found susceptible to *M. incognita* (Scotto La Massese 1984; Esmenjaud et al. 1997, 2009) showed this almond is susceptible to *M. floridensis* (Table 7-5). The resistance (R) gene to *M. javanica*, identified by Kochba and Spiegel-Roy (1975) and confirmed and designated *RMja* by Van Ghelder et al. (2010), might also control *M. arenaria*. With only few segregants (<30), it has been located on the LG7, i.e., closely linked to the *Ma* gene (Van Ghelder et al. (2010), using a set of polymorphic SSR markers selected as covering the entire eight LGs from the TxE reference *Prunus* map (Dirlewanger et al. 2004b) (Table 7-5). Moreover, given the total inability of the almond source to be vegetatively propagated, those few individuals segregating for *RMia* were indeed BC_1 hybrids between the resistance donor Alnem and easily-rooting susceptible Myrobalan plums, which allowed propagation of clonal material (cuttings) for an independent reliable evaluation of the same material to several RKN species.

7.4.1.2 Rootstock Breeding Scheme and Pyramiding Strategy Based on Molecular Markers

Breeding rootstocks for the deployment of durable resistance (R) to RKN in *Prunus* is a challenging task since their long developmental cycle highly increases the risk for resistance breaking and specific strategies have to be adapted to take into account the perennial status of the interaction. Major studies on resistance genetics have been focused on the model nematode gene *Mi* from tomato (Williamson et al. 1998, 2003, 2006) and the possible selection of virulent isolates under the resistance gene pressure has been today well documented (Castagnone-Sereno et al. 1996; Semblat et al. 2000; Castagnone-Sereno 2002). Thus the RKN resistance breeding strategy for *Prunus* rootstocks is based on pyramiding of the major genes *Ma, RMia* and *RMja* in order to confer a high-level, wide-spectrum and heat-stable resistance.

Characterization of the source material has been conducted through a program funded by EU (Esmenjaud 2004; Esmenjaud et al. 2009). The breeding program aims at the creation of rootstock material with a wide adaptation to biotic (RKN) and abiotic stresses suffered by the orchards of stone fruit species. New rootstocks will be based on interspecific hybrids involving i) Myrobalan plum (M) as the first parent and the donor of the complete-spectrum and high-level resistance gene *Ma* and ii) either peach (P), or almond (A) or almond-peach (AxP) as the second parent. Thus in this objective MxP [Myrobalan plum 'P.2175' x peach 'Nemared'], MxA[Myrobalan plum 'P.2175' x almond 'Grafi'] (the two-way hybrids), or Mx(AxP) [Myrobalan 'P.2175' x (almond 'Garfi'-peach 'Nemared')] (the three-way hybrids) have been created in order to pyramid the *Ma1* allele and *RMia* from 'Nemared' peach. A further step plans to cumulate all of the three R-genes in the Mx(PxA) hybrids of the type [Myrobalan plum 'P.2175' x (almond 'Alnem' x peach 'Nemared'] (Table 7-5). Rootstock material currently selected is the three-way hybrid [Myrobalan 'P.2175' x (almond 'Garfi' x peach 'Nemared')]. In this cross selection is also in progress for adaptative physiological characteristics putatively inherited from their parents: tolerance to waterlogging and rooting ability from Myrobalan plum, tolerance to drought and chlorosis from almond, graft compatibility with peach from peach (Table 7-6).

From our current observations, all resistance alleles of major RKN genes described here are completely dominant and can be pyramided at the heterozygous state. Indeed the combination of these three sources of resistance in a single rootstock would allow the control of each of three predominant RKN species, i.e., *M. incognita, M. arenaria,* and *M. javanica*, by at least two dominant resistance genes. In this way, *M. incognita* is controlled by *Ma* and *RMia, M. arenaria* by *Ma* and *RMia* (and hypothetically by *Mja*) and *M. javanica* by *Ma* and *RMja*. This pyramiding strategy should offer an unprecedented guarantee of stable and durable RKN resistance in rootstock material (Table 7-5). MAB for durable RKN resistance in *Prunus* rootstocks developed hereafter lies on the three major genes *Ma, RMia* and *RMja*.

7.4.1.2.1 MAB for *Ma*

MAS for RKN resistance using *Ma* has been developed since 1999 (Lecouls et al.) as a support to this project. The first results dealt with the complete-spectrum resistance conferred by *Ma* for which the heterozygous status of the source accessions P.2175 (allele *Ma1*) and P.2980 (allele *Ma3*) rendered possible a BSA approach based on F_1 R x S segregating progenies (*Ma/ma* x *ma/ma*). This approach allowed to develop SCAR markers, codominant in Myrobalan plum and successively designed from RAPD (SCAL19; 0.9 cM; Lecouls et al. 1999, 2004) and AFLP (SCAFLP2; cosegregating; Claverie et al. 2004a; Lecouls et al. 2004) fragments. The polymorphism of these two coupling-phase SCAR markers tightly linked to *Ma* was evaluated within diverse diploid *Prunus* accessions. The alleles SCAL19 and SCAFLP2 were null in three apricot cultivars ('Moniqui', 'Luizet' and 'Stark Early Orange') representative of the genetic diversity of this species. These results suggest that apricot, reported as resistant to *M. arenaria, M. incognita* and *M. javanica* (Nyczepir and Esmenjaud 2008) but for which genetics has been poorly studied (Esmenjaud et al. 2009), and Myrobalan plum, although belonging both to the subgenus *Prunophora*, might possess two different resistance systems. Alleles were also not recovered from the *Amygdalus* RKN resistance sources 'Nemared', 'Alnem 1' and 'GF. 557'. Then benefiting from the sequences of two bacterial artificial chromosomes (BACs) carrying, respectively, the *Ma1* (BAC 76H19) and *ma1* (BAC 40K9) alleles in P.2175, two codominant microsatellites SSR6 and SSR12, flanking the R-locus in an interval of less than 1 cM, were designed and validated for MAS (Esmenjaud and Dirlewanger 2007; Bus et al. 2009). The *Ma* gene has recently been cloned as the first member of a cluster of three TIR-NBS-LRR genes grouped in a 32 kb distance. It is the longest resistance gene cloned up-to-now (2048 aa) with a huge repeated post-LRR region (1088 aa). Two specific SCAR markers NSCAFLP2 and CT3-4N, located within the *Ma* genomic sequence, respectively, in the first gene intron and in the post-LRR region, are now available for MAB (Claverie et al. 2011).

7.4.1.2.2 MAB for *RMia*

Research efforts are also in progress for MAS of the *RMia* gene. SCAR markers OPAP4 and OPS14a established from RAPDs by Yamamoto and Hayashi (2002) for the *Mia* and *Mja* genes in the resistance source Juseitou have been shown to be linked to *RMia* in Nemared and Shalil material using a BSA approach from *RMia*-segregating progenies of two three-way crosses (Claverie et al. 2004a) (Table 7-5). One EST sequence 'EST 30M23' from the TxE map was then shown to be more closely linked to *RMia* than Yamamoto and Hayashi's markers (D. Esmenjaud, pers. com., INRA Sophia-Antipolis 2009) and could be located on the Lovell peach genome sequence (http://www.rosaceae.org/node/355) on scaffold 2 (LG2) at 4.2 Mb from the upper part of the scaffold sequence. This location definitively confirmed previous data on the telomeric position of the *RMia* gene (Claverie et al. 2004a). SSR markers were then designed from the online sequence and used for chromosome walking on each side of the gene. Using recombinant individuals, microsatellites SSR9 and SSR25, were shown to flank the gene respectively in centromeric (5.5 Mb) and telomeric positions (4.5 Mb), in an approx. 1 Mb physical interval. Thus the *RMia* gene is located in 1 Mb out of the 22.5 MB of scaffold 2 (i.e., a genetic interval of 4–5 cM). Several candidate genes belonging to the common RGA family lay in this interval. Further studies will be focused on the detection of recombination events occurring between both flanking markers in order to continue the chromosome walking from each marker position in the direction of the *RMia* locus. In this interval at least one codominant SSR or SCAR marker will be designed and validated. Sizes of the markers will be chosen to allow a duplex amplification procedure for *RMia* and *Ma* simultaneous genotyping.

7.4.1.2.3 MAB for *RMja*

MAB of *RMia* is just beginning as SSR markers covering the *Prunus* reference map placed *RMja* on LG7 in the same region as *Ma* and *Rjap* and thus showed its independence from the *RMia* gene (LG2) of the botanically closer peach (Van Ghelder et al. 2010). The limiting factor for the detection of markers closely linked to the gene is the very low number of segregating individuals currently available (<30). From the new segregating progenies produced, it will soon be possible to confirm or reject the hypothesis that *Ma* and *RMja* are in the same locus of LG7. MAS of *RMia* will benefit from its location in the *Ma* region in which many markers are already available and will only have to be evaluated for their polymorphism between the parental material used for the production of segregants. When the segregating crosses are obtained from almond parents, the high variability of this species should facilitate the identification of markers. The high interspecific polymorphism between Myrobalan and almond should also guarantee the detection of specific markers for each of the genes even though they are closely linked.

7.4.1.2.4 Perspectives of MAB for Adaptation to Drought

A population of 101 individuals from the 3-way hybrid [P.2175 x (Garfi x Nemared)22] (Tables 7-5 and 7-6) has been mapped and the two parental maps are available (Dirlewanger et al. 2004a). This population should be used for a first approach of the genetics of adaptation to drought and subsequently for detecting QTLs linked to this physiological trait (Cochard et al. 2008).

7.4.2 The Peach Fruit Breedomics Project

Special molecular breeding perspectives are also being developed in Europe through the Fruit Breedomics program (7th framework program) focused on peach as a model species for *Prunus*. Overall objectives of this program include in priority development of MAB to some predominant pests and pathogens in pre-breeding material that will serve as source material for further projects (Tables 7-3 and 7-4). For such objectives several genes or QTLs have been retained including resistances to peach green aphid and to powdery mildew. As previously reported, durable resistance to the green aphid will be based on the major genes *Rm1* from Weeping peach (Monet et Massonié 1994) and *Rm2* from Rubira (Lambert and Pascal 2011), respectively, They will be associated with the two QTLs (MP.SD-3.1 and MP.SD-5.1) from *P. davidiana* whereas durable resistance to powdery mildew will be based on the *Vr1 and Vr2* gene associated with two other QTLs also inherited from *P. davidiana* (PM.SD-6.1 and PM.SD-8.1). Common peach fruit traits for MAB (Table 7-3) will include the genes *D* (acid/non acid fruit flesh), *S* (flat fruit shape), *G* (glabrous fruit skin), *Y* (yellow fruit flesh) and *F-M* (freestone-melting flesh).

7.5 Limitations and Prospects of MAB

The long intervals between two successive cycles of *Prunus* crops will still be a serious limiting factor to impede further improvement of *Prunus* material. Nevertheless in perennial fruit varieties and most of the rootstock material are clones, a trait that allows a more thorough observation and identification than seed material and compensates partly the fact that reproduction cycles are long. However attempts for fast breeding technology are being initiated in fruit crops with the first results obtained in apple in 2009 (Flachowsky et al.). This approach is based on a transgenic cultivar obtained by introducing a MADS box gene from birch (*BpMADS4*) that permitted flowering of apple seedlings in a few weeks after sowing. In *Prunus* first successful adaptation of this early flowering 'Fastrack Breeding' technique has been developed in plum by Srinivasan in 2008 by transforming 'BlueByrd' plum with poplar *Flowering Locus T1* (*FT1*) gene (Scorza et al. 2011). This work will be described further in this chapter in the 'Transgenic Breeding' section. We will evoke hereafter the concept of genomic selection and will illustrate it through a precursor approach for peach breeding using *P. davidiana* as a donor source of major genes and QTLs. To overcome its traditional limiting factors, MAB will also rely more and more on an optimized approach of exploitation of biodiversity that are the genome-wide association (GWA) mapping based on linkage disequilibrium (LD) studies.

7.5.1 Genomic Selection

Clear limitations of MAB appear in the putative abilities to breed efficiently for quantitative traits contributed by a high number of loci with small effects. An interesting and precursor study has been conducted on *P. davidiana* to assess its potential for introgression as the donor of resistance factors to several peach pests and diseases, by comparative mapping (Foulongne et al. 2003). The three related mapping populations (F_1, F_2 and BC_2) generated for this species (see Section 7.3.3 for crosses) were used. Using SSR anchor loci from the *Prunus* map (Aranzana et al. 2003c) illustrated good colinearity between the three maps and this later reference map. Comparative mapping also revealed that homologous recombination occurred normally between *P. davidiana* and the *Prunus persica* genome, what confirmed the closeness of both species. Higher recombination rates were generally observed between *P. davidiana* and *P. persica* than between *P. amygdalus* and *P. persica*. Thus the three maps were confirmed to provide useful tools for QTL detection and MAS, as well as for assessing efficiency of the breeding scheme to introgress *P. davidiana* genes into peach varieties. In particular these studies provided confidence in QTL detection of powdery mildew resistance in peach x *P. davidiana* crosses. This introgression procedure of the polygenic resistance to powdery mildew from wild *P. davidiana* into peach in the light of molecular data (mapping and QTL detection) was re-examined in 2004 (Foulongne et al. 2004). Three alternative scenarios were compared by Monte-Carlo simulations: 1) random selection of F_1 hybrids, 2) selection on phenotypic data, and 3) selection on molecular markers. Simulations were performed in agreement with the genetic structure of the experimental population and the linkage mapping data available. For the selection of resistance loci, the selection with molecular markers appeared to be more efficient than the phenotypic selection, that itself was more efficient than random selection. Additional information about the genetic control of the resistance, collected after the selection step demonstrated the importance of preserving wild alleles over the whole genome. It showed that when little is known about the QTLs controlling the character to be introgressed, which is the case in fruit breeding, molecular markers have to be used to monitor both the selection of target loci and the preservation of wild alleles on the rest of the genome.

These studies prefigured the concept of genomic selection arisen recently that consists in the integration in the selection procedure of all the information acquired at the genome scale. This selection will rely on models for predicting the global selective value of the individuals. Phenotyping the individuals will thus not mainly aim at selecting them but at updating the prediction models. The new challenge will be to optimize valorization from the knowledge of the genome and from data on variability for the traits of interest in order to accelerate breeding performances. Such an objective is quite new for plant breeding and *a fortiori* in the case of stone fruit crops. It is generally agreed that many breeding programs will now aim at creating new cultivars with durable resistance to predominant pests and diseases together with an increased challenge for adaptation to environments that will be characterized by the global climatic change. If introgressing and pyramiding resistance factors to bio-agressors in the genetic background of a crop variety will be an obliged first step, the second step should be to select for varieties that will have the ability to maintain quantitatively and qualitatively their yield despite the climatic change. Genomic selection will have to be prioritarily developed in this second step.

7.5.2 Genome-wide Association Mapping

Another challenge will be to develop association mapping approaches at the scale of the genome. This new method is based on LD studies as stated by the preliminary results from Aranzana et al. (2010). This work, considering a large set of SSR markers (50 SSRs evenly distributed along the eight linkage groups of the *Prunus* reference map) in a collection of 224 peach cultivars from Europe and North America, supported previous data (Aranzana et al. 2003b) indicating a relatively low level of genetic variation. Results also indicated a strong subpopulation structure and to some extent surprisingly a high level of LD conservation, which may have been caused in part by the fact that most cultivars examined in the study originated from the small set of founders used by the early US breeding programs. These data anticipate the development of tools for genome-wide surveys of variability that will evolve from SSR markers to SNP markers allowing for association genetics studies in *Prunus* crops. This is one of the goals of the EU Breedomics project for peach that should complement these initial data with new phenotypic information together with genome-wide distributed SNP-marker data. Further LD based association studies will lead to the identification of chromosomal regions (major genes, QTL) affecting the phenotypic variation of both the traits of interest and linked molecular markers. It is expected that large SNP-based genotyping will make GW mapping a major tool of *Prunus* breeding.

7.6 Transgenic Breeding

7.6.1 Progress in Transformation of Prunus

Huge efforts have been paid to the objective of mastering plant transformation techniques as a tool for scientific studies (and in particular to elucidate the complex plant-pathogen interactions) and/or for breeding.

In comparison with other crops including many perennials of the Rosaceae family, transgenesis has been poorly developed in stone fruit crops (Petri and Burgos 2005; Srinivasan et al. 2005). Nevertheless like in most other tree species, *Prunus* crops have an enormous advantage in that, after successful integration of the gene of interest in a transgenic line, vegetative propagation, as the most classical method of multiplying fruit trees, provides unlimited production of the transgenic cultivar. Another favorable argument is that very few cultivars have a wide range of production in the national and/or international markets, what would enhance their commercial valorization. Such an example could be apricot, peach and plum cultivars that would express transgenic resistance to PPV. All investigations report *Agrobacterium tumefaciens*-mediated transformation of both mature and juvenile tissues but successful transformation and regeneration of commercial cultivars are limited to a few genotypes or to seedlings (see Petri and Burgos 2005 and Srinivasan et al. 2005 for reviews).

7.6.1.1 Apricot

Sharka appears to be the most important viral disease in *Prunus* and affects apricot very severely. Since Laimer da Camara Machado et al. (1992) showed that transformation of immature embryos of apricot with the sharka coat protein gene (cpPPV) resulted in an increased resistance of the plant to viral disease, transformation studies have sharka resistance as their main objective. In order to avoid the use of seed-derived material and thus to keep the traits of the clonal cultivar, Escalettes et al. (1994) attempted transformation of leaf tissue from mature apricot cultivars but regenerated only chimeric plants. Further studies also described transformation of commercial apricot cultivars without regenerating transformed plants (Burgos and Alburquerque 2003). Successful regeneration of transgenic plants from leaf explants was reported only in 2008 with a limited rate (Petri et al. 2008b) and later with a higher rate (Petri et al. 2008a). Because of the presence of marker genes conferring antibiotic resistance in transgenic plants is a major reason for their public non-acceptance, recent studies in the same group aimed at developing marker-free transformants. Combining a regeneration-promoting *ipt* gene and site-specific recombination allowed the elimination of marker genes together with an increased rate of apricot transformation (Lopez-Noguera et al. 2009).

7.6.1.2 Peach

Agrobacterium-mediated transformation of peach explants including leaf segments, immature embryos, and embryogenic calli was first reported by Scorza et al. (1990). A biolistic transformation of peach embryo callus tissue was described by Ye et al. (1994) but no transgenic shoots were produced. There is only one report of stable transgenic peach plant production. To produce transgenic peach plants, Smigocki and Hammerschlag (1991) used shooty mutant strains of *A. tumefaciens* with an *ipt* gene for cytokinin synthesis and an auxin synthesis gene interrupted by transposon *tn5*. These transformed plants with higher endogenous cytokinin levels produced shoots on a medium with low concentrations of growth regulators and had a marked branching and shorter internodes associated with delayed senescence. Transformation of zygotic embryo explants of stored seeds with a strain of *A. tumefaciens* with a GFP plasmid induced a higher number of transformation events (Perez-Clemente et al. 2004). Various *A. tumefaciens* strains, reporter genes and types of juvenile tissues excised from seeds were also tested to optimize gene transfer into peach explants (Padilla et al. 2006). Somatic embryos were regenerated *in vitro* from one to 3.0 mm long zygotic embryos excised from 25 peach and nectarine genotypes while adventitious shoot regeneration was observed from five peach genotypes (Srinivasan and Scorza 2007). However, no transgenic plants were regenerated from these somatic embryos or adventitious shoots. Recent work showed that although adventitious shoot regeneration has been achieved from juvenile or seedling leaf explants, it remains difficult to regenerate from leaves of mature plants (Zhou et al. 2010). However, efficient and routine regeneration of transgenic peaches is yet to be realized (Srinivasan et al. 2005).

7.6.1.3 Almond

In this species transformation studies was initiated with the objective of reducing the self-incompatibility problems. Transformation assays by Archilletti et al. (1995) and Ainsley et al. (2001) permitted only the production of transgenic calli from leaf explants. Miguel and Oliveira (1999) reported the first successful recovery of transgenic almond plants from leaf explants. Selection strategies were then thoroughly improved using positive or negative selection-based protocols (Ramesh et al. 2006) or the use of acetosyringone (Costa et al. 2006). To improve regeneration rate of transgenic adventitious shoots, recent histological studies identified two main stages of organogenesis. An integrated strategy based on differential gene expression in both of these stages has been designed to identify those key genes involved in organogenesis (Santos et al. 2009).

7.6.1.4 Cherry

Several regeneration systems have been described for sour cherry (Mante et al. 1989; Dolgov and Firsov 1999; Tang et al. 2000, 2002), sweet cherry

(Bhagwat and Lane 2004), and black and wild cherries (Hammatt and Grant 1998) but they did not produce stable transformation of cherry cultivars. Other studies (Druart et al. 1998; Gutierrez-Pesce et al. 1998; Dolgov and Firsov 1999) also could not generate transgenic cherry plants with a normal phenotype. Transgenic plants of cherry rootstock cultivar transformed with *A. rhizogenes* showed abnormal phenotypes (Gutierrez-Pesce et al. 1998). Transformation of a sour-sweet hybrid, with *A. tumefaciens* regenerated GUS positive putative transformants and successful integration of transgene was not demonstrated (Dolgov and Firsov 1999). Additional efforts to transform leaf explants of sour cherry (*Prunus cerasus* L.) scion cv. Montmorency and rootstock cv. Gisela 6 (*P. cerasus* × *P. canescens*) with *A. tumefaciens* proved successful transgene integrations (Song and Sink 2006). More recently transformation, regeneration, and rooting protocol to develop transgenic black cherry (*P. serotina*), a native *Prunus* species commercially valued for lumber production, has been developed from leaf explants using *A. tumefaciens*-mediated transformation to produce plants with the agamous (AG) gene (Liu and Pijut 2010).

7.6.1.5 Plum

A modification of the hypocotyl slice transformation technique described for *P. domestica* (Mante et al. 1991) allowed the transformation of two Japanese plum (*P. salicina*) cultivars, 'Angeleno' and 'Larry Anne', with efficiencies of 0.8 and 0.3%, respectively and the presence of the transgene was confirmed in plants after 18 months of growth in the greenhouse (Urtubia et al. 2008). Efficiency of recovery of transgenic *P. domestica* plum from transformed hypoctyl slices was dramatically improved to 42% by adding 2,4-dichlorophenoxy acetic acid in the cocultivation medium (Petri et al. 2008c). Quality of plum fruits was improved by ectopically expressing antisense peach ACC oxidase in the fruits (Callahan and Scorza 2007). Transformation in the domestic plum (*P. domestica*) will be detailed in the next paragraph for its effective application to PPV resistance.

7.6.2 The Example of Transgenic P. domestica—Plum for PPV Resistance

Plum (*P. domestica*) transformation has successfully produced PPV resistant plum. Scorza et al. (1994) regenerated the first transgenic plum plants from hypocotyl slices, transformed with the coat protein (CP) gene of plum pox virus (PPV-CP), the NPTII selection gene and the GUS reporter gene. Western analysis showed that transgenic plants produced amounts of CP which are generally correlated with amounts of detected mRNA. Based on observations and analyses over three years including two dormancy cycles, one out of five transgenic clones (C5), was found to be resistant to PPV infection whether inoculated by aphids or by chip budding. PPV could not be detected in any inoculated plants of the C5 clone by immunoblot or immunocapture-reverse transcriptase-PCR assays (Ravelonandro et al. 1997). The viral part of the binary construct contains the coat protein (CP) gene plus the 3′ non-translated region of the viral genome. This C5 clone displayed characteristics typical of post-transcriptional gene silencing (PTGS) as described in herbaceous species, with a complex multicopy transgene insertion characterized by a high level of transgene transcription in the nucleus and low levels of transgene mRNA in the cytoplasm (Scorza et al. 2001). These results on the blockage of PPV in transgenic plum C5 contributed to confirm the basic silencing mechanism by illustrating for the first time this phenomenon in a perennial plant (Ravelonandro et al. 2008).

PPV resistance in C5 plum was shown to be inherited when hybridized with the rootstock plum *P. spinosa* (Ravelonandro et al. 2001) and with the French plum cultivars 'Prunier d'Ente 303' and 'Quetsche 29061 (Ravelonandro et al. 2002). In *P. spinosa* hybrids, resistance assays of chip-bud inoculated plants following two cycles of growth and cold-induced dormancy demonstrated that the transgenic hybrid clones were symptomless, DAS-ELISA negative and that immune-blotting failed to detect PPV CP in inoculated transgenic hybrids. In French plum hybrids, histochemical GUS assays and molecular analyses revealed that almost 45% of the hybrids were transgenic, which suggests that C5 clone could be a useful transgenic PPV-resistant germplasm for classical breeding, Field trials were conducted in Poland (Hily et al. 2004; Malinowski et al. 2006), Romania (Ravelonandro et al. 2002) and Spain (Malinowski et al. 2006) and demonstrated resistance of C5 to several D and M strain PPV isolates. It appears that C5 clone is not infected when virus transmission naturally occurs by the aphid vectors and that it becomes infected, although only with mild symptoms on isolated shoots, when artificially inoculated with chip buds (Malinowski et al. 1998, 2006; Hily et al. 2004; Polak et al. 2008). Today this C5 plum represents a unique example of the use of genetic engineering for fruit crop improvement in *Prunus* spp. Regulation requirements for PPV in Chile constrained evaluation of C5 clone shoots, multiplied *in vitro* and micro-grafted onto *Prunus insititia* L. rootstock populations previously infected each with one of four Chilean PPV-D strains, under controlled conditions in a biosafety greenhouse (Wong et al. 2010). Symptoms' appearance, virus detection, and viral mRNA levels for the cylindrical inclusion (CI) and CP gene during three consecutive growing seasons showed complete resistance to all PPV isolates during the first two years in all of the C5 plants. In the third season, four out of ten C5 plants showed mild symptoms on leaves close to the union and low but detectable CI mRNA levels in the C5 scions. These data put forward the interest of a greenhouse test but also highlight the need for long term and field scale evaluations.

PTGS technique was further applied in plum to develop constructs with self-complementary sequences separated by an intron that produce "hairpin" RNA [intron-hairpin-RNA (ihpRNA)] structures (Hilly et al. 2007). ihpRNA constructs provided PPV resistance. The length of the PPV sequence introduced in the constructs was positively correlated with the frequency of transgenic-resistant plants and plants transformed with the full-length sequence produced a higher percentage of resistant lines.

7.6.3 Concluding Remarks

Transformation of stone fruit is still far from routine and additional research efforts are needed to provide the techniques that will allow successful introduction into elite cultivars of genes carrying the trait of interest or contributing to the adaptation of the plant to the biotic or abiotic stresses. In this field the major breakthrough is the successful transformation and regeneration of plants obtained recently from mature tissues for the apricot cultivar 'Helena' even though transformation rates are still low (Petri et al. 2008a, b). Recalcitrance to regenerate adventitious shoots from clonal tissues such as leaf explants is the primary obstacle in producing transgenic stone fruits; however recently Srinivasan et al. (2011) showed that transgenic plum plants ectopically expressing maize *KNOX1* (*knotted-like homeobox*) transcription factor gene produced regeneratively-competent leaves. Ninety six percent of the leaf explants excised from the maize *KNOX1* expressing plum shoots produced *in vitro* up to 38 adventitious shoots per leaf explant. Such *KNOX1* like regeneration inducing genes may be useful to make other stone fruit leaf explants regeneratively-competant to produce adventitious shoot regeneration from transformed cells.

Long (3 to 10 years) juvenile period is another bottleneck in stone fruit breeding in general. To circumvent this problem, Scorza et al. (2011) produced early flowering transgenic plums by transforming *P. domestica* plum with early flower inducing gene cloned from *Poplar trichocarpa* (Bohlenius et al. 2006). These early flowering transgenic plum lines when grown in pots in the greenhouse routinely produce flowers and fruits with fertile seeds within a year. By using these container-grown early flowering transgenic plum lines it is now possible to produce transgenic plum cultivar with gene of interest within 5 years, while the traditional breeding to produce a plum cultivar requires at least 15 years. This 'FasTrack' plum breeding technology is now successfully used to produce plum pox virus-resistant and high sugar containing California Dried Plums (Scorza et al. 2011; http://ucanr.org/sites/fastrack/). This is a typical example to show how transgenic breeding is an approach that might inflect the future fruit tree breeding by increasing their efficiency and effectiveness (Kean 2010). Nevertheless the development of genetically-modified organisms (GMO) is a matter of debate and in Europe it is up-to-now hampered by the public concern and thus prohibited by EU regulations (Directive 2001/18/EEC of the European Parliament and the Council of the European Union). Nevertheless fast breeding could be performed using transgenic lines carrying a single copy of the gene responsible for the trait (and thus segregating into transgenic and non-transgenic progenies). Such early-flowering lines may be used for the production of improved material by successive backcrosses with the material of interest, keeping for the last generation an additional selection step of material lacking the early-flowering gene and thus non-transgenic.

Transgenic plum trees that are highly resistant to Plum pox virus are one example of potential transgenic breeding and in the US they could be the first case of *Prunus* GMO release for growers who are concerned about the threat of PPV. The development and testing of this plum during 15 years in five countries has shown that it could bring significant benefit for the control of sharka disease while providing unique genetic material for use in conventional breeding programs (Hily et al. 2004; Malinowski et al. 2006; Polak et al. 2008). The C5 plum clone, renamed 'Honey Sweet', was granted deregulated status in June 2007 by the USDA Animal and Plant Health Inspection Service (Scorza et al. 2007) and was cleared by the US Food and Drug Agency (FDA) in January 2009.

References

Ainsley PJ, Collins GG, Sedgley M (2001) Factors affecting *Agrobacterium*-mediated gene transfer and the selection of transgenic calli in paper shell almond (*Prunus dulcis* Mill.) J Hort Sci Biotechnol 76: 522–528.

Aranzana MJ, Carbo J, Arus P (2003a) Using amplified fragment-length polymorphisms (AFLPs) to identify peach cultivars. J Am Soc Hort Sci 128: 672–677.

Aranzana MJ, Carbo J, Arus P (2003b) Microsatellite variability in peach *Prunus persica* (L.) Batsch: cultivar identification, marker mutation, pedigree inferences and population structure. Theor Appl Genet 106: 1341–1352.

Aranzana MJ, Pineda A, Cosson P, Dirlewanger E, Ascasibar J, Cipriani G, Ryder CD, Testolin R, Abbott A, King GJ, Iezzoni AF, Arus P (2003c) A set of simple-sequence repeat (SSR) markers covering the *Prunus* genome. Theor Appl Genet 106: 819–825.

Aranzana MJ, Abbassi EK, Howad W, Arus P (2010) Genetic variation, population structure and linkage disequilibrium in peach commercial varieties. BMC Genetics 11.

Aradhya MK, Weeks C, Simon CJ (2004) Molecular characterization of variability and relationships among seven cultivated and selected wild species of *Prunus* L. using amplified fragment length polymorphism. Sci Hort 103: 131–144.

Archilletti T, Lauri P, Damiano C (1995) Agrobacterium-mediated transformation of almond leaf pieces. Plant Cell Rep 14: 267–272.

Arulsekar S, Parfitt DE, Kester DE (1986) Comparison of isozyme variability in peach and almond cultivars. J Hered 77: 272–274.

Arus P, Messeguer R, Viruel M, Tobutt K, Dirlewanger E, Santi F, Quarta R, Ritter E (1994) The European *Prunus* Mapping Project—Progress in the almond linkage map. Euphytica 77: 97–100.
Arus P, Ballester J, Jauregui B, Joobeur T, Truco MJ, de Vicente MC (1998) The European *Prunus* mapping project: update on marker development in almond. Acta Hort 484: 325–330.
Arus P (2006) Molecular markers for variability assessment and map comparison in *Prunus*. In: Proc of the XIIth ISHS Symp on Apricot Culture and Decline, Vols 1 and 2. International Society Horticultural Science, Leuven 1, pp 173–180.
Audergon JM, Dosba F, Karayiannis I, Dicenta F (1994) Amélioration de l'abricotier pour la résistance à la sharka. EPPO Bull 24: 741–748.
Badenes ML, Martinez-Calvo J, Llácer G (2003) SEOPA-1 and GOLGI-2 apricot seedlings are resistant to plum pox virus. Hort Science 38: 135–137.
Ballester J, Boskovic R, Batlle I, Arus P, Vargas F, de Vicente MC (1998) Location of the self-incompatibility gene on the almond linkage map. Plant Breed 117: 69–72.
Ballester J, Socias i Company R, Arus P, De Vicente MC (2001) Genetic mapping of a major gene delaying blooming time in almond. Plant Breed 120: 268–270.
Bartolozzi F, Warburton ML, Arulsekar S, Gradziel TM (1998) Genetic characterization and relatedness among California almond cultivars and breeding lines detected by randomly amplified polymorphic DNA (RAPD) analysis. J Am Soc Hort Sci 123: 381–387.
Battistini S, Sansavini S (1991) Electrophoretic analysis of isozyme variability in apricot cultivars. J Genet Breed 45: 117–122.
Bhagwat B, Lane WD (2004) *In vitro* shoot regeneration from leaves of sweet cherry (*Prunus avium*) 'Lapins' and 'Sweetheart'. Plant Cell Tiss Org Cult 78: 173–181.
Bielenberg DG, Wang Y, Li ZG, Zhebentyayeva T, Fan SH, Reighard GL, Scorza R, Abbott AG (2008) Sequencing and annotation of the evergrowing locus in peach *Prunus persica* (L.) Batsch reveals a cluster of six MADS-box transcription factors as candidate genes for regulation of terminal bud formation. Tree Genet Genomes 4: 495–507.
Blenda AV, Wechter WP, Reighard GL, Baird WV, Abbott AG (2006) Development and characterisation of diagnostic AFLP markers in *Prunus persica* for its response to peach tree short life syndrome. J Hort Sci Biotechnol 81: 281–288.
Bliss FA, Arulsekar S, Foolad MR, Becerra V, Gillen AM, Warburton ML, Dandekar AM, Kocsisne GM, Mydin KK (2002) An expanded genetic linkage map of *Prunus* based on an interspecific cross between almond and peach. Genome 45: 520–529.
Bohlenius H, Huang T, Charbonnel-Campaa L, Brunner AM, Jansson S, Strauss SH, Nilsson O (2006) CO/FT regulatory module controls timing of flowering and seasonal growth cessation in trees. Science 312: 1040–1043.
Boskovic R, Tobutt KR (1996) Correlation of stylar ribonuclease zymograms with incompatibility alleles in sweet cherry. Euphytica 90: 245–250.
Botstein D, White RL, Skolnick M, Davis RW (1980) Construction of a Genetic-Linkage Map in Man Using Restriction Fragment Length Polymorphisms. Am J Hum Genet 32: 314–331.
Burgos L, Alburquerque N (2003) Ethylene inhibitors and low kanamycin concentrations improve adventitious regeneration from apricot leaves. Plant Cell Rep 21: 1167–1174.
Bus VGM, Esmenjaud D, Buck E, Laurens F (2009) Application of genetic markers in Rosaceous crops. In: Volta KM, Gardiner SE (eds) Genetics and Genomics of Rosaceae. Plant Genetics and Genomics: 6. Springer Science+Business Media, New York, pp 563–600.
Byrne DH (1990) Isozyme variability in 4 diploid stone fruits compared with other woody perennial plants. J Hered 81: 68–71.
Byrne DH, Littleton TG (1992) Characterization of isozyme variability in apricot. J Am Soc Hort Sci 114: 674–678.
Callahan A, Scorza R (2007) Effects of peach antisense ACC oxidase gene on plum fruit quality. In: Litz RE, Scorza R (eds) Proc of the Int Symp on Biotechnology of Temperate Fruit Crops and Tropical Species. Acta Hort 738: 567–573.

Casas AM, Igartua E, Balaguer G, Moreno MA (1999) Genetic diversity of *Prunus* rootstocks analyzed by RAPD markers. Euphytica 110: 139–149.

Castagnone-Sereno P (2002) Genetic variability of nematodes: a threat to the durability of plant resistance genes? Euphytica 124: 193–199.

Castagnone-Sereno P, Bongiovanni M, Palloix A, Dalmasso A (1996) Selection for *Meloidogyne incognita* virulence against resistance genes from tomato and pepper and specificity of the virulence/resistance determinants. Eur J Plant Pathol 102: 585–590.

Channuntapipat C, Wirthensohn M, Ramesh SA, Batlle I, Arus P, Sedgley M, Collins G (2003) Identification of incompatibility genotypes in almond (*Prunus dulcis* Mill.) using specific primers based on the introns of the S-alleles. Plant Breed 122: 164–168.

Chaparro JX, Werner DJ, Omalley D, Sederoff RR (1994) Targeted Mapping and Linkage Analysis of Morphological Isozyme, and Rapd Markers in Peach. Theor Appl Genet 87: 805–815.

Chen D, Juarez S, Hartweck L, Alamillo JM, Simon-Mateo C, Pérez JJ, Fernandez-Fernandez MR, Olszewski NE, Garcia JA (2005) Identification of secret agent as the O-GlcNac transferase that participates in plum pox virus infection. J Virol 79: 9381–9387.

Cipriani G, Lot G, Huang W-G, Marrazzo MT, Peterlunger E, Testolin R (1999) AC/GT and AG/CT microsatellite repeats in peach (*Prunus persica* (L.) Batsch): isolation, characterisation and cross-species amplification in *Prunus*. Theor Appl Genet 99: 65–72.

Clarke JB, Sargent DJ, Boskovic RI, Belaj A, Tobutt KR (2009) A cherry map from the inter-specific cross *Prunus avium* 'Napoleon' x P. nipponica based on microsatellite, gene-specific and isoenzyme markers. Tree Genet Genomes 5: 41–51.

Claverie M, Dirlewanger E, Bosselut N, Lecouls AC, Voisin R, Kleinhentz M, Lafargue B, Caboche M, Chalhoub B, Esmenjaud D (2004a) High resolution mapping and chromosome landing at the root-knot nematode resistance locus *Ma* from Myrobalan plum using a large-insert BAC DNA library. Theor Appl Genet 109: 1318–1327.

Claverie M, Bosselut N, Lecouls AC, Voisin R, Poizat C, Dirlewanger E, Kleinhentz M, Lafargue B, Laigret F, Esmenjaud D (2004b) Location of independent root-knot nematode resistance genes in plum and peach. Theor Appl Genet 108: 765–773.

Claverie M, Dirlewanger E, Bosselut N, Van Ghelder C, Voisin R, Kleinhentz M, Lafargue B, Abad P, Rosso MN, Chalhoub B, Esmenjaud D (2011) The *Ma* gene for complete-spectrum resistance to *Meloidogyne* spp. in *Prunus* is a TNL with a huge repeated C-terminal post LRR region. Plant Physiol 156: 779–792.

Cochard H, Barigah ST, Kleinhentz M, Eshel A (2008) Is xylem cavitation resistance a relevant criterion for screening drought resistance amongst *Prunus* species? J Plant Physiol 165: 976–982.

Campoy JA, Martinez-Gomez P, Ruiz D, Rees J, Celton JM (2010) Developing microsatellite multiplex and megaplex PCR systems for high-throughput characterization of breeding progenies and linkage maps spanning the apricot (*Prunus armeciaca* L.) genome. Plant Mol Biol Rep 28: 560–568.

Cook R, Evans K (1987) Resistance and tolerance. In: Brown RH, Kerry BR (eds) Principles and Practice of Nematode Control in Crops. Academic Press, Sydney, pp 179–231.

Costa MS, Miguel C, Oliveira MM (2006) An improved selection strategy and the use of acetosyringone in shoot induction medium increase almond transformation efficiency by 100-fold. Plant Cell Tiss Org Cult 85: 205–209.

Dabov S (1985) 'Malo Konare'—a new canning peach variety. Rastenievj Nauki 22: 84–87.

de Vicente MC, Truco MJ, Egea J, Burgos L, Arus P (1998) RFLP variability in apricot (*Prunus armeniaca* L.). Plant Breed 117: 153–158.

Decroocq V, Foulongne M, Lambert P, Le Gall O, Mantin C, Pascal T, Schurdi-Levraud V, Kervella J (2005) Analogues of virus resistance genes map to QTLs for resistance to sharka disease in *Prunus davidiana*. Mol Genet Genom 272: 680–689.

Dettori MT, Quarta R, Verde I (2001) A peach linkage map integrating RFLPs, SSRs, RAPDs, and morphological markers. Genome 44: 783–790.

Dicenta F, Martinez-Gomez P, Burgos L, Egea J (2000) Inheritance of resistance to *Plum pox potyvirus* (PPV) in apricot, *Prunus armeniaca*. Plant Breed 119: 161–4.

Dirlewanger E, Pronier V, Parvery C, Rothan C, Guye A, Monet R (1998) Genetic linkage map of peach [*Prunus persica* (L.) Batsch] using morphological and molecular markers. Theor Appl Genet 97: 888–895.

Dirlewanger E, Moing A, Rothan C, Svanella L, Pronier V, Guye A, Plomion C, Monet R (1999) Mapping QTLs controlling fruit quality in peach [*Prunus persica* (L.) Batsch]. Theor Appl Genet 98: 18–31.

Dirlewanger E, Pascal T, Zuger C, Kervella J (1996) Analysis of molecular markers associated with powdery mildew resistance genes in peach (*Prunus persica* (L.) Batsch) x *Prunus davidiana* hybrids. Theor Appl Genet 93: 909–919.

Dirlewanger E, Cosson P, Tavaud M, Aranzana MJ, Poizat C, Zanetto A, Arús P, Laigret F (2002) Development of microsatellite markers in peach [*Prunus persica* (L.) Batsch] and their use in genetic diversity analysis in peach and sweet cherry (*Prunus avium* L.). Theor Appl Genet 105: 127–138.

Dirlewanger E, Cosson P, Howad W, Capdeville G, Bosselut N, Claverie M, Voisin R, Poizat C, Lafargue B, Baron O, Laigret F, Kleinhentz M, Arus P, Esmenjaud D (2004a) Microsatellite genetic linkage maps of myrobalan plum and an almond-peach hybrid—location of root-knot nematode resistance genes. Theor Appl Genet 109: 827–838.

Dirlewanger E, Graziano E, Joobeur T, Garriga-Caldere F, Cosson P, Howad W, Arus P (2004b) Comparative mapping and marker-assisted selection in Rosaceae fruit crops. Proc Natl Acad Sci USA 101: 9891–9896.

Dirlewanger E, Capdeville G, Tauzin Y, Cosson P, Claverie J, Laigret F, Moing A (2008) A sweet cherry (*Prunus avium* L.) linkage map and its comparison to other *Prunus* species. In: Proc Vth Int Cherry Symp, Vols 1 and 2. International Society for Horticultural Science, Leuven 1, pp 115–126.

Dolgov SV, Firsov AP (1999) Regeneration and Agrobacterium transformation of sour cherry leaf discs. Acta Hort 484: 577–580.

Dondini L, Lain O, Geuna F, Banfi R, Gaiotti F, Tartarini S, Bassi D, Testolin R (2007) Development of a new SSR-based linkage map in apricot and analysis of synteny with existing *Prunus* maps. Tree Genet Genomes 3: 239–249.

Donoso JM, Aros D, Meneses C, Narvaez C, Infante R (2008) Genetic relationships in apricot (*Prunus armeniaca* L.) using SSR markers and their implications for breeding. J Food Agri Environ 6: 378–382.

Dosba F, Bernhard R, Zanetto A (1994) Importance des resources génétiques des *Prunus*. CR Acad Agri Fr 80: 45–57.

Druart P, Delporte F, Brazda M, Ugarte-Ballon C, da Camara Machado A, Laimer da Camara Machado M, Jacquemin J, Watillon B (1998) Genetic transformation of cherry trees. Acta Hort 468: 71–76.

Durham RE, Moore GA, Sherman WB (1987) Isozyme banding patterns and their usefulness as genetic markers in peach. J Am Soc Hort Sci 112: 1013–1018.

Egea J, Burgos P, Martinez-Gõmez, Dicenta F (1999) Apricot breeding for sharka resistance at CEBAS-CSIC, Murcia (Spain). Acta Hort 488: 153–157.

Escalettes V, Dahuron F, Ravelonandro M and Dosba F (1994) Utilisation de la transgenose pour l'obtention de pruniers et d'abricotiers exprimant le gene de la proteine capside du plum pox potyvirus. Bull OEPP/EPPO 24: 705–711.

Esmenjaud D (2004) Breeding for durable resistance to root-knot nematodes in perennials: a European initiative for *Prunus* rootstocks. In: Cook R, Hunt DJ (eds) Nematology Monographs and Perspectives 2. IV Int Congr of Nematology, Tenerife, Spain, June 8–13, 2002. Brill Press, Leiden, Netherlands, pp 279–287.

Esmenjaud D, Dirlewanger E (2007) Plum, In: Kole C (ed) Genome Mapping and Molecular Breeding in Plants. Vol 4: Fruits and Nuts. Springer, Heidelberg, pp 119–136.

Esmenjaud D, Minot JC, Voisin R (1996a) Effect of durable inoculum pressure and high temperature on root galling, nematode numbers and survival of Myrobalan plum genotypes (*Prunus cerasifera* Ehr.) highly resistant to *Meloidogyne* spp. Fund Appl Nematol 19: 85–90.

Esmenjaud D, Minot JC, Voisin R, Bonnet A, Salesses G (1996b) Inheritance of resistance to the root-knot nematode *Meloidogyne arenaria* in Myrobalan plum. Theor Appl Genet 92: 873–879.

Esmenjaud D, Minot JC, Voisin R, Pinochet J, Simard MH, Salesses G (1997) Differential response to root-knot nematodes in *Prunus* species and correlative genetic implications. J Nematol 29: 372–380.

Esmenjaud D, Voisin R, Van Ghelder C, Bosselut N, Lafargue B, Di Vito M, Dirlewanger E, Poëssel JL, Kleinhentz M (2009) Genetic dissection of resistance to root-knot nematodes *Meloidogyne* spp. in plum, peach, almond and apricot, from various segregating interspecific *Prunus* progenies. Tree Genet Genomes 5: 279–289.

Etienne C, Rothan C, Moing A, Plomion C, Bodenes C, Svanella-Dumas L, Cosson P, Pronier V, Monet R, Dirlewanger E (2002) Candidate genes and QTLs for sugar and organic acid content in peach *Prunus persica* (L.) Batsch. Theor Appl Genet 105: 145–159.

Fan S, Bielenberg DG, Zhebentyayeva TN, Reighard GL, Okie WR, Holland D, Abbott AG (2010) Mapping quantitative trait loci associated with chilling requirement, heat requirement and bloom date in peach (*Prunus persica*). New Phytol 185: 917–930.

Fargette M, Phillips MS, Block VC, Waugh R, Trudgill DL (1996) An RFLP study of relationships between species, populations, and resistance breaking lines of tropical *Meloidogyne*. Fund Appl Nematol 19: 193–200.

Fernández C, Pinochet J, Felipe A (1994b) Influence of temperature on the expression of resistance in six *Prunus* rootstocks infected with *Meloidogyne incognita*. Nematropica 23: 195–202.

Fernández C, Pinochet J, Esmenjaud D, Gravato-Nobre MJ, Felipe A (1995) Age of plant material influences resistance of some *Prunus* rootstocks to *Meloidogyne incognita*. HortScience 30: 582–585.

Fernandez-Fernandez MR, Camafeitat E, Bonay P, Mendez E, Albar JP, Garcia JA (2002) The capsid protein of a plant single-stranded RNA virus is modified by O-linked N-acetylglucosamine. J Biol Chem 277: 135–140.

Flachowsky H, Hanke MV, Peil A, Strauss SH, Fladung M (2009) A review on transgenic approaches to accelerate breeding of woody plants. Plant Breed 128: 217–226.

Foolad MR, Arulsekar S, Becerra V, Bliss FA (1995) A genetic map of *Prunus* based on an interspecific cross between peach and almond. Theor Appl Genet 91: 262–269.

Foulongne M, Pascal T, Pfeiffer F, Kervella J (2002) Introgression of a polygenic resistance to powdery mildew from a wild species *Prunus davidiana* into peach (*Prunus persica* (L.) Batsch), a case study of marker assisted selection in fruit tree. In: Proc 5th Intl Peach Symp. Vols 1 & 2. International Society Horticultural Science, Leuven 1, pp 259–265.

Foulongne M, Pascal T, Arus P, Kervella J (2003a) The potential of *Prunus davidiana* for introgression into peach *Prunus persica* (L.) Batsch assessed by comparative mapping. Theor Appl Genet 107: 227–238.

Foulongne M, Pascal T, Pfeiffer F, Kervella J (2003b) QTLs for powdery mildew resistance in peach x *Prunus davidiana* crosses: consistency across generations and environments. Mol Breed 12: 33–50.

Foulongne M, Pascal T, Pfeiffer F, Kervella J (2004) Re-examination of an introgression procedure of a polygenic resistance to powdery mildew from wild *Prunus davidiana* into peach in the light of molecular data and Monte-Carlo simulations. In: Proc of the XIth Eucarpia Symp on Fruit Breeding and Genetics, Vols 1 and 2. International Society Horticultural Science, Leuven 1, pp 849–854.

Gillen AM, Bliss FA (2005) Identification and mapping of markers linked to the *Mi* gene for root-knot nematode resistance in peach. J Am Soc Hort Sci 130: 24–33.

Granger AR, Clarke GR, Jackson JF (1993) Sweet cherry cultivar identification by leaf isozyme polymorphism. Theor Appl Genet 86: 458–464.

Greco N, Esmenjaud D (2004) Management strategies for nematode control in Europe. In: Cook R, Hunt DJ (eds) Nematology Monographs and Perspectives 2. IV Int Congr of Nematology, Tenerife, Spain, June 8–13, 2002, Brill Press, Leiden, pp 33–43.

Guillet-Bellanger I, Audergon JM (2000) Inheritance of the StarkEarly Orange apricot cultivar resistance to *Plum pox virus*. Acta Hort 550: 111–5.

Gutierrez-Pesce P, Taylor K, Muleo R, Rugini E (1998) Somatic embryogenesis and shoot regeneration from transgenic roots of cherry rootstock Colt (*Prunus avium* × *P. pseudocerasus*) mediated by pRi1855 T-DNA of *Agrobacterium rhizogenes*. Plant Cell Rep 17: 574–580.

Hagen LS, Chaib J, Fady B, Decroocq V, Bouchet JP, Lambert P, Audergon JM (2004) Genomic and cDNA microsatellites from apricot (*Prunus armeniaca* L.). Mol Ecol Notes 4: 742–745.

Hammatt N, Grant NJ (1998) Shoot regeneration from leaves of *Prunus serotina* Ehrh. (Black cherry) and *P. avium* L. (wild cherry). Plant Cell Rep 17: 526–530.

Handoo ZA, Nyczepir AP, Esmenjaud D, van der Beek JG, Castagnone-Sereno P, Carta LK, Skantar AM, Higgins JA (2004) Morphological, molecular and differential-host characterization of *Meloidogyne floridensis* n. sp. (Nematoda: Meloidogynidae), a root-knot nematode parasitizing peach in Florida. J Nematol 36: 20–35.

Hauck NR, Iezzoni AF, Yamane H, Tao R (2001) Revisiting the S-allele nomenclature in sweet cherry (*Prunus avium*) using RFLP profiles. J Am Soc Hort Sci 126: 654–660.

Hily JM, Scorza R, Malinowski T, Zawadzka B, Ravelonandro M (2004) Stability of gene silencing–based resistance to plum pox virus in transgenic plum (*Prunus domestica* L.) under field conditions. Transgen Res 13: 427–436.

Hily JM, Ravelonandro M, Damsteegt V, Bassett C, Petri C, Liu Z, Scorza R (2007) Plum pox virus coat protein gene Intron-hairpin-RNA (ihpRNA) constructs provide resistance to plum pox virus in *Nicotiana benthamiana* and *Prunus domestica*. J Am Soc Hort Sci 132: 850–858.

Hormaza JI (2001) Identification of apricot (*Prunus armeniaca* L.) genotypes using microsatellite and RAPD markers. In: Proc Intl Symp on Molecular Markers for Characterizing Genotypes and Identifying Cultivars in Horticulture. International Society Horticultural Science, Leuven 1, pp 209–215.

Hormaza JI (2002) Molecular characterization and similarity relationships among apricot (*Prunus armeniaca* L.) genotypes using simple sequence repeats. Theor Appl Genet 104: 321–328..

Horn R, Lecouls AC, Callahan A, Dandekar A, Garay L, McCord P, Howad W, Chan H, Verde I, Main D, Jung S, Georgi L, Forrest S, Mook J, Zhebentyayeva T, Yu YS, Kim HR, Jesudurai C, Sosinski B, Arus P, Baird V, Parfitt D, Reighard G, Scorza R, Tomkins J, Wing R, Abbott AG (2005) Candidate gene database and transcript map for peach, a model species for fruit trees. Theor Appl Genet 110: 1419–1428.

Howad W, Yamamoto T, Dirlewanger E, Testolin R, Cosson P, Cipriani G, Monforte AJ, Georgi L, Abbott AG, Arus P (2005) Mapping with a few plants: Using selective mapping for microsatellite saturation of the *Prunus* reference map. Genetics 171: 1305–1309.

Infante R, Martinez-Gomez P, Predieri S (2008) Quality oriented fruit breeding: Peach *Prunus persica* (L.) Batsch. J Food Agri Environ 6: 342–356.

Joobeur T, Viruel MA, de Vicente MC, Jauregui B, Ballester J, Dettori MT, Verde I, Truco MJ, Messeguer R, Batlle I, Quarta R, Dirlewanger E, Arus P (1998) Construction of a saturated linkage map for *Prunus* using an almond x peach F-2 progeny. Theor Appl Genet 97: 1034–1041.

Jauregui B (1998) Localizacion de marcadores moleculares ligados a caracteres agronomicos en un cruzamiento interespecifico almendro x melocotonero. Thesis de la Universidad de Barcelona (Facultad de Biologia), 183 p.

Kadkhodaei S, Shahnazari M, Nekouei MK, Ghasemi M, Etminani H, Imani A, Ariff AB (2011) A comparative study of morphological and molecular diversity analysis among cultivated almonds (*Prunus dulcis*). Aus J Crop Sci 5: 82–91.

Karyiannis I, Mainou A (1999) Apricot breeding in Greece for fruit quality and resistance to plum pox virus. Acta Hort 488: 111–117.

Kean S (2010) Besting Johnny Appleseed. Science 328: 301–303.

Kervella J, Pascal T, Pfeiffer F, Dirlewanger E (1998) Breeding for multiresistance in peach tree. Acta Hort 465: 177–184.

Kervella J, Foulongne M, Kraif R, Pascal T, Pfeiffer F, Pradier M, Moing A, Renaud C, Etienne C, Genard M, Quilot B, Lescourret F, Rothan C, Reich M (2002) *Prunus davidiana* and derived progenies: a valuable material for fruit quality studies. In: Proc of the 5th Int Peach Symp, Vols 1 and 2. International Society Horticultural Science, Leuven 1, pp 109–115.

Kester DE, Asay RN (1986) 'Hansen 2168' and 'Hansen 536': two new *Prunus* rootstock clones. HortScience 21: 331–332.

Kochba J, Spiegel-Roy P (1975) Inheritance of resistance to the root-knot nematode (*Meloidogyne javanica* Chitwood) in bitter almond progenies. Euphytica 24: 453–457.

Kochba J, Spiegel-Roy P (1976) Alnem 1, Alnem 88, Alnem 201 almonds: nematode-resistant rootstock seed source. HortScience 11: 270.

Lacis G, Rashal I, Ruisa S, Trajkovski V, Iezzoni AF (2009) Assessment of genetic diversity of Latvian and Swedish sweet cherry (*Prunus avium* L.) genetic resources collections by using SSR (microsatellite) markers. Sci Hort 121: 451–457.

Laimer da Camara Machado M, da Camara Machado A, Hanzer V, Weiss H, Regner F, Steinkelner H, Mattanovich D, Plail R, Knap E, Kalthoff B, Katinger H (1992) Regeneration of transgenic plants of *Prunus armeniaca* containing the coat protein gene of plum pox virus. Plant Cell Rep 11: 25–29.

Lalli D, Abbott A, Zhebentyayeva TN, Badenes ML, Damsteegt V, Polak J, Kirska B, Salava J (2008) A genetic linkage map for an apricot (*Prunus armeniaca* L.) BC1 population mapping *Plum pox virus* resistance. Tree Genet Genomes 4: 481–93.

Lambert P, Hagen LS, Arus P, Audergon JM (2004) Genetic linkage maps of two apricot cultivars (*Prunus armeniaca* L.) compared with the almond Texas x peach Earlygold reference map for *Prunus*. Theor Appl Genet 108: 1120–1130.

Lambert P, Dicenta F, Rubio M, Audergon JM (2007) QTL analysis of resistance to sharka disease in the apricot (*Prunus armeniaca* L.) Polonais' x 'Stark Early Orange' F1 progeny. Tree Genet Genomes 3: 299–309.

Lambert P, Pascal T (2011) Mapping of the *Rm2* gene conferring resistance to *Myzus persicae* in the peach cultivar "Rubira". Tree Genet Genomes: DOI 10.1007/s11295-011-0394-2.

Lamberti F (1979) Economic importance of *Meloidogyne* spp. in subtropical and mediterranean climates. In: Lamberti F, Taylor CE (eds) Root-Knot Nematodes (*Meloidogyne* Species): Systematics, Biology and Control. Academic Press, New York, pp 341–357.

Layne REC (1987) Peach rootstocks. In: Rom RC, Carlson RF (eds) Rootstocks for Fruit Crops. John Wiley, New York, pp 185–216.

Lecouls AC, Salesses G, Minot JC, Voisin R, Bonnet A, Esmenjaud D (1997) Spectrum of the *Ma* genes for resistance to *Meloidogyne* spp. in Myrobalan plum. Theor Appl Genet 95: 1325–1334.

Lecouls AC, Rubio-Cabetas MJ, Minot JC, Voisin R, Bonnet A, Salesses G, Dirlewanger E, Esmenjaud D (1999) RAPD and SCAR markers linked to the *Ma1* root-knot nematode resistance gene in Myrobalan plum (*Prunus cerasifera* Ehr.). Theor Appl Genet 99: 328–335.

Lecouls AC, Bergougnoux V, Rubio-Cabetas MJ, Bosselut N, Voisin R, Poessel JL, Faurobert M, Bonnet A, Salesses G, Dirlewanger E, Esmenjaud D (2004) Marker-assisted selection for the wide-spectrum resistance to root-knot nematodes conferred by the *Ma* gene from Myrobalan plum (*Prunus cerasifera*) in interspecific *Prunus* material. Mol Breed 13: 113–124.

Liu X, Pijut PM (2010) Agrobacterium-mediated transformation of mature *Prunus serotina* (black cherry) and regeneration of transgenic shoots. Plant Cell Tiss Org Cult 101: 49–57.

Lopez-Noguera S, Petri C, Burgos L (2009) Combining a regeneration-promoting ipt gene and site-specific recombination allows a more efficient apricot transformation and the elimination of marker genes. Plant Cell Rep 28: 1781–1790.

Lu ZX, Reighard GL, Baird WV, Abbott AG, Rajapakse S (1996) Identification of peach rootstock cultivars by RAPD markers. HortScience 31: 127–129.

Lu ZX, Sosinski B, Reighard GL, Baird WV, Abbott AG (1998) Construction of a genetic linkage map and identification of AFLP markers for resistance to root-knot nematodes in peach rootstocks. Genome 41: 199–207.

Lu ZX, Sossey-Alaoui K, Reighard GL, Baird WV, Abbott AG (1999) Development and characterization of a codominant marker linked to root-knot nematode resistance, and its application to peach rootstock breeding. Theor Appl Genet 99: 115–122.

Lu ZX, Reighard GL, Nyczepir AP, Beckman TG, Ramming DW (2000) Inheritance of resistance to root-knot nematodes (*Meloidogyne* sp.) in *Prunus* rootstocks. HortScience 35: 1344–1346.

Ma RC, Oliveira MM (2001) Molecular identification of S-genotypes of almond (*Prunus dulcis*). In: Proc of the Intl Symp on Molecular Markers for Characterizing Genotypes and Identifying Cultivars in Horticulture. International Society Horticultural Science, Leuven 1, pp 575–580.

Malinowski T, Zawadzka B, Ravelonandro M, Scorza R (1998) Preliminary report on the apparent breaking of resistance of a transgenic plum by chip bud inoculation of plumpox virus PPV-S. Acta Virol 42: 241–243.

Malinowski T, Cambra M, Capote N, Zawadzka B, Gorris MT, Scorza R, Ravelonandro M (2006) Field trials of plum clones transformed with the Plum pox virus coat protein (PPV-CP) gene. Plant Dis 90: 1012–1018.

Mante S, Scorza R, Cordts JM (1989) Plant-Regeneration from Cotyledons of *Prunus persica, Prunus domestica,* and *Prunus cerasus*. Plant Cell Tiss Org Cult 19: 1–11.

Mante S, Morgens P, Scorza R, Cordts J, Callahan A (1991) *Agrobacterium*-mediated transformation of plum (*Prunus domestica*) hypocotyls slices and regeneration of transgenic plants. Bio/Technology 9: 853–857.

Marandel G, Pascal T, Candresse T, Decroocq V (2009) Quantitative resistance to Plum pox virus in *Prunus davidiana* P1908 linked to components of the eukaryotic translation initiation complex. Plant Pathol 58: 425–435.

Marchese A, Tobutt KR, Caruso T (2005) Molecular characterisation of Sicilian *Prunus persica* cultivars using microsatellites. J Hort Sci Biotechnol 80: 121–129.

Marchese A, Tobutt KR, Caruso T (2006) The Sicilian peach (*Prunus persica* L. Batsch) germplasm: Evaluation of genetic diversity using SSRs. In: Proc of the VIth Intl Peach Symp. International Society Horticultural Science, Leuven 1, pp 135–138.

Martinez-Gomez P, Dicenta F (2000) Evaluation of resistance of apricot cultivars to a Spanish isolate of *Plum pox potyvirus* (PPV). Plant Breed 119: 179–81.

Martinez-Mora C, Rodriguez J, Cenis JL, Ruiz-Garcia L (2009) Genetic variability among local apricots (*Prunus armeniaca* L.) from the Southeast of Spain. Span J Agri Res 7: 855–868.

Marull J, Pinochet J, Felipe A, Cenis JL (1994) Resistance verification in *Prunus* selections to a mixture of 13 *Meloidogyne* isolates and resistance mechanisms of a peach-almond hybrid to *M. javanica*. Fund Appl Nematol 17: 85–92.

Massonie G, Maison R, Monet R, Grasselly C (1982) Resistance au puceron vert du pecher, *Myzus persicae* Sulzer (Homoptera: Aphididae) chez *Prunus persica* (L.) Batsch et d'autres especes de *Prunus*. Agronomie 2: 63–70.

Mnejja M, Garcia-Mas J, Audergon JM, Arus P (2010) *Prunus* microsatellite marker transferability across rosaceous crops. Tree Genet Genomes 6: 689–700.

Monnet R, Massonié G (1994) Déterminisme génétique de la résistance au puceron vert (*Myzus persicae*) chez le pêcher. Résultats complémentaires. Agronomie 2 : 177–182.

Mowrey BD, Werner DJ, Byrne DH (1990) Inheritance of isocitrate dehydrogenase, malate dehydrogenase, and Shikimate dehydrogenase in peach and peach x almond hybrids. J Am Soc Hort Sci 115: 312–319.

Nyczepir A, Esmenjaud D (2008) Peach nematodes. In: Wayne DR, Bassi D (eds) The Peach: Botany, Production and Uses. CAB Editions, New York, pp 505–535.

Okie WR, Beckman TG, Nyczepir AP, Reighard GL, Newell WC Jr, Zehr EI (1994) BY520-9, a peach rootstock for the southeastern United States that increases scion longevity. HortScience 29, 705–706.

Olukolu BA, Trainin T, Fan SH, Kole C, Bielenberg DG, Reighard GL, Abbott AG, Holland D (2009) Genetic linkage mapping for molecular dissection of chilling requirement and budbreak in apricot (*Prunus armeniaca* L.). Genome 52: 819–828.

Ortiz A, Renaud R, Calzada I, Ritter E (1997) Analysis of plum cultivars with RAPD markers. J Hort Sci 72: 1–9.

Padilla IMG, Golis A, Gentile A, Damiano C, Scorza R (2006) Evaluation of transformation in peach *Prunus persica* explants using green fluorescent protein (GFP) and beta-glucuronidase (GUS) reporter genes. Plant Cell Tiss Org Cult 84: 309–314.

Panda S, Martin JP, Aguinagalde I (2003) Chloroplast DNA study in sweet cherry cultivars (*Prunus avium* L.) using PCR-RFLP method. Genet Resour Crop Evol 50: 489–495.

Paran I, Michelmore RW (1993) Development of reliable PCR-based markers linked to downy mildew resistance genes in lettuce. Theor Appl Genet 85: 985–993.

Pascal T, Kervella J, Pfeiffer F, Sauge MH, Esmenjaud D (1998) Evaluation of the interspecific progeny *Prunus persica* cv. Summergrand x *Prunus.davidiana* for disease resistance and some agronomic features. Acta Hort 465: 185–192.

Pascal T, Pfeiffer F, Kervella J (2002) Preliminary observations on the resistance to sharka in peach and related species. In: Proc of the 5th Intl Peach Symp, Davis (CA), 8–11 July 2001, Vols 1 & 2, pp 699–706.

Pascal T, Pfeiffer F, Kervella J (2010) Powdery mildew resistance in the peach cultivar Pamirskij 5 is genetically linked with the Gr gene for leaf color. HortScience 45: 150–152.

Perez-Clemente RM, Perez-Sanjuan A, Garcia-Ferriz L, Beltran JP, Canas LA (2004) Transgenic peach plants (*Prunus persica* L.) produced by genetic transformation of embryo sections using the green fluorescent protein (GFP) as an *in vivo* marker. Mol Breed 14: 419–427.

Petri C, Burgos L (2005) Transformation of fruit trees. Useful breeding tool or continued future prospect? Transgen Res 14: 15–26.

Petri C, Lopez-Noguera S, Alburquerque N, Egea J, Burgos L (2008a) An antibiotic-based selection strategy to regenerate transformed plants from apricot leaves with high efficiency. Plant Sci 175: 777–783.

Petri C, Wang H, Alburquerque N, Faize M, Burgos L (2008b) *Agrobacterium*-mediated transformation of apricot (*Prunus armeniaca* L.) leaf explants. Plant Cell Rep 27: 1317–1324.

Petri C, Webb K, Hily JM, Dardick C, Scorza R (2008c) High transformation efficiency in plum (*Prunus domestica* L.): a new tool for functional genomics studies in *Prunus* spp. Mol Breed 22: 581–591.

Petri C, Scorza R (2010) Factors affecting adventitious regeneration from *in vitro* leaf explants of 'Improved French' plum, the most important dried plum cultivar in the USA. Ann Appl Biol 156: 79–89.

Pilarova P, Marandel G, Decroocq V, Salava J, Krska B, Abbott AG (2010) Quantitative trait analysis of resistance to plum pox virus in the apricot F1 progeny "Harlayne" x "Vestar". Tree Genet Genomes 6: 467–475.

Pinochet J, Anglés M, Dalmau E, Fernández C, Felipe A (1996a) *Prunus* rootstock evaluation to root-knot and lesion nematodes in Spain. J Nematol 28: 616–623.

Pinochet J, Fernández C, Alcañiz E, Felipe A (1996b) Damage by a lesion nematode, *Pratylenchus vulnus*, to *Prunus* rootstocks. Plant Dis 80: 754–757.

Polak J, Pivalova J, Kundu JK, Jokes M, Scorza R, Ravelonandro M (2008) Behaviour of transgenic Plum pox virus-resistant *Prunus domestica* L. clone C5 grown in the open field under a high and permanent infection pressure of the PPV-Rec strain. J Plant Pathol 90: 33–36.

Rajapakse S, Belthoff LR, He G, Estager AE, Scorza R, Verde I, Ballard RE, Baird WV, Callahan A, Monet M, Abbott AG (1995) Genetic linkage mapping in peach using morphological, RFLP and RAPD markers. Theor Appl Genet 90: 503–510.

Ramesh SA, Kaiser BN, Franks T, Collins G, Sedgley M (2006) Improved methods in Agrobacterium-mediated transformation of almond using positive (mannose/pmi) or negative (kanamycin resistance) selection-based protocols. Plant Cell Rep 25: 821–828.

Rammah A, Hirschmann H (1988) *Meloidogyne mayaguensis* n. sp. (*Meloidogynidae*), a root-knot nematode from Puerto Rico. J Nematol 20: 58–69.

Ramming DW, Cociu V (1991) Plum (*Prunus*). In: Moore JV, Ballington JR (eds) Genetic Resources of Temperate Fruit and Nut Crops. Acta Hort 290: 239–288.

Ramming DW, Tanner O (1983) Nemared peach rootstock. HortScience 18: 376.

Ravelonandro M, Scorza R, Bachelier JC, Labonne G, Levy L, Damsteegt V et al. (1997) Resistance of transgenic *Prunusdomestica* to plum pox virus infection. Plant Dis 81: 1231–1235.

Ravelonandro M, Briard P, Renaud R, Scorza R (2001) Transgene-based resistance to plum pox virus (Sharka disease) is transferred through interspecific hybridization in *Prunus*. In: Proc of the Intl Symp on Molecular Markers for Characterizing Genotypes and Identifying Cultivars in Horticulture. Montpellier, France, 6–8 March 2000, pp 569–574.

Ravelonandro M, Briard P, Monsion M, Scorza R, Renaud R (2002) Stable transfer of the plum pox virus (PPV) capsid transgene to seedlings of two French cultivars 'Prunier d'Ente 303' and 'Quetsche 2906', and preliminary results of PPV challenge assays. In: Proc of the 7th Intl Symp on Plum and Prune Genetics Breeding and Pomology, Plovdiv, Bulgaria, 20–24 August 2001, pp 91–96.

Ravelonandro M, Scorza R, Minoiu N, Zagrai I, Platon I (2002) Field tests of transgenic plums in Romania. Sănătatea Plant 6: 16–17.

Ravelonandro M, Briard P, Kundu J, Monsion M, Hily JM, Scorza R (2008) Silencing in Prunus: A natural defense developed by woody fruit trees in response to virus infection. In: Procs of the 20th Intl Symp on Virus and Virus-Like Diseases of Temperate Fruit Crops—Fruit Tree Diseases, Antalya, Turkey, 22–26 May 2006, pp 27–31.

Raz A, Stern RA, Bercovich D, Goldway M (2009) SFB-based S-haplotyping of apricot (*Prunus armeniaca*) with DHPLC. Plant Breed 128: 707–711.

Rehder A (1954) Manual of Cultivated Trees and Shrubs. 2nd edn. Dioscorides Press, Portland, OR 996 pp.

Ricciardi L, Girogio V, De Giovanni C, Lotti C, Gallotta A, Fanizza G (2002) The genetic diversity of apulian apricot genotypes (*Prunus armeniaca* L.) assessed using AFLP markers. Cell Mol Biol Lett 7: 431–436.

Rom RC, Carlson RF (1987) Rootstocks for Fruit Crops. John Wiley, New York 494 pp.

Rubio M, Pascal T, Bachellez A, Lambert P (2010) Quantitative trait loci analysis of Plum pox virus resistance in *Prunus davidiana* P1908: new insights on the organization of genomic resistance regions. Tree Genet Genomes 6: 291–304.

Rubio-Cabetas MJ, Lecouls AC, Salesses G, Bonnet A, Minot JC, Voisin R, Esmenjaud D (1998) Evidence of a new gene for high resistance to *Meloidogyne* spp. in Myrobalan plum (*Prunus cerasifera*). Plant Breed 117: 567–571.

Rubio-Cabetas MJ, Minot JC, Voisin R, Esmenjaud D (2001) Interaction of root-knot nematodes (RKN) and *Agrobacterium tumefaciens* in roots of *Prunus cerasifera*.: Evidence of the protective effect of the *Ma* RKN resistance against expression of crown gall symptoms. Eur J Plant Pathol 107: 433–441.

Ryan FJ, Ledbetter CA, Ramming DW, Palmquist DE, Bell DE, Peterson SJ (2001) Challenges in developing molecular markers for almond (*Prunus amygdalus*) and grape (*Vitis* species). In: Proc of the Intl Symp on Molecular Markers for Characterizing ecotypes and Identifying Cultivars in Horticulture. International Society for Horticultural Science, Leuven 1, pp 629–639.

Santos AM, Oliver MJ, Sanchez AM, Payton PR, Gomes JP, Miguel C, Oliveira MM (2009) An integrated strategy to identify key genes in almond adventitious shoot regeneration. J Exp Bot 60: 4159–4173.

Sanchez-Perez R, Martinez-Gomez P, Dicenta F, Egea J, Ruiz D (2006) Level and transmission of genetic heterozygosity in apricot (*Prunus armeniaca* L.) explored using simple sequence repeat markers. Genet Resour Crop Evol 53: 763–770.

Sasser JN (1979) Pathogenicity, host range and variability in *Meloidogyne* species. In: Lamberti F, Taylor CE (eds) Root-Knot Nematodes (*Meloidogyne* Species): Systematics, Biology and Control. Academic Press, New York, pp 257–268.

Sasser JN, Freckman DW (1987) A world perspective in nematology: The role of the society. In: Veech JA, Dickson DW (eds) Vistas on Nematology. Society of Nematologists, Hyattsville, Maryland, pp 7–14.

Sauge MH, Kervella J, Rahbe Y (1998) Probing behaviour of the green aphid *Myzus persicae* on resistant *Prunus* genotypes. Entomol Exp Appl 86: 223–232.

Sauge MH, Lacroze JP, Poessel JL, Pascal T, Kervella J (2002) Induced resistance by *Myzus persicae* in the peach cultivar Rubira. Entomol Exp Appl 102: 29–37.

Sauge MH, Pascal T, Lacroze JP, Pfeiffer F, Kervella J (2004) Mapping of a genetic factor of partial resistance to *Myzus persicae* in the wild peach *Prunus davidiana*, that impedes phloem sap ingestion by the aphid. In: Aphids in a New Millennium. INRA Edn, Versailles, France, pp 499–505.

Sauge MH, Lambert P, Pascal T (2011) QTLs linked to the feeding behaviour of the aphid Myzus persicae co-localize with host plant resistance factors. Heredity: DOI: 10.1038/hdy.2011.74.

Scotto La Massese C, Grasselly C, Minot JC, Voisin R (1984) Différence de comportement de 23 clones et hybrides de *Prunus* à l'égard de quatre espèces de *Meloidogyne*. Rev Nematol 7: 265–270 (in French).

Scorza R, Morgens PH, Cordts JM, Mante S, Callahan AM (1990) *Agrobacterium*-mediated transformation of peach (*Prunus persica* Batsch, L.) leaf segments, immature embryos, and long-term embryogenic callus. In Vitro Cell Dev Biol 26: 829–834.

Scorza R, Ravelonandro M, Callahan AM, Cordts JM, Fuchs M, Dunez J, Gonsalves D (1994) Transgenic plums (*Prunus domestica* L.) express the plum pox virus coat protein gene. Plant Cell Rep 14: 18–22.

Scorza R, Callahan AM, Levy L, Damsteegt V, Ravelonandro M (2001) Resistance to plum pox potyvirus in a transgenic woody perennial fruit tree, European plum (*Prunus domestica* L.) results from post-transcriptional gene silencing. In: Proc of the 18th Intl Symp on Virus & Virus-Like Diseases of Temperate Fruit Crops—Top Fruit Diseases, Canterbury, England, 9–15 July 2000, Vols 1 & 2. pp 425–430.

Scorza R, Melnicenco L, Dang P, Abbott AG (2002) Testing a microsatellite marker for selection of columnar growth habit in peach *Prunus persica* (L.) batsch. In: Proc of the 5th Intl Peach Symp, Davis (CA), 8–11 July 2001, Vols 1 and 2, pp 285–289.

Scorza R, Hily JM, Callahan A, Malinowski T, Cambra M, Capote N, Zagrai I, Damsteegt V, Briard P, Ravelonandro M (2007) Deregulation of plum pox resistant transgenic plum 'Honey Sweet'. In: Proc of the Intl Symp on Biotechnology of Temperate Fruit Crops and Tropical Species. International Society Horticultural Science, Leuven 1, pp 669–673.

Scorza R, Srinivasan C, Dardick C, Callahan A (2011) "FasTracking" Plum Breeding. USDA-ARS Agri Res 59(3): 16–17.

Semblat JP, Bongiovanni M, Wajnberg E, Dalmasso A, Abad P, Castagnone-Sereno P (2000) Virulence and molecular diversity of parthenogenetic root-knot nematodes, *Meloidogyne* spp. Heredity 84: 81–89.

Sherman WB, Lyrene, PM, Sharpe RH (1991) Flordaguard peach rootstock. HortScience 26: 427–428.

Smigocki AC, Hammerschlag FA (1991) Regeneration of plants from peach embryo cells infected with a shooty mutant strain of *Agrobacterium*. J Am Soc Hort Sci 116: 1092–1097.

Song G-Q, Sink KC (2005) Optimizing shoot regeneration and transient expression factors for *Agrobacterium tumefaciens* transformation of sour cherry (*Prunus cerasus* L.) cultivar Montmorency. Sci Hort 106: 60–69.

Song GQ, Sink KC (2006) Transformation of Montmorency sour cherry (*Prunus cerasus* L.) and Gisela 6 (*P. cerasus x P. canescens*) cherry rootstock mediated by *Agrobacterium tumefaciens*. Plant Cell Rep 25: 117–123.

Soriano JM, Vilanova S, Romero C, Llacer G, Badenes M (2005) Characterization and mapping of NBS-LRR resistance gene analogs in apricot (*Prunus armeniaca* L.). Theor Appl Genet 110: 980–989.

Soriano JM, Vera-Ruiz EM, Vilanova S, Martinez-Calvo J, Llacer G, Badenes ML, Romero C (2008) Identification and mapping of a locus conferring plum pox virus resistance in two apricot-improved linkage maps. Tree Genet Genomes 4: 391–402.

Sosinski B, Gannavarappu M, Hager LD, Beck LE, King GJ, Ryder CD, Rajapakse S, Baird WV, Ballard RE, Abbott AG (2000) Characterization of microsattelite markers in peach [*Prunus persica* (L.) Batsch]. Theor Appl Genet 97: 1034–1041.

Srinivasan C, Padilla IMG, Scorza R (2005) *Prunus* spp. Almond, apricot, cherry, nectarine, peach and plum . In: Litz RE (ed) Biotechnology of Fruit and Nut Crops. CAB International, Oxfordshire, UK, pp 512–542.

Srinivasan C, Scorza R (2007) The influence of genotype on the induction of somatic embryos from in vitro cultured zygotic embryos and adventitious shoot regeneration from cotyledons of peach and nectarine. Acta Hort 738: 691–696.

Srinivasan C, Liu, Z, Scorza R (2011) Ectopic expression of class 1 *KNOX1* genes induce adventitious regeneration and alter growth and development of tobacco (*Nicotiana tabacum* L) and European plum (*Prunus domestica* L.). Plant Cell Rep 30: 655–664.

Struss D, Ahmad R, Southwick SM, Boritzki M (2003) Analysis of sweet cherry (*Prunus avium* L.) cultivars using SSR and AFLP markers. J Am Soc Hort Sci 128: 904–909.

Tamura M, Ushijima K, Sassa H, Hirano H, Tao R, Gradziel TM, Dandekar AM (2000) Identification of self-incompatibility genotypes of almond by allele-specific PCR analysis. Theor Appl Genet 101: 344–349.

Tanksley SD, Young ND, Patterson AH, Bonierbale MW (1989) RFLP mapping in plant breeding: New tool for an old science. Bio/Technology 7: 257–264.

Tao R, Yamane H, Sassa H, Mori H, Gradziel TM, Dandekar AM, Sugiura A (1997) Identification of stylar RNases associated with gametophytic self-incompatibility in almond (*Prunus dulcis*). Plant Cell Physiol 38: 304–311.

Tang HR, Ren ZL, Krczal G (2000) Somatic embryogenesis and organogenesis from immature embryo cotyledons of three sour cherry cultivars (*Prunus cerasus* L.). Sci Hort 83, 109–126.

Tang HR, Ren ZL, Reustle G, Krczal G (2002) Plant regeneration from leaves of sweet and sour cherry cultivars. Sci Hort 93: 235–244.

Tautz D (1989) Hypervariability of simple sequences as a general source for polymorphic DNA markers. Nucl Acids Res 17: 6463–6471.

Testolin R, Marrazzo T, Cipriani G, Quarta R, Verde I, Dettori MT, Pancaldi M, Sansavini S (2000) Microsatellite DNA in peach [*Prunus persica* (L.) Batsch] and its use in fingerprinting and testing the genetic origin of cultivars. Genome 43: 512–520.

This P, Cuisset C, Boursiquot JM (1997) Development of stable RAPD markers for the identification of grapevine rootstocks and the analysis of genetic relationships. Am J Enol Vitic 48: 492–501.

Urtubia C, Devia J, Castro A, Zamora P, Aguirre C, Tapia E, Barba P, Dell'Orto P, Moynihan MR, Petri C, Scorza R, Prieto H (2008) *Agrobacterium*-mediated genetic transformation of *Prunus salicina*. Plant Cell Rep 27: 1333–1340.

Van Ghelder C, Lafargue B, Dirlewanger E, Ouassa A, Voisin R, Polidori J, Kleinhentz M, Esmenjaud D (2010) Characterization of the *RMja* gene for resistance to root-knot nematodes in almond: spectrum, location, and interest for Prunus breeding. Tree Genet Genomes 6: 503–511.

Vilanova S, Romero C, Abbott AG, Llacer G, Badenes ML (2003) An apricot (*Prunus armeniaca* L.) F2 progeny linkage map based on SSR and AFLP markers, mapping plum pox virus resistance and self-incompatibility traits. Theor Appl Genet 107: 239–247.

Viruel MA, Madur D, Dirlewanger E, Pascal T, Kervella J (1998) Mapping quantitative trait loci controllong peach leal curl resistance. Acta Hort 465: 79–97.

Vos P, Hogers R, Bleeker M, Reijans M, Vandelee T, Hornes M, Frijters A, Pot J, Peleman J, Kuiper M, Zabeau M (1995) AFLP—a New Technique for DNA-Fingerprinting. Nucl Acids Res 23: 4407–4414.

Wang Y, Georgi LL, Reighard GL, Scorza R, Abbott AG (2002) Genetic mapping of the evergrowing gene in peach [*Prunus persica* (L.) Batsch]. J Hered 93: 352–358.

Warburton ML, BecerraVelasquez VL, Goffreda JC, Bliss FA (1996) Utility of RAPD markers in identifying genetic linkages to genes of economic interest in peach. Theor Appl Genet 93: 920–925.

Warburton ML, Bliss FA (1996) Genetic diversity in peach (*Prunus persica* L batch) revealed by randomly amplified polymorphic DNA (RAPD) markers and compared to inbreeding coefficients. J Am Soc Hort Sci 121: 1012–1019.

Welsh J, McClelland M (1990) Fingerprinting genomes using PCR with arbitrary primers. Nucl Acids Res 18: 7213–7218.

Weinberger, JH, Marth PC, Scott DH (1943) Inheritance study of root-knot nematode resistance in certain peach varieties. Proc Am Soc Hort Sci 42: 321–325.

Williams JGK, Kubelik AR, Livak KJ, Rafalski JA, Tingey S (1990) DNA polymorphisms amplified by arbitrary primers are useful as genetic markers. Nucl Acids Res 18: 6531–6535.

Williamson VM (1998) Root-knot nematode resistance genes in tomato and their potential for future use. Annu Rev Phytopathol 36: 277–93.

Williamson VM, Gleason CA (2003) Plant–nematode interactions. Curr Opin Plant Biol 6: 327–333.

Williamson VM, Kumar A (2006) Nematode resistance in plants: the battle underground. Trends Genet 22: 396–403.

Wong W, Barba P, Alvarez C, Castro A, Acuna M, Zamora P, Rosales M, Dell'Orto P, Moynihan MR, Scorza R, Prieto H (2010) Evaluation of the resistance of transgenics C5 plum (*Prunus domestica* L.) against four Chilean plum pox virus isolates through micro-grafting. Chil J Agri Res 70: 372–380.

Yaegaki H, Shimada T, Moriguchi T, Hayama H, Haji T, Yamaguchi M (2001) Molecular characterization of S-RNase genes and S-genotypes in the Japanese apricot (*Prunus mume* Sieb. et Zucc.). Sex Plant Reprod 13: 251–257.

Yamamoto T, Hayashi T (2002) New root-knot nematode resistance genes and their STS markers in peach. Sci Hort 96: 81–90.

Yamamoto T, Shimada T, Imai T, Yaegaki H, Haji T, Matsuta N, Yamaguchi M, Hayashi T (2001) Characterization of morphological traits based on a genetic linkage map in peach. Breed Sci 51: 271–278.

Yamane H, Fukuta K, Matsumoto D, Hanada T, Mei G, Esumi T, Habu T, Fuyuhiro Y, Ogawa S, Yaegaki H, Yamaguchi M, Tao R (2009) Characterization of a novel self-compatible S-3 'haplotype leads to the development of a universal PCR marker for two distinctly originated self-compatible S haplotypes in Japanese apricot (*Prunus mume* Sieb. et Zucc.). J Jpn Soc Hort Sci 78: 40–48.

Ye XJ, Brown SK, Scorza R, Cordts J, Sanford JC (1994) Genetic-transformation of peach tissues by particle bombardment. J Am Soc Hort Sci 119: 367–373.

Zhou L, Kappel F, Wiersma PA, Hampson C, Bakkeren G (2005) Genetic analysis and DNA fingerprinting of sweet cherry cultivars and selections using amplified fragment length polymorphisms (AFLP). In: Proc of the IVth Intl Cherry Symp, Hood River, OR, 24–29 June 2001, Vols 1 & 2, pp 37–44.

Zhou H, Li M, Zhao X, Fan X, Guo A (2010) Plant regeneration from in vitro leaves of the peach rootstock 'Nemaguard' (*Prunus persica* x *P. davidiana*). Plant Cell Tiss Org Cult 101: 79–87.

Zhu M, Zhang XM, Zhang KC, Jiang LJ, Zhang LM (2004) Development of a simple molecular marker specific for detecting the self-compatible S4' haplotype in sweet cherry (*Prunus avium* L.). Plant Mol Biol Rep 22: 387–398.

8

Map-based Cloning of Single Gene Traits and Quantitative Traits

Renate Horn,[1,a,*] *Osama Sajer,*[1,b] *Daniel Esmenjaud,*[2]
Michel Claverie[3] and *Elisabeth Dirlewanger*[4]

ABSTRACT

In stone fruits, map-based cloning approaches to isolate genes for functional analyses have been successful, even in times of the tedious and time-consuming chromosome walking via contig construction using BAC libraries. Small genome sizes and diploid status of most *Prunus* species as well as the availability of genetic resources (*Prunus* reference genetic map based on SSR markers and framework physical map based on DNA fingerprinting of BAC libraries) were prerequisites for the success. In this chapter, map-based cloning strategies and the genetic resources utilized will be outlined. Genes for traits including *evergrowing*, self-incompatibility and *Meloidogyne* resistance were successfully isolated using this approach. With the peach genome sequence now available allowing an in silico walk in the region of interest, map-based cloning has become a very powerful tool for isolation of genes of unknown function in stone fruits. Ongoing and

[1]Institute of Biological Sciences, Department of Plant Genetics, University of Rostock, Albert-Einstein-Str. 3, 18051 Rostock, Germany.
[a]e-mail: *renate.horn@uni-rostock.de*
[b]e-mail: *osama.sajer@hotmail.de*
[2]Equipe Interactions Plantes-Nématodes, INRA, Centre de recherche Provence-Alpes-Côte d'Azur, UMR ISA (Institut Sophia Agrobiotech), INRA/CNRS/Université de Nice, 400 route des Chappes, BP 167, 06903 Sophia Antipolis Cedex, France; e-mail: *daniel.esmenjaud@sophia.inra.fr*
[3]CIRAD, UMR DAP, Avenue Agropolis, TA A-96/03, F–34398 Montpellier, Cedex 5, France; e-mail: *michel.claverie@cirad.fr*
[4]INRA, UR419, Unité de Recherches sur les Espèces Fruitières (UREF), B.P. 81, F-33140 Villenave d'Ornon, France; e-mail: *dirlewan@bordeaux.inra.fr*
*Corresponding author

newly initiated map-based cloning projects for single gene traits and genes underlying QTLs will be considerably accelerated owing to the availability of the genome sequence. However, fine-mapping in large segregating populations as well as the verification of candidate genes still represents a challenge.

Keywords: Genetic mapping, molecular markers, cloning, BAC libraries, transformation, mutants, peach genome sequence

8.1 Introduction

Map-based cloning (MBC) strategies have been successfully used in the stone fruits to narrow in on the target genes (Bielenberg et al. 2004; Romero et al. 2004; Boudehri et al. 2009; Esmenjaud et al. 2009; Sajer et al. 2012) and will be even more useful, now that the genome sequence of peach (*Prunus persica*) is available (http://www.jgi.doe.gov/; http://services.applied genomics.org/projects/drupomics/). However, there have been two limitations of this strategy. Firstly, MBC approaches have always been extremely time-consuming. Secondly, if the function of the gene of interest is unknown or no corresponding gene has been isolated from another species, either because it is not present in other species or has not been investigated so far, there are not many options open in stone fruits. Nevertheless, there are some advantages for map-based cloning approaches in stone fruits. Most *Prunus* species are diploid, peach as a model species within the Rosaceae has a small genome size and is rich in simple sequence repeat (SSR) markers, which showed a high transferability in the genus *Prunus* (Genome Database for Rosaceae, www.rosaceae.org; Mnejja et al. 2010). The high level of colinearity among genomes of different *Prunus* species (Dirlewanger et al. 2004) allowed the construction of a *Prunus* reference genetic and framework physical map (Dirlewanger et al. 2004; Zhebentyayeva et al. 2008), which proved to be very useful tools in MBC approaches. Since 1st April 2010 the peach genome sequence can be publicly accessed (http://www.jgi.doe.gov/), which will be considerably helpful for ongoing and future MBC approaches. Apart from simple genetic traits, polygenic traits or traits with large natural variation in the phenotype, which can only be addressed by quantitative trait loci (QTL) analyses, simply require MBC to identify the genes behind the trait (Peters et al. 2003). With the peach genome sequence and efficient marker technologies available, MBC has become a very powerful tool for isolation of genes in stone fruits.

8.2 Strategies: Chromosome Landing and Walking

The first step in MBC represents the identification of molecular markers tightly linked to the gene of interest. In the classical approach, those were then used as probes against large-insert genomic libraries to identify genomic clones near the target gene or containing the target gene (Tanksley et al. 1995). With the peach genome sequence available this step will in the future be substituted by homology searches against the peach genome sequence (http://www.jgi.doe.gov/), which considerably speeds up the process.

Prerequisites for MBC have been: (1) availability of large mapping populations, e.g., F_2, RILs, F_1BC_1 to use recombination events to develop high-density genetic maps around the target locus, (2) efficient marker technologies, and (3) large-insert genomic DNA libraries, e.g., bacterial artificial chromosome (BAC) libraries (Shizuya et al. 1992) or yeast artificial chromosome (YAC) libraries (Burke et al. 1987). In stone fruits, the latter is now obsolete with the access to the peach genome sequence. The availability of large segregating populations for fine-mapping poses a considerable problem for fruit trees due to the long generation times and the large field areas required. However, so far as marker technologies are concerned, a large number of SSR markers are now available in the genus *Prunus* (Cipriani et al. 1999; Sosinski et al. 2000; Cantini et al. 2001; Dirlewanger et al. 2002; Aranzana et al. 2003; Testolin et al. 2004; Mnejja et al. 2004, 2005; Howad et al. 2005; Verde et al. 2005), which allow the localization of genes on the *Prunus* reference map (Dirlewanger et al. 2004). Also, the amplified fragment length polymorphism (AFLP) technique (Vos et al. 1995), which is a high-throughput technology, not requiring any advance sequence information, has been successfully used in *Prunus* (Wang et al. 2002b; Claverie et al. 2004b; Olukolu et al. 2009; Fan et al. 2010; Sajer et al. 2012).

Large genomic libraries in form of BAC libraries are available for peach cultivars, e.g., Nemared and Lovell, as well as for apricot, *Prunus armeniaca,* (http://www.clemson. genomic.institute.edu/; Wang et al. 2001; Georgi et al. 2002; Georgi et al. unpublished; Vilanova et al. 2003b), which provide enough genome coverage (e.g., 8.8-fold for Nemared, 9.2-fold for Lovell) for MBC approaches, even though the average insert sizes of the BAC libraries are moderate with 50–95 kb. In addition, other BAC libraries have been constructed for the genotypes of interest (Claverie et al. 2004b; Boudehri et al. 2009). However, a binary BAC (BIBAC) library using, e.g. the vector BIBAC2 (Hamilton et al. 1996) has not been constructed in the genus *Prunus.* Using a binary vector for the construction of a large genomic library allows the immediate application of an identified BIBAC clone in transformation experiments using *Agrobacterium,* which could be used to verify candidate genes by complementation of mutants. However, a different system based on the transformation-competent artificial chromosome (TAC) vector pYLTAC7 has been established for peach and serves the same purpose (Liang et al. 2004). The average insert size is approximately 42 kb. As the

TAC library contains 41,472 recombinant clones this is equivalent to six-fold peach genome coverage (Liang et al. 2004).

With regard to the relationship of genetic maps with physical maps, 1 cM map distance represents about 420–450 kb in peach assuming an estimated haploid genome size (1C) of about 300 Mb (Baird et al. 1994) or the assembled peach genome sequence of about 220–230 Mb (http://www.jgi.doe.gov/) and a map size for the whole genome of about 519 cM–524 cM (based on the F_2-population of the interspecific cross between 'Texas' almond x 'Earlgold' peach; Dirlewanger et al. 2004; Howad et al. 2005, respectively).

In an MBC approach, the ideal case would be if markers very closely linked to the gene of interest directly hybridizes to the genomic clones carrying the target gene (chromosome landing, Tanksley et al. 1995). However, more realistic a chromosome walking approach has to be followed to isolate the gene. In this case, markers from both sides of the target gene would be used to identify BAC clones via colony hybridization or PCR screening of BAC pools. The positive BAC clones are needed to be arranged into contigs by BAC fingerprinting. BAC end sequences would then be used to develop probes for continuing hybridizations until a closed contig around the target gene is established, which is characterized by overlapping clones coming from both sides. This had been a tedious process.

However, MBC approaches using this classical approach have successfully led to the isolation of candidate genes in *Prunus*. But the availability of the peach genome sequence (http://www.jgi.doe.gov/) now allows an MBC strategy by homology searches of marker sequences, flanking the target gene, directly against the peach genome sequence. Development of new sequence-based markers, e.g., SSR markers closer to the target gene considerably speeds up the narrowing of the region of interest and thereby the identification of putative candidate genes. Chromosome walking, if necessary, can be performed based on the sequence by developing new markers more closely linked to the gene of interest narrowing down the region in a larger mapping population. The biggest challenge in the future will be the verification of potential candidate genes.

8.3 Traits and Genes Addressed by Map-based Cloning

8.3.1 Evergrowing Trait in Peach

In Mexico, a mutant phenotype of peach (Rodriguez et al. 1994, USDA P.I. 442380) was discovered, which is characterized by a continuous growth of the terminal shoot apices. The segregation analyses showed the *evergrowing* trait to be controlled by a single recessive nuclear gene (Rodriguez et al. 1994). *Evergrowing* peaches can be harvested twice a year in Mexico, but only once a year in temperate climates as the terminal buds get killed in low winter temperatures, whereas the lateral buds go into dormancy.

An MBC approach was started to isolate the gene responsible for this phenotype. Using the F_2-population derived from the cross 'Empress op op dwarf' x 'Evergrowing' (P.I. 442380) bulked segregant analyses (BSA; Michelmore et al. 1991) were performed using the amplified fragment length polymorphism (AFLP) technique for development of markers (Wang et al. 2002b). A map consisting of nine AFLP markers and the *evergrowing* (*evg*) locus was constructed using the software program Mapmaker (Lander et al. 1987). This map covered 79.3 cM (Wang et al. 2002b). The AFLP marker EAT/MCAC mapped 1 cM apart of the *evg* locus whereas the AFLP markers ETT/MCCA2 and EAT/MCTA were linked to the locus at distances of 5.3 cM and 6.7 cM, respectively. These three AFLP markers were cloned, sequenced and used as probes for colony hybridization against high density BAC filters of the peach library (Georgi et al. 2002). The AFLP markers ETT/MCCA2 and EAT/MCTA each identified four BAC clones, whereas EAT/MCAC only detected one BAC clone PpN18F12 (Wang et al. 2002a). SSR markers were derived from sequences of subclones from the BAC clones PpN18F12, PpN11B16, PpN70H22, PpN65B18 and PpN12D18. These SSR markers allowed the back mapping of the BAC clones detected by the AFLP markers EAT/MCAC and EAT/MCTA into the *evergrowing* region (Wang et al. 2002c). Chromosome walking was performed in both directions from the ends of BAC clone PpN18F12. The first step for this was the sequencing of the BAC ends using T7 and SP6 primers. Three types of probes were then used for colony hybridizations: single stranded DNA (ssDNA) probes from BAC SP6 ends, double stranded BAC end DNA fragments and overlapping oligonucleotide (Overgo) probes (Wang et al. 2002b). In total, 20 BACs were assembled into a contig at the location of the AFLP marker EAT/MCAC, very close to the *evg* locus (Wang et al. 2002a). This contig was developed by DNA fingerprinting, hybridization and the use of the Image 3.3 software package and the fingerprinting contig program FPC 4.7 (Soderlund et al. 2000). Three of the BAC clones from the *evg* contig (PpN089G02, PpN18F12 and PpN018G07) covered the whole target gene region (Bielenberg et al. 2004). Using the SP6 and T7 sequence information of the BAC ends from clone PpN089G02 and PpN018G07 the overlap with PpN018G07 could be verified. Sequencing of the BAC ends of PpN18G07 revealed a potential open reading frame with homology to a MADS-box type transcription factor in *Arabidopsis* (Georgi et al. 2003). Southern hybridization of *Hind*III and *Eco*RI digested genomic DNA of the wild type and *evergrowing* mutant proved the existence of a deletion of 42 kb affecting several genes in this region (Bielenberg et al. 2004). The BAC clones PpN089G02 and PpN018G07 that spanned the deletion in the mutant were subcloned and sequenced by primer walking as due to duplications in the *evg* region automated assembly of the sequences derived solely from BAC clone PpN018G17 with the Tn5 transposon insertion system (Epicenter, Madison, WI, USA) proved to be difficult (Bielenberg et al. 2008). The DNA sequence was assembled with the program Sequencher v 4.1.2 (Gene Codes Corp., Ann Arbor, MI, USA). For gene prediction, four programs were then used to identify potential exon-intron structures within the sequences: Fgenesh (Salamov and Solovyev 2000), GeneMark (Besemer and Brodovsky 2005), GenScan (Burge and Karlin 1997), and GlimmerHMM (Majoros et al. 2004). In total, 132 kb of the *evg* region were sequenced and annotated (Bielenberg et al. 2008). Nineteen genes, including six genes for MADS-box transcription factors, were predicted for the region. Four of these MADS-box genes are lost due to the 41,746 bp deletion in the *evergrowing* region (Bielenberg et al. 2008). However, all six MADS-box genes are not expressed in the mutant and represent candidate genes for the regulation of growth cessation and terminal bud formation in peach in response to dormancy-inducing conditions and were named dormancy-associated MADS-box (*DAM*) genes (Bielenberg et al. 2008). Using an MBC approach the potential candidate genes have been narrowed down to six genes, *DAM1* to *DAM6*. The relationship between physical distance to genetic distance in this region is only 10–35 kb/cM (Bielenberg et al. 2008), much lower than the average ratio.

To reduce the number of candidate genes, expression patterns of the *DAM* genes were analyzed under different seasonal and environmental conditions (Li et al. 2009). The *DAM* genes in peach followed at least four distinct expression patterns. *DAM1, DAM2,* and *DAM4* are temporally associated with seasonal elongation cessation and bud formation and represent the most likely candidate genes for control of the *evg* phenotype (Li et al. 2009). *DAM3, DAM5* and *DAM6* are all suppressed by exposure to chilling temperatures in the field as well as under controlled conditions (Jiménez et al. 2010). The expression patterns of *DAM5* and *DAM6* are consistent with a role as quantitative repressors of bud break (Jiménez et al. 2010). In phylogenetic studies, all six *DAM* genes grouped with the SVP/StMADS11 lineage of type II MADS-box genes (Jiménez et al. 2009). These MADS-box proteins bind to DNA as dimers or higher order complexes and are referred to as the MIKC-type due to their common structure of four domains: M domain, I region, K domain and C region (Alvarez-Buylla et al. 2000).

8.3.2 Map-based Cloning of QTLs for Chilling Requirement in Apricot and Peach

In commercial apricot production, severe yield losses occur due to unfulfilled chilling requirement (CR) timing the vegetative growth or flowering to avoid damages because of late winters or early spring frosts. Transition into dormancy is triggered by environmental cues such as day length, cold temperatures, or drought (Arora et al. 2003; Rhode et al. 2007). Using two apricot cultivars, 'Perfection' and 'A.1740', which represent high and low CR, a 2-way pseudo-testcross population consisting of 100 F_1 individuals was developed and screened for chilling requirement (Olukolu et al. 2009). A total of 1,059 markers (1,016 AFLP and 43 SSR markers) mapped on eight linkage groups, 176 AFLP and 12 SSR markers were bridge markers that mapped in both parents (Olukolu et al. 2009). The map lengths were 550.6 cM and 454.9 cM for the 'Perfection' and 'A.1740' maps, respectively. The quantitative trait loci (QTL) for CR were detected applying composite interval mapping (CIM, Zeng 1994) using the program Plab QTL version 1.2 (Utz and Melchinger 1996). A total of 12 unique QTLs were detected on six linkage groups at initial LOD thresholds of 9.44 and 8.46 for the 'Perfection' and 'A.1740' maps (Olukolu et al. 2009). Six main-effect QTLs (one on linkage group 1, 2, 7 and 8; two on LG 6) from the high CR parent 'Perfection' accounted for 58.5% ± 6.7% of the phenotypic variance. The eight main-effect QTLs (two on LG 1, 2 and 5, respectively; one on LG7 and LG8) explained even a total of 66.1% ± 5.8 % in the low CR parent 'A.1740' map. Two of the QTLs detected on LG 1 and 8 were common in both maps (Olukolu et al. 2009). The QTL on LG1 mapped at the location of the *evergrowing* locus in peach, which has been intensively studied and underlying candidate genes, especially six MADS-box genes, in this region and their expression have been characterized (Bielenberg et al. 2008; Li et al. 2009; Jiménez et al. 2010). Two other candidate genes were provided for the QTLs on LG7 and LG8 as the genomic sequence based SSR marker CPSCT042 and the EST-derived SSR marker EPDCU3454 mapped within these QTLs (Olukolu et al. 2009). CPSCT042 showed homology to the *MITOGEN-ACTIVATED PROTEIN KINASE7* (*MPK7*) and EPDCU3454 to *ABSCISIC ACID INSENSITIVE3* (*ABI3*). Utilizing the available *Prunus* genomic resources (Horn et al. 2005; Ogundiwin et al. 2008; Zhebentyayeva et al. 2008) as well as the *Prunus* genomic sequence (http://www.jgi.doe.gov/) additional candidate genes for CR in apricot underlying the QTLs will be analyzed.

In peach, CR was investigated together with heat requirement (HR) and blooming date (BD) (Fan et al. 2010). Twenty QTLs with additive effects were identified for these three traits in an F_2 population of the F_1 hybrid 'BY01p6245'. This F_1 hybrid originated from a cross of two genotypes having 'Contender' as high chilling female grandparent and 'Fla.92-2C' as low chilling male grandfather in their pedigree. The linkage map used for the QTL analyses was composed of 96 SSR markers, 30 AFLP markers and one phenotypic marker. In both investigated years (2008, 2009), four QTLs were detected for chilling requirement (qCR1a, qCR4b, qCR5, qCR7).

Two of these QTLs, qCR1a and qCR7, represented major QTLs explaining 40.5–44.8% and 17.8–24.9% of the phenotypic variants, respectively (Fan et al. 2010). Two QTLs were detected for heat requirement (qHR1 and qHR8), only qHR1 was recurrent. However, the full regression models for HR QTLs only explained 8.6% and 10.7% of the phenotypic variance in the two investigated years. For the blooming date (BD), 10 QTLs could be identified (Fan et al. 2010). Full regression models for BD QTLs accounted for 52% and 74.1% of the phenotypic variance. The three QTLs qCR1a, qHR1 and qBD1a overlap within a region characterized as the *evergrowing* region (Bielenberg et al. 2008).

Among all 20 QTLs for the three traits, only three BD QTLs and one CR QTL neither co-localized nor overlapped with any QTL for the other traits. But these QTLs explained only a small percentage of the variants or were not recurrent (Fan et al. 2010). The co-localization of the majority of the detected QTLs might indicate that the genes regulating these different traits (chilling requirement, heat requirement and blooming date) are tightly linked (Fan et al. 2010). The identification of these 20 QTLs represents the first step in map-based cloning of the genes underlying the QTLs (Fan et al. 2010). Depending on the size of the genome region spanned by the QTLs fine mapping will be required to reduce the region to look for putative candidate genes on the peach genome sequence level (Fan et al. 2010).

8.3.3 Map-based Cloning of Self-incompatibility Locus in Apricot

As many *Prunus* species in the Rosaceae family, apricot has a gametophytic self-incompatibility (SI) system controlled by a single highly polymorphic locus called S-locus (Burgos et al. 1997). In a number of species, the *S*-locus contains an *S-RNase* gene coding for a glycoprotein with ribonuclease activity expressed in the style (McCubbin et al. 2000) and adjacent to it an *S locus F-box (SFB)* gene encoding the interaction partner on the pollen side (Sijacic et al. 2004).

To clone the SI locus from apricot a combination of an MBC approach and a candidate gene (CG) approach was used. For mapping the SI locus, a genetic linkage map was constructed for apricot based on the F_2-population derived by self-pollination of an F_1 individual 'Lito', which originated from a cross between 'Stark Early Orange' and 'Tyrinthos' (Vilanova et al. 2003a). A total of 211 markers (180 AFLPs, 29 SSRs) together with two traits, self-incompatibility and sharka resistance, were assigned to 11 linkage groups covering 602 cM of the apricot genome. The SI locus mapped on linkage group 6 and the PVV resistance on linkage group 1 (Vilanova et al. 2003a). A BAC library of apricot cultivar 'Goldrich' (S_1S_2 haplotype) was constructed and colony hybridizations against high density BAC filters were performed with a PCR probe (C6–C5) using primers from conserved regions of a known wild cherry tree (*Prunus avium*) S-cDNA (Tao et al. 1999; Vilanova et al. 2003b). Eight positive BAC clones were detected and digested with the restriction enzymes *Hind*III and *Bam*HI (Vilanova et al. 2003b). The gel image was scanned, digitalized using an Image software program and assembled into a contig using FPC V6.2 (Soderlund et al. 2000). The contig spans approximately 180 kb of the chromosomal region containing the *S-RNase* gene and was subdivided into two classes (Vilanova et al. 2003b). Subclones of the BACs were again hybridized with C6–C5 PCR probe and all positive clones were sequenced (21 kb in total). The consensus sequences obtained from each BAC clone class were assigned to the haplotypes S_1 and S_2 of the cultivar 'Goldrich' (about 6.9 kb and 6.3 kb, respectively). In the S_2-haplotype, the physical distance between the S_2-*RNase* and SFB_2 alleles is only 2.9 kb. For the S_1-haplotype, the open reading frame for the S_1-*RNase* allele could be identified and the SFB_1 allele was directly amplified from the BAC clone and is located next to the S_1-*RNase* gene with a distance less than 49 kb (Romero et al. 2004). Expression analyses of S_1- and S_2-RNases showed that these are specifically expressed in style tissues, but not in pollen or leaves. In contrast, SFB_1 and SFB_2 are only expressed in pollen, but not in styles and leaves (Romero et al. 2004).

In other *Prunus* species, including almond, sweet and sour cherry and Japanese apricot, *S-RNase* genes have been identified by candidate gene approaches (Ushijima et al. 1998; Tao et al. 1999; Yaegaki et al. 2001; Yamane et al. 2001) as well as the corresponding *SFB* genes (Entani et al. 2003; Ushijima et al. 2003; Yamane et al. 2003).

The advantage of an MBC approach in comparison to a CG approach is that extensive information about the 5′-flanking region, necessary for the understanding of the expression on transcriptional level, can be obtained and analyzed regarding promoter elements (Romero et al. 2004).

8.3.4 Low Acidity in Peach Fruit Controlled by the D Locus

Several traits related to fruit quality are important to fruit breeders in response to consumer requirements. Acidity is an essential component of the organoleptic quality of fleshy fruits. For peach, consumers expect large and well-shaped fruits that are flavorful with high sugar content and low acidity. Although fruit pH is a trait whose phenotypic variation is continuous in peach, peach fruits can be grouped into two types according to their pH value: the normally-acid (pH below 4.0) and the low-acid phenotypes (pH above 4.0) (Yoshida 1970; Dirlewanger et al. 1998). A major dominant allele *D* (*D* for 'Doux' in French, mining sweet) from the honey-type peaches is responsible for low acidity (Monet 1979).

A progeny issued from a cross between two peach varieties, a low acid flat peach, 'Ferjalou Jalousia', and a normally-acid round nectarine,

'Fantasia', segregating for fruit quality traits, was used to map the *D* locus and additional QTLs controlling various physical and chemical components of peach fruit quality. Furthermore, QTLs for pH, titratable acidity (TA) and malic and citric acid contents mapped near the *D* locus (Dirlewanger et al. 1998, 1999). The first study was based on the analyses of 208 individuals. The *D* locus was mapped to the proximal end of linkage group 5 and co-localized with major QTLs involved in the control of fruit pH, titratable acidity and organic acid concentration and minor QTLs for sugar concentration.

In addition, a CG approach was carried out by studying 18 candidate genes involved in organic acid synthesis, degradation or vacuolar storage (Etienne et al. 2002a, b). Expression analyses of six selected CGs in fruits did not show a clear difference between the normally-acid and low-acid varieties (Etienne et al. 2002b). The genes showing a modification of their expression in the low-acid fruit compared to the normally-acid fruit were not located on linkage group 5, excluding their direct role in the control of acid content (Etienne et al. 2002b). More recently, additional CGs involved in fruit quality were mapped on this same segregating progeny, and none mapped in the region of the *D* locus (Le Dantec et al. 2010). The complexity of organic acids' metabolic pathways as well as the involvement of transporters, channels, and related proton pumps has hampered, so far, the identification of the gene(s) associated to the *D* locus using a CG approach. Thus, in order to investigate the molecular and physiological bases of fruit acidity in peach, a positional cloning strategy of the *D* locus was undertaken for the isolation of the gene(s) underlying this trait (Boudehri et al. 2009).

For the fine mapping of the *D* locus, a total of 1,878 F_2 additional individuals were obtained. In order to identify markers in the region of the *D* locus, the BSA-AFLP method was used. Four bulks were constituted for BSA with individuals chosen among the 208 F_2 used for the genetic linkage map; two with low-acid fruit (*D/D* or *D/d*) and two with normally-acid fruit (*d/d*). Individuals were selected according to juice pH and TA values measured during two years. Among the 1,024 AFLP primer combinations used to screen the bulks, 34 AFLP markers were mapped to the LG5, and the six nearest markers were transformed into codominant sequence characterized amplified region (SCAR) markers. These SCAR markers and additional SSR markers were used to genotype the complete F_2 progeny. A high-resolution map of the *D* locus was realized after genotyping and phenotyping recombinant individuals. Using these recombinant plants the *D* locus was delimited to a genetic interval of 0.4 cM (Boudehri et al. 2009).

A peach BAC library was constructed from the F_1 parent of the segregating population, heterozygous for the *D* locus. This BAC library contained about 52,000 clones with a mean insert size of approximately 90 kb corresponding to 15x coverage of the peach haploid genome (Boudehri et al. 2009). The screening of the BAC library with tightly linked markers allowed the construction of the physical map and the identification of BAC clones harboring the *D* locus. Two BAC clones, one for each allele, were sequenced. The sequence of the BAC clone containing the *D* allele was 98 kb and the one for the *d* allele was 78 kb long. A new set of markers was developed, which allowed the localization of the *D* locus in a 16 kb interval. The sequence of the dominant allele was highly homologous to the complete sequence of the peach genome. On the opposite, the *d* allele included a very high level of single nucleotide polymorphism (SNP). Within this two BAC sequences, 11 predicted genes were found and no recombinant individuals were able to eliminate any of them. Then, in order to identify the gene(s) responsible of the non-acid character, a differential gene expression analysis by transcriptome sequencing (RNASeq) of the two parents is in progress at the INRA Bordeaux (Dirlewanger and Le Dantec, personal communication). If positional candidate genes are also identified as expressional candidate genes they should be good candidates that will have to be validated by association mapping using peach collections.

8.3.5 Map-based Cloning Approach of the br Gene Responsible for the Pillar Growth Type in Peach

In peach as well as in other fruit trees, tree architecture has a major influence on yield and fruit quality as well as on the possibility of applying high density production systems and the mechanization of orchard operations (Miller and Scorza 2010). Trees showing columnar [also termed pillar or broomy] phenotype represent one growth type (Scorza 1984, 2005; Bassi et al. 1994; Scorza et al. 2006), which is of special interest for high density production systems. The columnar growth type of peach was first described in ornamental trees in Japan (Yamazaki et al. 1987). The narrow branch angle is the most striking feature of the pillar (PI) tree, which produces a crown diameter of only around 1.5–2.0 m whereas standard (ST) peach trees produce a canopy that reaches a diameter of 4.0 m or greater (Scorza 2005; Scorza et al. 2006). Heterozygotes show an intermediate, so called upright (UP) phenotype (Miller and Scorza 2010). In earlier molecular studies, the *br* gene, coming from a germplasm collection at Rutgers University, New Jersey (NJ) was investigated (Chaparro et al. 1994; Sosinski et al. 2000). Only two RAPD markers, OPG-13 and OPF-15, both at a distance of 8.2 cM, were identified (Chaparro et al. 1994) and the SSR marker pchgms1 mapped at 12.5 cM apart to the *br* gene (Sosinki et al. 2000). However, the NJ source for the pillar growth was linked to unfavorable traits regarding the fruit quality and the new source of the pillar trait from Italy promises to show markedly improved fruit quality.

An MBC approach was started to isolate the *br* gene responsible for the pillar growth in peach (Sajer et al. 2012). Cloning of the gene and understanding the function of the gene would reveal a fundamental process in the development of tree architecture that would be also interesting for other fruit trees.

To localize the *br* gene on the *Prunus* reference map (Dirlewanger et al. 2004) 104 SSR markers from different sources were screened for polymorphisms (Sajer et al. 2012). Forty-four polymorphic SSR markers representing each of the eight linkage groups of peach with at least three markers were used to map the *br* gene to the lower part of linkage group 2 (Sajer et al. 2012). The *br* gene mapped about in the middle between the SSR markers BPPCT004 and PceGA34. Screening 256 AFLP primer combinations by BSA identified 53 polymorphic primer combinations with 64 polymorphisms that were further analyzed in the F_2-population. Thirty-four AFLP markers and three SSR markers mapped together forming a linkage group that covered 92.9 cM (Sajer et al. 2012). Integration of the three SSR markers BPPCT004, PceGA34 and CPSCT023 into the AFLP map showed a colinear order of these markers in the pillar mapping population compared to the *Prunus* reference genetic map (Dirlewanger et al. 2004). The three cosegregating AFLP markers ETAM56_267A, ECAM61_426A und ETAM48_279A mapped at a distance of 3.8 cM to the *br* gene on one side, the AFLP marker ETGM61_291R with 4.8 cM on the other side. These four closest linked AFLP markers were successfully cloned and sequenced. Sequence-tagged site (STS)-markers, for easier use in marker-assisted breeding programs, were successfully derived from ETGM61_291R and ETAM48_279A (Sajer et al. 2012). BLAST searches against the peach genome (http://www.jgi.doe.gov/) placed the *br* gene in the region of 17–18 Mb on linkage group 2 (LG2). To confirm the localization on LG2, two SSR markers, pchgms128 and pchgms129, were derived from the peach genome sequence in the identified region. These two SSR markers co-segregated and mapped back to the *br* gene (Sajer et al. 2012).

Meanwhile, 1,024 AFLP primer combinations have been investigated by BSA. In total, 262 polymorphisms in 203 primer combinations were identified. The new linkage map obtained using the program Mapmaker consists of 48 markers, 45 AFLP markers and three SSR markers, and covers 85.7 cM. The closest marker EAAM40_261A is linked at a distance of 2.3 cM to the *br* gene. Seven additional AFLP markers mapping to both sides of the *br* gene were cloned and four of them also confirmed the location on LG2. The availability of the peach genome sequence (http://www.jgi.doe.gov/) now saves the need for the time-consuming identification of BAC clones and the formation of a closed contig around the target gene.

8.3.6 Map-based Cloning of QTLs for Plum Pox Virus Resistance in Apricot and Peach

Among all pathogens affecting stone fruits (mainly apricot, peach and plum), the sharka disease is the most devastating and most limiting for growing stone fruits in Europe and America (Cambra et al. 2006). This disease is caused by the plum pox virus (PPV), a potyvirus, that was first detected in Bulgaria one century ago but since then spread through the whole of Europe with the first report for Spain in 1984 (Llacer et al. 1985). Infection with the PPV transmitted by aphids results in fruits that are deformed, fail to ripen and cannot be sold on the market. The economic losses are estimated to be over 10 billion Euros over the last 30 years (Cambra et al. 2006). No valuable source of resistance to PPV has been found in peach, but polygenic resistance was described in Chinese wild peach (*Prunus davidiana*) clone 'P1908'. Using three connected progenies (F_1, F_2 and BC_2) derived from crosses between peach cv. 'Summergrand' and *P. davidiana* 'P1908' genetic maps were constructed and the resistance against PPV introduced (Foulongne et al. 2003). QTL analysis in the F_1-population identified six genomic regions on LG1, LG2, LG4, LG6 and LG7 (PPV-1.1 and PPV-1.2, PPV-2.1, PPV-4.1, PPV-6.1, PPV-7.1 (named according to the linkage group on which they were detected) related to PPV resistance in *P. davidiana* 'P1908' showing its multigenic character (Decroocq et al. 2005). Three analogs of virus resistance genes were found to be co-localized with two genomic regions for PPV resistance. QTL analyses in the F_2-population derived from the F_1-hybrid SD40 also implicated six genomic regions (PPV-1.a, PPV-1.b, PPV-4.a, PPV-5.a, PPV-6.a and PPV-7.a) in the resistance to the sharka disease (Marandel et al. 2009a). However, only four of the six QTLs were mapped in the F_1 and the F_2 population: PPV-1.1, PPV-4.1, PPV-6.1 and PPV-7.1 correspond to PPV-1.a, PPV-4.a, PPV-6.a and PPV-7a, respectively. The QTLs PPV-1.2 and PPV-2.1 (in the F_1-population) and PPV-1.b and PPV-5.a (in the F_2-population) were not observed in the other generation (Marandel et al. 2009a). As an involvement of translation initiation factors (eIF) for other resistances against Potyvirus had been described (reviewed in Robaglia and Caranta 2006), a candidate gene approach was also applied and five locations for the *eIF(iso)4G- like* genes, two on LG1 (one co-localized with PPV-1.a), one on the lower part of LG3, one on the upper part of LG6 (co-localized with PPV-6.a) and one on the lower part of LG7 were found. Interestingly, in *P. davidiana* both eukaryotic translation initiation complexes, *eIF4F* (composed of *eIF4E* and *eIF4G*) and *eIF(iso)4F* (composed of *eIF(iso)4E* and *eIF(iso)4G*) seem to be implicated in susceptibility (Marandel et al. 2009a).

Very few sources of PPV resistance have been identified in apricot originating from North American breeding programs (Martinez-Gomez and Dicenta 2000). The cultivar 'Harlayne' is regarded as the most resistant cultivar, followed by the resistant cv. 'Stark Early Orange' and its descendants 'Lito' and 'LE-3246', and finally 'Goldrich' showing an intermediate resistance (Ion-Nagy et al. 2006). All of these resistance sources have been used for QTL analyses regarding PPV resistance (Vilanova et al. 2003a; Lambert et al. 2007; Lalli et al. 2008; Sicard et al. 2008; Soriano et al. 2008; Marandel et al. 2009b; Pilarová et al. 2010). Consistently, all publications highlight a major contribution of the upper part of linkage group 1 to the sharka resistance. Recently, in the cross 'Harlayne' x 'Marlén' four distinct QTLs were detected, three named PPV-1H.a, PPV-1H.b, PPV-1H.c, mapping on LG1 and a fourth on LG3 (PPV-3H.a), individually explaining up to 39% of the phenotypic variance with the strongest effect from LG1 (Marandel et al. 2009b). Interestingly, candidate genes *eIF4E* and *eIF(iso)4G* mapped in the PPV-1H.b and PPV-3H.a intervals, respectively. Inoculating with distinct PPV strains, PPV-M and PPV-D, three and two QTLs were identified on LG1 in the cross 'Harlayne' x 'Vestar', respectively (Pilaróva et al. 2010), but none on LG3. Using the Meta QTL software (Veyrieras et al. 2007) the data of the six different studies (Lambert et al. 2007; Lalli et al. 2008; Sicard et al. 2008; Soriano et al. 2008; Marandel et al. 2009b) were compiled that resulted also in four meta QTLs for PPV resistance, three QTLs on LG1 and a putative fourth QTL on LG3 (Marandel et al. 2009b) with the different resistant sources contributing to them to different degrees. Several candidate genes mapped in the meta QTL intervals: *Cd83*, a homolog of the Argonaute protein AGO1, and *Cd66*, an RNA-directed RNA polymerase, mapped in MetaPPV-1.a; *Cd195*, corresponding to the translation initiation factor *eIF4E* in MetaPPV-1.b (Sicard et al. 2008; Marandel et al. 2009b). Eight expressed sequence tags were also identified in the region of Meta PPV-1.b, but no candidate gene was identified for Meta PPV-1.c.

However, the co-localization between resistance QTLs and eukaryotic translation initiation factors indicates an involvement of these translation initiation factors in PPV resistance coming from apricot (Sicard et al. 2008; Marandel et al. 2009b) and also *P. davidiana* (Decroocq et al. 2005; Marandel et al. 2009a) as has been reported for other resistances against Potyvirus (reviewed in Robaglia and Caranta 2006). In the future, integration of additional data and use of the peach genome sequence will allow cloning of the resistance genes against PPV in the mapped QTL regions (Marandel et al. 2009b).

Two high-density maps for 'Lito' and 'Goldrich' were developed using SSR markers in order to map the PPV resistance with higher accuracy in apricot (Ruiz et al. 2011). A new binary trait locus (BTL) analyses for PPV resistance then allowed to narrow down the marker intervals for PPV resistance to 7.3 cM and 5.9 cM for 'Lito' and 'Goldrich'. A set of primary BAC contigs surrounding the PPVres locus was identified by 71 overgo probes derived from flanking markers. Hybridization of these overgo probes against the apricot BAC library (Vilanova et al. 2003b) identified 870 single BAC clones of which 340 could be anchored on the map. High-information content fingerprinting (HICF) was used to build partial contigs. As a result, 14 primary contigs were assembled and confirmed by PCR to be localized within a region of approximately 30 cM in 'Goldrich' and 40 cM in 'Lito'. However, no overlapping of these contigs could be detected by PCR (Ruiz et al. 2011). A total of 300 BAC-end sequences were obtained of which 257 showed homology to LG1. Using the synteny to peach, the PPVres locus was placed by the apricot markers within an uncovered gap of 2.16 Mb between the markers Gol061 to Gol027 (Ruiz et al. 2011). According to the peach genome sequence this region contains 251 transcription units. Six of these transcription units showed significant homology to LRR receptor-like serine/threonine protein kinases (Martin et al. 2003), which represent known resistance genes in other species. These may now also be considered potential candidate genes for PPV resistance gene(s) (Ruiz et al. 2011).

8.3.7 The Ma Gene for Root-knot Nematode Resistance

Root-knot nematodes (RKN), *Meloidogyne* spp., are important pests of fruit tree crops and other perennial or annual plants. Three predominant RKN species including *M. incognita, M. javanica* and *M. arenaria* are present all over the world in tropical and Mediterranean climates (Abad et al. 2008). As nematicidal chemicals are progressively removed from the market, genetic control of *Meloidogyne* spp. is gaining interest (Cook and Evans 1987; Roberts 1992). In *Prunus* species, different ranges of resistance to RKN have been observed and corresponding genes used for rootstock breeding (Esmenjaud et al. 1997; see also Chapter 7 on Molecular Breeding).

Nematode juveniles were shown to penetrate the roots and emigrate from the roots in equivalent numbers in *Ma*-resistant and *Ma*-susceptible accessions, but they did not establish a feeding site nor develop in resistant accessions explaining the complete absence of galls in *Ma*-carrying accessions (Voisin et al. 1999). Histological studies have demonstrated that *Ma* resistant plants expressed a hypersensitive-like reaction (HLR). Nematode attacks, by disorganizing the meristematic apical tissues of *Ma*-accessions, induced the development of sub-terminal lateral roots replacing primary terminal apices and provided an active resistance reaction to HLR damage (Khallouk et al. 2011). MBC of *Ma* has then been conducted by an exclusive positional strategy using the source Myrobalan plum (*P. cerasifera*) accession 'P.2175' carrying the *Ma1* allele at the heterozygous stage (*Ma1/ma1*) in successive steps and in the validation procedure reported hereafter.

8.3.7.1 Identification of Markers by BSA and Development of SCAR and CAPS Markers

Molecular characterization of the *Ma* locus has begun by the detection of two reliable SCAR markers, SCAL19 and SCAFLP2, tightly linked to *Ma*. Both have been identified by BSA and are currently used for marker-assisted selection (MAS) (Lecouls et al. 1999, 2004). *Ma* was then mapped on the linkage group 7 of the reference European *Prunus* map (Joobeur et al. 1998; Aranzana et al. 2002) in a 2.7 cM interval between the SSR marker pchgms6 and SCAL19 (Claverie et al. 2004a). Additional markers in the *Ma* region were obtained by BSA of 320 AFLP primer combinations. From an original segregating population of 307 individuals, five AFLP markers tightly linked to the *Ma1* allele from 'P.2175' were obtained and mapped in the interval SCAL19-pchgms6 spanning the gene. Three of these markers were sequenced and transformed into SCAR or codominant cleaved amplified polymorphic sequence (CAPS) markers designated SCAFLP3, SCAFLP4 and SCAFLP5.

8.3.7.2 High Resolution Mapping, BAC Library Construction, Physical Contigs Spanning the Ma Region and Chromosome Landing

A total of 1,332 individuals, from 21 crosses segregating for *Ma1*, revealed 31 individuals recombining between SCAL19 and pchgms6 in the genetic interval of 2.7 cM encompassing the gene (Claverie et al. 2004a). The recombinant individuals were then genotyped with the markers SCAFLP2, SCAFLP3 and SCAFLP4 and RKN resistance tests allowed a finer location of the gene: *Ma* co-segregated with the SCAFLP2 marker and was separated from SCAFLP3 (proximal) and SCAFLP4 (distal) by a single and 11 recombination event(s), respectively. A large-insert BAC DNA library of Myrobalan plum 'P.2175', comprising 30,720 clones with a mean insert size of 145 kb and a 14–15x *Prunus* haploid genome coverage, was constructed and used to land on the *Ma* spanning interval with few BAC clones (Claverie et al. 2004b). As 'P.2175' is heterozygote for the gene, the resistant and susceptible physical contigs were constructed by PCR screening of the library with codominant markers. Additional microsatellite markers were designed from BAC subcloning or BAC end sequencing. Finally, in the resistant contig, a single 287-kb BAC clone (BAC 76H19) was shown to carry the *Ma* gene: this BAC clone contained two flanking markers on each side of the gene as well as two cosegregating markers (Claverie et al. 2004b).

8.3.7.3 Evidence of the Ma TNL Cluster as the Final Candidate Interval

Complete sequencing of BAC 76H19 allowed to place markers SCAFLP3, SCAFLP2 and SCAFLP4 on the sequence and to determine a physical interval of 174 kb containing *Ma* (Claverie et al. 2011). New SSR markers, manually designed within this 174-kb interval, reduced the interval containing the *Ma* gene to 70 kb flanked by the markers SSR2 and SCAFLP4. In this interval, four candidate genes, predicted by GeneScan and GeneMark.hmm, belonged to two distinct classes of resistance gene analogs: one lectin protein kinase (LecPK) and three toll/interleukin1 receptor (TIR)-NBS-LRR (TNL) class of genes. In the objective to more accurately locate the candidate R genes, a final round of mapping was undertaken. A total of 1,780 additional individuals were produced from several segregating crosses (*Ma/ma* x *ma/ma*) of which some carried the *RMia* gene from the Nemared peach, which can be discriminated from *Ma* by resistance assays using *M. floridensis*. In order to detect individuals recombining very closely to *Ma*, two flanking markers, SSR6 and SSR12, separated by 209 kb and located at 0.7 cM (nine recombinants) and 0.3 cM (four recombinants), respectively, were designed and used in a duplex amplification procedure. They allowed the detection of 19 new recombinants out of the 1,780 individuals. This, added to the 13 initial recombinants of the 1,332 initial individuals, represents a total of 32 (19 + 13) recombinants out of the 3,112 (1,332 + 1,780) meiosis in the SSR6-SSR12 interval, in agreement with the 1-cM genetic distance previously estimated. All recombinants and control individuals were then genotyped for nine SCAR or SSR markers designed within the 70-kb candidate interval. Five recombinant genotypes located within the interval of candidate genes were identified and evaluated for resistance to the species *M. floridensis*. Their resistance phenotype eliminated the candidate gene *LecPK* and reduced the final candidate region to the sole 32-kb sequence covering a cluster of three *TNL* genes.

8.3.7.4 Comparative Analysis of the Candidate Region in the Resistant and Susceptible BAC Clones

The BAC 40K9, 170 kb long and encompassing the susceptible allele of the *Ma* candidate region, (Claverie et al. 2004b), was sequenced. Three *TNLs* were identified in the susceptible haplotype of *Ma*, where sequence and structure comparison allowed the establishment of orthology relationships with those of the resistant haplotype (Claverie et al. 2011). Hereafter, *TNLs* have been designated *TNL1* to *TNL3* in the resistant haplotype and *tnl1* to *tnl3* in the susceptible haplotype. Intron/exon structures were predicted, and the complete cDNA sequence of *TNL1* was determined using specific and oligo(dT) primers and RACE-PCR. Sequences of the other *TNLs* were deduced from sequence alignments. For *TNL2*, a frameshift mutation (deletion of a single base) was detected in the NBS domain that makes it a pseudogene. *TNL3* had a C-terminal region excised after the seventh LRR repeat motif and was much shorter. The high divergence between *TNL1* and *tnl1*, added to the fact that *TNL3* was truncated and *TNL2* was a pseudogene, indicated that *TNL1* was the best candidate sequence for the *Ma* gene (Claverie et al. 2011). Validation of the *Ma* RKN gene using *A. rhizogenes* for plant transformation is a successful example of a complementation procedure based on this technique without the need for regeneration of the entire plant (Bosselut et al. 2011).

8.3.7.5 Summary of Characteristics of the TNL1/Ma Gene

Ma cloning was carried out by accurate fine genetic and physical mapping, comparative analysis of resistant and susceptible haplotypes sequences and finally by functional complementation of a susceptible *Prunus* accession. After the *Mi-1* gene from tomato (Milligan et al. 1998), the *Ma* gene is the second gene cloned for resistance to RKN and the first one in a perennial. This gene has the peculiarity to have been identified in a highly heterozygous and self-incompatible species from, moreover, a near-wild origin (Eremin 1978). The *Ma* gene belongs to the Toll/Interleukin1 Receptor (TIR)-NBS-LRR (*TNL*) class of genes instead of the CC (coil coiled)-NBS-LRR (*CNL*) family of *Mi-1*. Successful complementation by transformation via *A. rhizogenes* opens the way to the application of the hairy roots transformation in other plant species in order to move across species boundaries.

The *Ma* gene has the longest full-length cDNA size (2,048 deduced aa) of all *TNL* genes cloned up-to-now. Its post-LRR (PL) part is unique in that it represents 53% of the total cDNA length and is composed of five exons, with the three latter originating from recent duplications (Fig. 8-1). The presence of another allele for RKN resistance at the same locus in the Japanese plum (*Prunus salicina*) (Claverie et al. 2004a) suggests that resistance polymorphism is anterior to Myrobalan plum and Japanese plum species separation. Furthermore the *Ma* locus itself is probably also ancient: in the recently sequenced species *Prunus persica* (Peach genome V1.0, http://www.rosaceae.org/node/355), an orthologous locus (identified between the lectin-kinase and guanylate kinase genes flanking the *Ma* cluster) showed the same structure (the five PL exons) as *TNL1* from Myrobalan plum. This gene is probably the "direct" orthologous sequence of *TNL1*, which means that the particular structure of *Ma* is ancient and shared between *Prunus* species.

As *Ma* confers a complete-spectrum resistance to RKN and is not overcome under high inoculum pressure (Esmenjaud et al. 1996), it is highly probable that this gene recognizes a crucial effector or guards a virulence/

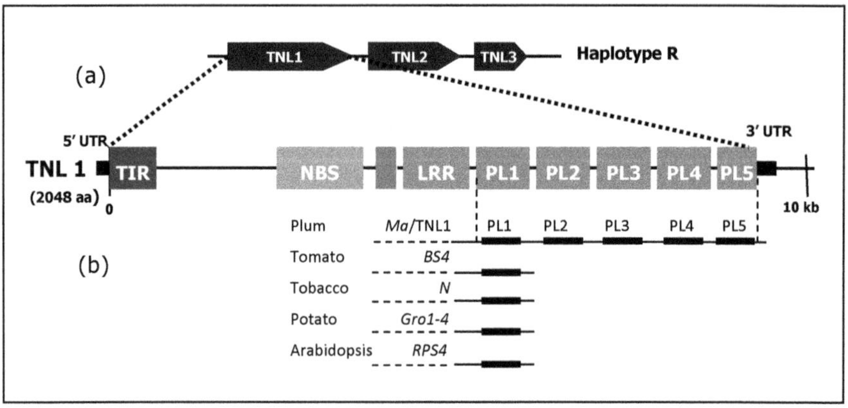

Figure 8-1 Structural characteristics of the *TNL1/Ma* gene (2048 aa).

a—Exon and intron structure (scale in kb) in the cluster of three *TNLs* (TIR-NBS-LRR) from accession P.2175. PL1 to PL5 are the C-terminal post-LRR exons (encompassing 1088 aa out of the 2048 aa of the entire deduced protein sequence). **b**—Structure alignment from each of the five PL exons of *TNL1/Ma1* with the post-LRR sequence (comprising a single exon) of four other cloned *TNL* genes [tomato/*BS4-MM* (Ballvora et al. 2001), tobacco/*N* (Whitham et al. 1994), potato/*Gro1-4* (Paal et al. 2004), *A. thaliana/RPS4* Col-0 (Gassmann et al. 1999)], illustrating that *Ma* is the longest R gene cloned up-to-now.

effector target that RKN species have in common. The spontaneous accession 'P.2175' originated from continental Romania. In the absence of co-evolution of Myrobalan plum and the apomictic RKN (*M. incognita, M. arenaria* and *M. javanica*), the wide-spectrum and high-level resistance conferred by the *Ma* gene may be best explained by an indirect interaction (guard hypothesis; Van der Biezen and Jones 1998; Dangl and Jones 2001) between the resistance gene product and the nematode avirulence factors. Alternatively, in the context of the decoy hypothesis proposed by Van der Hoorn and Kamoun (2008), at least one of the PL exons might mimic a conserved host target common to *Meloidogyne* spp. and unrelated plant pathogens, and lead to a direct interaction. In the *TNL1/Ma* gene, not only the PL domains but also the end of the LRR domain, and the gene regulatory region show extensive polymorphism. In this case, we may suppose that the 'guarded' host product or the host decoy is targeted by independent *Avr* products, from one or more pathogens species which are not apomictic RKN. Future research on RKN host targets will benefit from the sequencing of the *M. incognita* genome (Abad et al. 2008), which should provide new candidate avirulence genes and maybe nematode pathogenicity factors targeted to the gene.

8.4 Verification of Candidate Genes

8.4.1 Transformation Strategies in Prunus Species

Complementation of mutants and overexpression of candidate genes represent possibilities to verify a candidate gene identified by an MBC approach. However, both strategies require reliable and efficient transformation systems in the genus *Prunus*. Any attempt to transform *Prunus* species to confirm functions of the candidate genes isolated has been hampered by the fact that *Prunus* species, especially peach, have been proved to be recalcitrant in tissue culture due to the low regeneration potential.

8.4.1.1 Agrobacterium tumefaciens-based Transformation Protocols

Most promising for transformation experiments using *A. tumefaciens* seems to be European plum (*Prunus domestica*). A plum hypocotyl transformation method was developed by Mante et al. (1991), in which mature seeds of the breeding selection 'B70146' are surface sterilized and after removal of radicle and epicotyl the hypocotyl is sliced in three cross sections (0.5–1 mm). After co-cultivation with *A. tumefaciens* EHA101 containing the vector PCGN7001, which carries the genes for NPTII and GUS, and transfer to selective media, shoots were produced in 45% of the explants with 8 shoots per explant (Mante et al. 1991). Transgenic plum plants carrying the T-DNA could be regenerated. However, while the transformation protocol proved to produce reliable transgenic plum plants, improvement was still required to reduce the number of non-transformed plants that survived the selection (escapes) and to improve the rooting efficiency as well as the survival rate after transfer to the greenhouse (Gonzalez-Padilla et al. 2003). The transformation efficiency (based on the number of transgenic regenerants per inoculated explant x 100) ranged from 0.4 to 4.2% using the varieties 'Stanley' and 'Bluebyrd'. Early selection with 80 mg l^{-1} kanamycin and 300 mg l^{-1} Timentin proved to be superior to avoid escapes than late selection. Rooted shoots needed to be at least 1 cm in height to obtain a 70% rate of survival for the transfer to soil (Gonzalez-Padilla et al. 2003). Addition of 2,4-D during co-cultivation with *A. tumefaciens* significantly increased transformation efficiency up to 10 fold higher (42%) than the highest rate published up to then with an average efficiency of 25% of hypocotyl slices producing transgenic plants (Petri et al. 2008c). However, the whole process from co-cultivation of the hypocotyl slices with *Agrobacterium* to the regeneration of whole plants still takes approximately 19–28 weeks (Petri et al. 2008c).

Meanwhile this transformation protocol has been successfully applied to two Japanese plum varieties 'Angeleno' and 'Larry Anne' using GFP as reporter gene resulting in transformation efficiencies of 0.8 and 0.3% (Urtubia et al. 2008).

For apricot, a first protocol for *Agrobacterium*-mediated stable transformation using whole leaf explants of the cultivar 'Helena' was established with transformation efficiencies of up to 5.6% (Petri et al. 2008a, b).

In European plum, a reliable and efficient transformation protocol has been developed (Petri et al. 2008c) that is now available to verify candidate genes in their function by overexpression or knock-down via RNA silencing. Apple class I *KNOX* (*MdKN1* and *MdKN2*) and corn *KNOX1* genes were used to transform plum and tobacco (Srinivasan et al. 2011). Transgenic plum plants that expressed the *MdKN2* or corn *KNOX1* gene grew normally but *MdKN1* caused significant reduction in plant height, leaf shape and size and produced malformed curly leaves. A high frequency of adventitious shoot regeneration (96%) was observed of leaf explants excised from corn *KNOX1*-expressing transgenic plum shoots. This is the first report of high-frequency regeneration of adventitious shoots *in vitro* from leaves and stem internodial explants of a woody perennial fruit tree through the ectopic expression of *KNOX1* gene (Srinivasan et al. 2011).

8.4.1.2 Agrobacterium rhizogenes-based Transformation Protocols

A. rhizogenes-mediated transformation using hairy-roots and composite plants of *Prunus* (transformed roots and untransformed aerial part) represents an alternative, but is exclusively limited to functional analysis of root genes (Bosselut et al. 2011). The validation procedure for *Ma* resistance complementation by *TNL1* as candidate gene was based on a fragment of approximately 15.3 kb containing the 9.2-kb *TNL1* genomic sequence preceded by approximately 5.4 kb from the 5′ upstream region transformed into the RKN-susceptible *Prunus* accession 253 via *A. rhizogenes* (Claverie et al. 2011). Microplants, inoculated with agrobacteria above the base of the stem, produced roots of which the most vigorous were sectioned and grown in Petri dishes. Roots growing autonomously and considered as hairy roots were cultured, multiplied by division at each growth time period, sampled for DNA and RNA extraction, and inoculated for evaluation of resistance to different RKN species as described in Bosselut et al. (2011). Hairy roots of transformed accession 253 were resistant to the three predominant RKN species, *M. arenaria, M. incognita* and *M. javanica*, to the peach RKN *M. floridensis*, and to the *Mi-1* tomato uncontrolled RKN *M. mayaguensis* (= *M. enterolobii*). Effective transformation of the hairy roots from accession 253 with the *TNL1* gene (pKGW, TNL1-253) was confirmed by the amplification of the *nptII* gene from the vector pKGW and of within-gene

TNL1-specific genomic fragments (CT3-4N and NSCALF2) and then by the amplification of TNL1-specific cDNA fragments (CT3 and Am5'). Southern blot hybridization of genomic DNA with an *nptII* probe showed the presence of a single copy of the T-DNA in at least two of these transformants. Finally, hairy roots that were free of galls could be shown to express the allele-specific *TNL1* transcript fragments CT3 and Am5', whereas no *TNL1* transcription was detected from hairy roots that produced galls. Independently from the hairy root transformants, a supplemental complementation procedure was successfully performed using one composite plant (transformed roots and untransformed aerial part; Bosselut et al. 2011) of the same susceptible accession 253 evaluated against all five RKN species.

This work also demonstrated that *Prunus* can be transformed with relatively large DNA fragments as illustrated in the case for hairy roots transformed with an approx. 18.8 kb insert (Bosselut et al. 2011; Claverie et al. 2011).

8.4.2 Use of Heterologous Systems to Analyze Genes Isolated from Stone Fruits

As transformation experiments in the genus *Prunus* are still time-consuming, heterologous systems, e.g., *Arabidopsis thaliana* or *Nicotiana tabacum*, may represent an alternative to study the in vivo function of genes, which had been isolated from stone fruits, assuming that basic processes are conserved in dicotyledon plants. However, few reports are available.

The auxin influx into cells in *Arabidopsis* seems to be driven by an AUX/LAX family of plasma membrane permeases encoded by *AUX/LAX* genes (Kerr and Bennett 2007). Isolation of *AUX1*-related clones from wild cherry tree was started with PCR primers from conserved regions of the *Arabidopsis AUX1* gene. The PCR products were then used to design wild cherry tree specific primers for the isolation of a corresponding cDNA (Hoyerová et al. 2008). To characterize the gene *PaLAX1* from wild cherry, *N. tabacum* and *A. thaliana* were transformed with *PaLAX1*. Using these heterologous systems it could be demonstrated that the *PaLAX1* gene from wild cherry promotes the uptake of auxin into cells and as a putative auxin influx carrier affects the content and distribution of free endogenous auxin in transgenic plants (Hoyerová et al. 2008).

Pathogen-related protein-5 (PR-5) has been implicated in plant disease resistance and its antifungal activity has been demonstrated in some fruit species (Vigers et al. 1992; Tattersall et al. 1997). Pathogenesis related protein-5 (PR-5) cDNA from European plum was isolated and its expression studied in different cultivars varying in resistance to brown rot disease caused by the necrotrophic fungus *Monilina fructicola* (El-kereamy et al. 2011). Overexpressing this *PdPR5-1* gene in *Arabidopsis* increases the resistance to

Alternaria brassicola infection. Analysis of the transgenic *Arabidopsis* plants demonstrated that the *PR-5* gene from European plum in addition to its antifungal properties also has a possible role in activating other defense pathways including phytoalexin production (El-kereamy et al. 2011).

8.4.3 T-DNA Tagged Populations

An interesting new tool, especially for research on fruit quality traits, has become available by the establishment of a T-DNA tagged population of strawberry (Oosumi et al. 2010). Genomic DNA flanking the T-DNA insertions in 191 transgenic plants of the diploid strawberry *F. vesca* was analyzed. Thermal Asymmetric Interlaced (TAIL)-PCR (Liu et al. 1995) or hi-TAIL-PCR (Liu and Chen 2007) were used to amplify flanking regions adjacent to the pCAMBIA T-DNA. Tagged sequences for 126 of 191 transgenic plants (66%) had homology to plant ESTs including 85 *F. vesca* ESTs, 32 ESTs of other species within the Rosaceae family and 2 ESTs in fruit species (Oosumi et al. 2010). T-DNA integration in strawberry is not random but directed by sequence microsimilarities in the host genome. With a high throughput system for the systematic generation of a collection of T-DNA tagged mutant lines available (Oosumi et al. 2006), T-DNA tagging in *F. vesca* can be used as a model system for systematic reverse genetics research in the Rosaceae family (Oosumi et al. 2010). Hopefully in the future, a database based search in a well characterized T-DNA tagged *Fragaria vesca* population will be possible as is available in *Arabidopsis* (http://www.arabidopsis.org/).

8.4.4 RNAi Technology

The RNAi technology represents a powerful tool to investigate the function of a gene and to verify its involvement in the creation of a phenotype. The RNAi technology is based on the production of double stranded RNA, which via RNA silencing leads to the suppression of genes (Waterhouse et al. 1998; Smith et al. 2000; Schwab et al. 2006).

To obtain PPV resistant plants via an RNA silencing approach, plum trees with different transgene copy numbers of the PPV-coat protein (CP) were generated by transformation (Scorza et al. 1994; Ravelonandro et al. 2000). In particular, one transgenic clone C5 carried multiple copies of the transgene, produced low level of CP mRNA and no detectable level of CP. Transgenic clone C5 was highly resistant to PPV during four years of testing and showed all characteristics of post-transcriptional gene silencing (Ravelonandro et al. 1997; Scorza et al. 2001). The PPV-CP transgene proved to be methylated (Scorza et al. 2001).

The successful use of the RNAi technology could also be demonstrated with two intron-hairpin (ihp) constructs using peach phytoene desaturase (PDS) gene (Petri et al. 2008c). All independent PDS silenced plants showed albino phenotypes, but with variation. Regarding the accumulation of PDS mRNA the PDS silenced plants had reduced levels and at the same time PDS siRNA could be detected (Petri et al. 2008c). RNAi technology represents an interesting way to knock-down potential candidate genes to verify their function and their involvement in a phenotype. This technology has been proven to work in stone fruits as well.

8.5 Conclusions

Until recently, MBC in most plant species was a rather tedious, time-consuming and often unsuccessful approach, comparable to trying to reach the top of Mount Everest: although progress is initially easy, the final 100 kb or Meters, respectively, are increasingly difficult to conquer (Peters et al. 2003). However, in stone fruits MBC approaches to isolate genes for functional analyses have already been successful for the *evergrowing* locus, self-incompatibility genes and *Meloidogyne* resistance. Similar approaches for traits, e.g., low acidity and pillar growth or traits underlying QTLs like, e.g., chilling requirement, blooming date, heat requirement and sharka resistance resulted in map and genome positions and will allow the isolation of these genes in the future. Substituting the tedious chromosome walking using large insert genomic libraries to form closed contigs around the target gene by an in silico walking along the peach genome sequence will now accelerate on-going and new MBC projects. Fine mapping still remains a problem as it requires large mapping populations.

However, transformation systems using *A. rhizogenes* for root related traits and *A. tumefaciens* for European plum have been established at least to an efficiency that allows verification of candidate genes by overexpression of genes or a knock-down via RNAi technology. The possibility to use a heterologous system should be considered. Applying a TAC library might save time by avoiding the development of constructs (Liang et al. 2004). Due to the limited success of transformations in *Prunus* species complementation of mutants, occurring naturally or induced by gamma radiation is not yet a real option. However, mutations induced by gamma radiation (Sharafi and Motallebi-Azar 2011) will be helpful to understand the function of genes lying behind mutant phenotypes. The development of a T-DNA tagged *Fragaria* population (Oosumi et al. 2010) also represent an interesting possibility for investigating fruit related traits that cannot be investigated in T-DNA insertion or transposon deletion mutants of *Arabidopsis* (Krysan et al. 1999; Speulman et al. 1999).

Huge improvements in the availability of genetic resources (*Prunus* reference map, SSR markers, large genomic libraries, peach genome sequence, *Agrobacterium* transformation systems, T-DNA tagged population, radiation induced mutants) will considerably facilitate new and on-going MBC approaches in the future. However, fine mapping in large segregating populations and verification of candidate genes will still remain a challenge in stone fruits. Association mapping might be an alternative to link candidate genes to traits (Aranzana et al. 2010).

References

Abad P, Gouzy J, Aury J-M, Castagnone-Sereno P, Danchin EGJ, Deleury E, Perfus-Barbeoch L, Anthouard V, Artiguenave F, Blok VC, et al. (2008) Genome sequence of the metazoan plant-parasitic nematode *Meloidogyne incognita*. Nat Biotechnol 26: 909–915.

Alvarez-Buylla ER, Pelaz S, Lijegren SJ, Gold SE, Burgeff C, Ditta GS, de Pouplana LR, Martinez-Castilla L, Yanfsky MF (2000) An ancestral MADS-box gene duplication occurred before the divergence of plants and animals. Proc Natl Acad Sci USA 97: 5328–5333.

Aranzana MJ, Gracia-Mas J, Carbo J, Arus P (2002) Development and variability of microsatellite markers in peach. Plant Breed 121: 87–92.

Aranzana MJ, Abbassi EK, Howad W, Arús P (2010) Genetic variation, population structure and linkage disequilibrium in peach commercial varieties. BMC Genetics 11: 69.

Aranzana MJ, Pineda A, Cosson P, Dirlewanger E, Ascasibar J, Cipriani G, Ryder CD, Testolin R, Abbott A, King GJ, Iezzoni AF, Arús P (2003) A set of simple-sequence repeat (SSR) markers covering the *Prunus* genome. Theor Appl Genet 106: 819–825.

Arora R, Rowland LJ, Tanino K (2003) Induction and release of bud dormancy in woody perennials: a science comes of age. HortScience 38: 911–921.

Baird WV, Estager AS, Wells JK (1994) Estimating nuclear DNA content in peach and related diploid species using laser flow cytometry and DNA hybridization. J Am Soc Hort Sci 119: 1312–1316.

Ballvora A, Pierre M, van den Ackerveken G, Schornack S, Rossier O, Ganal M, Lahaye T, Bonas U (2001) Genetic mapping and functional analysis of the tomato *Bs4* locus governing recognition of the *Xanthomonas campestris* pv. *vescatoria* AvrBs4 protein. Mol Plant Microbe Interact 14: 629–638.

Bassi D, Dima A, Scorza R (1994) Tree structure and pruning response of six peach growth forms. J Am Soc Hort Sci 119: 378–382.

Besemer J, Borodovsky M (2005) GeneMark: web software for gene finding in prokaryotes, eukaryotes and viruses. Nucl Acids Res 33: W451–W454.

Bielenberg DG, Wang Y, Fan S, Reighard GL, Scorza R, Abbott AG (2004) A deletion affecting several gene candidates is present in the *evergrowing* peach mutant. J Hered 95: 436–444.

Bielenberg DG, Wang Y, Li Z, Zhebentyayeva T, Fan S, Reighard GL, Scorza R, Abbott AG (2008) Sequencing and annotation of the *evergrowing* locus in peach [*Prunus persica* (L.) Batsch] reveals a cluster of six MADS-box transcription factors as candidate genes for regulation of terminal bud formation. Tree Genet Genomes 4: 495–507.

Bosselut N, Van Ghelder C, Claverie M, Voisin R, Onesto JP, Rosso MN, Esmenjaud D (2011) *Agrobacterium rhizogenes*-mediated transformation of *Prunus* as an alternative for gene functional analysis in hairy-roots and composite plants. Plant Cell Rep 30: 1313–1326.

Boudehri K, Bendahmane A, Cardinet G, Troadec C, Moing A, Dirlewanger E (2009) Phenotypic and fine genetic characterization of the *D* locus controlling fruit acidity in peach. BCM Plant Biol 9: 59–72.

Burge C, Karlin S (1997) Prediction of complete gene structures in human genomic DNA. J Mol Biol 268: 78–94.

Burgos L, Ledbetter CA, Pérez-Tornero O, Ortin-Párraga F, Egea J (1997) Inheritance of sexual incompatibility in apricot. Plant Breed 116: 383–386.

Burke DT, Carle GF, Olson MV (1987) Cloning of large segments of exogenous DNA into yeast by means of artificial chromosome vectors. Science 236: 806–812.

Cambra M, Capote N, Myrta A, Llàcer G (2006) Plum pox virus and the estimated costs associated with sharka disease. EPPO Bull 36: 202–204.

Cantini C, Iezzoni AF, Lamboy WF, Boritzky M, Struss D (2001) DNA fingerprinting of tetraploid cherry germplasm using simple sequence repeats. J Am Soc Hort Sci 126: 205–209.

Chaparro JX, Werner DJ, Omalley D, Sederoff RR (1994) Targeted mapping and linkage analysis of morphological isozyme, and RAPD markers in peach. Theor Appl Genet 87: 805–815.

Cipriani G, Lot G, Huang WG, Marrazzo MT, Peterlunger E, Testolin R (1999) AC/GT and AG/CT microsatellite repeats in peach (*Prunus persica* (L.) Batsch): isolation, characterisation and cross-species amplification in *Prunus*. Theor Appl Genet 99: 65–72.

Claverie M, Bosselut N, Lecouls AC, Voisin R, Lafargue B, Poizat C, Kleinhentz M, Laigret F, Dirlewanger E, Esmenjaud D (2004a) Location of independent root-knot nematode resistance genes in plum and peach. Theor Appl Genet 108: 765–773.

Claverie M, Dirlewanger E, Cosson P, Bosselut N, Lecouls AC, Voisin R, Kleinhentz M, Lafargue B, Caboche M, Chalhoub B, Esmenjaud D (2004b) High-resolution mapping and chromosome landing at the root-knot nematode resistance locus *Ma* from Myrobalan plum using a large insert BAC DNA library. Theor Appl Genet 109: 1318–1327.

Claverie M, Dirlewanger E, Bosselut N, Van Ghelder C, Voisin R, Kleinhentz M, Lafargue B, Abad P, Rosso MN, Chalhoub B, Esmenjaud D (2011) The *Ma* gene for complete-spectrum resistance to *Meloidogyne* spp. in *Prunus* is a TNL with a huge repeated C-terminal post LRR region. Plant Physiol 156: 779–792.

Cook R, Evans K (1987) Resistance and tolerance. In: Brown RH, Kerry BR (eds) Principles and Practice of Nematode Control in Crops. Academic press, New York, pp 179–231.

Dangl JL, Jones JDG (2001) Plant pathogens and integrated defence responses to infection. Nature 411: 826–833.

Decroocq (V), Foulongne M, Lambert P, Le Gall O, Mantin C, Pascal T, Schurdi-Levraud V, Kervella J (2005) Analogues of virus resistance genes map to QTLs for resistance to sharka disease in *Prunus davidiana*. Mol Genet Genom 272: 680–689.

Dirlewanger E, Cosson P, Tavaud M, Aranzana MJ, Poizat C, Zanetto A, Arús P, Laigret F (2002) Development of microsatellite markers in peach [*Prunus persica* (L.) Batsch] and their use in genetic diversity analysis in peach and sweet cherry (*Prunus avium* L.). Theor Appl Genet 105: 127–138.

Dirlewanger E, Graziano E, Joobeur T, Garriga-Caldere F, Cosson P, Howad W, Arús P (2004) Comparative mapping and marker-assisted selection in Rosaceae fruit crops. Proc Natl Acad Sci USA 101: 9891–9896.

Dirlewanger E, Moing A, Rothan C, Svanella L, Pronier V, Guye A, Plomion C, Monet R (1999) Mapping QTL controlling fruit quality in peach (*Prunus persica* (L.) Batsch). Theor Appl Genet 98: 18–31.

Dirlewanger E, Pronier V, Parvery C, Rothan C, Guy A, Monet R (1998) Genetic linkage map of peach. Theor Appl Genet 97: 888–895.

El-kereamy A, El-sharkawy I, Ramamoorthy R, Taheri A, Errampalli D, Kumar P, Jayasankarr S (2011) *Prunus domestica* pathogenesis-related protein-5 activates the defense response pathway and enhances the resistance to fungal infection. PlosOne Vol 6, Issue 3, e17973.

Entani T, Iwano M, Shiba H, Che FS, Isogai A, Takayama S (2003) Comparative analysis of the self-incompatibility (S-) locus region of *Prunus mume*: identification of a pollen-expressed F-box gene with allelic diversity. Genes Cells 8: 203–213.

Eremin VG (1978) Genetic potential of species *Prunus cerasifera* Ehr. and its use in breeding. Acta Hort 74: 61–65.

Esmenjaud D, Minot JC, Voisin R (1996) Effect of durable inoculum pressure and high temperature on root galling, nematode numbers and survival of Myrobalan plum genotypes (*Prunus cerasifera* Ehr.) highly resistant to *Meloidogyne* spp. Theor Appl Genet 92: 873–879.

Esmenjaud D, Minot JC, Voisin R, Pinochet J, Simard MH, Salesses G (1997) Differential response to root-knot nematodes in *Prunus* species and correlative genetic implications. J Nematol 29:370–380.

Esmenjaud D, Voisin R, Van Gelder C, Bosselut N, Larfargue B, Di Vito M, Dirlewanger E, Poessel JL, Kleinhentz M (2009) Genetic dissection of resistance to root-knot nematodes *Meloidogyne* spp. in plum, peach, almond, and apricot from various segregating interspecific *Prunus* progenies. Tree Genet Genomes 5: 279–289.

Etienne C, Moing A, Dirlewanger E, Raymond P, Monet R, Rothan C (2002a) Isolation and characterization of six peach cDNAs encoding key proteins in organic acid metabolism and solute accumulation: involvement in regulating peach fruit acidity. Physiol Plant 114: 259–270.

Etienne C, Rothan C, Moing A, Plomion C, Bodenes C, Svanella-Dumas L, Cosson P, Pronier V, Monet R, Dirlewanger E (2002b) Candidate genes and QTLs for sugar and organic acid content in peach [*Prunus persica* (L.) Batsch]. Theor Appl Genet 105: 145–159.

Fan S, Bielenberg DG, Zhebentyayeva TN, Reighard GL, Okie WR, Holland D, Abbott AG (2010) Mapping quantitative trait loci associated with chilling requirement and bloom date in peach (*Prunus persica*). New Phytol 185: 917–930.

Foulonge M, Pascal T, Arús P, Kervella J (2003) The potential of *Prunus davidiana* for introgression into peach (*Prunus persica* L. Batsch) assessed by comparative mapping. Theor Appl Genet 107: 227–238.

Gassmann W, Hinsch ME, Staskawicz BJ (1999) The *Arabidopsis RPS24* bacterial-resistance gene is a member of the TIR-NBS-LRR family of the disease resistance genes. Plant J 20: 265–277.

Georgi LL, Wang Y, Yvergniaux D, Ormsbee T, Inigo M, Reighard G, Abbott AG (2002) Construction of a BAC library and its application to the identification of simple sequence repeats in peach [*Prunus persica* (L.) Batsch]. Theor Appl Genet 105: 1151–1158.

Georgi LL, Wang Y, Reighard GL, Mao L, Wing RA, Abbott AG (2003) Comparison of peach and *Arabidopsis* genomic sequences: fragmentary conservation of gene neighborhoods. Genome 46: 268–276.

Gonzalez-Padilla IM, Webb K, Scorza R (2003) Early antibiotic selection and efficient rooting and acclimatization improve the production of transgenic plum plants (*Prunus domestica* L.). Plant Cell Rep 22: 38–45.

Hamilton CM, Frary A, Lewis C, Tanksley SD (1996) Stable transfer of intact high-molecular-weight DNA into plant chromosomes. Proc Natl Acad Sci USA 93: 9975–9979.

Horn R, Lecouls AC, Callahan A, Dandekar A, Garay L, McCord P, Howad W, Chan H, Verde I, Main D, Jung S, Georgi L, Forrest S, Mook J, Zhebentyayeva TN, Yeisoo Y, Hye Ran K, Jesudurai C, Sosinski B, Arús P, Baird V, Parfitt D, Reighard GL, Scorza R, Tomkins J, Wing R, Abbott AG (2005) Candidate gene database and transcript map for peach, a model species for fruit trees. Theor Appl Genet 110: 1419–1428.

Howad W, Yamamoto T, Dirlewanger E, Testolin R, Cosson P, Cipriani G, Monforte AJ, Georgi L, Abbott AG, Arús P (2005) Mapping a few plants: using selective mapping for microsatellite saturation of the *Prunus* reference map. Genetics 171: 1305–1309.

Hoyerová K, Perry L, Hand P, Laňková M, Kocábek T, May S, Kottová J, Pačes J, Napier R, Zažímalová E (2008) Functional Characterization of *PaLAX1*, a Putative Auxin Permease, in Heterologous Plant Systems. Plant Physiol 146: 1128–1141.

Ion-Nagy L, Lansac M, Eyquard JP, Salvador B, Garia JA, Le Gall O, Hernould M, Schurdi-Levraud V, Decroocq V (2006) PPV long-distance movement is occasionally permitted in resistant apricot hosts. Virus Res 120: 70–78.

Jiménez S, Reighard GL, Bielenberg DG (2010) Gene expression of *DAM5* and *DAM6* is suppressed by chilling temperatures and inversely correlated with bud break rate. Plant Mol Biol 73: 157–167.

Jiménez S, Lawton-Rauh AL, Reighard GL, Abbott AG, Bielenberg DG (2009) Phylogenetic analysis and molecular evolution of the dormancy associated MADS-box genes from peach. BMC Plant Biol 9: 81–93.

Joobeur T, Viruel MA, De Vicente MC, Jauregui B, Ballester J, Dettori MT, Verde I, Troco MJ, Messeguer R, Battle I, Quarta R, Dirlewanger E, Arus P (1998) Construction of a saturated linkage map for *Prunus* using an almond x peach F2 progeny. Theor Appl Genet 97: 1034–1041.

Kerr ID, Bennett MJ (2007) New insight into the biochemical mechanisms regulating auxin transport in plants. Biochem J 401: 613–622.

Khallouk S, Voisin R, Van Ghelder C, Engler G, Amiri S, Esmenjaud D (2011) Histological mechanism of the resistance conferred by the *Ma* gene against *Meloidogyne incognita* in *Prunus*. Phytopathology 101: 945–951.

Krysan PJ, Young JC, Sussman MR (1999) T-DNA as an insertional mutagen in *Arabidopsis*. Plant Cell 11: 2283–2290.

Lalli D, Abbott A, Zhebentyayeva T, Badenes M, Damsteegt V, Polák J, Krska B, Salava J (2008) A genetic linkage map for an apricot (*Prunus armeniaca* L.) BC1 population mapping plum pox virus resistance. Tree Genet Genomes 4: 481–493.

Le Dantec L, Cardinet G, Bonet J, Fouché M, Boudehri K, Monfort A, Poëssel JL, Moing A, Dirlewanger E (2010) Development and mapping of peach candidate genes involved in fruit quality and their transferability and potential use in other Rosaceae species. Tree Genet Genomes 6: 995–1012.

Llácer G, Cambra M, Laviña A (1985) Detection of plum pox virus in Spain. EPPO Bull 15: 325–329.

Lambert P, Dicenta F, Rubio M, Audergon JM (2007) QTL analysis of resistance to sharka disease in apricot (*Prunus armeniaca* L.) 'Polonais' x 'Stark Early Orange' F_1 progeny. Tree Genet Genomes 3: 299–309.

Lander ES, Green P, Abrahamson J, Barlow A, Daley MJ, Lincoln SE, Newburg L (1987) MAPMAKER: An interactive computer package for constructing primary genetic linkage maps of experimental and natural populations. Genomics 1: 174–181.

Lecouls AC, Rubio-Cabetas MJ, Minot JC, Voisin R, Bonnet A, Salesses G, Dirlewanger E, Esmenjaud D (1999) RAPD and SCAR markers linked to the *Ma1* root-knot nematode resistance gene in Myrobalan plum (*Prunus cerasifera* Ehr.). Theor Appl Genet 99: 328–336.

Lecouls AC, Bergougnoux V, Rubio-Cabetas MJ, Bosselut N, Voisin R, Poessel JL, Faurobert M, Bonnet A, Salesses G, Dirlewanger E, Esmenjaud D (2004) Marker-assisted selection for the wide-spectrum resistance to root-knot nematodes conferred by the *Ma* gene from Myrobalan plum (*Prunus cerasifera*) in interspecific *Prunus* material. Mol Breed 13: 113–124.

Li Z, Reighard GL, Abbott AG, Bielenberg DG (2009) Dormancy-associated MADS genes from the *EVG* locus of peach [*Prunus persica* (L.) Batsch] have distinct seasonal and photoperiodic expression patterns. J Exp Bot 60: 3521–3530.

Liang FS, Zhang KC, Yu ZW, Yang JL, Zhang XM, Jin D, Wang B (2004) Construction, characterization, and screening of a transformation-competent artificial chromosome library of peach. Plant Mol Biol Rep 22: 37–48.

Liu YG, Chen Y (2007) High-efficiency thermal asymmetric interlaced PCR for amplification of unknown flanking sequences. Bio/Techniques 43: 649–656.

Liu YG, Mitsukawa N, Oosumi T, Whittier RF (1995) Efficient isolation and mapping of *Arabidopsis thaliana* T-DNA insert junctions by thermal asymmetric interlaced PCR. Plant J 8: 457–463.

Majoros WH, Pertea M, Salzberg SL (2004) TigrScan and GlimmerHMM: two open source ab initio eukaryotic gene-finders. Bioinformatics 20: 2878–2879.

Mante S, Morgens PH, Scorza R, Cordts JM, Callahan AM (1991) *Agrobacterium*-mediated transformation of plum (*Prunus domestica* L.) hypocotyl slices and regeneration of transgenic plants. Bio/Technology 9: 853–857.

Marandel G, Pascal T, Candresse T, Decroocq V (2009a) Quantitative resistance to plum pox virus in *Prunus davidiana* P1908 linked to components of the eukaryotic translation initiation complex. Plant Pathol 58: 425–435.

Marandel G, Slava J, Abbott A, Candresse T, Decroocq V (2009b) Quantitative trait loci meta-analysis of plum pox virus resistance in apricot (*Prunus armeniaca* L.): new insights on the organization and identification of genomic resistance factors. Mol Plant Pathol 10: 347–360.

Martin GB, Bagdanove AJ, Sess G (2003) Understanding the functions of plant disease resistance proteins. Annu Rev Plant Biol 54: 23–61.

Martinez-Gomez P, Dicenta F (2000) Evaluation of resistance of apricot cultivars to a Spanish isolate of plum pox virus (PPV). Plant Breed 199: 179–181.

McCubbin AG, Wang X, Kao TH (2000) Identification of self-incompatibility (S) locus linked pollen cDNA markers in *Petunia inflata*. Genome 43: 619–627.

Michelmore RW, Paran I, Kesseli RV (1991) Identification of markers linked to disease-resistance genes by bulked segregant analysis: A rapid method to detect markers in specific genomic regions by using segregating populations. Proc Natl Acad Sci USA 88: 9828–9832.

Miller S, Scorza R (2010) Response of two novel peach tree growth habits to in-row tree spacing, training system, and pruning: Effect on growth and pruning. J Am Pomol Soc 64: 199–217.

Milligan SB, Bodeau J, Yaghoobi J, Kaloshian I, Zabel P, Williamson VM (1998) The root knot nematode resistance gene *Mi* from tomato is a member of the leucine zipper, nucleotide binding, leucine-rich repeat family of plant genes. Plant Cell 10: 1307–1319.

Mnejja M, Gracia-Mas J, Howad W, Badenes ML, Arús P (2004) Simple-sequence repeat (SSR) markers of Japanese plum (*Prunus salicina* Lindl) are highly polymorphic and transferable to peach and almond. Mol Ecol Notes 4: 163–166.

Mnejja M, Gracia-Mas J, Howad W, Arús P (2005) Development and transportability across *Prunus* species of 42 polymorphic almond microsatellites. Mol Ecol Notes 5: 531–535.

Mnejja M, Garcia-Mas J, Audergon JM, Arús P (2010) *Prunus* microsatellite marker transferability across rosaceous crops. Tree Genet Genomes 6: 689–700.

Monet R (1979) Genetic transmission of the 'non-acid' character. Incidence on selection for quality. In: (INRA ed) Eucarpia Symposium Tree Fruit Breeding, Angers, France, pp 273–276.

Ogundiwin EA, Marti C, Forment J, Pons C, Granell A, Gradziel TM, Peace CP, Crisoto CH (2008) Development of Chill Peach genomic tools and identification of cold-responsive genes in peach fruit. Plant Mol Biol 68: 379–397.

Olukolu, BA, Trainin T, Fan S, Kole C, Bielenberg DG, Reighard GL, Abbott AG, Holland D (2009) Genetic linkage mapping for molecular dissection of chilling requirement and bud-break in apricot (*Prunus armeniaca* L.). Genome 52: 819–828.

Oosumi T, Gruszewski HA, Blischak LA, Baxter AJ, Wadl PA, Shuman JL, Veilleux RE, Shulaev V (2006) High-efficiency transformation of the diploid strawberry (*Fragaria vesca*) for functional genomics. Planta 223: 1219–1230.

Oosumi T, Ruiz-Rojas JJ, Veilleux RE, Dickerman A, Shulaev V (2010) Implementing reverse genetics in Rosaceae: analysis of T-DNA flanking sequences of insertional mutant lines in the diploid strawberry, *Fragaria vesca*. Physiol Plant 140: 1–9.

Paal J, Henselewski H, Muth J, Kesem K, Menéndez CM, Salamini F, Ballvora A, Gebhardt C (2004) Molecular cloning of the potato *Gro1-4* gene conferring resistance to pathotype Ro1 of the root cyst nematode *Globodera rostochiensis*, based on a candidate gene approach. Plant J 38: 285–297.

Peters JL, Cnudde F, Gerats T (2003) Forward genetics and map-based cloning approaches. Trends Plant Sci 8: 484–491.

Petri C, Lopez-Noguera S, Alburquerque N, Egea J, Burgos L (2008a) An antibiotic-based selection strategy to regenerate transformed plants from apricot leaves with high efficiency. Plant Sci 175: 777–783.

Petri C, Wang H, Alburquerque N, Faize M, Burgos L (2008b) *Agrobacterium*-mediated transformation of apricot (*Prunus armeniaca* L.) leaf explants. Plant Cell Rep 27: 1317–1324.

Petri C, Webb K, Hily JM, Dardick C, Scorza R (2008c) High transformation efficiency in plum (*Prunus domestica* L.): a new tool for functional genomics studies in *Prunus* spp. Mol Breed 22: 581–591.

Pilarova P, Marandel G, Decroocq V, Salava J, Krska B, Abbott AG (2010) Quantitative trait analyses of resistance to plum pox virus in the apricot F_1 progeny 'Harlayne' x 'Vestar'. Tree Genet Genomics 6: 467–475.

Ravelonandro M, Scorza R, Bachelier JC, Labonne G, Levy L, Damsteegt V, Callahan AM, Dunez J (1997) Resistance of transgenic *Prunus domestica* to plum pox virus infection. Plant Dis 81: 1231.

Ravelonandro M, Scorza R, Callahan A, Levy L, Jacquet C, Monsion M, Damsteegt V (2000) The use of transgenic fruit trees as a resistance strategy for virus epidemics: the plum pox (sharka) model. Virus Res 71: 63–69.

Remington DL, Unger MC, Purugganan MD (2001) Map-based cloning of quantitative trait loci: progress and prospects. Genet Res 78: 213–218.

Robaglia C, Caranta C (2006) Translation initiation factors: a weak link in plant RNA virus infection. Trends Plant Sci 11: 40–45.

Roberts PA (1992) Current status of the availability, development, and use of host plant-resistance to nematodes. J Nematol 24: 213–227.

Rodriguez J, Sherman WB, Scorza R, Wisniewski M, Okie WR (1994) Evergreen peach, its inheritance and dormant behavior. J Am Soc Hort Sci 119: 789–792.

Rohde A, Ruttnik T, Hostyn V, Sterck L, Van Driessche K, Boerjan W (2007) Gene epression during induction, maintenance, and release of dormancy in apical buds of poplar. J Exp Bot 58: 4047–4060.

Romero C, Vilanova S, Burgos L, Martínez-Calvo J, Vicente M, Llácer G, Badenes ML (2004). Analysis of the S-locus structure in *Prunus armeniaca* L. Identification of S-haplotype specific *S-RNase* and *F-Box* genes. Plant Mol Biol 56: 145–157.

Ruiz EMV, Soriano JM, Romeo C, Zhentyayeva T, Terol J, Zuraiga E, Llacer G, Abbott AG, Badenes ML (2011) Narrowing down the apricot plum pox virus resistance locus and comparative analysis within the peach genome syntenic region. Mol Plant Pathol 12: 535–547.

Sajer O, Scorza R, Dardick C, Zhebentyayeva T, Abbott A, Horn R (2012) Development of sequence-tagged site markers linked to the pillar growth type in peach [*Prunus persica* L. (Batsch)]. Plant Breed 131: 186–192.

Salamov AA, Solovyev VV (2000) Ab initio gene finding in *Drosophila* genomic DNA. Genome Res 10: 516–522.

Schwab R, Ossowski S, Riester M, Warthmann N, Weigel D (2006) Highly specific gene silencing by artificial MicroRNAs in *Arabidopsis*. Plant Cell 18: 1121–1133.

Scorza R (1984) Characterization of four distinct peach tree growth types. J Am Soc Hort Sci 109: 455–457.

Scorza R, Ravelonandro M, Callahan AM, Cordts JM, Fuchs M, Dunez J, Gonsalves D (1994) Transgenic plums (*Prunus domestica* L.) express the plum pox virus coat protein gene. Plant Cell Rep 14: 18–22.

Scorza R (2005) Theory and practice of genetically manipulating peach tree architecture. NY Fruit Quarterly 3: 27–31.

Scorza R, Callahan A, Levy L, Damsteegt V, Webb K, Ravelonandro M (2001) Post-transcriptional gene silencing in plum pox virus resistant transgenic European plum containig the plum pox potyvirus coat protein gene. Transgen Res 10: 201–209.

Scorza R, Miller S, Glenn DM, Okie WR, Tworkoski T (2006) Developing peach cultivars with novel tree growth habits. In: Infante R (ed) Proc 6th Intl Peach Symposium. Acta Hort (ISHS) 713: 61–64.

Sicard O, Marandel G, Soriano JM, Lalli DA, Lambert P, Salava J, Badenes ML, Abbott AG, Decroocq V (2008) Flanking the major plum pox virus resistance locus in apricot with co-dominant markers (SSRs) derived from candidate resistance genes. Tree Genet Genomes 4: 359–365.

Sijacic P, Wang X, Skirpan AL, Wang Y, Dowd PE, McCubbin AG, Huang S, Kao T (2004) Identification of the pollen determinant of S-RNase-mediated self-incompatibility. Nature 429: 302–305.

Sharafi Y, Motallebi-Azar AR (2011) Gamma irradiation influences on some biological traits in two almond (*Prunus amygdalus* L.) cultivars. J Med Plants Res 5: 255–258.

Shizuya H, Birren B, Kim UJ, Mancino V, Slepak T, Tachiiri Y, Simon M (1992) Cloning and maintenance of 300-kilobasepair fragments of human DNA in *Escherichia coli* using an F-factor based vector. Proc Natl Acad Sci USA 89: 8794–8797.

Smith NA, Singh SP, Wang MB, Stoutjesdijk PA, Green AG, Waterhouse PM (2000) Total silencing by intron-spliced hairpin RNAs. Nature 407: 319–320.

Soderlund C, Humphrey S, Dunhum A, French L (2000) Contigs built with fingerprints, markers and FPC 4.7. Genome Res 10: 1772–1787.

Soriano JM, Vera-Ruiz EM, Vilanova S, Martínez-Calvo J, Llácer G, Badenes ML, Romero C (2008) Identification and mapping of a locus conferring plum pox virus resistance in two apricot-improved linkage maps. Tree Genet Genomes 4: 391–402.

Sosinski B, Gannavarapu M, Hager LD, Beck LE, King GJ, Ryder CD, Rajapakse S, Baird WV, Ballard RE, Abbott AG (2000) Characterization of microsatellite markers in peach (*Prunus persica* L Batsch). Theor Appl Genet 101: 421–428.

Speulman E, Metz PLJ, van Arkel G, te Lintel Hekkert B, Stiekema WJ, Pereira A (1999) A two-component enhancer-inhibitor transposon mutagenesis system for functional analysis of the *Arabidopsis* genome. Plant Cell 11: 1853–1866.

Srinivasan C, Liu Z, Scorza R (2011) Ectopic expression of class 1 *KNOX* genes induce adventitious shoot regeneration and alter growth and development of tobacco (*Nicotiana tabacum* L) and European plum (*Prunus domestica* L). Plant Cell Rep 30: 655–664.

Tanksley SD, Ganal MW, Martin GB (1995) Chromosomal landing: a paradigm for map-based cloning in plants with large genomes. Trends Genet 11: 63–68.

Tao R, Yamane H, Sugiura A (1999) Molecular typing of S-alleles through identification, characterization and cDNA cloning for S-RNAses in sweet cherry. J Am Soc Hort Sci 124: 224–233.

Tattersall DB, van Heeswijck R, Hoj PB (1997) Identification and characterization of a fruit-specific, thaumatin-like protein that accumulates at very high levels in conjunction with the onset of sugar accumulation and berry softening in grapes. Plant Physiol 114: 759–769.

Testolin R, Messina R, Lain O, Marrazzo MT, Huang WG, Cipriani G (2004) Microsatellites isolated in almond from an AC-repeat enriched library. Mol Ecol Notes 4: 459–461.

Urtubia C, Devia J, Castro A, Zamora P, Aguirre C, Tapia E, Barba P, Dell'Orto, Moynhihan MR, Petri C, Scorza R, Prieto H (2008) *Agrobacterium*-mediated genetic transformation of *Prunus salicina*. Plant Cell Rep 27: 1333–1340.

Ushijima K, Sassa H, Dandekar AM, Gradziel TM, Tao R (2003) Structural and transcriptional analysis of the self-incompatibility locus of almond: identification of a pollen-expressed *F-box* gene with haplotype-specific polymorphism. Plant Cell 15: 771–781.

Ushijima K, Sassa H, Tao R, Yamane H, Dandekar AM, Gradziel TM, Hirano H (1998) Cloning and characterization of cDNAs encoding S-RNases from almond (*Prunus dulcis*). Primary structural features and sequence diversity of the S-RNases in Rosaceae. Mol Gen Gent 260: 261–268.

Utz HF, Melchinger AE (1996) PLABQTL: A program for composite interval mapping of QTL. J Agri Genom 2: 1–5.

Van der Biezen EA, Jones JD (1998) Plant disease-resistance proteins and the gene-for-gene concept. Trends Biochem Sci 23: 454–456.

Van der Hoorn RA, Kamoun S (2008) From Guard to Decoy: a new model for perception of plant pathogen effectors. Plant Cell 20: 2009–2017.

Verde I, Lauria M, Dettori MT, Vendramin E, Balconi C, Micali S, Wang Y, Marrazzo MT, Cipriani G, Hartings H, Testolin R, Abbott AG, Motto M, Quarta R (2005) Microsatellite and AFLP markers in the *Prunus persica* [L. (Batsch)] x *P. ferganensis* BC_1 linkage map: saturation and coverage improvement. Theor Appl Genet 111: 1013–1021.

Veyrieras JB, Goffinet B, Charcosset A (2007) MetaQTL: a package of new computational methods for the meta-analysis of QTL mapping experiments. BCM Bioinformat 8: 49.

Vigers AJ, Wiedermann S, Roberts WK, Legrand M, Selitrennikoff CP, et al. (1992) Thaumatin-like pathogenesis-related proteins are antifungal. Plant Sci 83 : 155–161.

Vilanova S, Romero C, Abbott AG, Llácer G, Badenes ML (2003a) An apricot (*Prunus armeniaca* L.) F_2 progeny linkage map based on SSR and AFLP markers, mapping plum pox virus resistance and self-incompatibility traits. Theor Appl Genet 107: 239–247.

Vilanova S, Romero C, Abernathy D, Abbott AG, Burgos L, Llacer G, Badenes ML (2003b) Construction and application of a bacterial artificial chromosome (BAC) library of *Prunus armeniaca* L. for the identification of clones linked to the self-incompatibility locus. Mol Genet Genom 269: 685–691.

Voisin R, Minot JC, Esmenjaud D (1999) Penetration, development and emigration of juveniles of the nematode *Meloidogyne arenaria* in Myrobalan plum (*Prunus cerasifera*) clones bearing the *Ma* resistance genes. Eur J Plant Pathol 105: 103–108.

Vos P, Hogers R, Bleeker M, Reijans M, van de Lee T, Hornes M, Frijters A, Pot J, Peleman J, Kuiper M, Zabeau M (1995) AFLP: a new technique for DNA fingerprinting. Nucl Acids Res 23: 4407–4414.

Wang Q, Zhang K, Qu X, Jia J, Shi J, Jin D, Wang B (2001) Construction and characterization of a bacterial artificial chromosome library of peach. Theor Appl Genet 103: 1174–1179.

Wang Y, Garay L, Georgi LL, Reighard GL, Scorza R, Abbott AG (2002a) Development of bacterial artificial chromosome contigs in the *evergrowing* gene region in peach [*Prunus persica* (L.) Batsch]. Acta Hort 592: 183–189.

Wang Y, Georgi LL, Reighard GL, Scorza R, Abbott AG (2002b) Genetic mapping of the *evergrowing* gene in peach [*Prunus persica* (L.) Batsch]. J Hered 93: 352–358.

Wang Y, Georgi LL, Zhebentyayeva TN, Reighard GL, Scorza R, Abbott AG (2002c) High-throughput targeted SSR marker development in peach (*Prunus persica*). Genome 45: 319–328.

Waterhouse PM, Grahm MW, Wang MB (1998) Virus resistance and gene silencing in plants can be induced by simultaneous expression of sense and antisense RNA. Proc Natl Acad Sci USA 95: 1359–13964.

Whitham S, Dinesh-Kumar SP, Choi D, Hehl R, Corr C, Baker B (1994) The product of the tobacco mosaic virus resistance gene *N*: similarity to Toll and Interleukin-1 receptor. Cell 78: 1101–1115.

Yaegaki H, Shimada T, Moriguchi T, Hayma H, Haji T, Yamaguchi M (2001) Molecular characterization of S-RNase genes and S-genotypes in Japanese apricot (*Prunus mume* Sieb. Et Zucc.) Sex Plant Reprod 13: 251–257.

Yamazaki K, Okabe M, Takahashi E (1987) New broomy flowering peach cultivars 'Terutebeni', 'Terutemomo' and 'Teruteshiro'. Bull Kanagawa Hort Exp Sta 34: 54–56.

Yamane H, Ikeda K, Ushijima K, Sassa H, Tao R (2003) A pollen-expressed gene for a novel protein with an F-box motif that is very tightly linked to gene for S-RNase of two species of cherry, *Prunus cerasus* and *P. avium*. Plant Cell Physiol 44: 764–769.

Yamane H, Tao R, Sugiura A, Hauck NR, Iezzoni AF (2001) Identification and characterization of S-RNases in tetraploid sour cherry (*Prunus cerasus*). J Am Soc Hort Sci 126: 661–667.

Yoshida M: Genetical studies on the fruit quality of peach varieties. I. Acidity (1970) Bull Hort Res Stn Jpn Ser A 9: 1–15.

Zhebentyayeva TN, Swire-Clark G, Georgi LL, Garay L, Jung S, Forrest S, Blenda AV, Blackmon B, Mook J, Horn R, Howad W, Arús P, Main D, Tomkins JP, Sosinski B, Baird WV, Reighard GL, Abbott AG (2008) A framework physical map for peach, a model Rosaceae species. Tree Genet Genomics 4: 745–756.

Zeng ZB (1994) Precision mapping of quantitative trait loci. Genetics 136: 1457–1468.

9

Genome Sequencing Initiative

Ignazio Verde,[1,*] *Elisa Vendramin,*[1] *Maria Teresa Dettori,*[1] *Sabrina Micali*[1] and *Bryon Sosinski*[2]

ABSTRACT

The genus *Prunus* encompasses many economically important stone fruit species. Peach is considered one of the best genetically characterized species in the Rosaceae, and it has distinct advantages that make it suitable as a model genome species for *Prunus* as well as for other species in the Rosaceae. In this context, an international consortium of scientists, The International Peach Genome Initiative (IPGI) has recently completed a first high quality draft sequence of the peach genome. The draft assembly (Peach V1.0) consists of an 8.5-fold coverage derived from whole genome shotgun Sanger sequences of the doubled haploid 'Lovell'. The assembly is arranged in 202 scaffolds larger than 1Kb, equalling approximately 227 Mbp rather than the 280–300 Mbp that was previously predicted. Eight pseudomolecules, named according to their corresponding linkage groups, comprising 99,3% of the peach genome were aligned using a network of *Prunus* linkage maps. Automated gene prediction and annotation produced 27,851 genes, 838 of which show alternative splicing giving a total of 28,689 predicted protein-coding genes. Studies were undertaken worldwide to investigate the genetic diversity within the species through the resequencing of different *Prunus* accessions. Millions of genetic markers, as SNPs and indels, were discovered allowing the development of high density SNP genotyping assays in peach and cherry.

Keywords: IPGI, Sanger WGS, NGS, variant discovery, SNP-chip

[1]CRA—Fruit Tree Research Center, via di Fioranello 52, 00134 Roma, Italy.
[2]Department of Horticultural Science, North Carolina State University, Raleigh, NC, USA.
*Corresponding author: *ignazio.verde@entecra.it*

9.1 Introduction

With the development of the earliest sequencing technologies (Maxam and Gilbert 1977; Sanger et al. 1977a) and the release of the first DNA sequence, that of the bacteriophage φ X174 in 1977 (Sanger et al. 1977b, 1978), foundations were laid for transition from the genetic to the genomic era of life science research. The ability of reading a nucleotide sequence has opened the way to study the secrets that lie within an organism's genome from many different perspectives at unprecedented levels of resolution. Whole genome sequencing has been regarded as a priority for the human species and for those representatives of animal and plant organisms of main importance. The possibility to obtain the DNA sequence of an organism on a whole genome scale provides a powerful tool to look into the complete catalog of genes, and their regulative elements and variation at the highest resolution possible. Similarly, sequencing of target DNA regions, or of the transcribed fraction of the genome, is a priority to identify genes involved in many different biological processes and mechanisms. Access to genetic variation lying within a given genomic region at a resolution as high as a single nucleotide and on a scale as wide as an entire germplasm collection, is crucial to explain and correlate phenotypic variation among individuals or populations of related species.

Efforts to sequence the peach genome can be traced back to 2005 with the initiatives engaged both in Italy and US to apply for grants aimed at producing a draft sequence of the species.

At the Plant and Animal Genome XV Meeting on January 2007, Jerry Tuskan from the Joint Genome Institute (JGI) announced the new roster of plants that would have been targeted for genome sequencing at JGI. Peach had been selected for an 8x whole genome shotgun (WGS) that would be completed by the end of 2009. Since the original genome sequencing announcement, an Italian consortium, in the frame of a national project (DRUPOMICS) supported by the Italian Ministry of Agriculture (MiPAAF), was established to join the US groups in the peach sequencing efforts. Since then the US-Italian consortium grew to gather a worldwide initiative, the International Peach Genome Initiative (IPGI), including also institutions from Spain, France and Chile (www.rosaceae.org/peach/genome).

9.2 Genomic Resources

The genus *Prunus* encompasses many economically important stone fruit species. Peach [*Prunus persica* L. (Batsch)] is by far the most important among them, being the third temperate fruit for production after apple and pear. Almond is also an important crop, the third for extension of cultivated areas, but only the fifth by production (data 2009 at FAO: http://faostat.fao.org/).

The economic importance of peach, together with its biological and genetic characteristics such as self-compatibility, a relatively short juvenile period and a small diploid genome, made this species a model for stone fruits. An advantage that turned to be very useful later, particularly in the development of molecular linkage maps since the species revealed low levels of variability, was the possibility to obtain interspecific crosses between some stone fruit species, both wild and cultivated.

Molecular genetic studies began later in trees than in herbaceous crops. At the end of the eighties in *Prunus* they were mainly restricted to peach and almond, which, in contrast to other stone fruits such as cultivated plums and sour cherries, share the characteristics of being diploid species with a low number of chromosomes ($n = 8$) and a small genome (estimated nuclear DNA content recently assessed at ~230 Mbp; Sosinski et al. 2010), less than twice that of *Arabidopsis thaliana* (Arumuganathan and Earle 1991). The knowledge available at the time for the genus, mainly represented by Mendelian genetic studies (Monet 1989) and the genetic resources like haploids, doubled haploids and sport mutants, had been mostly developed in *P. persica*.

The very first stone fruit map was published in 1994 (Chaparro et al.): it was a *P. persica* random amplified polymorphic DNA(RAPD) linkage map, molecular markers that were early abandoned lacking reproducibility. Already in 1995 another peach map (Rajapakse et al. 1995) and the first two almond maps (relative to a single F_1 segregating intraspecific population) followed (Viruel et al. 1995). Those maps contained robust codominant markers: restriction fragment length polymorphism (RFLPs developed from *Prunus* libraries) and isozymes. Fifteen year elapsed between the first map and the release of the complete sequence of the peach genome in 2010, and many more markers were developed allowing the constructions of several *Prunus* maps and of the *Prunus* reference map (TxE, Joobeur et al. 1998).

An overview of the resources employed to anchor and annotate the peach sequence will be presented in the following sections.

9.2.1 Segregating Populations and Linkage Maps

Saturated linkage maps, allowing the anchoring, ordering and orientation of the scaffolds, are crucial for reconstructing the sequence of each individual chromosome (pseudomolecule) of a genome.

Many *Prunus* maps are today available, covering most of the species included in the genus. Many can be viewed at the Genome Database for Rosaceae (GDR: http://www.rosaceae.org/).

At the beginning of the mapping efforts bi-parental intraspecific peach populations showed a low level of heterozygosity (peach is a self-compatible and predominantly autogamous species), making nearly impossible to detect sufficient polymorphisms to map in F_1s, as almost all the markers were monomorphic and did not segregate (Byrne 1990; Mnejja et al. 2010). Recently Aranzana et al. (2010) showed that the level of heterozygosity is widely variable in this species; values detected by microsatellite markers in 224 cultivars ranged from 0 to 0.80, with a mean value of 0.34.

In contrast other outcrossing stone fruit species, such as *P. dulcis* and *P. davidiana* proved to be extremely polymorphic (Mnejja et al. 2010). At that time, the problem of the peach low variability was overcome by mapping progenies obtained by interspecific crosses. In fact, peach can be easily crossed with phylogenetically near stone fruits such as *P. davidiana*, *P. ferganensis* and *P. dulcis* giving viable progenies with an enormous increase of genetic variability. Interspecific maps were obtained for all those crosses, (Joobeur et al. 1998; Dettori et al. 2001; Foulogne et al. 2003): the peach x almond map (TxE, Joobeur et al. 1998) was used to anchor the peach genome. This map is considered the reference map for the whole genus. Two more maps, a BC_1 *P. persica* x *P. ferganensis* (PxF) and a peach F_2 (Contender x Ambra, CxA), were employed to refine the Peach V 1.0 release.

The TxE map derives from a population of 88 individuals obtained by selfing a single peach x almond hybrid tree. This hybrid came from the cross of the almond cv. Texas and the peach cv. Earlygold. The first TxE map was the main result of the European project AIR3-CT93-1585. It represented a milestone in the development of *Prunus* genomics. During the project a set of transferable markers was developed and used to obtain maps in different *Prunus* species, thus providing anchor loci and a common terminology for the *Prunus* linkage groups and facilitating the construction of framework maps in other crosses. Most of the maps developed later contained a set of loci anchored with TxE, thus providing the possibility to perform comparisons between genomes of different species (Dirlewanger et al. 2004).

The first TxE map contained only codominant markers, 235 RFLPs and 11 isozymes. The RFLP probes came from genomic and cDNA libraries of almond, peach, *P. ferganensis*, cherry, plum and apple. This first map was later improved with the addition of 185 simple sequence repeats (SSRs), five sequence tagged sites (STSs) and 126 RFLPs (Aranzana et al. 2003; Dirlewanger et al. 2004), finally including 562 markers—361 RFLPs, 190 simple sequence repeats (SSRs) and 11 isozymes—spanning 519 cM with a high-density coverage of the *Prunus* genome (average 0.92 cM/marker). Dirlewanger et al. (2004) reported that the majority (87%) of the loci in the TxE map had known sequence.

As time passed the number of *Prunus* DNA and cDNA sequences increased, with hundreds of new unmapped SSR markers available; many were *Prunus* expressed sequence tags (ESTs), reposited at the Genome Database for the Rosaceae (GDR; http://www.rosaceae.org/). To exploit the possibility of using this major source of polymorphic markers Howad et al. (2005) applied to the TxE population a new mapping strategy proposed some years before by Vision (2000). They choose a subset of six individuals from the TxE population, based on the number and position of recombinational crossover sites detected in the framework map. This set allowed to map almost any marker to one of the 67 unique chromosome fragments, so-called bins (intervals), 7.8 cM on average in the *Prunus* genome. With this approach new markers can be mapped with lower precision, but greatly reducing the costs; the authors themselves published in that same article the "bin" position of 264 new SSR markers, thus bringing the number of mapped and bin-mapped TxE markers to 826. Bin-mapping proved an invaluable genomic resource in *Prunus*, in providing both the means for the further saturation of the 'reference map' and an excellent basis for marker searching in a specific region of interest for all the anchored species.

An updated version of the *Prunus* reference map (TxE), that includes SSRs, single nucleotide polymorphisms (SNPs) and RFLPs (with known sequence), was used to anchor, order and orientate the scaffold of the Peach v1.0 built. In Table 9-1 is reported the list of all the markers used.

Using this map it was possible to integrate the assembly produced by Arachne (Jaffe 2003) into eight chromosome-scale pseudomolecules, named according to the linkage group nomenclature adopted in *Prunus* (Joobeur et al. 1998). The eight pseudomolecules contained 96.1% of the released genome (Peach v 1.0, 215.9Mb out of 224.6 Mb); 3.9% (8.7 Mb) of the genome was unmapped and 85.6% had a known orientation along the pseudomolecules.

With the aim of refining this assembly, SNP and SSR markers were developed to univocally target regions belonging to unanchored and unoriented scaffolds. Those markers were mapped in TxE or in two intraspecific peach maps anchored with the *Prunus* reference map, PxF (Verde et al. 2005) and (CxA) F_2 map, (Eduardo et al. 2010).

The PxF map was published in 2001 (Dettori et al. 2001) and saturated in 2005 (Verde et al. 2005) with a total of 216 loci including 57 SSRs, and 78 RFLPs in eight linkage groups. This BC_1 population was obtained crossing a peach selection with a *P. ferganensis* accession. A single F_1 hybrid derived from this cross was then backcrossed to the peach parent. At the time of the cross, *P. ferganensis* was believed to be a different species from peach, even if doubts were raised by their morphological similarity. The genetic analysis of this cross showed that *P. ferganensis* was essentially homozygous and shared the same alleles with peach. More recently the resequencing of *P. ferganensis* (Verde et al. 2011) confirmed that this cross should be considered as an intraspecific one. The last map employed for the refinement of Peach v 1.0 is an updated version of CxA. The F_1 self-pollinated seedling that produced

Table 9-1 List of the markers used for anchoring the peach genome: type, name, source.

Type	Marker	Species	Library	Reference
SSR	BPPCT	*P. persica*	genomic library	Dirlewanger et al. 2002
	CPDCT	*P. dulcis*	genomic library	Mnejja et al. 2004
	CPPCT	*P. persica*	genomic library	Aranzana et al. 2002
	CPSCT	*P. salicina*	genomic library	Mnejja et al. 2004
	EPDCU	*Prunus dulcis*	cDNA	unpublished; developed from GenBank ESTs (accessions from BU572502 to BU575295)
	EPPB *	*P. persica*	cDNA (EST in GeneBank)	Dirlewanger 2006
	EPPCU (syn ECUPP)	*P. persica*	cDNA (EST in GeneBank)	unpublished; developed from GenBank ESTs (accessions from BU039022 to BU049005)
	EPPISF	*P. persica*	cDNA	Vendramin et al. 2007
	MA	*P. persica*	genomic library	Yamamoto et al. 2002
	M	*P. persica*	cDNA	Yamamoto et al. 2002
	Pchcms	*P. persica*	cDNA	Sosinski et al. 2000
	Pchgms	*P. persica*	genomic library	Sosinski et al. 2000
	PS	*P. avium*	genomic library	Cantini et al. 2001
	RPPG	*P. spersica*	Peach V1.0	Verde Unpublished
	UDP	*P. persica*	genomic library	Cipriani et al. 1999; Testolin et al. 2000
	UDA	*P. dulcis*	genomic library	Testolin et al. 2004
	UDAp	*P. armeniaca*	genomic library	Messina et al. 2004
STS	MD	*P. persica*	genes	Yamamoto et al. 2005
SNPs, InsDels	RosCOS (Rosaceae Conserved Orthologous Sequences)	*Rosaceae*	EST	Cabrera et al. 2009
	CG (Candidate Genes)	*Prunus*	EST	Illa et al. 2011
RFLP*	AC	*P. dulcis*	cDNA	Viruel et al. 1995; Joobeur et al. 1998
	AG	*P. dulcis*	genomic DNA	Viruel et al. 1995; Joobeur et al. 1998
	CC	*P. avium*	cDNA	Joobeur et al. 1998
	FG	*P. ferganensis*	genomic DNA	Joobeur et al. 1998; Dettori et al. 2001
	PC	*P. persica*	cDNA	Joobeur et al. 1998
RFLPs/ known genes	Ltp2, Pgl1, Pru1, Ole1, Omt1	*P. dulcis*	cDNA	Viruel et al. 1995; Joobeur et al. 1998

*Some of the RFLPs of this libraries were sequenced

the F_2 population (about 300 seedlings) was in the panel of resequenced peaches (Verde et al. 2011) so it was possible to saturate the map by choosing and mapping evenly spaced SNP markers obtained by aligning the next generation sequencing (NGS) reads against the peach genome. With this approach it was possible to target unsolved issues in TxE (i.e., unoriented scaffold in Peach v 1.0) due to the high power of resolution of this map.

As a result of all the refinements, 99.3% (223.1 Mb out of 224.6 Mb) of the assembled sequence is now anchored to chromosomes, and 96.2% is also orientated. All these refinements will be included in a future release of the peach genome. They will be visualized in a dedicated track available on the Gbrowses.

9.2.2 Expressed Sequence Tags (ESTs) and Bacterial Artificial Chromosome (BAC) Libraries

The expressed sequence tag (EST) sequences of cDNA libraries constructed from various tissues and in various conditions serve as a crucial tool in the development of markers for genetic mapping, in the development of probes for gene expression studies or in direct mapping to genomic clones or chromosomes, and in gene prediction when large genomic sequences are available. For the *Prunus* genus thousands of ESTs are available, that have been developed in worldwide labs from different tissues and in different conditions (Table 9-2). The *Prunus* unigene set (*Prunus* Assembly V4) was created to reduce ESTs redundancy and to obtain longer transcripts. These transcripts were computationally annotated using BLAST pairwise comparison against protein databases, searching for SSR motifs, mining for SNPs, and other functional characterizations. The descriptions of the *Prunus* Assembly V4 are reported in Table 9-3. All the information about the *Prunus* ESTs are available at the GDR (http://www.rosaceae.org/; Jung et al. 2008) and at the ESTree websites (http://www.itb.cnr.it/estree/; Lazzari et al. 2008) that represent the only existing online resources dedicated to *Prunus* EST analysis.

Bacterial artificial chromosomes (BACs) are the vectors of choice for large insert libraries; they can allocate fragment up to 300 kb. Compared with the main alternative, yeast artificial chromosomes (YAC), they are easy to handle, and the clones are stable and less likely to be chimeric. Importantly, DNA purification and sequencing from BACs are simple and straight forward. BAC libraries provide a platform for the creation of physical maps and complete genome sequencing, which are powerful tools for large-scale gene discovery and elucidation of gene function and regulation (Shizuya et al. 1992; Woo et al. 1994; Zhang and Wu 2001; Farrar and Donnison 2007).

Table 9-2 Statistics on the *Prunus* ESTs available.

EST Libraries	
Number of ESTs available	89166
Number of Libraries	39
Number of Tissues	21
Number of Development Stages	20
Species	
Prunus armeniaca	15105
Prunus avium	21
Prunus avium x cerasus x canescens	89
Prunus cerasus	1255
Prunus dulcis	3864
Prunus domestica	54
Prunus persica	70939

Table 9-3 Data outline of the *Prunus* unigene set.

Prunus Assembly V4 Summary	
Number of ESTs available after filtering	89166
Average Length	568
Number of Contigs (CAP3 Assembly)	9179
Average Length of Contigs	919
Number of Singlets	15128
Number of Putative Unigenes	24307

Two complementary large-insert genomic libraries were developed in peach. 'The Nemared' BAC library, consists of 44,160 BAC clones with an average insert size of 70 kb giving 8.8-fold genome coverage. The second library is based on a haploid strain (Plov 2-1N) derived from the peach rootstock Lovell and consists of 34,560 clones with an average insert size of 80 kb providing a 9.2-fold genome coverage. These two libraries were used to develop a genome-wide framework physical map of peach that contains 2,138 contigs composed of 15,655 clones (Zhebentyayeva et al. 2008). BAC ends from the haploid strain was also sequenced in the frame of the IPGI to assist the assembly of the peach genome.

9.3 Sequencing Strategies: Methods for Genome Sequencing

From the beginning of the human genome project, in 1990, until now many advances were made in sequencing technologies. The implementation of the Sanger sequencer from one capillary to 96 parallel runs allowed the human genome project to finish two years in advance (International Human Genome Consortium 2004). Nowadays the innovation of the sequence-by-synthesis technology (or next generation sequencing, NGS) that dramatically reduces the cost and the time to generate sequences agrees to sequence a genome in a very little time (not including the huge bioinformatics efforts) and at reduced cost (Metzker 2010). In this scenario different genomes of model plant species and crops have been sequenced (Feuillet et al. 2011) with mainly two different strategies: BAC-by-BAC approach (also known as hierarchical shotgun sequencing) and the whole-genome shotgun strategy (WGS).

The BAC-by-BAC approach was the first method applied and consists of different steps: (a) Obtain a large collection of BAC clones, (b) Construct a physical map, (c) Select a minimum tiling path, (d) Sequence each clone in the path with shotgun, (e) Assemble, (f) Anchor the scaffolds using linkage and physical maps.

Once a BAC library is obtained the next step is to construct the physical map. A physical map is a linear ordering of a set of clones encompassing one or more chromosomes. Physical maps can be generated by first partially digesting clones with restriction enzymes and then detecting the clone overlaps by matching the lengths of the fragments produced by digestion (Bozdag et al. 2008). A minimum-cardinality set of overlapping clones that spans the region represented by the physical map is called minimal tiling path (MTP). These strategies produce a huge logical puzzle (Fig. 9-1). The individually mapped clones of the MTP are then subjected to shotgun sequencing. DNA from each BAC clone is fragmented randomly into smaller pieces that are either cloned into a plasmid and sequenced with Sanger sequencing or directly via NGS technologies. The sequences are then assembled looking for the sequence portions that overlap to form contiguous sequences (contigs) that reconstruct the entire insert allocated in each BAC vector. At the end all the contigs are ordered and orientated based on the genetic map information to obtain a finished sequence.

The advantage of this approach is essentially related to knowing the position of the clones (MTP) which simplifies assembly procedure and lowers the level of coverage required. On the other hand, the construction of the physical map is very time consuming and expensive in terms of money and human efforts. Using the BAC-by-BAC approach, in 2000, the first whole sequence of a plant genome, *Arabidopsis thaliana*, was obtained (Arabidopsis Genome Initiative 2000). The *Arabidopsis* genome size is 125

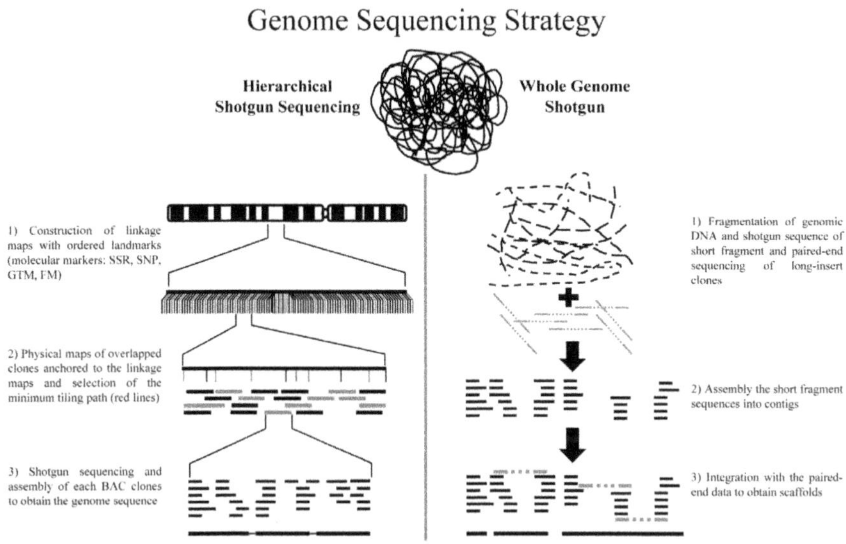

Figure 9-1 Hierarchical shotgun and whole genome shotgun sequencing strategies.

Color image of this figure appears in the color plate section at the end of the book.

Mb and the final coverage obtained is 15-folds. Five years later the first crop in human feed, the rice (*Oryza sativa* ssp. *japonica*), was sequenced using the hierarchical approach (International Rice Genome Sequencing Project 2005). This species is more than three times larger than *A. thaliana*, 389 Mb. The rice genome and the *A. thaliana* one are the only finished plant genome sequences; they are a gold standard with less than one error per 100 kb and where each replicon is assembled into a single contiguous sequence (Chain et al. 2009).

Crop species show a high percentage of repetitive elements in the genome; this fraction is responsible for the variation of the plant genome sizes (Bennetzen et al. 1998). In maize 37,000 genes have been predicted, this value is very close to the rice (32,000 genes predicted), but the size of these two genomes is very different: 2,600 Mb for the *Z. mays* cv. B73 (Schnable et al. 2009) and 389 Mb for *O. sativa*. Moreover, in plants an important factor affecting the genome size is the polyploidy that can cause erroneous merging during the assembly.

An alternative approach to overcome the problems related to repeats, transposable elements and polyploidy of plant genomes is gene enrichment. In this category are gathered together different techniques: EST, methylation filtration (MF), high cot (HC) sequencing, Transposon insertion-site sequencing. All these procedures select, in different manners, the genome part composed of the coding regions (gene space, Rabinowicz 2007), trying

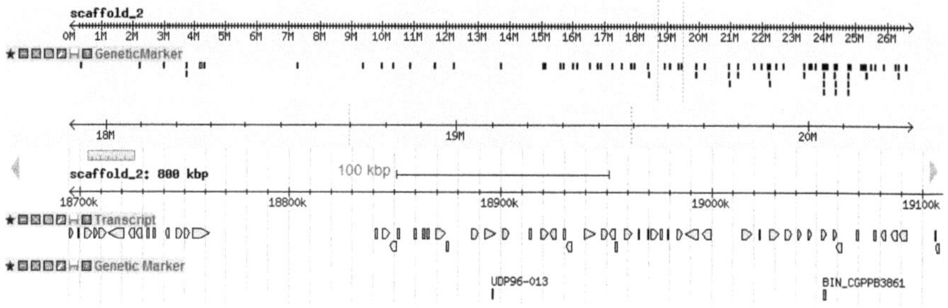

Figure 9-2 *Prunus* persica—JGI v1.0 genome browse: Pseudochromosome survey, localization of all genetic marker used to anchor and orientate the sequence, zoom view of 200 kb showing transcripts and genetic marker in this region.

Color image of this figure appears in the color plate section at the end of the book.

to identify a relatively complete gene catalog for a species. The EST approach was broadly used in all the species, thanks to its high efficiency to discover genes, and to give information on the expression and alternative splicing events in different tissue or in the same tissue at different stages. Billions of ESTs are collected in many databases (National Center of Biotechnology Information—NCBI, European Molecular Biology Laboratory—EMBL), which give significant support to annotate sequences individuating the coding part of a genome. A large catalog of EST enables good gene prediction and annotation. As mentioned before, for the Rosaceae two dedicated websites collect the ESTs produced from different stone fruit tissues: the Genome Database for Rosaceae, and the ESTree database.

EST sequences are very efficient for gene discovery, but, nontranscribed flanking regulatory sequences as well as intron sequences that are excluded from EST data contain important information that can only be retrieved by sequencing genomic DNA. Methylation filtration (MF), high cot (HC) sequencing, transposon insertion-site sequencing are strategies that provide this genomic information while minimizing the amount of repetitive genome portion. These approaches have been successfully used in maize (Emberton et al. 2011), sorghum (Jones-Rhoades et al. 2006), onion (Jakne et al. 2008) and wheat (Lamoureux et al. 2005).

The most frequently used and most cost-effective sequencing approach is the whole-genome shotgun (WGS, Fig. 9-1) that involves the assembly of sequences in a random genome-wide manner. Seventeen out of the 26 sequenced plant genomes were obtained by WGS. This strategy consists essentially of three principal steps: (a) Shearing the genomic DNA into random fragment of defined size that are subcloned into plasmid vectors or directly sequenced via NGS technologies; (b) Sequencing clones, to produce a high redundant coverage; (c) Assembling shotgun reads into contigs using various computational methods to produce a consensus sequence.

The fragment assembly is a conceptually simple procedure that generates longer sequences by detecting overlapping fragments. If the fragment assembly could be done perfectly, the genome sequencing would be a simple problem. On the other hand, there are extensive repetitive sequences in all genomes, which can easily mislead the fragment assembly procedure (Kim and Tang 2007).

A method to overcome the production of misassembled contigs, a consequence of the repeats presence, is to sequence both ends of a fragment, generating two sequences per fragment. Since the size of the fragment is known, we identify the approximate distance between two sequences. This technique named paired-end was developed by Roach et al. (1995). One strategy to use the pair-end information effectively is to assemble contigs as accurate as possible by detecting potentially misassembled contigs and then utilize the pair-end information using only contigs that are likely to be assembled correctly. The pair-end information is also used to join contigs into longer layouts called supercontings or scaffolds.

After the assembly, scaffolds must be anchored, ordered and oriented properly along the chromosomes; this can be done using saturated linkage maps. Markers (SSRs, SNPs, RFLPs) can be placed on the WGS scaffolds using ePCR or BLASTn procedures allowing the correct assignment of supercontigs along the chromosomes. The order of the markers within a scaffold resolves the orientation along the chromosome.

The increasing of the efficiency of the NGS technology in terms of read lengths, base-calling and cost reduction, enables to sequence the crop genomes composed from hundreds to thousands of mega bases. This is possible thanks to the high coverage that can be obtained and the capability to apply the mate-pair procedure allowing clearing the hurdle related to the repeats and transposable elements.

The critical step of the WGS strategy, at the moment, is the computational efforts needed. For the first time in the biological science history the technology is more powerful than our capacity to analyze the data. In this scenario bioinformatics is the leading actor, indeed the assembly pipelines affect all the downstream studies, first of all the gene prediction and annotation. For a sequencing project is essential to think about the information technology (IT) platforms to handle the huge amount of data produced in a single run of a NGS apparatus ($\cong$ 600 GB). In silico lab must be constructed by computing clusters, storage facilities, personal computers with the power to manage the raw data produced during the run in term of storage and analysis: the informatics knowledge is crucial to handle the IT platforms. The pyramid is upset due to the fact that a little number of scientists can produce dozens of terabyte of data but a lot of informaticists and bioinformaticists are essential to analyze these results and answer to biological questions. This is the very bottleneck of the WGS strategy coupled with NGS and applied to plant genomes.

P. persica is considered one of the genetically better characterized species in the Rosaceae, and it has distinct advantages that make it suitable as a model genome for *Prunus* as well as for other species of the family. While some *Prunus* species, such as cultivated plums and sour cherries, are polyploid, peach is a diploid with $n = 8$ and has a comparatively small genome currently estimated to be ~220-230 Mb, based upon the Peach v 1.0 assembly, instead of the 280–300 Mb previously predicted (Baird et al. 1994). Peach has a relatively short juvenility period of 2–3 years compared to most other fruit tree species that require 6–10 years. In addition, a number of genes for important traits have been genetically described in *P. persica*, including genes controlling flower and fruit development, tree growth habit, dormancy, cold hardiness, and disease and pest resistance (Dirlewanger et al. 2004; Peace et al. 2005; Boudehri et al 2009; Esmenjaud et al. 2009; Olukolu et al. 2009; Fan et al. 2010; Lambert and Pascal 2011).

For all these reasons the IPGI decided to sequence the peach genome with a WGS strategy using data from Sanger sequencing. The first draft assembly, Peach v 1.0, was publicly released on April 1st 2010 and is available for all the scientific community at: http://www.rosaceae.org/peach/genome; www.phytozome.net/peach;http://http://services.appliedgenomics.org/fgb2/iga/prunus_public/gbrowse/prunus_public/Peach v 1.0 was produced from the DNA of a doubled haploid generated by the genome duplication of a natural haploid from the cv. Lovell. Genes and intervening DNA are "fixed" or identical for all alleles and both chromosomal copies of the genome. Its doubled haploid nature was confirmed by the evaluation of >200 SSRs, and has facilitated a highly accurate and consistent assembly of the peach genome (Zhebentyayeva et al. 2008).

The assembly is arranged in 202 scaffolds covering 227.3 Mb of the genome with a coverage of 8.54-folds. These scaffolds have been selected through a screening analysis for contamination and integrated into eight pseudomolecules using 827 markers from an updated version of the *Prunus* reference map (Dirlewanger et al. 2004; Howad et al. 2005) to form the final release; the pseudochromosome molecules are named according to their corresponding linkage groups (Joobeur et al. 1998). To date in the refined version 99.3% (223.1 Mb) of the peach genome of the contig sequence has been anchored to the reference map while 96.2% (216.1 Mb) of it has been also orientated (Sosinski et al. 2010).

The gene prediction and annotation was performed constructing transcript assemblies, with 88,500 peach ESTs and 424,656 ESTs belonging to the Rosaceous family, using the software PASA applying the repeated sequences filter. Gene models were predicted by homology-based predictors, mainly by FGENESH+ with the addition of GenomeScan if FGENESH+ produced no model at the locus. Predicted genes were UTR-extended and/or improved by PASA. Final gene set was made from gene selection based on ESTs support or peptide homology support subjected to filtering of repeats/transposable elements. Automated gene prediction and annotation produced 27,851 genes, 838 of which show alternative splicing giving a total of 28,689 predicted protein-coding genes (Sosinski et al. 2010). About 200 genes belonging to carbohydrate metabolism, carotenoids/ABA volatile compounds, cell wall metabolism, hormone pathways, cold induced and resistance genes have been manually annotated.

In order to support the bioinformatics gene prediction and to define the peach transcriptional landscape, an RNA-seq approach was performed on pollen and three bulks obtained pooling the mRNA from fruit (mesocarp, epicarp and endocarp at different ripening stages), leaf (shoot and fully expanded leaf), seed (embryo and cotyledon). A total number of 352 million of pair-end reads 75bp long (23.6 Gbp) was obtained, with an average of 44 million per tissue, ranging from 24 in the root to 62 million in the fruit. After reads alignment on the peach reference genome, the accuracy of bioinformatics gene annotation appeared improved: previously un-annotated genes and 3′- or 5′-untranslated regions (UTR) have been identified and the exon-intron boundaries precisely defined (Fig. 9-3). Moreover the dynamics of transcription and transcript processing, in particular alternative isoforms expressed within a single tissue or differentially expressed between different tissues, have been studied and both transcriptionally active and inactive regions have been identified at whole genome level (Vendramin et al. 2010).

The relatively small genome size is probably due to the fact that only 62.3Mb (27.4%) of the genome is constituted by repeats. Comparison of the genome with the plant section of RepBase disclosed that only 19.7Mb (8.7%) is similar to other plant repeats, while *de novo* identification of repeats within peach proved that a good portion of repeats is present in high number: 40.3 Mb (17.7%) were the result of a repeats detection with ReAS successively projected onto the peach genome and 27.4Mb (12.1%), detected with LTR_finder, are formed exclusively by LTR transposable elements (Sosinskif et al. 2010)

This improved high-quality draft genome sequence (Chain et al. 2009) can become the reference genome for the Rosaceae family providing access to the gene catalog, to the regulatory elements that control their function and a framework for understanding genomic variation. Furthermore it is a prerequisite resource for all the whole genome approaches, that can facilitate to fully understand the roles of genes implied in all the processes that determine the quality and the health of crop productions.

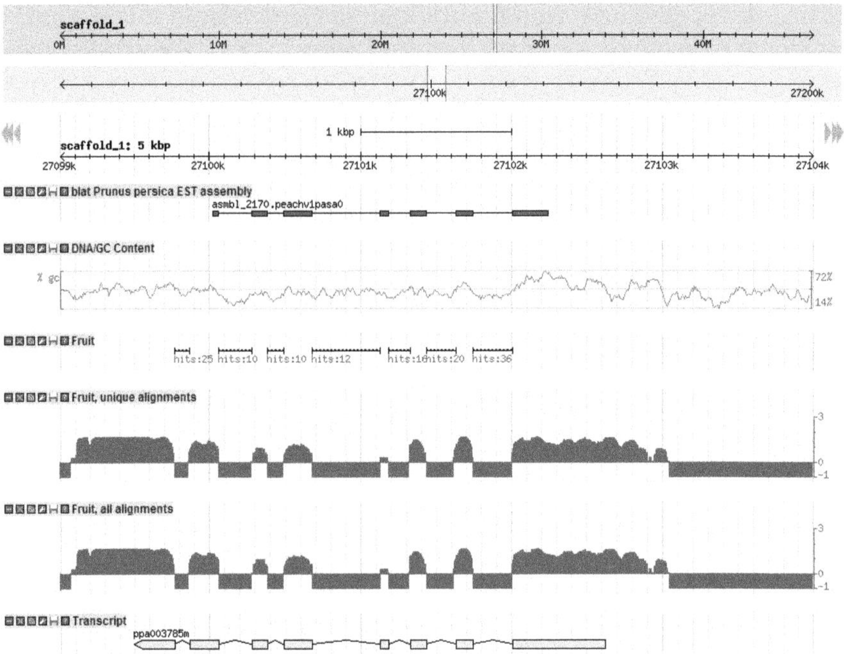

Figure 9-3 Peach v1.0 genome browse: Pseudochromosome survey, zoom view of 3000 kb showing EST assembly, DNA/GC content, PASA aligned ESTs, localization of RNA-seq reads produced from fruit, transcript and exon-intron boundaries.

Color image of this figure appears in the color plate section at the end of the book.

9.4 Advances in Sequencing Technology (NGS) and Genome Based Molecular Marker Discovery and Applications in *Prunus*

Most of the early genome and transcriptome sequencing projects were carried out on instruments performing a semi-automated version of the Sanger biochemistry, also known as 'dideoxy sequencing' or 'chain termination' method. The method is based on the use of a mix of normal and modified nucleotides, dideoxynucleotides containing a hydrogen group on the 3′ carbon instead of a hydroxyl group (OH) whose integration into the nascent DNA strand causes the interruption of the elongation step of the sequencing reaction. It involves the production of a population of fragments of different length, each beginning with the same primer sequence and ending with a different labeled nucleotide, which are eventually resolved electrophoretically for the nucleotide sequence decodification. Cost and throughput have always been the key limiting factors of all sequencing efforts and industry has always been on the run to issue new technologies to overcome them. Many different improvements have been introduced into the original Sanger protocol such as the use of fluorescently labeled dideoxynucleotides in spite of ^{32}P labeling system (Smith et al. 1986; Prober et al. 1987) and the adoption of the capillary gel electrophoresis (Luckey et al. 1990; Swerdlow and Gesteland 1990). These improvements together with the availability of robotic platforms have had significant impact both on the throughputs and the automatization of the sequencing reactions, leading to current read lengths of up to 1,000 bps at an error rate as low as 0.001% per base (Ewing and Green 1998). Accuracy is measured using Phred, an algorithm that provides a quality numerical value that is an estimate of the probability of error for each base-call in a raw read (Ewing et al. 1998). The Sanger method is considered the golden standard for both the sequence length and the high-quality raw data over most of the read length and has been extensively used in human and plant sequencing projects.

Over the past few years novel sequencing platforms have become available, which were developed with completely different chemistries than the Sanger one to respond to the needs of automatization, throughput and cost. The so called 'next generation sequencing' (NGS) technologies are revolutionizing genomic research reducing the cost of sequencing by over 10,000 times during the last ten years (Feuillet et al. 2011) while enabling in just a matter of days the production of a quantity of sequence data that would have taken months to generate with a classic Sanger sequencing approach (Shendure and Ji 2008).

The common feature of these methods is that they are 'massively parallel', meaning that the number of sequence reads generated from a single experiment is greater than that obtained with modern Sanger sequencers by many orders of magnitude. One of the main advantages of these platforms is that sequence data are generated from a pool of identical copies amplified from a single DNA fragment, thus bypassing vector cloning and colony picking which are time consuming steps necessary in the Sanger method. A limiting factor is that the very high throughput is achieved at the expense of length and accuracy of the individual reads when compared to Sanger sequencing, even though continuous advances both in the chemistry and in the technology are greatly reducing the gap. These platforms produce DNA sequences which range in size from 25 to up to 1,000 bases and generate from several hundreds of millions to several billions bases per runs. Less degree of accuracy is however compensated by very high depth of sequence coverage (Hutchinson 2007). The huge amount of data generated by these systems (over a gigabase per run) in the form of short reads represents a bioinformatics challenge requiring both the development of computer algorithms and efficient software to perform fast and accurate data analyses, as well as infrastructural platforms able to support computational efforts and data storage (Metzeker 2010).

There are many commercially available technologies, collectively referred to as 'second generation', while a 'third generation' of platforms is still underway or has only recently entered the market. Currently, the most predominantly NGS platforms used in plant research projects are essentially the Roche/454 system, the Solexa/Illumina sequencing system and the SOLiD™ system.

The Roche/454 GS FLX System workflow involves the generation of a single-stranded template DNA library followed by an emulsion-based clonal amplification of each fragment (300 to 800 bases long) of the library. Data are generated via sequencing-by-synthesis in the presence of a reaction mixture containing sulfurylase and luciferase. Incorporation of a base into the growing chain releases a pyrophosphate group with consequent emission of light detected by a CCD camera. Among the NGS platforms the GS FLX System produces the longest reads, up to 1 kb (with an average length of 700bp) with the latest advances in sequencing chemistry and an output of up to 700Mb/run. On the other hand the Roche/454 sequencing performance in the presence of homopolymer stretches is lower in comparison to the other NGS technologies.

The Solexa/Illumina sequencing-by-synthesis technology supports massively parallel sequencing using terminator nucleotides similar to those of the Sanger method but featuring a proprietary reversible chemistry that enables detection of single bases as they are incorporated into growing DNA strands. Solexa/Illumina sequencing technology produces reads of up to 150bp and generates up to 100Gb of data/run.

Both the Roche/454 and the Solexa/Illumina technology support paired-end libraries for efficient sequence assembly, and a barcoded-based multiplexing system providing a recognizable sequence tag for sequencing different libraries together.

The AB SOLiD™ system is based on a massively parallel sequencing by ligation chemistry where a set of four fluorescently labeled di-base probes compete for ligation to the sequencing primer. Specificity of the di-base probe is achieved by interrogating every 1st and 2nd base in each ligation reaction. Sequences are obtained by measuring serial ligation of a given oligonucleotide to the sequencing primer. The system produces reads of up to 75bp with an output of up to 20Gb/day and supports paired-end and multiplexing approaches.

The Life Technologies/Ion Torrent sequencer has more recently entered the market with a technology that couples clonal amplification of DNA libraries with a miniaturized pH meter able to detect variation in the pH of a solution after the release of a hydrogen ion subsequent to the incorporation of a nucleotide into a DNA strand. Depending on the type of semiconductor sequencing chip, the system produces an output from more than 10 Mb to over 1 Gb with a read-length from 100 to 400 bases.

Driven by the need to further lower sequencing cost while increasing base call accuracy, a 'third generation' of new platforms is being developed essentially based on single molecule sequencing technologies.

The Helicos Biosciences/HeliScope™ Sequencer has been the first commercial single molecule platform, based on a high-performance image analysis software able to identify and extract millions of nucleotide incorporations in near real time.

The Pacific Biosciences/PacBio*RS* system, recently on the market, performs, monitors and analyzes single molecule realtime (SMRT™) sequencing reactions. The system uses advanced collection optics to record light pulses emitted as a byproduct of nucleotide incorporation and proprietary algorithms translate each pulse into an A, C, G or T base call with its own set of quality metrics.

At Oxford Nanopore Technologies a system is being developed that uses an electronics-based platform, GridION, to sequence and analyze the DNA with improvements in yield and cost through scalability.

Whichever technology is being exploited for issuing new sequencing platforms, the common goals are to obtain better performances in terms of throughput and time coupled with base calling error rates comparable or even better than the Sanger method at reduced costs.

The great potential of NGS systems has prompted a huge variety of applications in plant research comprising development of SNP-based markers both with the aid of a sequenced genome as in *Arabidopsis* (Weber et al. 2007) and *Medicago* (Cheung et al. 2006) and in the absence of a reference genome as in maize (Barbazuk et al. 2007) and *Eucalyptus* (Novaes et al. 2008); genome-wide profiling of epigenetic modifications and chromatin structure as in *Arabidopsis* (Cokus et al. 2008; Lister et al. 2008); genome wide identification and analysis of microRNAs and small RNAs as in rice (Zhu 2008; Zhou et al. 2008). One of the major applications of NGS platforms, which promises to gain more and more importance in the future, is the high-throughput sequencing of mRNAs both for sampling transcript populations from plant tissues (Weber et al. 2007) and for gene expression studies. The so called RNA-Seq technology is challenging and replacing microarray-based platforms for its better performance both in terms of reliability and accuracy in transcript quantification (from analog to digital measure, Shendure 2008). For instance in maize, the Roche/454 technology applied for 3'-UTR profiling was able to resolve gene- and allele-specific trascripts (Eveland 2008).

The NGS technologies coupled with the growing number of genome sequences provide new opportunities to redesign strategies for more effective genome analysis and powerful investigation of variants discovery (as SNPs, small InDels and structural variants) through the whole genome resequencing of unrelated and divergent accessions.

In species with a reference genome available whole genome resequencing via NGS platforms of unrelated accessions may be applied for variant discovery aligning the reads to the reference genome.

Multiplex SNP genotyping allows the simultaneous high-throughput interrogation of hundreds of thousands of loci with high measurement precision at a cost that enables large-scale studies. After the release of the Sanger-based draft assembly Peach v1.0, (http://www.rosaceae.org/peach/genome), a number of research projects were started based on the application of NGS sequencing platforms primarily aiming at genome-wide variant discovery. Genome-wide variant detection within a species is the initial step towards linking genotypic variation and phenotype. The genetic markers obtained from the discovered variants can play an important role allowing cost effective marker-assisted selection strategies, whole genome fingerprinting, association studies, map based gene cloning and population-based analyses.

One of the main limitations to the dissection of economically important traits in crop plants has been the lack of a large number of genetic markers for the development of high-density and high-throughput assays for genome wide analysis, in particular association studies. The genetic dissection of quantitative traits is quite complicated and requires large number of molecular markers and multiplex genotyping. To date in *Prunus* several QTLs have been localized in large chromosomal regions by using bi-parental population and a limited number of molecular markers, such as SSRs and RFLPs (Dirlewanger et al. 1999; Etienne et al. 2002; Verde et al. 2002; Quilot et al. 2004).

The availability of the peach genome as reference and the worldwide resequencing initiatives aimed at the identification of large number of SNPs allow the development of such platform also for stone fruits making possible genome wide association studies (GWAS) in this important group of fruit species. In this section we discuss the ongoing studies aimed at the identification of variants (SNPs, InDels, structural variants) in *Prunus* species as well as the genome wide survey of peach SSRs identified in the frame of the IPGI. We will also report the high-throughput platforms available for these species.

9.4.1 Simple Sequence Repeats in the Peach Assembly

SSRs or microsatellites are a class of molecular markers abundant and well dispersed in all genomes. The availability of genome sequences allows the identification of genome wide survey of microsatellites. About 32 thousands of SSRs with a length higher than 20 nt have been identified in the peach genome. SSRs can be visualized on the peach genome browser at the following url: http://services.appliedgenomics.org/fgb2/iga/*Prunus*_public/gbrowse/*Prunus*_public/.

9.4.2 Variants Discovery in *Prunus*

The availability of the high quality draft sequence of the peach genome has speeded up several SNP detection initiatives in *Prunus*.

In peach a whole genome resequencing effort was carried out in Italy, in the frame of the DRUPOMICS project within the IPGI consortium, to investigate nucleotide variation in the peach germplasm and detect useful SNPs using the Illumina Genome Analyzer IIx. A panel of peach accessions has been selected for their broad genetic variability according to previous molecular analysis. The chosen accessions also display a wide range of geographical origins and phenotypical variations. In total 28 peach accessions were resequenced producing about 19.5 Gb of sequences, corresponding to 3–4 fold coverage per each accession using paired-end reads (2x100bp) and indexing methodology. Almost a million of SNPs with a minimum allele frequency (MAF) of 10% were discovered (Verde et al. 2011), ~800 thousands in intergenic regions and ~200 thousand in genic regions (CDS, UTR, introns).

In the frame of the IPGI a set of 13 *Prunus* accessions including peach, almond and *P. davidiana* was resequenced at higher coverage (20–30x) using Illumina paired-end reads (2 x 100bp) in order to deeply investigate single individual variations including structural variants such as the movement of transposable elements, large insertions/deletions, segmental duplications and inversions (Scalabrin et al. 2011).

In the frame of the RosBREED project, an USDA-funded multi-institutional and trans-disciplinary project, a panel of 21 peach accessions, founders of the modern US germplasm, was resequenced using Illumina paired-end reads (2 x 80bp) with a coverage of about 1 x for each accession allowing the identification of thousands of SNPs (Bassil et al. 2011).

Reduced representation library (RRL) was used in Spain at IRTA (P. Arús, personal communication) to resequence eight peach accessions from the European gene-pool using a Roche 454 GS FLX platform, allowing the discovery of thousands of SNPs.

The extensive synteny individuated among *Prunus* species (Dirlewanger et al. 2004) makes it possible to use the peach genome sequence as reference for other *Prunus* related species. In the frame of the RosBREED project, a panel of 24 cherry (16 sweet and 8 tart cherries) accessions was resequenced using an Illumina platform with an average coverage of 2x. The alignment of the reads to the peach reference sequence (Peach v1.0) allowed the identification of about 2 millions of SNPs (Bassil et al. 2011).

All the above mentioned initiatives allowed the discovery of large amount of SNPs both in peach and in cherry making possible the design of high-density assays that include thousands of SNPs distributed throughout the genome.

9.4.3 Highthrouput SNP Assays in Prunus

The ability to simultaneously genotype hundreds of thousands of SNPs in a single assay has recently become feasible due to innovative combinations of assay and array platform multiplexing. The Infinium Illumina's Bead Array Technology platform is an extremely high throughput SNP genotyping system that allows the identification and scoring of up to 2.5 million SNPs per single DNA sample (Gunderson et al. 2009). It is based on 3-micron silica beads that self-assemble in microwells on planar silica slides. When randomly assembled on this substrate, the beads have a uniform spacing of ~5.7 microns. Each bead is covered with hundreds of thousands of copies of a specific oligonucleotide that act as the capture sequences in one of the Illumina's assays. The availability of such platform coupled with the growing number of SNP discovered in *Prunus* opens new opportunities to develop a high density, genome-wide SNP assay with the power to detect genomic segments harboring inter-individual DNA sequence variation. The availability of such a tool will foster the application of GWAS in *Prunus* and in the future will give the opportunity to apply genome selection in stone fruits.

A high density SNP genotyping assay was recently released and commercialized for peach by Illumina using the SNPs discovered in the above mentioned resequencing initiatives. More than 1 million of peach SNPs were filtered using two main criteria: MAF higher than 0.2, Illumina design score (http://www.illumina.com/downloads/Illumina_Assay_Design_Tool.pdf) higher than 0.9 to have a final number of 9,000 SNPs on the beadchip. Considering the reduced size of the peach genome (~ 230 Mb) the 9,000 peach SNP chip can be considered a high density one having on average a SNP every 25 kb. An important criterion for a whole genome genotyping assay is the uniform distribution of the SNPs across the genome, as this greatly facilitates finding associations between markers and phenotypes. All the chosen SNPs but one were localized and well spaced along the eight refined pseudomolecules of Peach v1.0 built that represent 99.3% of the peach genome. Thanks to the availability of a large and well spaced number of SNPs in genic regions only genic SNPs were included in the chip. Moreover, only Infinium type II SNPs (A/C; A/G; T/C; T/G) were included in the chip because this platform requires only a single bead type on the chip.

Similarly, in cherry a SNP genotyping assay was recently released. The cherry SNPs were filtered as described in peach and 6000 well spaced ones were positioned on the chip.

The availability of the peach and cherry SNP genotyping assays will enhance genetic studies in these important fruit species. It will be now possible to conduct studies aimed at the molecular dissection of complex traits of economic importance using GWAS and to perform genome selection in a near future.

References

Arabidopsis Genome Initiative (2000) Analysis of the genome sequence of the flowering plant *Arabidopsis thaliana*. Nature 408: 796–81.

Aranzana MJ, Garcia-Mas J, Carbò J, Arús P (2002) Development and variability analysis of microsatellite markers in peach. Plant Breed 121: 87–92.

Aranzana MJ, Pineda A, Cosson P, Dirlewanger E, Ascasibar J, Cipriani G, Ryder CD, Testolin R, Abbott A, King GJ, Iezzoni AF, Arús P (2003) A set of simple-sequence repeat (SSR) markers covering the *Prunus* genome. Theor Appl Genet 106: 819–825.

Aranzana MJ, Abbassi el-K, Howad W, Arús P (2010) Genetic variation, population structure and linkage disequilibrium in peach commercial varieties. BMC Genetics 11: 69.

Arumuganathan K, Earle E (1991) Nuclear DNA content of some important plant species. Plant Mol Biol Rep 9: 208–218.

Baird WV, Estager AS, Wells J (1994) Estimating nuclear DNA content in peach and related diploid species using laser flow cytometry and DNA hybridization. J Am Soc Hort Sci 119: 1312–1316.

Barbazuk WB, Emrich SJ, Chen HD, Li L, Schnable PS (2007) SNP discovery via 454 transcriptome sequencing. Plant J 51: 910–918 .

Bassil NV, Gilmore B, Main D, Peace C, Mockler T, Wilhelm L, van de Weg E, Davis T, Chagne D, Gardiner S, Crowhurst R, Verde I, Sosinski B, Morgante M, Arus P, Velasco R, Troggio T, Cestaro A, Fazio G, Norelli J, Rees J, Iezzoni A (2011) New age Of DNA marker discovery in the Rosaceae. In: Plant Anim Genomes XIX Conf, Jan 15–19, San Diego, CA, Abstr W 252.

Bennetezen JL, SanMiguel P, Chen M, Tikhonov A, Francki M, Avramova Z (1998) Grass genomes. Proc Natl Acad Sci USA 95: 1975–1978.

Boudehri K, Bendahmane A, Cardinet G, Troadec C, Moing A, Dirlewanger E (2009) Phenotypic and fine genetic characterization of the D locus controlling fruit acidity in peach. BMC Plant Biol 9: 59.

Bozdag S, Close T J, Lonardi S (2008) Computing the minimal tiling path from a physical map by integer linear programming. ALG IN BIOINF, Volume 5251/2008, 148–161, DOI: 10.1007/978-3-540-87361-7.

Byrne DH (1990) Isozyme variability in four diploid stone fruits compared with other woody perennial plants. J Hered 81: 68–71.

Cabrera A, Kozik A, Howad W, Arús P, Iezzoni A, van der Knaap E (2009) Development and bin mapping of a Rosaceae Conserved Ortholog Set (COS) of markers. BMC Genomics 10: 562.

Cantini C, Iezzoni AF, Lamboy WF, Bortizki M, Struss D (2001) DNA fingerprinting of tetraploid cherry germplasm using simple sequence repeats. J Am Soc Hort Sci 126: 205–209.

Chaparro JX, Werner DJ, O'Malley D, Sederoff RR (1994) Targeted mapping and linkage analysis of morphological isozyme, and RAPD markers in peach. Theor Appl Genet 87: 805–815.

Cipriani G, Lot G, Huang WG, Marrazzo MT, Peterlunger E, Testolin R (1999) AC/GT and AG/CT microsatellite repeats in peach [*Prunus persica* (L.) Batsch]: isolation, characterization and cross-species amplification in *Prunus*. TheorAppl Genet 99: 65–72.
Chain PSG, Grafham DV, Fulton RS, FitzGerald MG, et al. (2009) Genome project standards in a new era of sequencing. Science 326: 236–237.
Cheung F, Haas BJ, Goldberg SMD, May GD, XiaoY, Town CD (2006) Sequencing *Medicagotruncatula* expressed sequenced tags using 454 life sciences technology. BMC Genom 7: 272–282.
Cokus SJ, Feng S, Zhang X, Chen Z, Merriman B, Haudenschild CD, Pradhan S, Nelson SF, Pellegrini M, Jacobsen SE (2008) Shotgun bisulphite sequencing of the *Arabidopsis* genome reveals DNA methylation patterning. Nature 452: 215–219.
Dettori MT, Quarta R, Verde I (2001) A peach linkage map integrating RFLPs, SSRs, RAPDs, and morphological markers. Genome 44: 783–790.
Dirlewanger E, Moing A, Rothan C, Svanella L Pronier V, Guye A, Plomion C, Monet R (1999) Mapping QTL controlling fruit quality in peach (*Prunus persica* (L.) Batsch). Theor Appl Genet 98: 18–31.
Dirlewanger E, Cosson P, Tavaud M, Aranzana J, Poizat C, Zanetto A, Arus P, Laigret F (2002) Development of microsatellite markers in peach [*Prunus persica* (L.) Batsch] and their use in genetic diversity analysis in peach and sweet cherry (*Prunusavium* L.). Theor Appl Genet 105: 127–138.
Dirlewanger E, Graziano E, Joobeur T, Garriga-Calderé F, Cosson P, Howad W, Arús P (2004) Comparative mapping and marker-assisted selection in Rosaceae fruit crops. Proc Natl Acad Sci USA 101: 9891–9896.
Eduardo I, Pacheco I, Chietera G, Bassi D, Pozzi C, Vecchietti A, Rossini L. (2011) QTL analysis of fruit quality traits in two peach intraspecific populations and importance of maturity date pleiotropic effect. Tree Genet Genomes 7: 323–335.
Emberton J, Ma J, Yuan Y, SanMiguel P, Bennete JL (2005) Gene enrichment in maize with hypomethilated partial restriction (HMPR) libraries. Genome Res 15: 1441–1446.
Esmenjaud D, Voisin R, Van Ghelder C, Bosselut N, Lafargue B, Di Vito M, Dirlewanger E, Poëssel J L, Kleinhentz M (2009) Genetic dissection of resistance to root-knot nematodes *Meloidogyne* spp. in plum, peach, almond, and apricot from various segregating interspecific *Prunus* progenies. Tree Genet Genomes 5: 279–289.
Etienne C, Rothan C, Moing A, Plomion C, Bodenes C, Dumas LS, Cosson P, Pronier V, Monet R, Dirlewanger E (2002) Candidate genes and QTL for sugar and organic acid content in peach (*Prunus persica* (L.) Batsch). Theor Appl Genet 105: 145–159.
Eveland AL, McCarty DR and Koch KE (2008) Transcript profiling by 3'-untranslated region sequencing resolves expression of gene families. Plant Physiol 146: 32–44.
Ewing B, Green P (1998) Base-calling of automated sequencer traces using phred. II. Error probabilities. Genome Res 8: 186–194.
Ewing B, Hillier L, Wendl MC, Green P (1998) Base-calling of automated sequencer traces using Phred. I. Accuracy assessment. Genome Res 8: 175–185.
Fan S, Bielenberg DG, Zhebentyayeva TN, Reighard GL, Okie WR, Holland D, Abbott AG (2010) Mapping quantitative trait loci associated with chilling requirement, heat requirement and bloom date in peach (*Prunus persica*). New Phytol 185: 917–930.
Farrar K, Donnison IS (2007) Construction and screening of BAC libraries made from *Brachypodium* genomic DNA. Nat Protoc 2: 1661–1674.
Feuillet C, Leach JE, Rogers J, Schnable PS, Eversole K (2011) Crop genome sequencing: lessons and rationales. Trends Plant Sci 16: 77–88.
Foulongne M, Pascal T, Arus P, Kervella J (2003) The potential of *Prunusdavidiana* for introgression into peach [*Prunus persica* (L.) Batsch] assessed by comparative mapping. TheorAppl Genet 107: 227–238.
Gunderson KL (2009) Whole-genome genotyping on bead arrays. Meth Mol Biol 529: 197–213.

Howad W, Yamamoto T, Dirlewanger E, Testolin R, Cosson P, Cipriani G, Monforte AJ, Georgi L, Abbott AG, Arús P (2005) Mapping with a few plants: using selective mapping for microsatellite saturation of the Prunus reference map. Genetics 171: 1305–1309.

Hutchison CA (2007) DNA sequencing: bench to bedside and beyond. Nucl Acids Res 35: 6227–6237.

Illa E, Eduardo I, Audergon JM, Barale F, Dirlewanger E, Li X, Moing A, Lambert P, Le Dantec L, Gao Z, Poessel JL, Pozzi C, Rossini L, Vecchietti A, Arus P, Howad W (2011) Saturating the *Prunus* (stone fruits) genome with candidate genes for fruit quality. Mol Breed DOI: 10.1007/s11032-010-9518-x.

International Human Genome Consortium (2004) Finishing the euchromatic sequence of the human genome. Nature 431: 931–945.

International Peach Genome Initiative (2010) www.rosoceae.org/peach/genome.

International Rice Genome Sequencing Project (2005) The map-based sequence of the rice genome. Nature 436: 793–800.

Jaffe D B, Butler J, Gnerre S, Mauceli E, Lindblad-Toh K, Mesirov JP, Zody MC, Lander ES (2003) Whole-genome sequence assembly for mammalian genomes: Arachne 2. Genome Res 13: 91–96.

Jones-Rhoades MW, Bartel DP, Bonnie Bartel (2006) MicroRNAs and their regulatory roles in plants.Annu Rev Plant Biol 57: 19–53.

Joobeur T, Viruel MA, de Vicente MC, Jáuregui B, Ballester J, Dettori MT, Verde I, Truco MJ, Messeguer R, Batlle I, Quarta R, Dirlewanger E, Arús P (1998) Construction of a saturated linkage map for *Prunus* using an almond x peach F_2 progeny. Theor Appl Genet 97: 1034–1041.

Jakne J,Meyer JDF, Suzuki G, McCallum J, Cheung F, Town CD, Havey MJ (2008) Pilot sequencing of onion genomic DNA reveals fragments of transposable elements, low gene densities, and significant gene enrichment after methyl filtration. Mol Genet Genom 280: 287–292.

Jung S, Staton M, Lee T, Blenda A, Svancara R, Abbott AG, Main D (2008) GDR (Genome Database for Rosaceae): integrated web-database for Rosaceae genomics and genetics data. Nucl Acids Res 36: D1034–D1040.

Kims S, Tang H (2007) Genome Sequencing Technology and Algorithms. 1st Artech House, Norwood, MA, USA © ISBN:1596930942 9781596930940.

Lambert P, Pascal T (2011) Mapping *Rm2* gene conferring resistance to the green peach aphid (*Myzuspersicae*Sulzer) in the peach cultivar "Rubira®". Tree Genet Genomes 7: 1057–1068.

Lamoureux D, Peterson DG, Li W, Fellers JP, Gill BS (2005) The efficacy of Cot-based gene enrichment in wheat (*Triticumaestivum* L.). Genome 48: 1120–1126.

Lazzari B, Caprera A, Vecchietti A, Merelli I, Barale F, Milanesi L, Stella A, Pozzi C (2008) Version VI of the ESTreedb: an improved tool for peach transcriptome analysis. BMC Bioinformatics 9 (Suppl 2): S9.

ListerR,O'MalleyRC,Tonti-Filippini J, GregoryBD,Berry CC, Millar AH, Ecker JR (2008) Highly integrated single-base resolution maps of the epigenome in *Arabidopsis*. Cell 133: 523–536.

Luckey JA, Drossman H, Kostichka AJ, Mead DA, D'Cunha J, Norris TB, Smith LM (1990) High speed DNA sequencing by capillary electrophoresis. Nucl Acids Res 18: 4417–4421.

Maxam AM, Gilbert W (1977) A new method for sequencing DNA. ProcNatl Acad Sci USA 74: 560–564.

Metzker ML (2010) Sequencing technologies—the next generation. Nat Rev Genet 11: 31–46.

Messina R, Lain O, Marrazzo M T, Cipriani G, Testolin R (2004) New set of micro satellite loci isolated in apricot. Mol Ecol Notes 4: 432–434.

Mnejja M, Garcia-Mas J, Howad W, Badenes M L, Arús P (2004) Simple-sequence repeat (SSR) markers of Japanese plum (*Prunussalicina* Lindl.) are highly polymorphic and transferable to peach and almond. Mol Ecol Notes 4: 163–166.

Mnejja M, Garcia-Mas J, Audergon JM, Arús P (2010) *Prunus* microsatellite marker transferability across rosaceous crops. Tree Genet Genomes 6: 689–700.

Monet R (1989) Peach genetics: past, present and future. Acta Hort 254: 49–53.

Novaes E, Drost DR, Farmerie WG, Pappas GJ Jr, Grattapaglia D, Sederoff RR, Kirst M (2008) High-throughput gene and SNP discovery in *Eucalyptus grandis*, an uncharacterized genome. BMC Genomics 9: 312.

Olukolu B, Trainin T, Fan S, Kole C, Bielenberg D, Reighard G, Abbott A, Holland D (2009) Genetic linkage mapping for molecular dissection of chilling requirement and budbreak in apricot (*Prunusarmeniaca* L.). Genome 52: 819–828.

Peace CP, Crisosto CH, Gradziel TM (2005) Endopolygalacturonase: a candidate gene for free stone and melting in peach. Mol Breed 16: 21–31.

Prober JM, Trainor GL, Dam RJ, Hobbs FW, Robertson CW, Zagursky RJ, Cocuzza AJ, Jensen A, Baumeister K (1987) A system for rapid DNA sequencing with fluorescent chain-terminating dideoxynucleotides. Science 238: 336–341.

Quilot B, Wu BH, Kervella J, Génard M, Foulongne M, Moreau K (2004) QTL analysis of quality traits in an advanced backcross between *Prunus persica* cultivars and the wild relative species *P. davidiana*. Theor Appl Genet 109: 884–897.

Rabinowicz PD (2007)Plant genomic sequencing using gene-enriched libraries. Chem Rev 107: 3377–3390.

Rajapakse S, Bethoff LE, He G, Estager AE, Scorza R, Verde I, Ballard RE, Baird WV, Callahan A, Monet R, Abbott AG (1995) Genetic Linkage mapping in peach using morphological, RFLP and RAPD markers. Theor Appl Genet 90: 503–510.

Roach JC, Boysen C, Wang K, Hood L (1995) Pairwise end aequencing: A unified approach to genome mapping and sequencing. Genomics 26: 345–353.

Sanger F, Nichklen S Coulson AR (1977a) DNA sequencing with chain-terminating inhibitor. ProcNatlAcadSci USA 74: 5463–5467.

Sanger F, Air GM, Barrell BG, Brown NL, Coulson AR, Fiddes JC, Hutchinson CA III, Slocombe PM, Smith M (1977b) Nucleotide sequence of bacteriophage φ X 174 DNA. Nature 265: 687–695.

Sanger F, Coulson AR, Friedmann T, Air GM, Barrell BG, Brown NL, Fiddes JC, Hutchinson CA III, Slocombe P, Smith M (1978) Nucleotide sequence of bacteriophage φX174 DNA. J Mol Biol125: 225–246.

Scalabrin S, Policriti A, Nadalin F, Pinosio S, Cattonaro F, Vendramin E, Aramini V, Verde I, Bassi D, Pirona R, Rossini L, Testolin R, Morgante M (2011) A catalog of molecular diversity of *Prunus*germplasm gathered from aligning NGS reads to the peach reference sequence: bioinformatic approaches and challenges. Acta Hort (in press).

Schnable PS,Ware D, Fulton R S, SteinJ C, Wei F, Pasternak S et al (2009) The B73 maize genome: complexity, diversity and dynamics. Science 326: 1112–1115 .

Shendure J, Ji H (2008) Next-generation DNA sequencing. Nat Biotechnol 26: 1135–1145.

Shendure J (2008) The beginning of the end for microarrays? Nat Meth 5: 585–587.

Shizuya H, Birren B, Kim UJ, Mancino V, Slepak T, Tachiiri Y, Simon M (1992) Cloning and stable maintenance of 300-kilobase-pair fragments of human DNA in *Escherichia coli* using an F-factor-based vector. Proc Natl Acad Sci USA 89: 8794–8797.

Smith LM, Sanders JZ, Kaiser RJ, Hughes P, Dodd C, Connell CR, Heiner C, Kent SB, Hood LE (1986) Fluorescence detection in automated DNA sequence analysis. Nature 321: 674–679.

Sosinski B, Gannavarapu M, Hager LD, Beck LE, King GJ, Ryder CD, Rajapakse S, Baird WV, Ballard RE, Abbott AG (2000) Characterization of microsatellite markers in peach [*Prunus persica* (L.) Batsch]. Theor Appl Genet 101: 421–428.

Sosinski B, Verde I, Morgante M, Rokhsar D, and The International Peach Genome Initiative (IPGI) (2010) A first draft of the peach genome sequence and its use for genetic diversity analysis in peach. 5th IntlRosaceae Genome Conf, Stellenbosch, South Afroca, Nov 14–17.

Swerdlow H, Gesteland R (1990) Capillary gel electrophoresis for rapid, high resolution DNA sequencing. Nucl Acids Res 18: 1415–1419.

Testolin R, Marrazzo T, Cipriani G, Quarta R, Verde I, Dettori MT, Pancaldi M, Sansavini S (2000) Microsatellite DNA in peach (*Prunus persica* L. Batsch) and its use in fingerprinting and testing the genetic origin of cultivars. Genome 43: 512–520.

Testolin R, Messina R, Lain O, Marrazzo MT, Huang WG, Cipriani G (2004) Microsatellites isolated in almond from an AC-repeat enriched library. Mol Ecol Notes 4: 459–461.

Vendramin E, Dettori MT, Giovinazzi J, Micali S, Quarta R, Verde I (2007) A set of EST-SSRs isolated from peach fruit transcriptome and their transportability across *Prunus* species. Mol Ecol Notes 7: 307–310.

Vendramin E, Scalabrin S, Felice N, Del Fabbro C, Aramini V, Dettori M T, Quarta R, Cattonaro F, Morgante M, Verde I (2010) Whole-genome trascriptional landscape definition and gene annotation improvement in peach by RNA-seq on Illumina platform.Proc XLVIII Ital Soc of Agri Genet –SIFV- SIGA Joint Meeting Matera, Italy, 27/30 Sept.

Verde I, Quarta R, Cerdrola C, Dettori MT (2002) QTL analysis of agronomic traits in a BC1 peach population. Acta Hort 592: 291–297.

Verde I, Lauria M, Dettori MT, Vendramin E, Balconi C, Micali S, Wang Y, Marrazzo MT, Cipriani G, Hartings H, Testolin R, Abbott AG, Motto M, Quarta R (2005) Microsatellite and AFLP markers in the *Prunus persica* [L. (Batsch)]· x *P.ferganensis* BC1 linkage map: saturation and coverage improvement. Theor Appl Genet 111: 1013–1021.

Verde I , Cattonaro F, Scalabrin S, Vendramin E , Policriti A, Del Fabbro C, Quarta R, Aramini V, Dettori MT, Rokhsar DS, Sosinski B, Morgante M (2011) Whole genome resequencing reveals a reduced nucleotide variability in peach. In: Plant Anim Genomes XIX Conf, Jan 15–19, San Diego, CA, Abstr W 249.

Viruel MA, Messeguer R, de Vicente MC, Garcia-Mas J, Puigdomènech P, Vargas F, Arùs P (1995) A linkage map with RFLP and isozyme markers for almond. Theor Appl Genet 91: 964–971.

Vision TJ, Brown DG, Shmoys DB, Durrett RT, Tanksley SD (2000) Selective mapping: a strategy for optimizing the construction of high-density linkage maps. Genetics 155: 407–420.

Weber APM, Weber KL, Carr K, Wilkerson C, Ohlrogge JB (2007) Sampling the *Arabidopsis* transcriptome with massively parallel pyrosequencing. Plant Physiol 144: 32–42.

Woo SS, Jiang J, Gill BS, Paterson AH, Wing RA (1994) Construction and characterization of a bacterial artificial chromosome library of Sorghum bicolor. Nucl Acids Res 22: 4922–4931.

Zhang HB, Wu C (2001) BAC as tools for genome sequencing. Plant Physiol Biochem 39: 195–209.

Zhebentyayeva TN, Swire-Clark G, Georgi LL, Garay L, Jung S, Forrest S, Blenda AV, Blackmon B, Mook J, Horn R, Howad W, Arús P, Main D, Tomkins JP, Sosinski B, Baird WV, Reighard GL, Abbott AG (2008) A framework physical map for peach, a model Rosaceae species. Tree Genet Genomes 4: 745–756.

Zhou X, Sunkar R, Jin H, Zhu JK, Zhang W (2008) Genome-wide identification and analysis of small RNAs originated from natural antisense transcripts in *Oryza sativa*. Genome Res 19: 70–78.

Zhu JK (2008) Epigenome sequencing comes of age. Cell 133: 395–397.

Yamamoto T, Mochida K, Imai T, Shi YZ, Ogiwara I, Hayashi T (2002) Microsatellite markers in peach [*Prunus persica* (L.) Batsch] derived from an enriched genomic and cDNA libraries. Mol Ecol Notes 2: 298–301.

Yamamoto T, Yamaguchi M, Hayashi T (2005) An integrated genetic linkage map of peach by SSR, STS, AFLP and RAPD J Jpn Soc Hort Sci 74: 204–213.

10

Comparative Genetics and Genomics Initiatives

Daniel J. Sargent,[1,*] *Sook Jung*[2,a] and *Dorrie Main*[2,b]

ABSTRACT

This chapter describes the developments in comparative genomics in *Prunus* that have occurred over the past two decades. It charts the progress of research beginning with the development of the first genetic linkage maps for *Prunus,* constructed from RFLP, isoenzyme and microsatellite markers, followed by the mapping of transferable, gene-specific and EST markers, and describes the evolution of the interspecific reference map for the genus. The development of linkage maps using transferable markers permitted interspecific comparisons to be made in *Prunus,* demonstrating the genomes of all the diploid species are essentially colinear. The chapter then describes map-based comparative genomics performed between *Prunus* and other Rosaceous genera, which have highlighted the conservation of synteny between Rosaceous genomes and the proposal of a hypothetical ancestral genome structure for the family. It then summarizes the studies performed between *Prunus* and more distantly-related genomes (*Arabidopsis,* cotton and poplar). The chapter concludes with a look towards the anticipated release of a whole genome sequence for *P. persica,* which will permit comparative genomics studies to be performed at a level of resolution that was not previously possible, and promises to reveal insights into the nature of genome evolution that has led to the extraordinary phenotypic diversity apparent between species within the genus.

Keywords: linkage mapping, marker development, *Prunus,* genome sequencing, evolutionary genetics, Rosaceae

[1]Fondazione Edmund Mach Istituto Agrario S. Michele all'Adige, Via E. Mach, 1 38010 S. Michele all'Adige, Italy.

[2]Department of Horticulture and Landscape, Washington State University, Pullman, WA 99164, USA.

[a]e-mail: *sook_jung@wsu.edu*

[b]e-mail: *dorrie@wsu.edu*

*Corresponding author: *dan.sargent@iasma.it*

10.1 Introduction

The genus *Prunus* is a member of the Rosaceae, the third-most economically-important plant family (Dirlewanger et al. 2002), and contains a large number of commercially valuable fruit and nut tree species, including almond, apricot, cherry, peach and plum, as well as a number of prized ornamental species. The genus is a member of the Spiraeoideae subfamily of Rosaceae, which contains other economically important genera including *Malus* (the apples) and *Pyrus*(the pears). The base chromosome number for the Spiraeoideae is $x = 9$, however there are many deviations from this, and *Prunus* forms part of the Amygdaleae tribe, which has a base chromosome number of x=8. The majority of the *Prunus* species are diploid, including peach (*P. persica*; $2n = 2x = 16$) which, with a genome size of ~280 Mbp/C, has been adopted as the model organism for the genus (Shulaev et al. 2008), and has been the subject of a recent whole genome sequencing initiative (International Peach Genome Sequencing Initiative 2010; www.rosaceae.org/peach/genome).

Due to the diversity of edible fruits and nuts, as well as ornamental tree-species, *Prunus* is a genus of profound commercial value, and as such, molecular genetics and genomics resources have been developed to assist with the breeding and development of fruit, nut and ornamental varieties in the genus. The molecular genetics and genomics resources that have been amassed have permitted comparisons to be made between the genomes of many *Prunus* species, between the genome of *Prunus* and those of other members of the Rosaceae, as well as between *Prunus* and more distantly related plant families (Georgi et al. 2003; Dirlewanger et al. 2004; Jung et al. 2006, 2008, 2009; Vilanova et al. 2008; Illa et al. 2011).

Comparative genomics includes the identification of conserved markers with a high degree of sequence homology within the genomes of more than one organism and a comparison of the relative positions of such markers in relation to each other between the genomes. The conservation of the positions of a number of such markers allows inferences to be made regarding the marker loci, based on knowledge of those loci in another organism. Such knowledge is especially useful for breeding, where molecular markers associated with agronomic traits are used to increase the efficiency of the breeding process, but it can also help to answer more profound questions about the nature and processes involved in genome evolution. Often a locus conserved between two organisms may perform a specific role in one species, but may have an altered but related function in another. Through comparative genomics within *Prunus*, the Rosaceae family and beyond, knowledge of the positions of loci associated with major genes and quantitative trait loci (QTL) will facilitate the advancement of our understanding of such loci in other species and genera.

Whole-genome molecular genetics investigations in *Prunus* began with the formation of the *Prunus* genome mapping group in 1991 and from it, the aim of constructing genetic maps of molecular markers and agronomic characters for apricots, peaches, almonds and cherries (Tobutt and Nicoll 1992). Since that time, *Prunus* genetics and genomics resources have developed rapidly, from marker characterization, to genetic linkage map construction using sequence-characterized transferable genetic markers, culminating recently in the release of the whole genome sequence for peach *P. persica* (*Prunus* Genome Sequencing Initiative 2010; www.rosaceae.org/peach/genome). These resources have contributed to a growing body of comparative genetics and genomics knowledge for the genus and are bringing new insights into the nature of genome evolution in Rosaceae and beyond.

10.2 Marker Development and Initial Mapping Experiments

10.2.1 Arbitrary Markers in Linkage Map Development

Initial linkage mapping analyses were performed in peach (*P. persica*) using random amplified polymorphic DNA (RAPD) markers (Chaparro et al. 1994). In their study, a linkage map consisting of 83 RAPD markers, one isoenzyme and four morphological characters was described, spanning 15 of the 16 expected linkage groups for a map constructed of dominant markers in an F_1 progeny, and covering 396 cM. Following that investigation, a linkage map containing 10 linkage groups and covering a total genetic distance of 503.3 cM was developed for sweet cherry (*P. avium*) using 87 RAPD markers and two isoenzymes (Stockinger et al. 1996). Whilst these linkage maps covered a large portion of two economically important *Prunus* genomes, they were composed primarily of arbitrary molecular markers, which are difficult to transfer between progenies and are thus of limited use for comparative mapping studies.

10.2.2 Transferable Marker Systems in Linkage Mapping

Other more readily transferable marker systems were also used in the initial linkage mapping studies within *Prunus*, including isoenzymes, to determine linkage relationships in both almond (*P. dulcis*) (Arus et al. 1994) and cherry (*P. avium* and *P. nipponica*) (Boskovic et al. 1997; Boskovic and Tobutt 1998). A number of other linkage maps were developed, including a map of peach containing 270 markers comprising 92 RAPDs and 115 amplified fragment length polymorphisms (AFLPs) as well as 50 restriction fragment length polymorphisms (RFLPs) (Dirlewanger et al. 1998). Other maps reported around this period included a map of an F_1 almond progeny comprising seven isoenzymes and 120 RFLPs (Viruel et al. 1995). These maps, and others, were the first to contain a significant number of highly transferable molecular markers that if applied to mapping populations of other species and genera, could be used for comparative analyses.

10.2.3 Prunus Microsatellite Development

Following the development of isoenzyme and RFLP markers and their application to genetic investigations in *Prunus*, microsatellites began to be developed for the genus. Microsatellites, often termed simple sequence repeats (SSRs) have advantages over RFLPs and isoenzymes for linkage map construction in a species because, as they are sequence-characterized PCR-based markers, they are less labor-intensive to use experimentally, and thus quicker to screen on large numbers of individuals. Microsatellites were initially developed for peach (Cipriani et al. 1999; Sosinski et al. 2000) and shown to be transferable to other *Prunus* species, making them useful for synteny studies within *Prunus*. Since then, SSRs have been reported for many *Prunus* species, including apricot (Lopes et al. 2002; Messina et al. 2004), almond (Xu et al. 2004) and cherry (Clarke and Tobutt 2003), and these markers have permitted the elucidation of the extent of conservation of genome structure and organization within the genus *Prunus* (Dirlewanger et al. 2004).

10.3 The Development of the *Prunus* ReferenceLlinkageMap

10.3.1 Development of a Framework Map

Joobeur et al. (1998) developed a linkage map from an interspecific F_2 cross between the almond (*P. armeniaca*) variety 'Texas' and the peach (*P. persica*) variety 'Earlygold' (herein referred to as T×E). The resultant progeny of n=75 individuals was mapped initially using a total of 246 markers (11 isoenzymes and 235 RFLPs) and a linkage map covering a total of 491 cM and spanning the eight expected *Prunus* linkage groups was produced. The progeny size of this population was increased to n=82 and an additional 96 microsatellite loci were added by Aranzana et al. (2003). Subsequently, Dirlewanger et al. (2004) added a further 220 markers including 89 SSRs, five sequence-tagged site (STS) markers and 126 RFLPs and established the T×E linkage map as the reference map for the genus *Prunus* (Fig. 10-1). The map of Dirlewanger et al. (2004) comprised a total of 562 molecular markers covering 519 cM, and due to the sequence-characterized nature of all of the markers employed in its construction, it provided the ideal platform from which to perform comparative mapping studies.

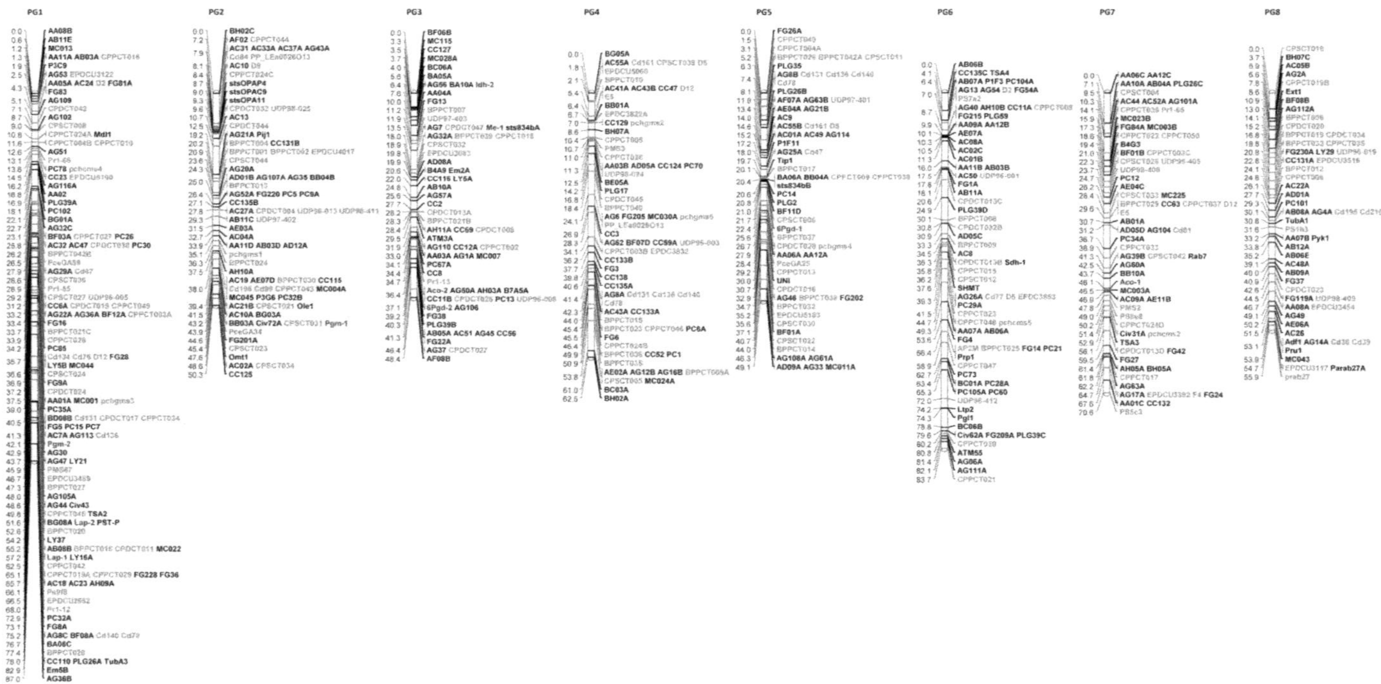

Figure 10-1 The *Prunus* reference linkage map 'Texas' × 'Earlygold' (T×E) based on the most current dataset for markers scored in the full T×E progeny submitted to the Genome Database for Rosaceae (Jung et al. 2008) of 602 markers. The Map depicts the eight *Prunus* linkage groups (PG1-PG8) and the locations of markers mapped to each. Microsatellites are given in red, RFLP markers in black, RGAs in green, isoenzymes in blue and STS markers are given in black underlined. Genetic distances are given in centiMorgans to one decimal place.

Color image of this figure appears in the color plate section at the end of the book.

10.3.2 Bin-mapping, Transferable Markers and Map Saturation

In recent years, genetic mapping resources for the T × E progeny have been extended further, with the development of a selective- or bin-mapping set comprising just six progeny (Howad et al. 2005), which divided the T × E reference map into 67 mapping bins. The development of this resource has enabled the cost-effective, rapid and accurate location of genetic markers to the *Prunus* genome, albeit at the expense of precision in marker ordering within mapping bins. Nevertheless, the bin set permitted the location of a further 264 SSR loci by Howad et al. (2005), 158 markers, including 109 gene-specific markers by Ogundiwin et al. (2009) and 613 Rosaceous conserved orthologous set (RosCOS) markers by Cabrera et al. (2009) (Fig. 10-2). Since its inception and subsequent adoption as the *Prunus* reference map, in excess of 600 molecular markers have been mapped in the full T × E progeny, more than 1,000 markers have been located by bin-mapping, whilst in excess of 2,300 markers have been associated with positions on the T × E linkage map via a physical map framework (Zhebentyayeva et al. 2008). The location of an abundance of conserved, transferable molecular markers, including RFLPs, gene-specific and Ros COS markers to the *Prunus* reference linkage map has permitted comparative mapping studies to be performed between *Prunus* and other members of the Rosaceae, as well as more distantly-related plant species (Dirlewanger et al. 2004; Vilanova et al. 2008; Jung et al. 2009; Illa et al. 2011).

10.4 Comparative Mapping in the Genus *Prunus*

The genus *Prunus* consists of numerous economically important stone fruit species, including apricot *(P. armeniaca)*, sweet cherry (*P. avium*), sour cherry (*P. cerasus*), peach (*P. persica*), plum (*Prunusdomestica*), almonds (*P. dulcis*) and many ornamental species such as Japanese cherry (*P. serrulata*), as well as wild species containing agronomically important traits, such as *P. davidiana*. Using mainly SSRs, which have proven to be highly polymorphic in addition to being transferable between species within the genus, the genomes of *Prunus* species have been extensively compared (Dirlewanger et al. 2004; Dondini et al. 2007; Clarke et al. 2009; Illa et al. 2009; Ogundiwin et al. 2009).

In the most comprehensive study to date, (Dirlewanger et al. 2004) analyzed markers on a total of 16 published linkage maps of diploid *Prunus* species containing a minimum of 28 transferable, sequence characterized markers per linkage map common to the *Prunus* T×E reference map. Results from these marker-based linkage analyses demonstrated a very high degree of conservation of synteny. The authors concluded that the genomes of all diploid *Prunus* species investigated, including almond, apricot, cherry and

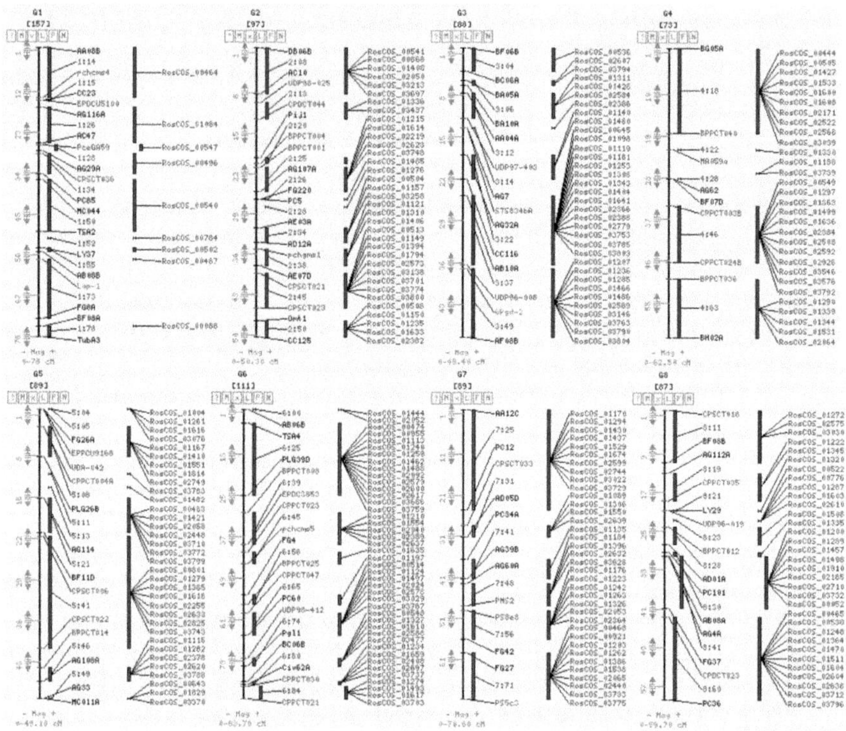

Figure 10-2 The *Prunus* T×E bin map of Howad et al. (2005) showing the bin positions of the 613 RosCOS markers of Cabrera et al. (2009). Black vertical lines depict the eight T×E linkage groups, red vertical lines depict the individual mapping bins, delineated by their flanking markers on the T×E linkage map (numbered 1:14 to 8:60). Marker names highlighted in red are the 613 RosCOS markers of Cabrera et al. (2009), linked by black lines to the mapping bin in which they were located. Figure adapted from the CMap viewer on the Genome Database for Rosaceae (Jung et al. 2008).

Color image of this figure appears in the color plate section at the end of the book.

myrobalan plum, as well as the wild species *P. davidiana* and *P. feragensis* were essentially colinear and thus could be treated as a single genetic entity. Figure 10-3 depicts the syntenic comparison between linkage group 1 (PG1) of the T × E reference map and PG1 of the cherry linkage map of (Clarke et al. 2009) showing the high degree of colinearity between the two genomes.

10.5 Comparative Mapping in the Rosaceae

10.5.1 Initial Comparative Mapping

Initially, comparative mapping was performed between *Prunus* and other members of the Rosaceae using modest numbers of transferable

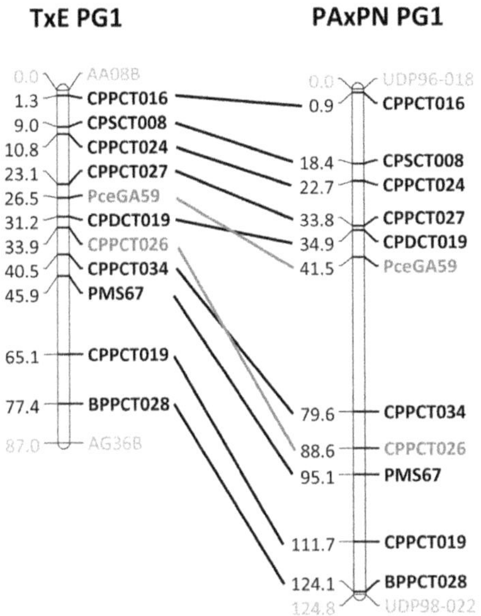

Figure 10-3 Syntenic comparison between linkage group 1 (PG1) of the T×E reference map and the cherry linkage map of Clarke et al. (2009) showing the high degree of colinearity between the common markers mapped in the two genomes. The two non-colinear markers, PceGA59 and CPPCT026 are displayed in red, non-common markers defining linkage group ends are presented in grey.

Color image of this figure appears in the color plate section at the end of the book.

genetic markers. Employing a set of 24 RFLP markers and six isoenzymes, Dirlewanger et al. (2004) compared the F_1 apple (*Malus pumila*) ($2n=2x=34$) linkage map developed from a cross between the cultivars 'Prima' and 'Fiesta' (P×F) (Maliepaard et al. 1998) to the *Prunus* reference map. Their investigation identified syntenic relationships between three of the eight T×E linkage groups, and seven of the 17 linkage groups of the P×F map (Fig. 10-4). Markers mapping to the distal end of linkage group 1 of the T×E map (PG1) were homologous to markers mapping to *Malus* linkage group 8 (MG8) and the order of markers on the linkage groups of the two was generally conserved, whilst the proximal end of PG1 was syntenic to MG13 and MG16, two groups that had been shown to be homeologous previously by Maliepaard et al. (1998). Additionally, PG3 showed clear conservation of synteny with two further homeologous *Malus* linkage groups, MG9 and MG17, whilst PG4 showed clear conservation of synteny in marker order with the homeologous *Malus* groups MG5 and MG10. From this initial work, the authors concluded that there was a high level of conservation of synteny between the genomes of *Prunus* and *Malus*, and due to the recent whole genome duplication of the *Malus* genome, a single

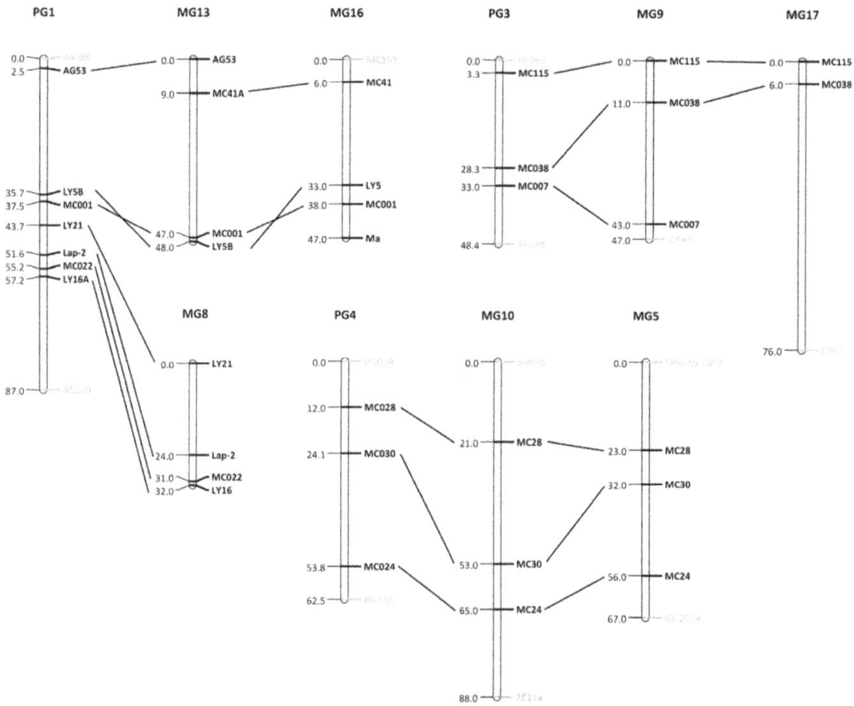

Figure 10-4 Syntenic comparison between T×E linkage groups PG1, PG3 and PG4 and their homologues on the *Malus* 'Prima' × 'Fiesta' (P×F); MG8, MG13, and MG16, MG9 and MG17, and MG5 and MG10. Comparisons were based on 24 RFLP markers and six isoenzymes by Dirlewanger et al. (2004). Only syntenic relationships are shown. Marker positions are given in centiMorgans. Markers defining linkage group ends that do not form part of the comparison are given in grey.

Prunus linkage group usually corresponded to two homeologous *Malus* linkage groups. They also concluded that the long *Prunus* chromosome represented by PG1 may have been formed from a fusion/fission event between two chromosomes represented by extant non-homeologous linkage groups in *Malus*.

Following genomic comparisons using RFLP and isoenzyme markers, a number of studies were performed using PCR-based markers that had been developed from both *Malus* and *Fragaria* (strawberry), which were applied to comparative mapping between *Prunus* and these other members of the Rosaceae. In one of these studies, primer pairs for intron-flanking 'functional' markers were developed from 24 *Fragaria* genes (Sargent et al. 2007), four of which mapped to linkage groups of a map of the interspecific cherry cross *P. avium* × *P. nipponica* (PA×PN). Homologous markers were mapped to PG1 of *Prunus* and *Fragaria* linkage group 2 (FG2) of the diploid *Fragaria* ($2n=2x=14$) reference map (FV×FN) (Sargent et al. 2006), and markers mapping to *Fragaria* linkage group FG6 were located on two *Prunus* linkage groups (PG3 and PG6). This study was followed by a more extensive investigation of intron-flanking markers developed from *Malus* gene sequences deposited in the EMBL database (Sargent et al. 2009). Primer pairs were designed for 27 intron-containing *Malus* genes, 85% of which were transferable to *Prunus* and *Fragaria*. A total of ten genes mapped to a linkage map of a *Malus* mapping progeny (F×T), whilst ten were bin-mapped in *Prunus* (T×E) and ten were bin-mapped in *Fragaria* (FV×FB). Comparisons of markers mapped in *Prunus* and the other two genera revealed support for previously undetermined relationships between homeologous *Malus* linkage groups MG7, MG1 and PG2, between MG6, MG14 and PG5, whilst a further putative relationship was identified between MG4 and PG6. This study also provided evidence of a strong syntenic relationship between PG2 and FG7 through the mapping of three markers to each of those groups, and tentative relationships between PG4 and FG3, and PG5 and FG5.

These initial fragmentary results however did not contain sufficient data from all the *Prunus*, *Fragaria* and *Malus* linkage groups to draw conclusions as to the nature of genome evolution in the Rosaceae across more than just a few linkage groups, or syntenic blocks within linkage groups.

10.5.2 Whole Genome Comparisons in Rosaceae

10.5.2.1 Prunus-Fragaria Comparison

By exploiting the relatively large number of RFLP markers mapped in the full *Prunus* T×E progeny, and the high level of transferability of RFLP markers between genera, Vilanova et al. (2008) were able to determine the syntenic relationships conserved between the genomes of *Prunus* and *Fragaria* through a comparison of 71 transferable markers, mapped to all the linkage groups of the T×E map and the diploid *Fragaria* FV×FB reference map of Sargent et al. (2006). This investigation was the first to employ markers spanning the entire *Prunus* genome and gave the first insights into the degree of conservation of whole genome structure between two distantly-related (Potter et al. 2007) and morphologically diverse genera within the Rosaceae.

The 71 anchor markers (40 RFLPs, 21 ESTs, eight gene-specific markers and two SSRs) of Vilanova et al. (2008) were distributed evenly across the eight *Prunus* and the seven *Fragaria* linkage groups. Their relative map positions showed that in general, chromosomal structure between *Prunus* and *Fragaria* was highly conserved given the ~29 million years since they diverged from a common ancestor, and that the majority of markers mapping to a single *Prunus* linkage group were located to just one or two *Fragaria* groups (Fig. 10-5). In some cases, such as PG5 and FG5, conservation

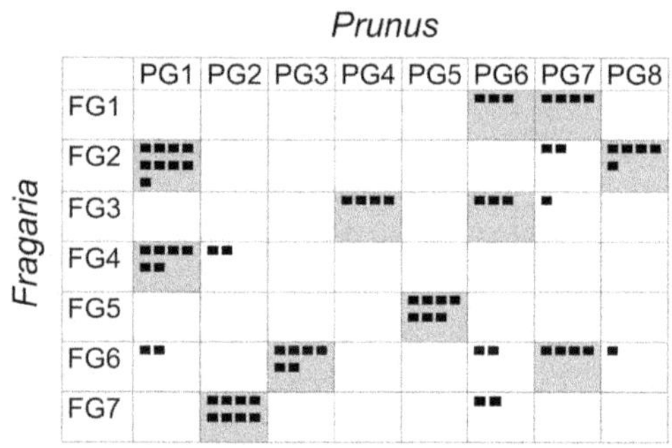

Figure 10-5 Summary of the synteny revealed between *Prunus* and *Fragaria* reference maps through a comparison of 71 sequence-characterized markers by Vilanova et al. (2008). The number of markers from each linkage group corresponding to linkage groups in the other genus are indicated by a black dot. Cells containing three or more markers are marked with a grey background. Adapted from Vilanova et al. (2008).

of synteny was almost complete with just a single inversion required to place all markers in the same order on both maps. There was less conservation between the genomes as a whole however and a total of 36 chromosomal rearrangements were required to place the markers that were compared into the same conformation in both genomes. Marker coverage and resolution across the two genomes was sufficient however for Vilanova et al. (2008) to propose an ancestral Rosaceae genome structure comprising nine ancestral chromosomes, the base chromosome number for the Spiraeoideae.

10.5.2.2 Prunus-Fragaria Comparison to the Malus Genome Sequence

More recently, Illa et al. (2011) have exploited the release of the whole genome sequence of the apple cultivar 'Golden Delicious' (Velasco et al. 2010) to perform a whole genome synteny comparison between *Prunus, Malus* and *Fragaria*. In their investigation, the sequences of 1,473 markers, 1,321 of which were mapped to the T×E reference map, were used as queries for BLAST against the *Malus* genome sequence. Of the 1,013 sequences (214 SSRs, 72 RFLPs, 182 ESTs and 545 RosCOS markers) that were significantly conserved in *Malus*, 784 were common to *Prunus* and *Malus* and 129 were conserved between *Prunus, Malus* and *Fragaria*, providing a high-resolution genome-wide data set for comparison between the three genera.

The analysis of the positions of the 129 common markers identified clear, conserved syntenic blocks that were common to all three genomes, with a single syntenic block in *Prunus* corresponding to one or two syntenic regions in *Fragaria*, consistent with the findings of Vilanova et al. (2008), and two or four syntenic regions in apple, consistent with the recent whole genome duplication in the *Malus* genome (Velasco et al. 2010). The identification of highly conserved syntenic regions between the three genera permitted the reconstruction of a hypothetical ancestral genome for the Rosaceae containing nine chromosomes (x=9) (Fig. 10-6), which was in accordance with the ancestral chromosomal arrangements suggested by Vilanova et al. (2008) and with the report of Velasco et al. (2010), who proposed a model for the genome evolution of *Malus* via polyploidization from a single ancestral genome donor. The data from the comparison of the 784 markers common to *Prunus* and *Malus* confirmed the hypothesis of a single ancestral genome donor for *Malus* as the arrangement of syntenic blocks within the *Prunus* genome did not contain rearrangements that predated genome duplication in *Malus* which would have been conserved in an ancestral genome donor from a Prunoideae lineage.

A pair-wise comparison of the arrangement of syntenic blocks in *Prunus, Fragaria* and *Malus* to the proposed ancestral chromosome configuration showed that they differed by three, four and eight major translocations respectively, consistent with a uniform rate of evolution since the genera diverged from a common ancestor and a whole genome duplication during the evolution of the extant *Malus* genome. A pair-wise comparison of all the rearrangements between the 129 common markers revealed a higher number of inversions between *Prunus* and *Fragaria* (27), than between *Prunus* and *Malus* (seven per diploid genome), consistent with the evolutionary relatedness, and therefore distances between the three genera (Potter et al. 2007). However, when all markers common to *Prunus* and *Malus* were considered, a total of 43 inversions were observed. The data therefore suggests that small scale genomic inversions, which are not detectable when the positions of a modest number of evenly-spaced genetic markers are compared, may have played an important role in speciation within Rosaceae. Whole genome macro- and micro-synteny studies exploiting whole genome sequence data for *Prunus* and other Rosaceae may reveal much about genome evolution in the family that has remained hitherto undetected due to the level of resolution at which studies have been performed to date.

10.6 BAC and Physical Map Development and BAC Sequencing

High molecular weight genomic bacterial artificial chromosome (BAC) libraries and BAC-based physical maps are a useful precursor for genome-wide comparative genomics investigations and are essential for the development of high density maps for fine mapping of specific genomic locations, targeted sequencing and sequence-based micro synteny studies. In *Prunus*, BAC libraries have been developed for a number of species, including a 22,000 clone library with an average insert size of 95 kb

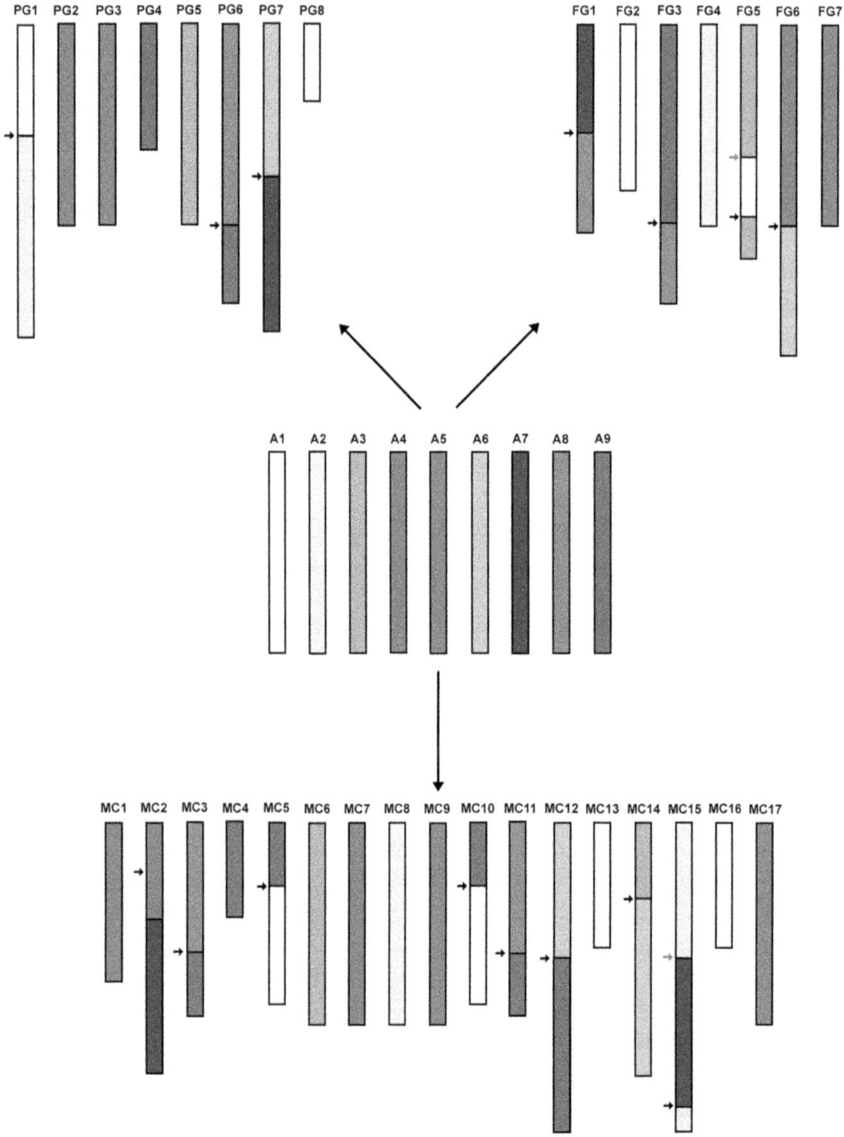

Figure 10-6 The reconstruction of a hypothetical ancestral Rosaceae genome containing nine chromosomes (A1–A9) through a comparison of the major structural differences between the *Malus*, *Prunus* and *Fragaria* genomes (Illa et al. 2011). Syntenic regions were identified through a comparison of 129 common orthologous markers. Extant *Malus* chromosomes and *Prunus* and *Fragaria* linkage groups are coloured according to the hypothetical ancestral chromosomes. Breakpoints indicating fusion-fission events are indicated with arrows. Adapted from (Illa et al. 2011).

Color image of this figure appears in the color plate section at the end of the book.

and thus an estimated 7× coverage of the peach genome (Wang et al. 2001), a 52,000 clone library of peach with an average insert size of 90 kb (15× genomic equivalent) (Boudehri et al. 2009), a library of myrobalan plum (*P. cerasifera*) containing 30,720 clones with an average insert size of 145 kb (15× genomic equivalent) (Claverie et al. 2004) and a 101,376 clone library of apricot (*P. armeniaca*) containing an average insert size of 64 kb (22× genomic equivalent) (Vilanova et al. 2003).

For the purpose of physical map development, a BAC library was initially constructed from DNA of the peach variety Nemared digested with the restriction enzyme *Hind*III. The library consisted of 44,160 clones with an average insert size of 60 kb, and thus an estimated 8× coverage of the haploid peach genome (Georgi et al. 2002). This was subsequently complemented by the development of an additional library of the doubled haploid clone of the peach variety Lovel (Plov2-1N) which was digested with the restriction enzyme *Sau*3A1 and consisted of 34,560 clones with an average insert size of 80 kb and therefore an effective 9.2× coverage of the peach haploid genome (Zhebentyayeva et al. 2008). The Nemared library was surveyed using 33 probes derived from AFLP and RFLP markers, many of which had previously been genetically mapped in *Prunus*, and which hybridized to a total of 152 BAC clones (Georgi et al. 2002). Additionally, one BAC clone, PpN31C7 containing a genomic fragment of 48,844 bp was completely sequenced and analyzed for SSR and gene content, providing the first extensive genome sequence data-set for *Prunus* that could be used for comparative genomics investigations (Georgi et al. 2002).

Following its development and initial characterization, a further 141 probes derived from previously mapped RFLP, SSR and AFLP markers were hybridized to 679 BAC clones (an average of 5.5 clones per probe) from the Nemared BAC library (Horn et al. 2005) to create an initial framework physical map. A further 1,304 peach ESTs were then hybridized to 4,983 clones of the Nemared library, 147 of which were anchored using the genetically-mapped probes, creating transcript map of peach fruit ESTs. The Nemared and Plov2-1N libraries were subsequently used to create a high density BAC-based physical map for peach (Zhebentyayeva et al. 2008), through the assembly of 15,655 BAC clones (4.3× genomic equivalent) fingerprinted with high information content fingerprinting (HICF) into 2,138 contigs covering an estimated physical distance of 303 Mb. To this framework, in excess of 2,300 ESTs and cDNA probes were hybridized, as well as 165 markers that had been previously mapped to the T×E reference map, enabling 45 Mb of BAC contigs to be genetically anchored to locations on the eight *Prunus* linkage groups of the T×E map (Zhebentyayeva et al. 2008) (Fig. 10-7). In addition to peach ESTs, candidate genes representing NBS-LRR, kinase, transmembrane domain classes, as well as, pathogen response proteins and resistance-associated transcription factors were

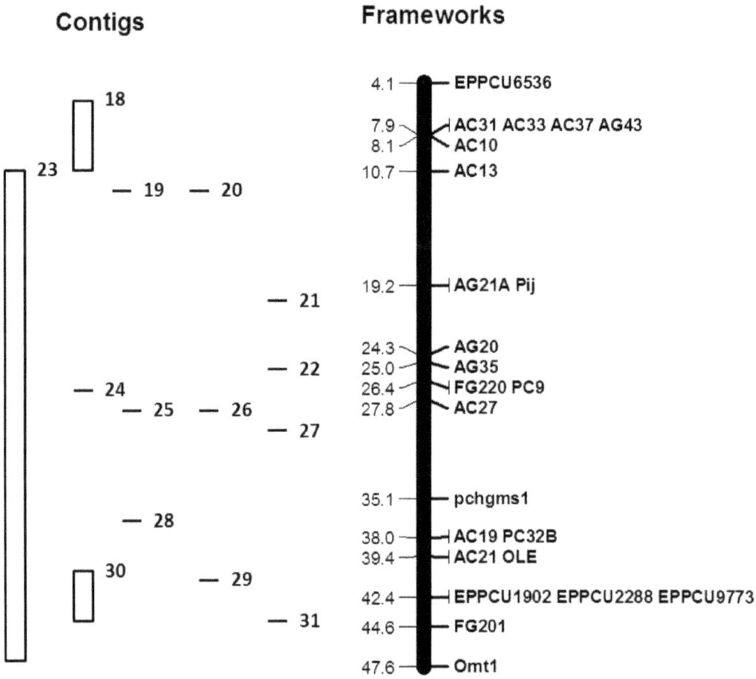

Figure 10-7 Schematic representation of chromosome 2 of the peach physical map, detailing the 14 BAC contigs comprised of 249 BAC clones, and the 24 markers used to anchor the contigs to the *Prunus* reference map. Figure reproduced from the Genome Database for Rosaceae (Jung et al. 2008); www.rosaceae.org.

hybridized to a peach BAC library and mapped (Lalli et al. 2005) by using the peach physical map database and the Genome Database for Rosaceae (Jung et al. 2008). A resistance map for *Prunus* was generated that contains 42 map locations for putative resistance regions distributed among seven of the eight linkage groups of TxE map (Lalli et al. 2005).

The physical map, and its associated anchored EST sequences and resistance gene homologs, the sequence of many of which are conserved between distantly related species including *Arabidopsis, Medicago* and *Populus* (Jung et al. 2009), are a valuable genetic resource for genome-wide marker comparisons and synteny studies not only within the Rosaceae, but also between *Prunus* and more distantly-related plant families, and have assisted in the elucidation of the patterns of evolution that have occurred since *Prunus* diverged from its common ancestor with the model plant species for which genome sequences are available.

10.7 Comparative Genomics: Rosaceae and Beyond

The availability of peach EST sequences anchored to *Prunus* genetic maps and to the peach physical map (Horn et al. 2005) allowed an assessment of the degree of conservation of synteny between the *Prunus* genome and other model genome sequences to be made (Jung et al. 2006, 2009). An analysis of the positions of 475 peach ESTs that were anchored to *Prunus* reference genetic maps to the positions of homologous sequences on the *Arabidopsis* genome revealed regions of microsynteny between seven linkage groups of the *Prunus* reference map and all five *Arabidopsis* chromosomes. An additional 1,097 peach ESTs that were anchored to 431 BAC contigs of peach physical map and their *Arabidopsis* homologs were also used in the analysis. However, the syntenic regions identified were small comprising only three or four conserved gene or EST pairs. Many blocks in peach mapped to multiple sites in the *Arabidopsis* genome that had previously been reported as paralogous blocks.

To further analyze the evolutionary relationships between the *Arabidopsis* and *Prunus* genomes, conserved syntenic regions were identified between the *Prunus* reference map and the pseudo-ancestral *Arabidopsis* genome in which all the recently duplicated paralogous blocks had been merged (Blanc et al. 2003). The comparison revealed examples of *Prunus* genome blocks that were syntenic to a single region in the ancestral *Arabidopsis* genome, but two different regions in the current *Arabidopsis* genome. Thus the gene content was more conserved between the *Prunus* genome and the ancestral *Arabidopsis* genome and different subsets of genes were found in the two regions of the current *Arabidopsis* genome, supporting the hypothesis that there has been a large-scale genome duplication in the evolutionary history of *Arabidopsis*, followed by selective gene loss. With the availability of whole genome sequences for fellow rosid I members *Populus* and *Medicago*, the degree of conserved synteny between *Prunus* and the three model genomes, *Populus*, *Medicago*, and *Arabidopsis* (rosid II), was investigated (Jung et al. 2009). In this analysis, eight *Prunus* BAC sequences along with 3,233 sequences that were mapped to the *Prunus* reference map were used. The map-anchored *Prunus* sequences comprised 1,093 ESTs or markers that were mapped directly to the *Prunus* reference map and a further 2,140 ESTs that were associated with 1,500 genetically-anchored BAC contigs of the peach physical map.

All of the eight *Prunus* BAC sequences compared revealed extensive conservation of synteny with the *Populus* and *Medicago* genomes, but not with the *Arabidopsis* genome. The gene order was well conserved within the syntenic regions identified between *Prunus* BACs, *Populus* and *Medicago*, without any sign of translocation. The syntenic regions detected were up to 477 kb in *Populus* (Fig. 10-8). The two syntenic regions

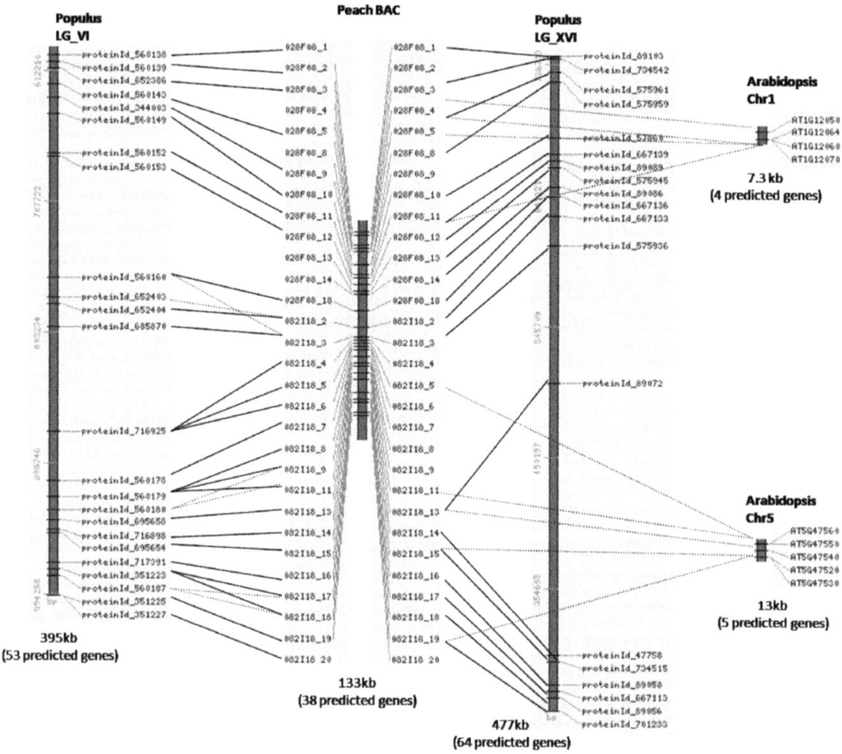

Figure 10-8 Conserved syntenic regions between the two overlapping peach BACs and the *Poplulus* or *Arabidopsis* genomes. When peach sequences match more than one genes in *Populus*, the matches with lower E value (lower PID when the E value is the same) are depicted with dotted lines. The matches between the peach sequences and the *Arabidopsis* genes are depicted with dotted lines. Only the predicted genes showing homology with genes in other genomes, not the intervening genes, are depicted. The numbers on the left side of the bar stand for base pair positions in the *Prunus* BACs, the *Populus* linkage groups, or the *Arabidopsis* genome. The lengths of the syntenic regions and the total numbers of predicted genes in the regions are given below the bar. Adapted from Jung et al. (2009).

detected between *Arabidopsis* and the eight BACs were much shorter, with neither exceeding 10 kb (Fig. 10-8). In most cases, two distinct syntenic regions were detected in *Populus* and *Medicago*, which corresponded with the proposed whole genome duplication events in the lineages of these species since they diverged from their common ancestor with *Prunus*. The analysis of map-anchored *Prunus* sequences and their *Populus* and *Arabidopsis* homologs identified many small syntenic blocks with several gene pairs between *Prunus* and *Populus* or *Arabidopsis*. The same blocks in *Prunus* however, detected more than twice the number of syntenic regions in *Arabidopsis* than in *Populus*. The data supported previous reports that concluded that *Arabidopsis* and *Populus* went through two and one whole genome duplication events respectively, and that a genome triplication event occurred before the species diverged from a common ancestor (Tang et al. 2008). Thus, the comparative studies performed with *Prunus* provided insights into the nature of genome evolution within rosid II and between rosid I and rosid II, and for the first time revealed the extent of the conservation of microsynteny between *Prunus* and its closest relatives for which whole genome reference sequences were available.

10.8 Sequencing of the *Prunus* Genome and the Implications for Comparative Genomics

A recent jointly-funded international initiative (the International Peach Genome Initiative; IPGI), consisting of members in countries throughout the world, and led by representatives from the US, Italy (Drupomics), Spain and Chile, have sequenced the complete genome of the doubled haploid peach cultivar 'Lovell', using whole genome shot-gun sequencing. Employing traditional long-read dideoxynucleotide chain-terminator sequencing (Sanger sequencing), the consortium have sequenced ~99% of the peach genome to 7.7× depth of coverage in 202 sequencing scaffolds greater than 1 kbp in length. Using markers located to the T×E genetic map, more than 92% of this sequence has been anchored and orientated and eight pseudo-molecules representing the eight *Prunus* chromosomes have been produced. The full v1.0 peach genome sequence contains a total of 227 Mbp of genome sequence, and 27,851 predicted genes have been identified that code for a total of 28,689 predicted proteins (International peach genome sequencing initiative 2010; www.rosaceae.org/peach/genome) providing a robust tool for sequence-based synteny studies with *Prunus*.

10.9 Future Expectations

Molecular genetics and genomics in *Prunus* have progressed rapidly over the past 20 years with the development of numerous sequence-characterized markers (Cipriani et al. 1999), tens of thousands of ESTs (Horn et al. 2005), high density, saturated genetic linkage maps (Dirlewanger et al. 2004), a physical mapping framework for the genus (Zhebentyayeva et al. 2008), and the elucidation of the conservation of synteny within *Prunus* and between *Prunus* and a number of other genomes of important Rosaceous crops and model plant species (Jung et al. 2009; Illa et al. 2011). The sequencing of the peach genome and thus the creation of a high quality reference sequence for *Prunus*, along with rapid advances in short read sequencing technologies will enable unprecedented resolution in future comparative genomics investigations between *Prunus* and the genome sequences of other genera, but also between the genomes of the many *Prunus* species. The high degree of colinearity between diploid *Prunus* genomes (Dirlewanger et al. 2004) has meant that the causal genomic factors for the vast morphological diversity between *Prunus* species have yet to be elucidated. With the relatively low cost per nucleotide of whole genome re-sequencing, the high quality peach reference genome sequence and the well-equipped bioinformatics resources available for *Prunus*, it will now be possible to conduct whole genome re-sequencing for many *Prunus* species and to perform genome-wide micro-synteny studies within the genus to elucidate the genetic basis of the morphological diversity that characterises these economically-important species. Additionally, the peach genome sequence will be invaluable for whole genome synteny studies at the DNA sequence level between *Prunus* and other sequenced Rosaceous genomes such as that of the wild strawberry *Fragaria vesca* (206 Mbp) (Shulaev et al. 2011), and that of the cultivated apple *Malus* × *domestica* (742 Mbp) (Velasco et al. 2010), as well as more distantly-related species such as grape *Vitis vinifera* (505 Mbp) (Velasco et al. 2007), and *Arabidopsis thaliana* (125 Mbp) (Kaul et al. 2000).

References

Aranzana MJ, Pineda A, Cosson P, Dirlewanger E, Ascasibar J, Cipriani G, Ryder CD, Testolin R, Abbott A, King GJ, Iezzoni AF, Arus P (2003) A set of simple-sequence repeat (SSR) markers covering the Prunus genome. Theor Appl Genet 106: : 819–825.

Arus P, Messeguer R, Viruel M, Tobutt K, Dirlewanger E, Santi F, Quarta R, Ritter E (1994) The European Prunus Mapping Project—Progress in the almond linkage map. Euphytica 77: 97–100.

Blanc G, Hokamp K, Wolfe KH (2003) A recent polyploidy superimposed on older large-scale duplications in the Arabidopsis genome. Genome Res 13: 137–144.

Boskovic R, Tobutt KR (1998) Inheritance and linkage relationships of isoenzymes in two interspecific cherry progenies. Euphytica 103: 273–286.

Boskovic R, Tobutt KR, Nicoll FJ (1997) Inheritance of isoenzymes and their linkage relationships in two interspecific cherry progenies. Euphytica 93: 129–143.

Boudehri K, Bendahmane A, Cardinet G, Troadec C, Moing A, Dirlewanger E (2009) Phenotypic and fine genetic characterization of the D locus controlling fruit acidity in peach. BMC Plant Biol 9.

Cabrera A, Kozik A, Howad W, Arus P, Iezzoni AF, van der Knaap E (2009) Development and bin mapping of a Rosaceae Conserved Ortholog Set (COS) of markers. BMC Genomics 10: Article No 562.

Chaparro JX, Werner DJ, Omalley D, Sederoff RR (1994) Targeted mapping and linkage analysis of morphological, isozyme, and RAPD markers in Peach. Theor Appl Genet 87: 805–815.

Cipriani G, Lot G, Huang WG, Marrazzo MT, Peterlunger E, Testolin R (1999) AC/GT and AG/CT microsatellite repeats in peach [*Prunus persica* (L.) Batsch]: isolation, characterisation and cross-species amplification in *Prunus*. Theor Appl Genet 99: 65–72.

Clarke JB, Sargent DJ, Boskovic RI, Belaj A, Tobutt KR (2009) A cherry map from the inter-specific cross *Prunus avium* 'Napoleon' x *P. nipponica* based on microsatellite, gene-specific and isoenzyme markers. Tree Genet Genomes 5: 41–51.

Clarke JB, Tobutt KR (2003) Development and characterization of polymorphic microsatellites from Prunus avium 'Napoleon'. Mol Ecol Notes 3: 578–580.

Claverie M, Dirlewanger E, Cosson P, Bosselut N, Lecouls AC, Voisin R, Kleinhentz M, Lafargue B, Caboche M, Chalhoub B, Esmenjaud D (2004) High-resolution mapping and chromosome landing at the root-knot nematode resistance locus Ma from Myrobalan plum using a large-insert BAC DNA library. Theor Appl Genet 109: 1318–1327.

Dirlewanger E, Cosson P, Tavaud M, Aranzana MJ, Poizat C, Zanetto A, Arus P, Laigret F (2002) Development of microsatellite markers in peach [*Prunuspersica* (L.) Batsch] and their use in genetic diversity analysis in peach and sweet cherry (*Prunus avium* L.). Theor Appl Genet 105: 127–138.

Dirlewanger E, Graziano E, Joobeur T, Garriga-Caldere F, Cosson P, Howad W, Arus P (2004) Comparative mapping and marker-assisted selection in Rosaceae fruit crops. Proc Natl Acad Sci USA 101: 9891–9896.

Dirlewanger E, Pronier V, Parvery C, Rothan C, Guye A, Monet R (1998) Genetic linkage map of peach [*Prunus persica* (L.) Batsch] using morphological and molecular markers. Theor Appl Genet 97: 888–895.

Dondini L, Lain O, Geuna F, Banfi R, Gaiotti F, Tartarini S, Bassi D, Testolin R (2007) Development of a new SSR-based linkage map in apricot and analysis of synteny with existing Prunus maps. Tree Genet Genomes 3: 239–249.

Georgi LL, Wang Y, Reighard GL, Mao L, Wing RA, Abbott AG (2003) Comparison of peach and Arabidopsis genomic sequences: fragmentary conservation of gene neighborhoods. Genome 46: 268–276.

Georgi LL, Wang Y, Yvergniaux D, Ormsbee T, Inigo M, Reighard G, Abbott AG (2002) Construction of a BAC library and its application to the identification of simple sequence repeats in peach [*Prunus persica* (L.) Batsch]. Theor Appl Genet 105: 1151–1158.

Horn R, Lecouls AC, Callahan A, Dandekar A, Garay L, McCord P, Howad W, Chan H, Verde I, Main D, Jung S, Georgi L, Forrest S, Mook J, Zhebentyayeva T, Yu YS, Kim HR, Jesudurai C, Sosinski B, Arus P, Baird V, Parfitt D, Reighard G, Scorza R, Tomkins J, Wing R, Abbott AG (2005) Candidate gene database and transcript map for peach, a model species for fruit trees. Theor Appl Genet 110: 1419–1428.

Howad W, Yamamoto T, Dirlewanger E, Testolin R, Cosson P, Cipriani G, Monforte AJ, Georgi L, Abbott AG, Arus P (2005) Mapping with a few plants: Using selective mapping for microsatellite saturation of the Prunus reference map. Genetics 171: 1305–1309.

Illa E, Lambert P, Quilot B, Audergon JM, Dirlewanger E, Howad W, Dondini L, Tartarini S, Lain O, Testolin R, Bassi D, Arus P (2009) Linkage map saturation, construction, and comparison in four populations of Prunus. J Hort Sci Biotechnol 84: 168–175.

Illa E, Sargent D, J, Lopez Girona E, Bushakra J, Cestaro A, Crowhurst R, Pindo M, Cabrera A, Van der Knapp E, Iezzoni A, Gardiner S, Velasco R, Arus P, Chagné D, Troggio M (2011) Comparative analysis of rosaceous genomes and the reconstruction of a putative ancestral genome for the family. BMC Evol Biol 11: 9.

Joobeur T, Viruel MA, de Vicente MC, Jauregui B, Ballester J, Dettori MT, Verde I, Truco MJ, Messeguer R, Batlle I, Quarta R, Dirlewanger E, Arus P (1998) Construction of a saturated linkage map for Prunus using an almond x peach F-2 progeny. Theor Appl Genet 97: 1034–1041.

Jung S, Jiwan D, Cho I, Lee T, Abbott A, Sosinski B, Main D (2009) Synteny of Prunus and other model plant species. BMC Genomics 10.

Jung S, Main D, Staton M, Cho IH, Zhebentyayeva T, Arus P, Abbott A (2006) Synteny conservation between the Prunus genome and both the present and ancestral Arabidopsis genomes. BMC Genomics 7: (14 April 2006).

Jung S, Staton M, Lee T, Blenda A, Svancara R, Abbott A, Main D (2008) GDR (Genome Database for Rosaceae): integrated web-database for Rosaceae genomics and genetics data. Nucl Acids Res 36: D1034–D1040.

Kaul S, Koo HL, Jenkins J, Rizzo M, Rooney T, Tallon LJ, Feldblyum T, Nierman W, Benito MI, Lin XY, Town CD, Venter JC, Fraser CM, Tabata S, Nakamura Y, Kaneko T, Sato S, Asamizu E, Kato T, Kotani H, Sasamoto S, Ecker JR, Theologis A, Federspiel NA, Palm CJ, Osborne BI, Shinn P, Conway AB, Vysotskaia VS, Dewar K, Conn L, Lenz CA, Kim

CJ, Hansen NF, Liu SX, Buehler E, Altafi H, Sakano H, Dunn P, Lam B, Pham PK, Chao Q, Nguyen M, Yu GX, Chen HM, Southwick A, Lee JM, Miranda M, Toriumi MJ, Davis RW, Wambutt R, Murphy G, Dusterhoft A, Stiekema W, Pohl T, Entian KD, Terryn N, Volckaert G, Salanoubat M, Choisne N, Rieger M, Ansorge W, Unseld M, Fartmann B, Valle G, Artiguenave F, Weissenbach J, Quetier F, Wilson RK, de la Bastide M, Sekhon M, Huang E, Spiegel L, Gnoj L, Pepin K, Murray J, Johnson D, Habermann K, Dedhia N, Parnell L, Preston R, Hillier L, Chen E, Marra M, Martienssen R, McCombie WR, Mayer K, White O, Bevan M, Lemcke K, Creasy TH, Bielke C, Haas B, Haase D, Maiti R, Rudd S, Peterson J, Schoof H, Frishman D, Morgenstern B, Zaccaria P, Ermolaeva M, Pertea M, Quackenbush J, Volfovsky N, Wu DY, Lowe TM, Salzberg SL, Mewes HW, Rounsley S, Bush D, Subramaniam S, Levin I, Norris S, Schmidt R, Acarkan A, Bancroft I, Quetier F, Brennicke A, Eisen JA, Bureau T, Legault BA, Le QH, Agrawal N, Yu Z, Martienssen R, Copenhaver GP, Luo S, Pikaard CS, Preuss D, Paulsen IT, Sussman M, Britt AB, Selinger DA, Pandey R, Mount DW, Chandler VL, Jorgensen RA, Pikaard C, Juergens G, Meyerowitz EM, Theologis A, Dangl J, Jones JDG, Chen M, Chory J, Somerville MC, Ar Gen I (2000) Analysis of the genome sequence of the flowering plant *Arabidopsis thaliana.* Nature 408: 796–815.

Lalli DA, Decroocq V, Blenda AV, Schurdi-Levraud V, Garay L, Le Gall O, Damsteegt V, Reighard GL, Abbott AG (2005) Identification and mapping of resistance gene analogs (RGAs) in Prunus: a resistance map for Prunus. Theor Appl Genet 111: 1504–1513.

Lopes MS, Sefc KM, Laimer M, Machado AD (2002) Identification of microsatellite loci in apricot. Mol Ecol Notes 2: 24–26.

Maliepaard C, Alston FH, van Arkel G, Brown LM, Chevreau E, Dunemann F, Evans KM, Gardiner S, Guilford P, van Heusden AW, Janse J, Laurens F, Lynn JR, Manganaris AG, den Nijs APM, Periam N, Rikkerink E, Roche P, Ryder C, Sansavini S, Schmidt H, Tartarini S, Verhaegh JJ, Vrielink-van Ginkel M, King GJ (1998) Aligning male and female linkage maps of apple (*Malus pumila* Mill.) using multi-allelic markers. Theor Appl Genet 97: 60–73.

Messina R, Lain O, Marrazzo MT, Cipriani G, Testolin R (2004) New set of microsatellite loci isolated in apricot. Mol Ecol Notes 4: 432–434.

Ogundiwin EA, Peace CP, Gradziel TM, Parfitt DE, Bliss FA, Crisosto CH (2009) A fruit quality gene map of Prunus. BMC Genomics 10: Article No: 587.

Potter D, Eriksson T, Evans RC, Oh S, Smedmark JEE, Morgan DR, Kerr M (2007) Phylogeny and classification of Rosaceae. Plant Syst Evol 266: 5–43.

Sargent D, J, Clarke J, Simpson DW, Tobutt KR, Arus P, Monfort A, Vilanova S, Denoyes-Rothan B, Rousseau M, Folta KM, Brassil NV, Battey NH (2006) An enhanced microsatellite map of diploid *Fragaria.* Theor Appl Genet 112: 1349–1359.

Sargent DJ, Marchese A, Simpson DW, Howad W, Fernandez-Fernandez F, Monfort A, Arus P, Evans KM, Tobutt KR (2009) Development of "universal" gene-specific markers from Malus spp. cDNA sequences, their mapping and use in synteny studies within Rosaceae. Tree Genet Genomes 5: 133–145.

Sargent DJ, Rys A, Nier S, Simpson DW, Tobutt KR (2007) The development and mapping of functional markers in Fragaria and their transferability and potential for mapping in other genera. Theor Appl Genet 114: 373–384.

Shulaev V, Korban SS, Sosinski B, Abbott AG, Aldwinckle HS, Folta KM, Iezzoni A, Main D, Arus P, Dandekar AM, Lewers K, Brown SK, Davis TM, Gardiner SE, Potter D, Veilleux RE (2008) Multiple models for Rosaceae genomics. Plant Physiol 147: 985–1003.

Shulaev V, Sargent DJ, Crowhurst RN, Mockler RN, Veilleux RE, Folkerts O, Delcher AL, Jaiswal P, Liston A, Mane SP, Burns P, Mockaitis K, Davis TM, Slovin JP, Bassil N, Hellens RP, Evans C, Jensen RV, Allan AC, Michael TP, Setubal JC, Celton JM, Rees DJG, Williams K, Holt SH, Dickerman A, Ruiz Rojas JJ, Chatterjee M, Liu B, Silva H, Meisel L, Filichkin AAS, Velasco R, Troggio M, Viola R, Borodovsky M, Ashman TL, Aharoni A, Bennetzen J, Dharmawardhana P, Elser J, Raja R, Priest HD, Bryant DW, Fox SE, Givan SA, Naithani S, Christoffels A, Salama DY, Carter J, Lopez Girona E, Zdepski A, Wang W, Kerstetter

RA, Salzberg SL, Schwab W, Korban SS, Davik J, Monfort A, Denoyes-Rothan B, Arus P, Mittler R, Flinn B, Folta KM (2011) Whole Genome Sequencing and Analysis of Diploid Strawberry, Fragaria vesca, a Compact and Versatile Perennial Rosaceous Fruit. Nat Genet 43: 109–116.

Sosinski B, Gannavarapu M, Hager LD, Beck LE, King GJ, Ryder CD, Rajapakse S, Baird WV, Ballard RE, Abbott AG (2000) Characterization of microsatellite markers in peach [*Prunus persica* (L.) Batsch]. Theor Appl Genet 101: 421–428.

Stockinger EJ, Mulinix CA, Long CM, Brettin TS, Iezzoni AF (1996) A linkage map of sweet cherry based on RAPD analysis of a microspore-derived callus culture population. J Hered 87: 214–218.

Tang HB, Bowers JE, Wang XY, Ming R, Alam M, Paterson AH (2008) Perspective—Synteny and collinearity in plant genomes. Science 320: 486–488.

Tobutt KR, Nicoll RJ (1992) Joint Meeting of the European-Prunus-and-Apple-Genome-Mapping-Groups, East-Malling, Uk, May 1991. Fruit Var J 46: 263–265.

Velasco R, Zharkikh A, Affourtit J, Dhingra A, Cestaro A, Kalyanaraman A, Fontana P, Bhatnagar SK, Troggio M, Pruss D, Salvi S, Pindo M, Baldi P, Castelletti S, Cavaiuolo M, Coppola G, Costa F, Cova V, Dal Ri A, Goremykin V, Komjanc M, Longhi S, Magnago P, Malacarne G, Malnoy M, Micheletti D, Moretto M, Perazzolli M, Si-Ammour A, Vezzulli S, Zini E, Eldredge G, Fitzgerald LM, Gutin N, Lanchbury J, Macalma T, Mitchell JT, Reid J, Wardell B, Kodira C, Chen Z, Desany B, Niazi F, Palmer M, Koepke T, Jiwan D, Schaeffer S, Krishnan V, Wu C, Chu VT, King ST, Vick J, Tao Q, Mraz A, Stormo A, Stormo K, Bogden R, Ederle D, Stella A, Vecchietti A, Kater MM, Masiero S, Lasserre P, Lespinasse Y, Allan AC, Bus V, Chagne D, Crowhurst RN, Gleave AP, Lavezzo E, Fawcett JA, Proost S, Rouze P, Sterck L, Toppo S, Lazzari B, Hellens RP, Durel CE, Gutin A, Bumgarner RE, Gardiner SE, Skolnick M, Egholm M, Van de Peer Y, Salamini F, Viola R (2010) The genome of the domesticated apple (*Malus* x *domestica* Borkh.). Nat Genet 42: 833–839.

Velasco R, Zharkikh A, Troggio M, Cartwright DA, Cestaro A, Pruss D, Pindo M, FitzGerald LM, Vezzulli S, Reid J, Malacarne G, Iliev D, Coppola G, Wardell B, Micheletti D, Macalma T, Facci M, Mitchell JT, Perazzolli M, Eldredge G, Gatto P, Oyzerski R, Moretto M, Gutin N, Stefanini M, Chen Y, Segala C, Davenport C, Dematte L, Mraz A, Battilana J, Stormo K, Costa F, Tao QZ, Si-Ammour A, Harkins T, Lackey A, Perbost C, Taillon B, Stella A, Solovyev V, Fawcett JA, Sterck L, Vandepoele K, Grando SM, Toppo S, Moser C, Lanchbury J, Bogden R, Skolnick M, Sgaramella V, Bhatnagar SK, Fontana P, Gutin A, Van de Peer Y, Salamini F, Viola R (2007) A high quality draft consensus sequence of the genome of a heterozygous grapevine variety. PLOS One 2: 18.

Vilanova S, Romero C, Abernathy D, Abbott AG, Burgos L, Llacer G, Badenes ML (2003) Construction and application of a bacterial artificial chromosome (BAC) library of *Prunus armeniaca* L. for the identification of clones linked to the self-incompatibility locus. Mol Genet Genom 269: 685–691.

Vilanova S, Sargent DJ, Arus P, Monfort A (2008) Synteny conservation between two distantly-related Rosaceae genomes: *Prunus* (the stone fruits) and *Fragaria* (the strawberry). BMC Plant Biol 8: 12.

Viruel MA, Messeguer R, Devicente MC, Garciamas J, Puigdomenech P, Vargas F, Arus P (1995) A linkage map with RFLP and isozyme markers for almond. Theor Appl Genet 91: 964–971.

Wang Q, Zhang K, Qu X, Jia J, Shi J, Jin D, Wang B (2001) Construction and characterization of a bacterial artificial chromosome library of peach. Theor Appl Genet 103: 1174–1179.

Xu Y, Ma RC, Xie H, Liu JT, Cao MQ (2004) Development of SSR markers for the phylogenetic analysis of almond trees from China and the Mediterranean region. Genome 47: 1091–1104.

Zhebentyayeva TN, Swire-Clark G, Georgi LL, Garay L, Jung S, Forrest S, Blenda AV, Blackmon B, Mook J, Horn R, Howad W, Arus P, Main D, Tomkins JP, Sosinski B, Baird WV, Reighard GL, Abbott AG (2008) A framework physical map for peach, a model Rosaceae species. Tree Genet Genomes 4: 745–756.

11

Functional Genomics: Transcriptomics

Livio Trainotti,[1,*] *Stefano Cagnin,*[1,2] *Claudio Forcato,*[2]
Claudio Bonghi,[3] *Amit Dhingra,*[4] *Tyson Koepke,*[4] *Loreto Prat,*[5]
Jonathan Maldonado[5] and *Herman Silva*[5,*]

ABSTRACT

Transcriptomics approaches in the *Prunus* genus started to be developed since the beginning of the new century. In few years, a set of tools have been developed and used, mainly in peach, apricot and almond. Transcriptomics tools have been primarily used to investigate fruit ripening and post-harvest physiology, but also disease resistance and flower transition. At the beginning of the second decade of the century, more than 100,000 ESTs are available in public databases, the majority of which were obtained from peach fruit. This repertoire has been used for digital expression analyses of transcriptome changes associated with fruit ripening and the appearance of chilling-induced post-harvest fruit disorders. The microarrays developed on peach ESTs have been extensively used mainly to investigate fruit biology but are soon going to be replaced by genome-wide platforms based on the recently released peach genome sequence. Transcriptome profiling based on second generation DNA sequencing (SGS) technologies are also expected

[1]Department of Biology, University of Padova, Via U. Bassi 58/B, 35121 Padova, Italy.
[2]CRIBI Biotechnology Centre, University of Padova, Via U. Bassi 58/B, 35121 Padova, Italy.
[3]Department of Environmental Agronomy and Crop Science, University of Padova, Agripolis, Legnaro (PD) 35020, Italy.
[4]Department of Horticulture and Landscape Architecture and Molecular Plant Sciences Graduate Program, Washington State University, PO BOX 646414, Pullman, WA 99164, USA.
[5]Millennium Nucleus in Plant Cell Biotechnology and Functional Genomics & Bioinformatic Lab, Departamento de Producción Agrícola, Facultad de Ciencias Agronómicas, Universidad de Chile, Av. Santa Rosa 11315, 8820808 La Pintana, Casilla 1004, Santiago, Chile.
*Corresponding authors: *livio.trainotti@unipd.it; hesilva@uchile.cl*

to have a tremendous impact in *Prunus* genomics. Pilot experiments conducted in several laboratories show that the peach genome sequence may be used for SGS approaches also in other *Prunus* species.

Many molecular markers have been developed from ESTs and used to construct transcriptome maps in order to find the genetic determinants of Mendelian and quantitative traits. Transcriptomics is expected to speed up the finding of functional molecular markers to be used in marker-assisted programs for the development of new varieties across *Prunus* species. The use of information obtained with the new generation sequencing technologies will allow crossing information and looks for differential expression into different varieties within the same species as well as different ones.

Keywords: EST, microarray, cDNA-AFLP, eQTL, second-generation sequencing (SGS), RNA-seq, *Prunus*

11.1 Transcriptomics Tools Developed in Stone fruits

The beginning of the modern transcriptomics era in *Prunus* spp. might be traced back to the first report describing expressed sequence tag (EST) production from peach, dating back to year 2000 (Hayama et al. 2000), when about 200 sequences were described, obtained from libraries constructed from RNA of three different fruit developmental stages. Besides EST production, several differential display approaches in different species have been undertaken with peach (*P. persica*) being investigated primarily. By the time that this chapter has been written, the whole peach genome sequence was available (Sosinski et al. 2010) as the main achievement of the activities of several laboratories that in the last decade faced genomics in order to expand the molecular tools needed to tackle the many biological questions underneath the agronomical traits in the *Prunus* genus.

11.1.1 Sequence-independent Transcriptomics Tools

Several genomics techniques have been used to investigate transcriptome variations during development or as a response to external stimuli.

The cDNA-AFLP (amplified fragment length polymorphism) technique has been used to identify almond genes that are differentially expressed in response to dehydration (Campalans et al. 2001). Plantlets, either obtained from germinating seeds or grown *in vitro* and exposed to dehydration or treated with abscisic acid (ABA), were used as starting material for the RNA isolations. The cDNA-AFLP experiments allowed the identification of 40 differential bands, of which thirteen could be confirmed by reverse northern. Sequence analysis allowed the characterization of five cDNAs, whose differential expression was validated by northern and semi-quantitative RT-PCR. Clone IVC2, named PaCYP1 and encoding a cysteine protease, showed similar induction after drought stress in eight cultivars differing for drought adaptation, thus showing no correlation with the dehydration tolerance character.

Mesocarp transcripts of the nectarine *slow ripening* (*slr*) selection have been compared with those of Fantasia, a mid-term climacteric variety, at three different developmental stages (Ziliotto et al. 2005). Almost 1,400 transcript-derived fragments (TDFs) have been identified, indicating a large alteration of the *slr* mesocarp transcriptome. The *slr* genotype produces fruits blocked in their development at the S3 stage. Thus, several Fantasia TDFs are related to senescence, a genetic program not entered by *slr* fruits. On the other hand, several *slr* TDFs are involved in DNA replication and cytokinesis, confirming that S3 *slr* fruits are still in a juvenile stage.

An extensive cDNA-AFLP analysis has been carried out in apricot, where RNAs from fruit at six developmental stages and from leaf and developing seeds, used as control for fruit specificity, allowed the isolation of 265 differentially expressed bands (Geuna et al. 2005). A consistent part (125) of the differentially expressed bands were cloned and sequenced, allowing the identification of genes involved in cell wall, sugar, lipid, organic acid and hormonal metabolisms, besides those (51 sequences) involved in hormonal signaling and signal transduction. A substantial agreement in the differential expression measured by cDNA-AFLP and northern blot analysis has been shown for two selected mRNAs, a xyloglucan endotransglycosylase (XET) and a lipoxygenase (LOX), thus pinpointing the reliability of the method applied to discover ripening related genes. Nevertheless, the comparison of the expression profiles measured by cDNA-AFLP and northern blots evidenced that the success of the technique largely depends on the sample selection. Indeed, being cDNA-AFLP an amplification-based technique, the identification of differentially expressed transcripts largely dangles on the degree of differential transcription. Only genes whose mRNAs accumulate at substantially different levels can be highlighted. Those genes that are only moderately differentially expressed are missed by this approach because the amplification steps, brought to (almost) saturation, flatten amplicons differences among samples. This was evident also in Geuna's work (Geuna et al. 2005), in which the expression of XET was detected since stage 2 by cDNA-AFLP but only from stage 4 by northern and that of LOX was clear in stage 4 by cDNA-AFLP but barely detectable by northern. This intrinsic limit hides the discovery of genes whose transcriptional modulation, albeit being very subtle, might be functionally relevant, as, for example, those coding for transcription factors (Tadiello et al. 2009).

The impact of phytoplasma infection on the transcriptome of apricot leaf was evaluated by means of the mRNA differential display (DD) technique

(Carginale et al. 2004). Leaves of plants infected with European stone fruit yellows (ESFY) phytoplasma belonging to ribosomal subgroup 16SrX-B showed a significant change in expression of several genes. Among the four genes identified by sequence analysis, three were up-regulated and one down-regulated by the presence of the pathogen. Among the up-regulated ones, there was also a metallothionein (MT), a gene family with members up-regulated also during ripening (Geuna et al. 2005). Even thought the information of a possible role of the identified genes during the infection was only correlative, the DD approach demonstrated its validity as a possible tool toward the identification of the metabolic pathway(s) altered by the phytoplasma infection.

11.1.2 Production and in-silico Transcript Profiling of ESTs

Since the very early beginning, the available ESTs have been used to both develop molecular markers and transcriptomics tools. Initially, simple sequence repeats (SSRs) (Yamamoto et al. 2002) were developed followed by single nucleotide polymorphisms (SNPs; Ogundiwin et al. 2009). Meanwhile, nylon macroarrays were the first transcriptomics tools (Trainotti et al. 2003; González-Agüero et al. 2008) followed by glass microarrays (ESTree Consortium 2005). Unfortunately, the generation of *Prunus* ESTs based on Sanger sequencing has been very limited compared to other species (Table 11-1); thus peach is only 54th for number of ESTs (78,633) available in public databases (http://www.plantgdb.org/prj/ESTCluster/progress.php) while apricot, the second species of the genus for produced ESTs (15,032), only 181st (slight differences in numbers may depend from the different databases queried, see Table 11-1). The low amount of available ESTs limited the possibility of serious approaches of digital gene expression profiling but some interesting work was done for the comparison of vegetative and fruit samples and, among the latter, between skin and mesocarp (Vecchietti et al. 2008). Regarding the fruit, also the effect of different post-harvest regimes on fruit quality has been investigated (Vizoso et al. 2009). In the first example, EST frequency analysis (EFA) was calculated among mesocarp at two developmental stages (S3 and S4), skin of S4 fruit and shoot for 391 transcripts. A k-mean cluster analysis of these abundantly transcribed transcripts allowed the identification of nine clusters, the first of which contains mRNAs abundant in the shoot. The two clusters representing transcripts more abundant in the skin contained sequences coding for lipid transfer proteins, the major protein allergen (which, besides being the most abundant in ripe fruit, seems to be specific of the genus) and a 9-cis-epoxycarotenoid dioxygenase (NCED), a key enzyme in abscisic acid (ABA) synthesis. Furthermore, transcripts encoding proteins involved in synthesis of volatiles were more

Table 11-1 Plant Transcriptomes. A selection of EST resources available from two main public repositories. *Prunus* species are highlighted in bold.

		Plant Transcripts Assemblies			Plant Gene Indices		
Common name	Specie	ESTs + cDNAs	Clusters	Singletons	ESTs + cDNAs	Clusters	Singletons
Thale-cress	*Arabidopsis thaliana*	682040	27983	120385	1104399	49009	63818
Tomato	*Solanum lycopersicum*	251814	21523	32268	302789	28167	24335
Black cottonwood	*Populus trichocarpa*	89243	12687	18395	432027	50563	50313
Rice	*Oryza sativa*	1204150	49870	197646	1365381	82184	119958
Wine grape	*Vitis vinifera*	313050	21627	40880	382717	34634	47495
Papaya	*Carica papaya*	--	--	--	74675	9890	31827
Cucumber	*Cucumis sativus*	5254	631	3299	--	--	--
Maize	*Zea mays*	1038677	64601	232592	1698645	112156	202931
Sorghum	*Sorghum bicolor*	203641	20714	28218	209434	23442	22583
Cultivated apple	*Malus* x *domestica*	250228	26757	50847	319428	34945	29013
Soybean	*Glicine max*	352750	36399	80566	1356732	73178	63996
Purple false brome	*Brachypodium distachyon*	20324	2785	8359	127579	18209	11051
Woodland strawberry	*Fragaria vesca*	45085	4825	8624	--	--	--
Peach	***Prunus persica***	**65979**	**6596**	**21845**	**78633**	**9633**	**16454**
Apricot	***Prunus armeniaca***	**15032**	**2372**	**3571**	--	--	--
Sweet almond	***Prunus dulcis***	**3886**	**224**	**1711**	--	--	--
Sour cherry	***Prunus cerasus***	**1252**	**115**	**824**	--	--	--
Sweet orange	*Citrus sinensis*	93451	11061	18222	202570	26081	72817
Potato	*Solanum tuberosum*	220473	26280	54792	239559	31972	30358

Source = http://plantta.jcvi.org/cgi-bin/plantta_release.pl, http://compbio.dfci.harvard.edu/tgi/plant.html

frequently cloned from the skin, confirming that the peel is central in peach flavor production. Ripening related transcripts, such as those encoding the enzymes for ethylene biosynthesis or cell wall degrading enzymes, were abundant both in mesocarp and skin at S4. Cluster 6 contained ripening related genes expressed mainly in mesocarp at S3, thus confirming that not only cell wall related genes are expressed before the ethylene climacteric peak (Trainotti et al. 2003).

With the aim of identifying genes related to disorders caused by cold storage, more than 41,000 ESTs have been obtained from four different libraries, each one constructed from fruit either ripe or unripe either with or without the post-harvest treatment (Vizoso et al. 2009). The statistical analysis of the EST population by pair-wise comparisons highlighted 197 transcripts that significantly differed in abundance in at least two conditions. The analysis revealed that ripening has a stronger effect on the peach transcriptome than the cold treatment, as transcriptomes of ripe and unripe fruit, either treated or not, were more similar than those of treated and untreated ones, either ripe or unripe. Besides giving further support to the role played by cell wall enzymes and ethylene, the cluster analysis of differentially expressed genes highlighted the relevance of those associated with oxidative stress and metabolic imbalances occurring mainly in the organelles (mitochondria and plastids), but also at the level of the plasma membrane. The findings opened new fields for future research not only because they added new candidates to the list of genes/metabolic processes causing post-harvest diseases but also because many of the sequenced clones represented full-length transcripts flanked by lox-P sites and thus ready for further functional analyses, as they may be easily recombined in compatible expression vectors.

In apricot, more than 13,000 ESTs have been generated from mesocarp at three developmental stages (green, half-ripe and ripe) but only the 11,043 containing a 3′ end were used for in silico northern (Grimplet et al. 2005). This analysis identified 1,793 clusters (contigs) among which only four accounted for up to more than 5% of the whole transcriptome. The two more abundant transcripts code for a putative allergen and for an abscisic stress-related protein, orthologs to those observed in peach (Vecchietti et al. 2008). This collection, although far from saturation, allowed the identification of differentially expressed genes, many of which code for cell wall proteins or belong to defense and stress responses, besides 1-aminocyclopropane-1-carboxylate (ACC) oxidase (ACO), encoding the enzyme of the last step in ethylene biosynthesis. Different statistical significances were attributed to three classes of differentially expressed genes and the expression of genes belonging to the most robust ones was confirmed by quantitative RT-PCR.

Many of the *Prunus* ESTs have been produced to identify genetic molecular markers, such as SSRs and SNPs. This has been the case for ESTs produced from peach (Jung et al. 2004; Horn et al. 2005; Vendramin et al. 2007; Ogundiwin et al. 2008), almond (Jiang and Ma 2003; Jung et al. 2004), apricot (Grimplet et al. 2005), cherry and ume (Li et al. 2010) and their use, not pertinent to this chapter, as source of molecular markers is discussed elsewhere in this book.

11.1.3 Microarray Transcript Profiling

Transcripts encoding proteins related to the cell wall metabolism were the first to be investigated by means of nylon macroarray hybridizations (Trainotti et al. 2003). Although the platform contained only 32 cDNAs, they were selected among one of the first peach EST collection, consisting of more than 800 putative unigenes (contigs) obtained by random sequencing of clones from a pre-climacteric fruit library. Besides giving a general overview of the many genes directly or indirectly involved in peach softening, this report was the first to show, at the molecular level, that softening starts before ripening proper. Moreover, it was there speculated that structural proteins, acting as molecular tethers, might hold in place the remnants of the cell wall polymers, thus ensuring mesocarp softening without the lysis of its cells.

The second and last macroarray of which there are reports in the literature has been developed with 847 EST clones produced at Clemson University (Jung et al. 2004) and used to investigate transcriptional changes occurring during chilling injury (González-Agüero et al. 2008). The use of this platform allowed the identification of 106 differentially expressed genes in juicy and woolly fruits. These results are probably biased by the probe selection and that might explain the lower number of differentially expressed genes identified by the same research group using EST transcription profiling, which substantially confirmed macroarray-derived findings (Vizoso et al. 2009).

µPEACH1.0 (ESTree Consortium 2005) and ChillPeach (Ogundiwin et al. 2008) are the two main microarray platforms that have been developed from *Prunus* derived ESTs. µPEACH1.0 consists of 4,806 80-mer oligonucleotide probes, each spotted twice together with controls on glass slides. It has been extensively used by members of the ESTree Consortium but also by other researches as it was commercially available from Operon (now Eurofins MWG Operon). The transcript set used to design the probes was based on both in house developed ESTs and those publicly available in the NCBI databank, but was, nonetheless, biased towards genes more actively transcribed in the mesocarp. Nevertheless, being ripening and fruit storage are the main interests of the researchers using this tool, it allowed relevant finding in these fields. The comparison between S3 and S4 fruits highlighted a dramatic change in the mesocarp transcriptome, with hundreds of genes up- and down-regulated (Trainotti et al. 2006; Trainotti et al. 2007). Among the differentially transcribed genes, many belong to the cell wall and chloroplast compartments. Auxin and ethylene treatments of pre-climacteric fruits allowed assigning a new and independent role to auxin during ripening of this climacteric fruit (Trainotti et al. 2007). The role of ethylene and auxin during ripening has been confirmed also by transcriptome profiling of fruit treated with the ethylene inhibitor 1-methylcyclopropene (1-MCP; Ziliotto et al. 2008). This intense auxin-ethylene cross-talk during mesocarp ripening does not seem to be exclusive of peach, since the up-regulation of genes belonging to the auxin domain has been documented also in apricot, by both in silico northern (Grimplet et al. 2005) and microarray profiling (Manganaris et al. 2011). Interestingly, the work on apricot (Manganaris et al. 2011) shows that μPEACH1.0 can be used also in species other than peach. On the other hand, peach transcriptome changes occurring during endocarp development could be better assayed using an apple platform than μPEACH1.0 (Dardick et al. 2010). Indeed, 80% of the probes on the 15,720 oligonucleotide apple microarray (Schaffer et al. 2007) could detect peach mRNAs, thus allowing to asses transcriptional changes of about 12,700 peach genes, many more than those corresponding to the 4,806 probes on μPEACH 1.0. Also the effect of jasmonates (JAs), applied to fruit on the tree at the S3 stage, on the transcriptome of ripening mesocarp has been investigated using μPEACH1.0 (Ziosi et al. 2008). Transcriptome profiling revealed a profound effect of JAs, which repressed ethylene and auxin-related genes and, as a consequence, also ripening, while inducing stress and defense-related genes, keeping fruit cell in a more metabolically active state.

μPEACH2.0 has been developed with the goal of underpinning the discovery of genes involved in aroma and flavor production (Pirona R, Vecchietti A, Lazzari B, Caprera A, Malinverni R, Consolandi C, Severgnini M, De Bellis G, Chietera G, Rossini L, Pozzi C, unpublished results). It is based on the 12k Combimatrix platform and contains 4,776 oligonucleotides probes, 1,951 of which are also present on μPEACH1.0. The other probes are derived from 9 ESTs libraries (S3, S4, PP_Lea. Pp-S4, S3II, Lf, PP_YEa, UD_S4, Skin; ESTreedb version V).

A combination of suppression subtractive hybridization (SSH) and microarray hybridization has been employed to identify genes associated with bud dormancy release in peach (Leida et al. 2010). Four SHH cDNA libraries have been prepared from dormant and non-dormant buds and from germinating buds of two genotypes differing for the chilling requirement needed to release dormancy. This approach has the advantage that only the 364 differentially expressed clones, which probably are the biologically relevant ones, were sequenced and further analyzed. Overlapping of the clones reduced the number of relevant genes down to 101, among which are MADS-box transcription factors and genes known as dormancy factors in other species.

Chilling injury (CI) and diseases arising in peach fruit that underwent post-harvest cold storage are among the main problems for the peach industry. The ChillPeach cDNA microarray platform contains 4,261 UniGene derived probes selected from an EST collection prepared from mesocarp tissues of two full-sib progeny contrasting for CI susceptibility (Ogundiwin et al. 2008). As regards the fruit, only a pilot experiment has been published with this platform. It has been shown that at least 516 genes were up-regulated by the CI treatment in peach mesocarp, thus showing the usefulness of both the microarray platform and of the segregating population. This research also highlighted that not all the CI-modulated genes in peach behave as their *Arabidopsis* orthologs, with some behaving in the opposite way, probably because of the specific features of the mesocarp tissue.

11.2 Transcriptome QTL Mapping

There are many traits in *Prunus* that are inherited in a quantitative fashion and several mapping populations have been described and used to map several QTLs (Boudehri et al. 2009; Ogundiwin et al. 2009; Eduardo et al. 2010). Since the early development of EST collections, candidate genes have been selected and mapped on several maps (Horn et al. 2005). The selection of the candidate genes to be mapped was based either on the putative function deduced by sequence homology with genes characterized in other biological systems or on the expression levels of genes in a given biological process, such as fruit ripening or pathogen infection. This selection scheme allowed mapping 11 ripening-related and 13 cold-responsive genes on the Pop-DG and TxE maps (Ogundiwin et al. 2009). Nevertheless, systematic expression data of genes comprised in the regions spanning the QTL have not yet been published. As a QTL locus may span hundreds of genes, several kinds of data are needed to select the right candidate gene(s) that may contribute to the character. Therefore, the segregation of the expression levels in the individuals of the mapping population may help to reveal the molecular mechanisms underneath the QTL. Thus, the finding of expression QTLs (eQTLs) might lead to the identification of the loci/genes responsible for the QTL (Hansen et al. 2008). Furthermore, eQTL data may help to shed light on the complex regulatory networks that link molecular variation at multiple loci with the physiological and phenotypic variation that is observed in natural populations or in breeding germplasm collections. The availability of the peach genome will speed up the development of new high throughput transcriptomics tools, making the construction of eQTL maps easier. Similar maps have been produced in other systems for a variety of traits using different kinds of populations (West et al. 2007; Chen et al. 2010; Hammond et al. 2011). Although not yet explored in *Prunus*, microarray expression profiling of tomato fruits selected from only six different introgression lines of a mapping population allowed to associate the higher Brix found in two lines with the coordinated up-regulation of enzymes of sucrose mobilization and respiration (Baxter et al. 2005).

11.3 Target Deliverables and Current Status

The availability of the peach genome poses new challenges to transcriptomics in this species but also in strictly related ones. New technologies are being applied and some, as RNA sequencing (RNAseq) with different platforms (see Section 11.5), have already been successfully used in peach, almond and apricot (Martínez-Gómez et al. 2011). Microarray hybridizations are still going to be used as they can still stand the competition for data quality and prices with the new DNA/RNA sequencing technologies, especially if large amount of data have to be produced, as in eQTL mapping experiments.

A new microarray platform, named µPEACH3.0, has recently been developed and is going to be used in the near future. It is based on a custom-made Agilent Sure Print G3 platform with a 8 x 60k format. The probe selection has been carried out on 30,113 predicted transcripts. The 28,689 officially released transcripts (Sosinski et al. 2010) have been implemented by adding 1,424 new predictions (Forcato et al., unpublished data). 454 and Solexa RNAseq experiments have been used to implement the new in silico transcript predictions. Each of the eight arrays on each slide contains 29,800 peach probes, each printed twice. As for other steps of the genomics research, also transcriptomics is facing the challenge of how to handle data and to retrieve meaningful information. Bioinformatics will thus be crucial for any functional genomics project and, with this in mind, dedicated tools are going to be developed to make fruitful the use of the µPEACH3.0 platform.

As a "proof of concept", a pilot experiment has been carried out by mapping the transcript levels measured by sequencing a peach mesocarp cDNA library prepared from mesocarp at stage S3. Mapping this data on the peach reference genome allowed identifying regions highly transcribed in mesocarp soon before fruit ripening (Fig. 11-1). In the near future, expression data, either produced by RNAseq or by microarray, will be used to build similar maps in order to highlight DNA regions that could be hotspot of transcription activity in a given physiological, pathological or developmental process.

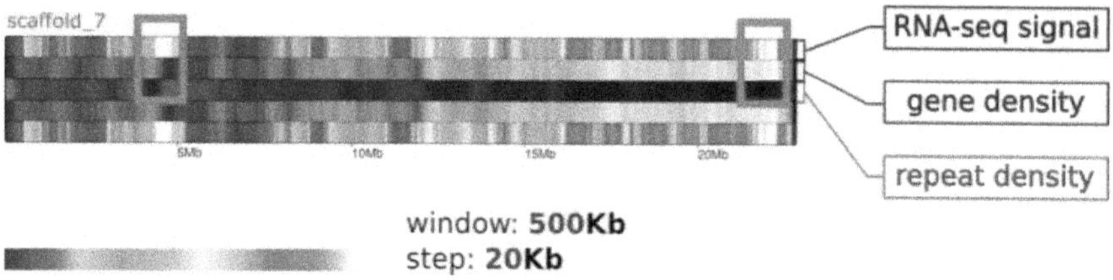

Figure 11-1 Transcript landscape of genes expressed in fruit at stage 3 and mapping on scaffold 7 of the draft of the peach genome. Transcript levels have been measured by counting 16,113,430 Solexa RNAseq reads. 79% could be successfully mapped on the draft of the peach genome. The top and bottom lines (in color, identical) show the RNAseq signal (see scale on the bottom left of the figure, with blue meaning low and white high expression). Gene density is given on the second and fourth lines (grey scale, dark meaning low and white high gene density). In the middle is depicted the repeat density (black means low and purple high), because repeated sequences give a high expression signal (see red window on the left). Islands of highly expressed genes may be considered real if the repeat density is low (red window on the right).

Color image of this figure appears in the color plate section at the end of the book.

11.4 Use of Related Species Resources

One of the advantages of using functional genomics analysis is the capability of comparing data among different species. One of the resources being used on the last couple of years is massive gene expression. The use of microarrays, SAGE, MPSS as well as transcriptome sequencing has allowed the Rosaceae research community to improve their analysis. The sequencing of the peach (International Peach Sequencing Consortium, Sosinsky et al. 2010), apple (Velasco et al. 2010) and strawberry (Shulaev et al. 2011) genomes (all members of the Rosaceae family) are helping to understand genes functionality as well as metabolic pathways. Among the *Prunus* genomes that are sequenced or in the process, we have peach (http://www.rosaceae.org) and sweet cherry (http://genomics.wsu.edu/pages/researchcherry/index.html). The amount of available data increases daily, therefore making easier analysis of gene expression. Today is possible to compare different transcriptomes and look for expression levels as well as genome sequences to identify gene families and orthologs. Table 11-2 shows the total list of plant genomes sequenced to date. There are 14 published genomes sequenced and several in progress. Most of the plant genomes have been sequenced in the last five years. The amount of plant sequences has increased dramatically (NCBI), allowing labs to share information and use it for functional genomics analysis. One of the analyses widely used is transcriptome sequencing. This technology uses the power of massive sequencing to identify a high number of genes being expressed in a particular condition. Table 11-1 shows a comparative analysis among the different number of input and output sequences available for the Plant

Table 11-2 Sequenced plant genomes.

Common name	Specie	Number of chromosomes	Size (Mbp)	Gene number (number of loci)	Paper	Year of Publication	Centre	Link
Thale-cress	*Arabidopsis thaliana*	5	135	27416	http://www.ncbi.nlm.nih.gov/pubmed/11130711	2000	The Arabidopsis Information Resource	http://www.arabidopsis.org
Tomato	*Solanum lycopersicum*	12	950	36974	http://www.ncbi.nlm.nih.gov/pubmed/18629226	2005	International Tomato Genome Sequencing Project; solgenomics.net	http://solgenomics.net/chado/organism.pl?organism_id=1
Black cottonwood	*Populus trichocarpa*	19	403	40668	http://www.sciencemag.org/content/313/5793/1596.abstract	2006	DOE Joint Genome Institute (JGI)	http://genome.jgi-psf.org/poplar/poplar.home.html
Rice	*Oryza sativa*	12	372	40838	http://www.ncbi.nlm.nih.gov/pubmed/17145706	2006	The MSU Rice Genome Annotation Project Database and Resource is a National Science Foundation project	http://rice.plantbiology.msu.edu
Wine grape	*Vitis vinifera*	19	487	26346	http://www.nature.com/nature/journal/v449/n7161/abs/nature06148.html	2007	Consortium for Grapevine Genome Characterization	http://www.genoscope.cns.fr/spip/Vitis-vinifera-e.html
Papaya	*Carica papaya*	9	372	27332	http://www.nature.com/nature/journal/v452/n7190/full/nature06856.html	2008	the papaya Genome sequencing Consortium	http://asgpb.mhpcc.hawaii.edu/papaya

Table 11-2 contd....

Table 11-2 contd....

Common name	Specie	Number of chromosomes	Size (Mbp)	Gene number (number of loci)	Paper	Year of Publication	Centre	Link
Cucumber	*Cucumis sativus, line 9930*	7	367	26682	http://www.nature.com/ng/journal/v41/n12/full/ng.475.html	2009	The cucumber Genome Initiative	http://cucumber.genomics.org.cn/page/cucumber/index.jsp
Maize	*Zea mays line B73*	10	2500	80000	http://www.sciencemag.org/content/326/5956/1112.short	2009	The Maize Genome Sequencing Project	http://www.maizesequence.org
Sorghum	*Sorghum bicolor*	10	730	34496	http://www.nature.com/nature/journal/v457/n7229/full/nature07723.html	2009	DOE-JGI Community Sequencing Program (CSP)	http://www.phytozome.org/sorghum.php
Cultivated apple	*Malus x domestica*	17	742,3	63541	http://www.nature.com/ng/journal/v42/n10/full/ng.654.html	2010	IASMA research center	http://genomics.research.iasma.it
Soybean	*Glicine max*	20	975	66153	http://www.nature.com/nature/journal/v463/n7278/full/nature08670.html	2010	DOE-JGI Community Sequencing Program (CSP)	http://www.soybase.org
Purple false brome	*Brachypodium distachyon*	5	272	25532	http://www.nature.com/nature/journal/v463/n7282/full/nature08747.html	2010	Brachypodium distachyon project	http://www.brachypodium.org
Woodland strawberry	*Fragaria vesca*	7	240	34809	http://www.nature.com/ng/journal/v43/n2/full/ng.740.html	2011	Strawberry Genome Sequencing Consortium	https://strawberry.plantandfood.co.nz

Peach	*Prunus persica*	8	220–230	27864	--		The International Peach Genome Initiative (IPGI)	http://www.phytozome.org/peach.php
Sweet orange	*Citrus sinensis*	9	319	25376	--		Citrus Genomics initiative	http://www.citrusgenomedb.org
Cucumber	*Cucumis sativus, line Gy14*	7	203	21491	--		The cucumber genome project and collaborators	http://genome.jgi-psf.org/cucumber/cucumber.home.html
Potato	*Solanum tuberosum*	12	844	39031	http://www.nature.com/nature/journal/v475/n7355/full/nature10158.html	2011	The Potato Genome Sequencing Consortium (PGSC)	http://www.potatogenome.com

Transcript Assemblies and Plant Gene Indexes (http://plantta.jcvi.org/cgi-bin/plantta_release.pl; http://compbio.dfci.harvard.edu/tgi/plant.htm; accessed September 30, 2011). Tables 11-1 and 11-2 show most of the information that it is available for plants. So far just one *Prunus* genome has been sequenced (peach); however analysis of different transcriptomes will allow to have a better idea of gene conservation between species. Several genes involved in biotic and abiotic stress in peaches were positioned in clusters through the genome after doing analysis of different genomes/transcriptomes (Lee Meisel and Herman Silva, unpublished).

11.5 Impact of Advances in Sequencing Technology

11.5.1 Introduction

While non-sequencing based transcriptomic methods such as microarrays produce meaningful results, they require a priori knowledge of the genes being examined. As advances in sequencing technologies have improved genomics research since 2005, the field of transcriptomics has also been significantly enhanced. The transcriptomic methods previously discussed in this chapter rely either on model system data or sequences developed through Sanger sequencing. Since the advances discussed here are still recent, stone fruit research using the newer methods have yet to be published; therefore, this section focuses on the technological improvements pertaining to transcriptomic research and approaches in other species are presented to provide insight on the future direction of transcriptomics.

11.5.2 Sequencing Technologies and Their Applications for Transcriptomics

During the past decade, sequencing technologies have developed that far exceed the capabilities of Sanger sequencing for many applications. These new systems, or next generation sequencers, have recently been renamed second generation sequencers with the advent of third generation machines.

11.5.2.1 1st Generation

Sanger Sequencing: Incorporation of capillary electrophoresis in Sanger sequencing that uses the chain termination method was the first major advancement in technology. This enhancement reduced the number of reactions per sequence from 4 to 1 and enabled higher throughput than sequencing gels by running the DNA through a capillary tube yielding the flow gram and sequence output. Though the newer technologies have much higher throughput, the high length and quality of Sanger reads remain unmatched. Transcriptomic approaches using Sanger sequencing are limited due to the necessity for individual reactions per gene being investigated.

Sequencing of cDNA libraries has been the primary method to develop ESTs and full-length cDNA sequences. When entire cDNA libraries are sequenced, most of the transcriptome from a tissue is identified. There are, however, limitations of cloning that may result in the loss of non-clonable genes or generate representation bias during colony picking for subsequent sequencing. Since genes have different levels of expression, it can be costly and time consuming to thoroughly investigate a library to the depth required for identification of genes expressed at low levels. This approach was employed on three types of peaches that resulted in the identification of 10,847 unique sequences or genes. This information was further used to identify 1,819 simple sequence repeats (SSRs) that could be used in molecular genetics studies in *Prunus* (Vecchietti et al. 2009). Another method based on Sanger sequencing is differential display (Liang and Pardee 1998) which helps to identify genes that are expressed differently among the tested samples. This method was used in stone fruits for unraveling *Phytoplasma* induced gene expression changes in *Prunus armeniaca* (Carginale et al. 2004).

11.5.2.2 2nd Generation

Second-generation sequencing (SGS) technologies started becoming available in 2005 and have initiated the bulk of the advancements in sequencing based transcriptomics. 454, Illumina, and SOLiD are the three original SGS technologies and each uses a sequencing by synthesis methodology. Though Ion Torrent is a much more recently developed system, the majority of the workflow and technology is very similar to 454 including sequencing by synthesis so it is included here as an SGS technology. The SGS technologies generally have the common themes of creating cDNA, PCR amplifying the cDNA, and sequencing by synthesis. Bias can be introduced during both the cDNA preparation and PCR amplification but have been generally discounted due to a lack of alternatives. This potential bias also increases the need for confirmatory qRT-PCR on any quantitative sequencing results pertaining to a given gene. The high-throughput of these SGS methods has led to easier deciphering of splice junctions, discovery of novel transcripts, quantitative analyses and identification of allelic variation. The methods behind the technologies give each platform unique advantages and disadvantages when comparing their utility for transcriptome analysis and these are described below in order of decreasing sequence read length generated by individual platforms.

454: 454 sequencing is a pyrosequencing method where photo-capture of the light emitted as a result of the nucleotide addition is analyzed and converted into sequence. One million reads with an average length of 400 bases are being produced using the Titanium chemistry, and reads of ~800bp have recently been commercialized. Limitations in the light capture create potential issues with sequencing homopolymer (presence of 7 or more A, T, G or C) runs which can cause open reading frames to be disrupted if not resolved properly. The long read lengths of 454, >400bp, are useful for developing de novo transcriptome assemblies. The major drawback of using 454 for transcriptomic analyses is the relatively low number of reads produced per run.

One of the first uses of 454 for transcriptomics examined prostate cancer cell lines and found 10,000 genes expressed as predicted, 25 novel splicing events and an additional 10,000 novel ESTs with average reads of 102 bp (Bainbridge et al. 2006). The same year, a method for using 454 to sequence the 5′ end of transcripts was developed as an improvement on SAGE based techniques and tested on maize (Gowda et al. 2006). In this experiment, multiple translation start sites were identified in the sequenced transcripts though only 34% of the sequences mapped back to the maize genome at the time of their analysis. Later, a quantitative transcriptome profiling method was developed by selectively sequencing the 3′ UTR (Eveland et al. 2008). In their work, Eveland et al. (2008) identified 14,822 transcripts and found expression differences between the wild type and their mutant as well as expression differences between alleles with up to 98% sequence identity in maize. During the past year, sequencing of the phloem RNAs from ash species yielded 58,673 ESTs of which only 55% had homology in the non-redundant GenBank database (Bai et al. 2011). This study also identified 1,272 SNPs and 980 SSRs of which 7 out of 25 tested SSR loci were polymorphic in the test species. Deep Super SAGE, an ultra-deep sequencing version of Super SAGE, was developed and used to examine chickpea gene expression after salt stress (Molina et al. 2011). This process identified 469 genes that were differentially regulated in both roots and nodules of chickpea 2 hours after salt treatment. Using the 3′UTR strategy, 2,243 SNPs were identified in sweet cherry with many being polymorphic in a screen of some major parental varieties (Koepke et al. unpublished).

Ion Torrent: Similar to the 454 workflow, the Ion Torrent sequencing platform creates up to five million reads of 200 bp using pH detection on a semiconductor instead of an enzymatic cascade to produce light (http://www.iontorrent.com/). Transcriptome and small-RNA sequencing can be completed on this system and the smaller system allows individual projects to be run without requiring coordination with other labs to fill the larger output systems runs. Detailed transcriptome analysis protocols and citations are not yet readily available since the platform became publically accessible in December 2010.

Illumina: The newest Illumina platform, the High-Seq 2000, produces up to six billion reads of 100 bp for paired end sequencing. A major difference between Illumina and Ion Torrent is the bridge PCR performed on the sequencing flow cell which yields clusters of similar sequences compared to the emulsion PCR for amplification with the other SGSs. This platform sequence by synthesis through many cycles of addition of one dye incorporated nucleotide, detection and cleavage of the dye. The transition from GAII to High-Seq 2000 has nearly increased the output 10-fold and further increases in read length are expected. This high sequencing depth typically yields very high confidence in the consensus sequences from Illumina data.

A quantitative method for sequencing cDNA to estimate gene expression was developed and first tested on human liver and kidney RNA samples and showed high correlation to a microarray on the same samples (Marioni et al. 2008). This method has been widely applied and was recently used to dissect transcriptome variations during berry development of grapes (Zenoni et al. 2010). RNA-seq of the grape berries revealed that 6,695 of 17,324 expressed genes exhibit stage specific expression including stage specific alternate splicing of 210 genes. The project of Zenoni et al. also identified 85,870 SNPs in addition to the quantitative profile of gene expression for the berry stages tested. Illumina based RNA-seq has also been modified to specifically sequence miRNAs. One miRNA project found that 249 miRNAs, 236 previously known, are expressed in safflower leaf, seed, and petal and that 116–133 were differentially expressed when comparing the tissues (Li et al. 2011). An alternative to the traditional RNA to cDNA to sequencing paradigm of transcriptome sequencing is flowcell reverse-transcription sequencing, FRT-seq (Mamanova et al. 2010). As the name suggests, the RNA is bound to the flowcell and reverse transcribed into DNA on the flow cell instead of prior to binding. FRT-seq showed a stronger correlation between samples from the same library than the standard RNA-seq protocol.

SOLiD: The SOLiD platform creates a unique file format for the 700 million reads of up to 75 bp which can be obtained in a single run (http://www.appliedbiosystems.com/). Called sequencing by ligation, the ligation of an eight base probe with the first two nucleotides dye labeled to the growing strand, detection of incorporation and the last three bases cleaved is cycled with five different primers to read each base with its 5′ and 3′ neighbor. Since each base in a read is sequenced twice, the accuracy is claimed to be higher than other systems.

A method for transcriptome sequencing, short quantitative random RNA sequencing (SQR-RNA seq), that identified 31–37% of transcripts were not from known exons or predicted genes (Cloonan et al. 2008). SOLiD sequencing of total small RNAs and immunoprecipitated small RNAs from human embryonic stem cells yielded the identification of 818 new miRNAs including many that were less conserved than the consensus miRNA pattern (Goff et al. 2009). For single cell transcriptomic investigations, a protocol was developed that amplifies the transcripts of a cell enough for use with SOLiD's RNA-seq application (Tang et al. 2010). Comparing to microarrays, this method identified 5,260 more genes in the single cell and found 1,753 new splice sites. Additionally, 8–19% of genes known to have multiple isoforms were found to have multiple isoforms expressed in the oocyte or blastomere. These data also revealed that knockout lines of miRNA processing genes showed large expression differences in a majority of transcripts compared to the wild type.

Combinations: Since each system is unique in its strengths and weaknesses, combinations have been applied to supplement each other. Using Illumina and SOLiD, grape transcripts were examined and 401 transcripts showing C to U RNA editing events were identified with 28% of these having significant tissue specific expression differences (Picardi et al. 2010). Compared to the individual platform results, the combined data included a lower percentage of false positives. Using both 454 and Illumina for non-overlapping samples, mitochondrial RNA expression was profiled in mosquito revealing species and stage specific differences but not sex specific changes in expression (Neira-Oviedo et al. 2011).

11.5.2.3 3rd Generation

All the previously available sequencing platforms rely on cDNA preparation and PCR amplification prior to sequencing. Both of these steps introduce biases in interpretation of data related to quantitative expression. Third-generation sequencing (TGS) systems are based on single molecule sequencing. This allows the direct sequencing of RNA molecules, thus removing the cDNA preparation and amplification steps that could introduce bias into in the results. These newest methods aim to use electron microscopy to visualize the actual molecular sequence but are not yet commercialized. The different TGS technologies are reviewed more generally by Shadt (2010) including how they compare to SGS.

PacBio: The PacBio system uses single-molecule real time (SMRT) technology to obtain a maximum of 750,000 reads of ~1kb in total length by anchoring the polymerase to a plate to visualize which dye incorporated bases are added (http://www.pacificbiosciences.com/). While direct RNA sequencing is still in development, single molecule sequencing of cDNAs will not prevent the RNA to cDNA errors but will eliminate the amplification step of SGS thereby reducing the potential for errors in the sequences. The potential for sequencing the RNA directly may reduce bias or loss during cDNA preparation. Research using the PacBio platform for transcriptomics has not yet been published.

Helicos: Helicos offers a ligation free, amplification free process to sequence cDNAs where up to billions of 25–35bp reads are generated in a single run with this true single molecule sequencing (tSMS; http://www.helicosbio.com/). The technology relies on incorporating a fluorescently labeled nucleotide, detecting the addition through light capture, removing the label and repeating. This capture of light from single base additions eliminates the homopolymer problems seen with 454's light capture. Removing bias from ligation and amplification is proposed to improve the quantitative prowess of this method for transcriptome profiling.

To perform transcriptomics on the Helicos machine, single-molecule sequencing direct gene expression (smsDGE) was used on first strand cDNA from yeast (Lipson et al. 2009). smsDGE demonstrated the ability to highly correlate to microarray data and identify both SNPs and splice variants. The next development for Helicos was direct RNA sequencing (DRS) as reported where the cDNA synthesis used for smsDGE was bypassed (Ozsolak et al. 2009). Later, low quantity RNA sequencing (LQ-RNA-seq) was tested on 400ng and 250pg of starting RNA and yielded very similar results between the samples (Ozsolak et al. 2010).

11.5.3 Conclusions

Overall, the sequencing advances of the past decade and the continued progress in sequencing technology are drastically changing the field of transcriptomics. As costs continue to diminish and throughput increases in the coming years, the ability for stone fruit researchers to conduct transcriptomic analyses will facilitate a new level of understanding for important biological processes in these species. With these advances, generating quality transcriptomic data will no longer be the bottleneck in the research pipeline, as shown in other systems (Table 11-3). Data analysis, however, is progressing more slowly and each sequencing platform has its own error types that need to be handled separately.

11.6 Application of Functional Genomics in Genomics-assisted Breeding

Functional genomics encompasses genome scale approaches to define gene function and gene networks at the transcript or protein level. It is also being used in the parlance of testing gene function by transgenic or

Table 11-3 Sequencing platforms used in different organisms for the analysis of transcriptomics.

Transcriptomic Method	Platform	Organism	Citation
Differential display	Sanger	apricot	(Carginale et al. 2004)
EST sequencing	Sanger	peach	(Vecchietti et al. 2009)
RNA-seq	454	human	(Bainbridge et al. 2006)
5'RATE	454	maize	(Gowda et al. 2006)
3' UTR-seq	454	maize	(Eveland et al. 2008)
RNA-seq	454	ash	(Bai et al. 2011)
deepSuperSAGE	454	chickpea	(Molina et al. 2011)
RNA-seq	Illumina	human	(Marioni et al. 2008)
RNA-seq	Illumina	grape	(Zenoni et al. 2010)
miRNA-seq	Illumina	safflower	(Li et al. 2011)
FRT-seq	Illumina	human	(Mamanova et al. 2010)
SQR-RNA-seq	SOLiD	mouse	(Cloonan et al. 2008)
miRNA-seq/Chip-seq	SOLiD	human	(Goff et al. 2009)
Single cell RNA-seq	SOLiD	mouse	(Tang et al. 2010)
RNA-seq	Illumina/ SOLiD	grape	(Picardi et al. 2010)
RNA-seq	454/Illumina	mosquito	(Neira-Oviedo et al. 2011)
smsDGE	Helicos	yeast	(Lipson et al. 2009)
Direct RNA seq	Helicos	yeast	(Ozsolak et al. 2009)
LQ-RNAseq	Helicos	yeast	(Ozsolak et al. 2010)

mutation-generated perturbation of gene expression. Functional genomics is a powerful tool that could be integrated in breeding at various stages. Functional genomics approaches can be used to generate useful molecular markers to increase the efficiency of variety development in a classical breeding program and discover new genes of interest (Martínez-Gómez et al. 2003). Recently, a new approach to reduce the juvenility using transgenic intermediates in members of Rosaceae has been demonstrated (Kean 2010). This is an important development where deployment of functionally validated molecular markers along with early flowering transgenic intermediates holds the potential to accelerate stone fruit breeding.

Progeny selection using molecular markers is particularly important in *Prunus*, as it is in other perennial crops, since they have a long juvenile period. Development of a new variety in peach may take 12 to 15 years, in plums, apricots and almond 13 to 15 years and cherries 14 to 16 years in the confines of a classical breeding program. The use of molecular markers decreases the resources needed to evaluate the segregating progeny (Infante et al. 2011). On the other hand, the markers are also necessary when the expression of the gene is recessive or when the traits are difficult to evaluate such as resistance to abiotic and biotic stress (Luby and Shaw 2001; Scorza 2001).

In the literature there are some examples of molecular markers, microsatellites or SSRs used in peach for genetic study and/or to study quality fruit traits. Among these we can find developmental stage, fruit weight, maturity time, ripening time, skin color, soluble solids, sucrose and titratable acidity (Cipriani et al. 1999; Testolin et al. 2000; Aranzana et al. 2002; Dirlewanger et al. 2002, 2009; Yamamoto et al. 2001, 2002, 2005; Etienne et al. 2002; Verde et al. 2002; Infante et al. 2011). In Japanese plum we can find their use in genetic study (Mnejja 2004); in apricot in genetic study and quality fruit traits (fruit weight, ripening time, skin color soluble solids; Lopes et al. 2002; Hagen et al. 2004; Messina et al. 2004; Ruiz et al. 2008); in almond in genetic study and quality fruit traits (double kernel, shell weight, kernel weight and taste, productivity, ripening time and shell hardiness; Mnejja et al. 2005; Testolin et al. 2004; Sanchez-Perez et al. 2007); in black cherry (Downey and Iezzoni 2000) and in sweet cherry in genetic study and fruit weight (Clarke and Tobutt 2003; Struss et al. 2003; Vaughan and Russell 2004; Zhang et al. 2009; Klagges, Silva and Meisel, unpublished).

The progress in the introduction of useful genes into a *Prunus* species for high-throughput functional genomics will depend on the development of a reliable and efficient transformation system to study stable transformation or functional characterization of promoters in transient transformation assays. In the literature there are examples of stable transformation in almond only for selectable marker genes (Aldwinckle and Malnoy 2009) and its efficiency depends on genotypes and explants tissues, having a transformation efficiency between 0.1 to 2.3% (Miguel and Oliveira 1999; Ainsley et al. 2002; Costa et al. 2006; Ramesh et al. 2006). In apricot also there was some reports of stable transformation for marker genes, with transformation efficiency between 2.1 to 8.5% (Lamier da Camara Machado et al. 1992; Petri et al. 2004; Lopez-Noguera et al. 2006). In sour cherry a report of stable transformation with selectable marker shows differences in the transformation efficiency depending on the genotype used. The range was between 3.1 to 3.9% (Song and Sink 2007). In peach we find reports of stable transformation (Smigochi and Hammerschlag 1991; Perez-Clement et al. 2004; Padilla et al. 2006) and transient transformation (Honda et al. 2006; Spolaore et al. 2001), nevertheless none of them were used to study gene function and the transformation efficiency was less than 4%. In the case of plum, there are many examples of stable transformation (Mante et al. 1991; Scorza et al. 1994; Petri et al. 2008; Urtubia et al. 2008) and transient transformation (Nagel et al. 2008; Tittarelli et al. 2009). This *Prunus* species is the only member with an example of a commercial transgenic variety (Honey Sweet). This variety carries a gene that confers resistance to PPV (plant pox virus) (Scorza et al. 2007). The development of functional genomics in association with the studies of geneticists, molecular biologists and breeders should increase the efficiency of stone fruit breeding research allowing the generation of more interesting varieties adapted to different conditions and consumers.

11.6.1 Development of Robust Molecular Markers Using Gene Expression

So far only few reports describe the study of stone fruit development in peach where microarray approach was used to study change in gene expression profiles (Dardick et al. 2010). Another comprehensive study in peach used EST transcript profiling under different post-harvest conditions to identify candidate genes associated with fruit quality (Vizoso et al. 2009). Similarly, fruit development in apple, another important member of Rosaceae, has been extensively studied using microarrays (Soglio et al. 2009; Botton et al. 2011). While no molecular markers, genetic or expression-based, have been developed yet, it is only a matter of time that molecular markers could be derived from such studies. This statement is supported by a specific gene expression study relating to anthocyanin accumulation in fruits. The transcript levels of the anthocyanin biosynthetic genes were compared in red and white-fleshed apples along with a transcription factor, MdMYB10 (Espley et al. 2007). There was a strong correlation between anthocyanin accumulation and MdMYB10 expression. Further investigations revealed that in the red-fleshed varieties, the presence of five direct tandem repeats of 23bp in length were responsible for ectopic production of anthocyanin (Espley et al. 2009). The above mentioned example provides a case study where expression of a gene led to the development of a molecular marker that can now be deployed in a breeding program to eliminate white fleshed fruit if the program is selecting for red-fleshed progeny. In the previous case, expression of the gene also produces a distinct phenotype that is excessive anthocyanin accumulation in foliage and stems. Such gene-based markers can be of vital use where the phenotype is visible only when the progeny sets fruit. One such example is the use of allelic differences of the endopolygalacturonase gene as an indicator of free stone or melting flesh trait in peach (Peace et al. 2005). Based on the correct allelic association of the gene with the desired trait, progeny can be segregated or eliminated at the seedling stage.

11.6.2 Functional Genomics to Establish Robust Marker-trait Relationships

Model plant systems like *Arabidopsis* and rice have been the largest beneficiary of functional genomics approaches where gene perturbations have been used to establish gene-trait relationships (Hilson et al. 2003; Kant et al. 2008; Tsuchida-Mayama et al. 2010). The advantage of gene perturbations, which includes ectopic expression and gene silencing, is that it provides a direct cause and effect relation between the genotype and phenotype. Within Rosaceae, stone fruits have remained recalcitrant to genetic transformation. However, the efficiency of transformation remains low resulting in lack of utilization of reverse genetics approaches in stone fruits. First, there is a need to develop efficient and reproducible regeneration systems for stone fruits.

11.6.3 The Future of Functional Genomics and Molecular Markers in Stone Fruit Breeding

It has been often proposed that molecular markers can reduce the time to develop new varieties in perennial crops. However, no tangible examples exist. With the recent development of early flowering transgenic plants in stone fruits, molecular markers will be vital to speed up the introgression of desirable traits delivering the promise of time savings. Ralph Scorza's group has recently developed early flowering plums that continue to produce flowers soon after regeneration (Scorza 2011). Similar early flowering transgenic plants have been reported previously in other members of Rosaceae, as apple and pear (Flachowsky et al. 2007; Matsuda et al. 2009). The FasTrack Breeding concept is based on the following experimental premise. An elite cultivar in which a wild type desirable trait needs to be introgressed is transformed with an early or continually flowering gene. The resulting transgenic plants produce pollen in year 2 after the initiation of the experiment and it is used to pollinate the wild type accession. Resulting seeds are expected to have 50% of the genome from the wild type parent and 50% from the elite cultivar. The transgenic seeds are germinated, pollen obtained from them and backcrosses are made with the elite cultivar. Thereafter three more backcrosses are made with the elite cultivar to introgress the desirable trait. All progeny with the early or continually flowering transgene are eliminated since the desirable trait and the transgene are expected to segregate. This cycle can be accomplished within five years compared to 16 years if a 4-year flowering cycle is considered for the plant in a conventional breeding program (Scorza 2011). As is evident, in such an accelerated process, it is very important to have a robust molecular marker that will enable elimination of progeny that do not have the desirable trait. This also necessitates the need for molecular markers with a strong gene-trait relationship. As discussed in section 5, gene expression studies can provide such markers.

In summary, functionally validated molecular markers are an excellent tool to increase efficiency in a conventional stone fruit breeding program.

With the use of continually flowering transgenic intermediates, molecular markers hold even more potential to save resources in terms of finances and time.

Acknowledgements

Research in LT and CB laboratories is founded by grants of MIUR and Ministry of Agriculture. Research on cherry SNPs in AD Lab is supported by the Washington Tree Fruit Research Commission. TK would like to acknowledge support from NIH Protein Biotechnology Graduate Training Program and ARCS Fellowship. HS is founded by Millennium Nucleus in Plant Cell Biotechnology-ICM P06-065-F and INNOVA/CORFO 07CN13PBT167.

References

Ainsley PJ, Collins GG, Sedgley M (2002) Factors affecting *Agrobacterium*-mediated gene transfer and the selection of transgenic calli in paper shell almond (*Prunus dulcis* Mill). J Hort Sci Biotechnol 76: 522–528.

Aldwinckle H, Malnoy M (2009) Plant regeneration and transformation in the Rosaceae. Transgen Plant J 3(1): 1–39.

Aranzana MJ, García-Mas J, Carboa J, Arús P (2002) Development and variability analysis of microsatellite markers in peach. Plant Breed 121: 87–92.

Bai X, Rivera-Vega L, Mamidala P, Bonello P, Herms DA, Mittapalli O (2011) Transcriptomic signatures of ash (*Fraxinus* spp.) phloem. PLoS ONE 6(1): e16368. doi: 10.1371/journal.pone.0016368.

Bainbridge MN, Warren RL, Hirst M, Romanuik T, Zeng T, Go A, Delaney A, Griffith M, Hickenbotham M, Magrini V, Mardis ER, Sadar MD, Siddiqui AS, Marra MA, Jones SJ (2006) Analysis of the prostate cancer cell line LNCaP transcriptome using a sequencing-by-synthesis approach. BMC Genomics 7: 246.

Baxter CJ, Sabar M, Quick WP, Sweetlove LJ (2005) Comparison of changes in fruit gene expression in tomato introgression lines provides evidence of genome-wide transcriptional changes and reveals links to mapped QTLs and described traits. J Exp Bot 56: 1591–1604.

Botton A, Eccher G, Forcato C, Ferrarini A, Begheldo M, Zermiani M, Moscatello S, Battistelli A, Velasco R, Ruperti B, Ramina A (2011) Signaling pathways mediating the induction of apple fruitlet abscission. Plant Physiol 155: 185–208.

Boudehri K, Bendahmane A, Cardinet G, Troadec C, Moing A, Dirlewanger E (2009) Phenotypic and fine genetic characterization of the D locus controlling fruit acidity in peach. BMC Plant Biol 9: 59.

Campalans A, Pagès M, Messeguer R (2001) Identification of differentially expressed genes by the cDNA-AFLP technique during dehydration of almond (*Prunu samygdalus*). Tree Physiol 21: 633–643.

Carginale V, Maria G, Capasso C, Ionata E, La Cara F, Pastore M, Bertaccini A, Capasso A (2004) Identification of genes expressed in response to phytoplasma infection in leaves of *Prunus armeniaca* by messenger RNA differential display. Gene 332: 29–34.

Chen X, Hackett CA, Niks RE, Hedley PE, Booth C, Druka A, Marcel TC, Vels A, Bayer M, Milne I, Morris J, Ramsay L, Marshall D, Cardle L, Waugh R (2010) An eQTL analysis of partial resistance to *Puccinia hordei* in barley. PLoS ONE 5: e8598.

Cipriani G, Lot, G, Huang W, Marrazzo M, Peterlunger E, Testolin R (1999) AC/GT and AG/CT microsatellite repeats in peach [*Prunus persica* (L.) Batsch]: isolation, characterization and cross-species amplification in *Prunus*. Theor Appl Genet 99: 65–72.

Clarke JB, Tobutt KR (2003) Development and characterization of polymorphic microsatellites from *Prunus avium* 'Napoleon'. Mol Ecol Notes 3: 578–580.

Cloonan N, Forrest AR, Kolle G, Gardiner BB, Faulkner GJ, Brown MK, Taylor DF, Steptoe AL, Wani S, Bethel G, Robertson AJ, Perkins AC, Bruce SJ, Lee CC, Ranade SS, Peckham HE, Manning JM, McKernan KJ, Grimmond SM (2008) Stem cell transcriptome profiling via massive-scale mRNA sequencing. Nat Meth 5: 613–619.

Costa MS, Miguel C, Oliveira MM (2006) An improved selection strategy and the use of acetosyringone in shoot induction medium increase almond efficiency by 100-fold. Plant Cell Tiss Org Cult 85: 205–209.

Dardick C, Callahan A, Chiozzotto R, Schaffer R, Piagnani MC, Scorza R (2010) Stone formation in peach fruit exhibits spatial coordination of the lignin and flavonoid pathways and similarity to Arabidopsis dehiscence. BMC Biology 8: 13.

Dirlewanger E, Cardinet G, Boudehri K, Renaud C, Monllor S, Illa E, Howad W, Arús P, Croset C, Poëssel J L, Maucourt M, Deborde C, Moing A (2009) Detection of QTLs controlling major fruit quality components in peach within the European project ISAFRUIT. Acta Hort 814(2): 533–538.

Dirlewanger E, Cosson P, Tavaud M, Aranzana MJ, Poizat C, Zanetto A, Arus P, Laigret F (2002) Development of microsatellite markers in peach [*Prunus persica* (L.) Batsch] and their use in genetic diversity analysis in peach and sweet cherry (*Prunu savium* L.). Theor Appl Genet 105: 127–138.

Downey SL, Iezzoni AF (2000) Polymorphic DNA markers in black cherry (*Prunus serotina*) are identified using sequences from sweet cherry, peach, and sour cherry. J Am Soc Hort Sci 125: 76–80.

Eduardo I, Pacheco I, Chietera G, Bassi D, Pozzi C, Vecchietti A, Rossini L (2010) QTL analysis of fruit quality traits in two peach intraspecific populations and importance of maturity date pleiotropic effect. Tree Genet Genomes 7: 323–335.

Espley RV, Brendolise C, Chagne D, Kutty-Amma S, Green S, Volz R, Putterill J, Schouten HJ, Gardiner SE, Hellens RP, Allan AC (2009) Multiple repeats of a promoter segment causes transcription factor autoregulation in red apples. Plant Cell 21: 168–183.

Espley RV, Hellens RP, Putterill J, Stevenson DE, Kutty-Amma S, Allan AC (2007) Red colouration in apple fruit is due to the activity of the MYB transcription factor, MdMYB10. Plant J 49: 414–427.

ESTree Consortium (2005) Development of an oligo-based microarray (μPEACH 1.0) for genomics studies in peach fruit. Acta Hort 682: 263–268.

Etienne C, Rothan C, Moing A, Plomion C, Bodénès C, Svanella-Dumas L, Cosson P, Pronier V, Monet R, Dirlewanger E (2002) Candidate genes and QTLs for sugar and organic acid content in peach [*Prunus persica* (L.) Batsch]. Theor Appl Genet 105: 145–159.

Eveland AL, McCarty DR, Koch KE (2008) Transcript profiling by 3'-untranslated region sequencing resolves expression of gene families. Plant Physiol 146: 32–44.

Flachowsky H, Peil A, Sopanen T, Elo A, Hanke V (2007) Overexpression of BpMADS4 from silver birch (*Betula pendula* Roth.) induces early-flowering in apple (*Malus* x *domestica* Borkh.). Plant Breed 126: 137–145.

Geuna F, Banfi R, Bassi D (2005) Identification and characterization of transcripts differentially expressed during development of apricot (*Prunus armeniaca* L.) fruit. Tree Genet Genomes 1: 69–78.

Goff LA, Davila J, Swerdel MR, Moore JC, Cohen RI, Wu H, Sun YE, Hart RP (2009) Ago2 immunoprecipitation identifies predicted microRNAs in human embryonic stem cells and neural precursors. Plos One 4(9): e7192.

González-Agüero M, Pavez L, Ibáñez F, Pacheco I, Campos-Vargas R, Meisel LA, Orellana A, Retamales J, Silva H, González M, Cambiazo V (2008) Identification of woolliness response genes in peach fruit after post-harvest treatments. J Exp Bot 59: 1973–1986.

Gowda M, Li H, Alessi J, Chen F, Pratt R, Wang GL (2006) Robust analysis of 5′-transcript ends (5′-RATE): a novel technique for transcriptome analysis and genome annotation. Nucl Acids Res 34(19): e126. doi: 10.1093/nar/gkl522.

Grimplet J, Romieu C, Audergon J, Marty I, Albagnac G, Lambert P, Bouchet J, Terrier N (2005) Transcriptomic study of apricot fruit (*Prunus armeniaca*) ripening among 13,006 expressed sequence tags. Physiol Plant 125: 281–292.

Hagen LS, Chaib J, Fady B, Decroocq V, Bouchet JP, Lambert P, Audergon JM (2004) Genomic and cDNA microsatellites from apricot (*Prunus armeniaca* L.). Mol Ecol Notes 4: 742–745.

Hammond JP, Mayes S, Bowen HC, Graham NS, Hayden RM, Love CG, Spracklen WP, Wang J, Welham SJ, White PJ, King GJ, Broadley MR (2011) Regulatory hotspots are associated with plant gene expression under varying soil phosphorus (P) supply in *Brassica rapa*. Plant Physiol 156: 1230–1241.

Hansen BG, Halkier BA, Kliebenstein DJ (2008) Identifying the molecular basis of QTLs: eQTLs add a new dimension. Trends Plant Sci 13: 72–77.

Hayama H, Shimada T, Yamamoto T, Iketani H, Matsuta N, Yoshioka H, Hayashi T (2000) Characterization of randomly obtained cDNAs from peach fruits at various developmental stages. J Jpn Soc Hort Sci 69: 183–185.

Hilson P, Small I, Kuiper MTR (2003) European consortia building integrated resources for Arabidopsis functional genomics. Curr Opin Plant Biol 6: 426–429.

Honda C, Moriguchi T (2006) High Gus expression in protoplasts isolated from immature peach fruits. Sci Hort 109: 244–247.

Horn R, Lecouls A-C, Callahan A, Dandekar A, Garay L, McCord P, et al. (2005) Candidate gene database and transcript map for peach, a model species for fruit trees. Theor Appl Genet 110: 1419–1428.

Infante R, Martinez-Gomez P, Predieri S (2011) Breeding for fruit quality in *Prunus*. In: Jenks M, Bebeli P (eds) Breeding for Fruit Quality. Wiley-Blackwell, Oxford, pp 201–229.

Jiang Y-Q, Ma R-C (2003) Generation and analysis of expressed sequence tags from almond (*Prunus dulcis* Mill.) pistils. Sex Plant Reprod 16: 197–207.

Jung S, Jesudurai C, Staton M, Du Z, Ficklin S, Cho I, Abbott A, Tomkins J, Main D (2004) GDR (Genome Database for Rosaceae): integrated web resources for Rosaceae genomics and genetics research. BMC Bioinformatics 5: 130.

Kant P, Gordon M, Kant S, Zolla G, Davydov O, Heimer YM, Chalifa-Caspi V, Shaked R, Barak S (2008) Functional-genomics-based identification of genes that regulate Arabidopsis responses to multiple abiotic stresses. Plant Cell Environ 31(6): 697–714.

Kean S (2010) Besting Johnny appleseed. Science 328: 301–303.

Laimer da Camara Machado M, da Camara Machado A, Hanzer V, Weiss H, Regner F, Steinkellner H, Mattanovich D, Plail R, Knapp E, Kalthoff B, Katinger H (1992) Regeneration of transgenic plants of *Prunus armeniaca* containing the coat protein gene of plum pox virus. Plant Cell Rep 11: 25–29.

Leida C, Terol J, Martí G, Agustí M, Llácer G, Badenes ML, Ríos G (2010) Identification of genes associated with bud dormancy release in *Prunus persica* by suppression subtractive hybridization. Tree Physiol 30: 655–666.

Li H, Deng H (2010) Systems genetics, bioinformatics and eQTL mapping. Genetica 138: 915–924.

Li X, Shangguan L, Song C, Wang C, Gao Z, Yu H, Fang J (2010) Analysis of expressed sequence tags from *Prunus mume* flower and fruit and development of simple sequence repeat markers. BMC Genetics 11: 66.

Li H, Dong Y, Sun Y, Zhu E, Yang J, Liu X, Xue P, Xiao Y, Yang S, Wu J, Li X (2011) Investigation of the microRNAs in safflower seed, leaf, and petal by high-throughput sequencing. Planta 233(3): 611–619.

Liang P, Pardee AB (1998) Differential display—A general protocol. Mol Biotechnol 10(3): 261–267.

Lipson D, Raz T, Kieu A, Jones DR, Giladi E, Thayer E, Thompson JF, Letovsky S, Milos P, Causey M (2009) Quantification of the yeast transcriptome by single-molecule sequencing. Nat Biotechnol 27: 652-U105.

Lopes MS, Sefc KM, Laimer M, Da Camaro Machado A (2002) Identification of Microsatellite loci in apricot. Mol Ecol Notes 2: 24–26.

Lopez-Noguera S, Petri C, Burgos L (2006) Production of marker-free transgenic plants after transformation of apricot cultivars. Acta Hort 717: 225–227.

Luby J, Shaw D (2001) Does marker-assisted selection makes dollars and sense in a fruit breeding program? HortScience 36: 872–879.

Mamanova L, Andrews RM, James KD, Sheridan EM, Ellis PD, Langford CF, Ost TW, Collins JE, Turner DJ (2010) FRT-seq: amplification-free, strand-specific transcriptome sequencing. Nat Meth 7(2): 130–132.

Manganaris GA, Rasori A, Bassi D, Geuna F, Ramina A, Tonutti P, Bonghi C (2011) Comparative transcript profiling of apricot (*Prunus armeniaca* L.) fruit development and on-tree ripening. Tree Genet Genomes 7: 609–616.

Mante S, Morgens PH, Scorza R, Cordts JM, Callahan AM (1991) *Agrobacterium*-mediated transformation of plum (*Prunus domestica* L.) hypocotyl slices and regeneration of transgenic plants. Bio/Technology 9: 853–857.

Marioni JC, Mason CE, Mane SM, Stephens M, Gilad Y (2008) RNA-seq: An assessment of technical reproducibility and comparison with gene expression arrays. Genome Res 18(9): 1509–1517.

Martínez-Gómez P, Crisosto CH, Bonghi C, Rubio M (2011) New approaches to Prunus transcriptome analysis. Genetica 139: 755–769.

Martínez-Gómez P, Sozzi G, Sánchez-Pérez R, Rubio M, Gradziel TM (2003) New approaches to *Prunus* tree crop breeding. Food Agri Environ 1: 52–63.

Matsuda N, Ikeda K, Kurosaka M, Takashina T, Isuzugawa K, Endo T, Omura M (2009) Early flowering phenotype in transgenic pears (*Pyrus communis* L.) expressing the CiFTgene. J Jpn Soc Hort Sci 78: 410–416.

Messina R, Lain O, Marrazzo T, Cipriani G, Testolin R (2004) New set of microsatellite loci isolated in apricot. Mol Ecol Notes 4: 432–434.

Miguel CM, Oliveira MM (1999) Transgenic almond (*Prunus dulcis* Mills) plants obtained by Agrobacterium-mediated transformation of leaf explants. Plant Cell Rep 18: 387–393.

Mnejja M J, Garcia-Mas W, Howad L, Arús P (2005) Development and transportability across Prunus species of 42 polymorphic almond microsatellites. Mol Ecol Notes 5: 531–535.

Mnejja MJ, Garcia-Mas W, Howad L, Badenes ML, Arús, P (2004) Simple-sequence repeat (SSR) markers of Japanese plum (*Prunus salicina* Lindl.) are highly polymorphic and transferable to peach and almond. Mol Ecol Notes 4: 163–166.

Molina C, Zaman-Allah M, Khan F, Fatnassi N, Horres R, Rotter B, Steinhauer D, Amenc L, Drevon JJ, Winter P, Kahl G (2011) The salt-responsive transcriptome of chickpea roots and nodules via deep Super SAGE. BMC Plant Biol 11: 31.

Nagel A, Schnabel G, Petri C, Scorza R (2008) Generation and characterization of transgenic plum lines expressing the gastrodia antifungal protein. HortScience 43: 1514–1521.

Neira-Oviedo M, Tsyganov-Bodounov A, Lycett GJ, Kokoza V, Raikhel AS, Krzywinski J (2011) The RNA-Seq approach to studying the expression of mosquito mitochondrial genes. Insect Mol Biol 20: 141–152.

Ogundiwin E, Martí C, Forment J, Pons C, Granell A, Gradziel T, Peace C, Crisosto C (2008) Development of ChillPeach genomic tools and identification of cold-responsive genes in peach fruit. Plant Mol Biol 68: 379–397.

Ogundiwin E, Peace C, Gradziel T, Parfitt D, Bliss F, Crisosto C (2009) A fruit quality gene map of *Prunus*. BMC Genomics 10: 587.

Ozsolak F, Goren A, Gymrek M, Guttman M, Regev A, Bernstein BE, Milos PM. (2010) Digital transcriptome profiling from attomole-level RNA samples. Genome Res 20: 519–525.

Ozsolak F, Platt AR, Jones DR, Reifenberger JG, Sass LE, McInerney P, Thompson JF, Bowers J, Jarosz M, Milos PM (2009) Direct RNA sequencing. Nature 461: 814–818.

Padilla IMG, Golis A, Gentile A, Damiano C, Scorza R (2006) Evaluation of transformation in peach *Prunus persica* explants genes. Plant Cell Tiss Org Cult 84: 309–314.
Peace CP, Crisosto CH, Gradziel TM (2005) Endopolygalacturonase: A candidate gene for Freestone and Melting flesh in peach. Mol Breed 16: 21–31.
Perez-Clement RM, Perez-Sanjuan A, Garcia-Ferriz L, Beltran JP and Canas A (2004) Transgenic peach plants (*Prunus persica* L.) produced by genetic transformation of embryo sections using green fluorescent protein (GFP) as an *in vitro* marker. Mol Breed 14: 419–417.
Petri C, Alburquerque N, Garcia-Castillo S, Egea J, Burgos I (2004) Factors affecting gene transfer efficiency to apricot leaves during early *Agrobacterium*-mediated transformation steps. J Hort Sci Biotechnol 79: 704–712.
Petri C, Webb K, Hily JM, Dardick C, Scorza R (2008) High transformation efficiency in plum (*Prunus domestica* L.) a new tool for functional genomics studies in *Prunus* spp. Mol Breed 22: 581–591.
Picardi E, Horner DS, et al. (2010) Large-scale detection and analysis of RNA editing in grape mtDNA by RNA deep-sequencing. Nucl Acids Res 38: 4755–4767.
Ramesh SA, Kaise BN, Francks T, Collins G, Sedgley, M (2006) Improved methods in Agrobacterium-mediated transformation of almonds using positive (mannose pmi) or negative (Kan resistance) selection based protocols. Plant Cell Rep 25: 821–828.
Ruiz D, Lambert P, Audergon JM, Bureau B, Gouble B, Reich M, Renard CM, Dondini L, Tartarini S, Adami M, Gennari F, Cervellati C, De Franceschi P, Sansavini S, Bassi D, Testolin R (2008) Identification of QTLs for fruit quality traits in apricot. Acta Hort 862: 587–590.
Sánchez-Pérez R, Howad W, Dicenta F, Arús P, Martínez-Gómez P (2001) Mapping major genes and quantitative trait loci controlling agronomic traits in almond. Plant Breed 126: 310–318.
Schadt EE, Turner S, Kasarskis A (2010) A window into third-generation sequencing. Hum Mol Genet 19: 227–40.
Schaffer RJ, Friel EN, Souleyre EJF, Bolitho K, Thodey K, Ledger S, Bowen JH, Ma J-H, Nain B, Cohen D, Gleave AP, Crowhurst RN, Janssen BJ, Yao J-L, Newcomb RD (2007) A genomics approach reveals that aroma production in apple is controlled by ethylene predominantly at the final step in each biosynthetic pathway. Plant Physiol 144: 1899–1912.
Scorza R (2011) FasTracking Plum Breeding: USDA-ARS (http://www.ars.usda.gov/is/AR/2011/mar11/breeding0311.htm).
Scorza R (2001) Progress in tree improvement thought molecular genetics. HortScience 36: 855–857.
Scorza R, Hily JM, Callahan A, Matinowski T, Cambra M, Capote N, Zagra I, Damsteegt V, Briard P, Ravelonandro M (2007) Deregulation of plum pox resistant transgenic plum 'Honey Sweet'. Acta Hort 738: 669–673.
Scorza R, Ravelonandro M, Callahan A, Cordts JM, Fuchs M, Dunez J, Gonsalves D (1994) Transgenic plums (*Prunus domestica* L.) express the plum pox virus coat protein gene. Plant Cell Rep 14: 18–22.
Shulaev V, Sargent DJ, Crowhurst RS, Mockler T, Veilleux RE, Folkerts O, et al. (2010) The genome of woodland strawberry (*Fragaria vesca*). Nat Genet 43: 109–116.
Smigocki AC, Hammerschlag FA (1991) Regeneration of plants from peach embryo cells infected with a shoot mutant strain of *Agrobacterium*. J Am Soc Hort Sci 116: 1092–1097.
Soglio V, Costa F, Molthoff JW, Weemen-Hendriks WMJ, Schouten HJ, Gianfranceschi L (2009) Transcription analysis of apple fruit development using cDNA microarrays. Tree Genet Genomes 5: 685–698.
Song GK, Sink KC (2007) Transformation of cherry: *Prunus cerasus* L. Montmorency and *P. cerasus* x *P. canescens* Gisela 6 mediated by *Agrobacterium tumefaciens* and a two-step selection system. Acta Hort 738: 683–689.
Sosinski B, Verde I, Morgante M, Rokhsar D (2010) The International Peach Genome Initiative. A first draft of the peach genome sequence and its use for genetic diversity analysis in peach. In: 5th Intl Rosaceae Genomics Conf, Nov 2010, Stellenbosch, South Africa, O46.

Spolaore S, Trainotti L, Casadoro G (2001) A simple protocol for transient gene expression in ripe fleshy fruit mediated by *Agrobacterium*. J Exp Bot 52: 845–850.

Struss D, Ahmad R, Southwick S (2003) Analysis of sweet cherry (*Prunus avium* L.) cultivars using SSR and AFLP markers. J Am Soc Hort Sci 128: 904–909.

Tadiello A, Pavanello A, Zanin D, Caporali E, Colombo L, Rotino GL, Trainotti L, Casadoro G (2009) A PLENA-like gene of peach is involved in carpel formation and subsequent transformation into a fleshy fruit. J Exp Bot 60: 651–661.

Tang F, Barbacioru C, Nordman E, Li B, Xu N, Bashkirov VI, Lao K, Surani MA (2010) RNA-Seq analysis to capture the transcriptome landscape of a single cell. Nat Protoc 5: 516–535.

Testolin R, Marrazzo T, Cipriani G, Quarta R, Verde I, Dettori T, Pancaldi M, Sansavini S (2000) Microsatellite DNA in peach (*Prunus persica* L. Batsch) and its use in fingerprinting and testing the genetic origin of cultivars. Genome 43: 512–520.

Testolin R, Messina R, Lain O, Marrazzo MT, Huang WG, Cipriani G (2004) Microsatellites isolated in almond from an AC-repeat enriched library. Mol Ecol Notes 4: 459–461.

Tittarelli A, Santiago M, Morales A, Meisel L, Silva H (2009) Isolation and functional characterization of cold-regulated promoters, by digitally identifying peach fruit cold-induced genes from large EST dataset. BMC Plant Biol 9: 121–136.

Trainotti L, Bonghi C, Ziliotto F, Zanin D, Rasori A, Casadoro G, Ramina A, Tonutti P (2006) The use of microarray μPEACH1.0 to investigate transcriptome changes during transition from pre-climacteric to climacteric phase in peach fruit. Plant Sci 170: 606–613.

Trainotti L, Tadiello A, Casadoro G (2007) The involvement of auxin in the ripening of climacteric fruits comes of age: the hormone plays a role of its own and has an intense interplay with ethylene in ripening peaches. J Exp Bot 58: 3299–3308.

Trainotti L, Zanin D, Casadoro G (2003) A cell wall-oriented genomic approach reveals a new and unexpected complexity of the softening in peaches. J Exp Bot 54: 1821–1832.

Tsuchida-Mayama T, Nakamura H, Hakata M, Ichikawa H (2010) Rice transgenic resources with gain-of-function phenotypes. Breed Sci 60: 493–501.

Urtubia C, Devia J, Castro A, Zamora P, Aguirre C, Tapia E, Barba P, Dell'Orto P, Moynihan MR, Petri C, Scorza R, Prieto H (2008) *Agrobacterium*-mediated genetic transformation of *Prunus salicina*. Plant Cell Rep 27: 1333–1340.

Vaughan S, Russell K (2004) Characterization of novel microsatellites and development of multiplex PCR for large-scale population studies in wild cherry, *Prunus avium*. Mol Ecol Notes 4: 429–431.

Vecchietti A, Lazzari B, Ortugno C, Bianchi F, Malinverni R, Caprera A, Mignani I, Pozzi C (2008) Comparative analysis of expressed sequence tags from tissues in ripening stages of peach (*Prunus persica* L. Batsch). Tree Genet Genomes 5: 377–391.

Velasco R, Zharkikh A, Affourtit J, Dhingra A, Cestaro A, Kalyanaraman A, et al. (2010) The genome of the domesticated apple (*Malus × domestica* Borkh.). Nat Genet 42: 833–841.

Vendramin E, Dettori MT, Giovinazzi J, Micali S, Quarta R, Verde I (2007) A set of EST-SSRs isolated from peach fruit transcriptome and their transportability across *Prunus* species. Mol Ecol Notes 7: 307–310.

Verde I, Quarta R, Cedrola C, Dettori MT (2002) QTL analysis of agronomic traits in a bc1 peach population. Acta Hort (ISHS) 592: 291–297.

Vizoso P, Meisel L, Tittarelli A, Latorre M, Saba J, Caroca R, Maldonado J, Cambiazo V, Campos-Vargas R, Gonzalez M, Orellana A, Silva H (2009) Comparative EST transcript profiling of peach fruits under different post-harvest conditions reveals candidate genes associated with peach fruit quality. BMC Genomics 10: 423.

West MAL, Kim K, Kliebenstein DJ, van Leeuwen H, Michelmore RW, Doerge RW, Clair DAS (2007) Global eQTL mapping reveals the complex genetic architecture of transcript-level variation in *Arabidopsis*. Genetics 175: 1441–1450.

Yamamoto T, Mochida K, Imai T, Shi Y, Ogiwara I, Hasashi T (2002) Microsatellite markers in peach (*Prunus persica* L. Batsch) derived from an enriched genomic and cDNA libraries. Mol Ecol Notes 2: 298–301.

Yamamoto T, Shimada T, Imai T, Yaegaki H, Haji T, Matsuta N, Yamaguchi M, Hayashi T (2001) Characterization of morphological traits based on a genetic linkage map in peach. Breed Sci 51: 271–278.

Yamamoto T, Yamaguchi M, Hasashi T (2005) An integrated genetic linkage map of peach by SSR, STS, AFLP and RAPD. J Jpn Soc Hort Sci 74: 204–213.

Zenoni S, Ferrarini A, Giacomelli E, Xumerle L, Fasoli M, Malerba G, Bellin D, Pezzotti M, Delledonne M (2010) Characterization of transcriptional complexity during berry development in *Vitis vinifera* using RNA-Seq. Plant Physiol 152: 1787–1795.

Zhang G, Sebolt AM, Sooriyapathirana S, Wang D, CAM Bink M, Olmstead J, Iezzoni A (2009) Fruit size QTL analysis of an F_1 population derived from a cross between a domesticated sweet cherry cultivar and a wild forest sweet cherry. Tree Genet Genomes 6 (1): 25–36.

Ziliotto F, Begheldo M, Rasori A, Bonghi C, Ramina A, Tonutti P (2005) Molecular and genetic aspects of ripening and qualitative traits in peach and nectarine fruits. Acta Hort 682: 237–246.

Ziliotto F, Begheldo M, Rasori A, Bonghi C, Tonutti P (2008) Transcriptome profiling of ripening nectarine (*Prunus persica* L. Batsch) fruit treated with 1-MCP. J Exp Bot 59: 2781–2791.

Ziosi V, Bonghi C, Bregoli AM, Trainotti L, Biondi S, Sutthiwal S, Kondo S, Costa G, Torrigiani P (2008) Jasmonate-induced transcriptional changes suggest a negative interference with the ripening syndrome in peach fruit. J Exp Bot 59: 563–573.

12

Proteomics and Metabolomics

María V. Lara[1] and *María F. Drincovich*[1,*]

ABSTRACT

"Proteomics" and "metabolomics" are the most recently developed "omic" technologies used to describe biological systems. They are used to identify the entire suite of proteins and metabolites of a sample, developing the so-called *protein profiling* and *metabolic profiling*, respectively. Both technologies have been largely improved over the last years and make use of the same steps: experimental design, protein/metabolite extraction and preparation, protein/metabolite separation, protein/metabolite identification and data analysis and integration. Even though these technologies have been largely used in plants, their application to stone fruits has been just recently performed. In the present chapter, the basics of "proteomics" and "metabolomics" are briefly described; as well as their application to stone fruits. Specifically, the results obtained when applying these technologies to the analysis of the development, ripening and postharvest of peach fruit have lead to the identification of key proteins and metabolites involved in, e.g., the tolerance to chilling injury. Moreover, the results obtained indicate particular metabolic programs for peach fruit, which are different when comparing to other types of fruits. Therefore, the extension of these studies to other stone fruits, as well as under different developmental or environmental situations, together with the interpretation and integration of the large datasets derived, will help in modeling the molecular networks related to traits of agronomic interest in stone fruits.

Keywords: networks, fruit improvement, fruit quality, storage condition, metabolic processes, stress response, ripening

[1]Center of Photosynthetic and Biochemical Studies (CEFOBI), Faculty of Pharmaceutical and Biochemical Sciences, Rosario National University, Suipacha 531, Rosario, Argentina.
*Corresponding author: *drincovich@cefobi-conicet.gov.ar*

12.1 Introduction

Among the "omic" technologies, "proteomics" and "metabolomics" are the most recently developed strategies used to describe biological systems. These technologies, when applied to the comprehensive screening of proteins and metabolites on a given cell, tissue or organ, and their comparison to the screening under different developmental or environmental conditions or in response to genetic perturbations, are useful to identify components involved in the gene-phenotype relationships. The identification of these components can be applied, among others, to molecular breeding, metabolic engineering or to the elucidation of stress response pathways. When applied to fruit, these components may also help in defining factors involved in fruit quality, such as flavor and color; as well as in the identification of biomarkers, health promoting compounds, or factors involved in better storage conditions, all of which have important implications for better fruit management and marketing. Although "proteomics" and "metabolomics" technologies have been widely applied to different plant species and fruits since several years ago, their application to stone fruits has been just recently performed. In this chapter, a brief summary of the basics of "proteomics" and "metabolomics" are described, followed by the analysis of the results obtained when applying these technologies to stone fruits, especially peach. A broader application of these technologies to different stone fruits, as well as under different developmental or environmental situations, is needed in order to obtain larger volume of data on protein and metabolite fluctuations that must be interpreted and integrated in a manner that could significantly contribute to fruit improvement. This data integration in stone fruit, with that already obtained on transcriptome, will facilitate the construction of networks and the modeling of key processes of this fruit, which is a challenge for their application on stone fruit improvement.

12.2 Proteomics

The term "proteomics" has been most ambiguously used to encompass the discipline of protein research which includes almost any aspect of protein expression, structure or function and even protein-protein and protein-ligand interactions. Nevertheless, it is now well established in the lexicon of plant biologists (Rose et al. 2004). Therefore, as defined here it relates to the systematic study of the full complement of the proteins expressed by the genome of a cell, a tissue, an organ, an organelle, or an organism at a specific point of time. Proteomics provides an essential bridge between the transcriptome and the metabolome (Renaut et al. 2006).

Proteomics studies can be divided into the *protein profiling* of biological material to separate, sequence and catalog as many proteins as possible so as to build a "map" identifying the entire suite of proteins of a sample, and in the *comparative proteomics*, which search for differences among different protein populations differing both in protein quality and quantity (Rose et al. 2004). Therefore, in this later approach, most commonly referred to as differential-expression proteomics, it is aimed at protein identification and relative quantitation (Kosová et al. 2011). Similarly to comparative DNA microarray profiling, comparative proteomics studies usually include analysis of the proteome of a same plant material in non-stress vs. stress conditions or the study of the response of two different genotypes with contrasting levels of tolerance to a given stress factor or the evaluation of proteins from wild type vs. mutant plants or the evaluation of tissues at different developmental stages, among others.

Even though the coining of the word "proteomics" was simultaneous with the onset of practicable mass spectrometry methods in the field, it was not the true birth of proteomics. A detailed historical review on proteomics from its birth to its current state is provided by Rabilloud and colleagues (2010). Following, there is a brief summary. While sodium dodecyl sulfate (SDS)-electrophoresis was introduced in the early 70s by Laemmli (1970), it was some years later that successful two-dimensional electrophoresis (2-DE) was achieved by coupling denaturing isoelectric focusing (IEF) to SDS-polyacrylamide gel electrophoresis (PAGE) (MacGillivray and Rickwood 1974; O'Farrell 1975). Since then, major efforts have been underway for several years to improve the study of proteins through 2-DE. With the introduction of the use of immobilized pH gradient (IPG) strips in the last 80s (Gorg et al. 1987) there was a real quantum leap in reproducibility in 2-DE (Rabilloud et al. 2010). But it was in the 90s that proteomics really took off by the introduction of mass spectrometry-based methods for protein identification and generating "peptide mass fingerprinting" (PMF) (Henzel et al. 1993).

12.2.1 Proteomics: The Basics

Since over the last years many reviews specifically dedicated to plant proteomics have been written, including conceptual and specific technical details, as well as describing online plant proteomics and mass spectrometry-related resources (Rossignol 2001; van Wijk 2001; Rabilloud 2002; Heazlewood and Millar 2003; Canovas et al. 2004; Rose et al. 2004; Bienvenut 2005; Rabilloud et al. 2010) it is not the purpose of this chapter to describe and review all these aspects, rather than to focus on the results obtained in fruit development and postharvest management, especially in stone fruits. Nevertheless a quick overview to the proteomic approach will be given.

Briefly, proteomics analysis involves the following steps: experimental design, protein extraction and preparation, protein separation, protein identification and data analysis and integration. Protein separation can be achieved by one- or two-dimensional electrophoresis (1-DE and 2-DE, respectively) or liquid chromatography (LC). Proteomic studies uses mass spectrometry (MS) based methods to profile the proteome. The most critical steps in any proteomics study nowadays are no longer in the electrophoretic procedure per se, but rather upstream and downstream, i.e., protein extraction and sample preparation, and image production and analysis (Rose et al. 2004; Rabilloud et al. 2010).

Protein extraction and preparation is a critical step, the protocol to be used should avoid post-extraction artifacts and non-proteinaceous contaminants such as polysaccharides, polyphenolics, pigments, nucleic acids, lipids and a broad array of secondary metabolites (Newton et al. 2004). Depending on the purpose of the research, a subcellular fractionation step may be included, in which case contamination with protein from other compartments must be avoided. In consequence, a wide range of protocols varying in solvents, buffers, physical treatments, precipitating agents and number of step have been developed.

Protein separation is most frequently achieved by 2-DE and in the case of comparative proteomics it has been the method of choice for several decades (Rabilloud 2002). Proteins are separated according to their charge in the first dimension by the use of immobilized pH gradient (IPG) strips, followed by a molecular mass based separation through SDS-PAGE in the second dimension. The main advantages of 2-DE are its robustness, its parallelism, and its unique ability to analyze complete proteins at high resolution with all their modifications (Rabilloud et al. 2010). The use of multiplexed electrophoresis by differentially labeling the samples with different flourophores before migration has greatly increased the reproducibility and precision of the technique (Unlu et al. 1997). The robustness has been tested thoroughly in inter- and intra-laboratory comparisons (Corbett et al. 1994; Blomberg et al. 1995; Challapalli et al. 2004). Modern equipment allows parallel running of 12 and up to 20 gels, enabling the running of biological replicates and thus increasing statistical confidence (Rabilloud et al. 2010). On the contrary, the main weakness of 2-DE is its very low efficiency in the analysis of hydrophobic proteins. In consequence, to study hydrophobic and/or membrane proteins the blue-native PAGE (BN-PAGE) has been developed (Schagger and von Jagow 1991).

After gel electrophoresis, subsequent protein staining—for which a range of stains with different sensitivity are available—and image acquisition and editing allow protein detection, annotation and quantification. Although the different software packages available have been largely improved, this step is still quite time-consuming and requires substantial manual intervention

(Rose et al. 2004). The introduction of the difference in-gel electrophoresis (DIGE, Ettan™ DIGE system, GE Healthcare, Buckinghamshire, UK) technique in 1997 by Unlu and colleagues represented a step forward facilitating comparisons of protein. By this approach the different protein extracts to be compared are covalent labeled with different fluorescent dyes, then mixed and separated on the same 2-DE gel. Images are acquired by scanning on a variable wavelength laser-based imaging system. 2D-DIGE broadens the dynamic range of the analysis by detecting more low abundant proteins without saturating the highly abundant ones and provides quantitative results (Renaut et al. 2006).

After identification of the differentially expressed spots, the next step involves protein identification by MS. For this purpose, selected spots containing proteins differentially expressed are picked and in-gel digested with a site-specific protease (commonly trypsin). Eluted peptides are then identified. Currently, protein identification is carried out based on spectra generation either by matrix-assisted laser desorption-ionization time-of-flight mass spectrometry (MALDI-TOF/MS) (Hillenkamp and Karas 1990) or/and electrospray ionization tandem mass spectrometry (ESI MS/MS) (Fenn et al. 1989). One of the fundamental differences between these methods is that MALDI is employed on samples in a solid state, whereas ESI is employed on samples in a liquid state. In the case of MALDI, the energy of a laser is absorbed by the matrix and causes the ionization of the analytes. MALDI is typically used to measure the masses of the peptides derived from the trypsinized parent protein spot, generating a "peptide mass fingerprint" (PMF). By MALDI-TOF/MS analysis, the identity of a protein is assigned through the in silico match of experimentally determined peptide PMF vs. predicted peptide masses of tryptic peptide fragments for every protein in the public databases, including equivalent translated genomic and expressed sequence tag (EST) databases, rather than actual amino acid sequence. Nevertheless, in some cases it is difficult to distinguish among members of protein families with a high degree of sequence similarity since they may generate indistinguishable PMFs. MALDI-TOF/PMF approach is very rapid, requires relatively little user expertise, can be automated and is tolerant of contaminants. Problems arise when there are no data regarding genome sequence or at least EST collections. In consequence, peptide analysis using de novo sequencing by ESI-MS/MS is the method of choice for unsuccessful PMF protein identification (Hirano et al. 2004). By this approach sequencing of the peptide by MS/MS yields amino acid sequences of selected tryptic peptides. ESI system ionization is based on a high electric potential applied to a capillary needle in which the sample is passing through. This produces a mist of highly charged droplets. After evaporation of the solvent, ions will reach the detector of the mass spectrometer (Gaskell 1997). In short, it includes ionization of a sample and separation based upon the mass-to-charge ratio (m/z) of the primary ions. Then an ion with a specific m/z value is selected, fragmented, and the fragment ions detected after passing through the second mass spectrometer. Therefore, a series of fragment ions differing by single amino acids are produced, allowing a portion of the peptide sequence, termed an "amino acid sequence tag", to be determined and used for database searching. When a peptide has been identified in the database, the theoretical fragmentation pattern can be predicted and compared with the observed MS/MS spectrum for assignment of other peaks that can validate the identification. This procedure can be repeated for every fragmented peptide in the sample. As it can be inferred, MS/MS analysis is technically more challenging, requires specialist training, has a slower throughput and is far more expensive.

At the present time, characterizing individual proteins by MS is often considered to be a straightforward exercise because of the high mass accuracy and sensitivity of modern mass spectrometers, which are now able to conduct a range of procedure combinations such as MALDI-TOF/TOF and tandem quadrupole/time-of-flight mass spectrometer with a matrix-assisted laser desorption/ionization source (MALDI-QqTOF) (Loboda et al. 2000; Bienvenut 2005).

When conventional approaches such as 2D-DIGE fail, the more recently one- or two-dimension liquid chromatography (LC) is a good replacement for the gel separation. In this respect, LC is becoming the favored method for the initial separation. For ESI and similar methods, the LC is coupled directly to the ion source, and the fractions are examined sequentially as they elute from the LC. For desorption methods, on the other hand, the most natural form of coupling is offline; and the technique is known as HPLC-MALDI-qTOF-MS (Krokhin et al. 2005).

12.2.2 The Benefits from Using Proteomics over Transcriptomics

As extended description by Rose and colleagues (2004), there are many arguments that favor the use of proteomic over transcriptomic analysis, even when the experimental manipulation is more complicated for the former case. Although correlation studies of proteome of different organs and transcriptome in *Arabidopsis* suggest that protein accumulation in this species is primarily regulated at transcriptional level (Baerenfaller et al. 2008), several studies have revealed poor correlations between changes in the abundance of specific mRNAs and their corresponding proteins. In consequence, mRNA levels not always are indicative of cognate protein abundances. Moreover, substantial regulation of cellular events can occur at the protein level with no apparent changes in mRNA abundance. Apart from regulation of protein abundance by transcription and translation, direct fine-tuned control of protein activity, protein abundance, or protein localization is achieved at the post-translational level resulting in a dramatic increase in protein complexity without a concomitant increase in gene expression (Schulze 2010). In addition, proteome reflects the expression of the molecules that more directly influence cellular biochemistry, rather than representing information of intermediates as the case of profiling the expression of mRNAs (Rose et al. 2004).

12.2.3 Proteomic Studies' Applications in Peach

Among Prunoids, peach (*Prunus persica* L. Batsch) is one of the genetic reference species not only because of its high economic value but also because of its genetic and genomic characteristics such as small genome size, taxonomic proximity to other important species, and availability of homozygous doubled haploids (Shulaev et al. 2008; Aranzana et al. 2010).

Although, significant improvements in sensitivity, mass accuracy, and fragmentation in recent years have led to widespread increasing number of very specialized studies, the use of proteomics in the Prunoids has been only recently adopted. Differential proteomics have been conducted mainly in peach fruit to elucidate molecular events involved in ripening as well as changes regarding different postharvest conditions (Chan et al. 2007; Borsani et al. 2009; Lara et al. 2009; Dagar et al. 2010; Nilo et al. 2010; Zhang et al. 2010). On the contrary, in bark tissues only one study has been conducted (Renaut et al. 2008).

12.2.3.1 The Ripening of Peach Fruit

During ripening, fleshy fruits are transformed into palatable products with modifications in color, texture, levels of sugars, organic acids and volatile compounds leading to changes in nutritional quality, flavor and aroma. Thus, during ripening fleshy fruit are prepared for human eating (Giovannoni 2004). The peach is a climacteric fruit which has its ripening process controlled by the production of ethylene. Other hormones have been implicated in this process as well (Trainotti et al. 2007).

Ripening encompasses a myriad of biochemical and physiological modifications (Borsani et al. 2009). The typical softening behavior of peach during the postharvest at ambient temperature is characterized by an initial stage of slow firmness decrease, followed by a rapid loss of firmness (melting stage) (Tonutti et al. 1996). The decline in fruit firmness during ripening is the result of the cell wall disassembly due to a series of modifications in polysaccharide components of the primary cell wall and middle lamella (Brummell 2006). The softening process correlates with ethylene production as well as with 1-aminocycloproane-1-carboxylate oxidase 1 (*PpACO1)* expression (Borsani et al. 2009). At a transcriptomic level, analysis conducted during the transition from the pre-climacteric to the climacteric phase in peach fruit has led to the characterization of the molecular components of fruit ripening process (Trainotti et al. 2006). In addition, a global approach by sequencing and analyzing the abundance of ESTs during ripening revealed the complex biological processes that take place during the postharvest, as illustrated by a significant raise in the transcript levels of 30 genes and a reduction in the expression of 39 genes in ripe fruits when compared to the unripe peach (Vizoso et al. 2009).

Proteomic analyses during postharvest ripening of peach fruits have also been conducted. About six hundred spots were detected on the 2D map carried out with soluble protein extracted from "Dixiland" peach fruit after harvest and during the postharvest ripening process during 7 days (Borsani et al. 2009), while about 560 spots were identified in "O'Henry" fruits at these stages (Nilo et al. 2010). By using 2D-DIGE followed by MALDI-TOF/TOF only five spots showed significant variation during the postharvest ripening process of "Dixiland" peach. Two polypeptides decreased in samples collected at 3 days after harvest (DAH) with respect to samples taken after harvest while two others spots decreased in 7 DAH in relation to 3 DAH. On the other hand, one spot was identified as only occurring in samples taken at 7 DAH. Fingerprinting mass analysis allowed the identification of a Ca^{2+}/calmodulin-dependent protein kinase and a eukaryotic initiation factor 4A-2, both showing reduced expression during the postharvest ripening process. In Table 12-1, the identified proteins are listed and their corresponding Genebank accession numbers are shown. Changes in these proteins during ripening may have profound effects in modification of a wide set of protein targets and in overall protein biosynthesis (Borsani et al. 2009).

In the case of "O'Henry" peach ripening after harvest during 6 days 34 proteins were detected by 2D-DIGE as differentially abundant during ripening, some of them are listed in Table 12-1. Up-regulation of endopolygalacturonase (endoPG), pectin methylesterase (PME) and two ACC oxidases agree with the softening of the fruit (Nilo et al. 2010). In accord with an increased pyruvate decarboxylase (PDC) activity measured in ripening "Dixiland" peach (Borsani et al. 2009); in "O'Henry" peach two PDCs were up-regulated (Nilo et al. 2010). Decreases in the levels of catalase (CAT), gluthatione S-transferase, and quinone oxidoreductase and increases in the levels of glyceraldehyde-3-P dehydrogenase, thioredoxin H, subunit F of NADH dehydrogenase, NADPH isocitrate dehydrogenase in ripe fruit (Nilo et al. 2010) reveals a deep alteration in the redox process, indicating that ROS metabolism, protein modulation by redox mechanism and energy supply are deeply altered during ripening. In this respect, abundant evidence supports the concept that ripening is an oxidative

Table 12-1 Summary of some of the most relevant proteins identified by proteomics as differentially expressed in peach fruit during the postharvest.

Condition	Cultivar	NCBI Acc N°	Protein identification	Reference
POSTHARVEST RIPENING AT AMBIENT CONDITIONS				
Decrease during the postharvest at 20°C (7 days)	"Dixiland"	116060059	Ca^{2+}/calmodulin-dependent Protein Kinase	Borsani et al. 2009
		116060059	Initiation Factor 4A-2	
Increase during the postharvest at 20°C (6 days)	"O'Henry"	CAJ84510	NADH dehydrogenase subunit F	Nilo et al. 2010
		AAL26915	Thioredoxin H	
		CAB57457	Pectin methylesterase	
		CAB39974	Glyceraldehyde-3-P dehydrogenase	
		CAA54449	1-aminocyclopropane-1-carboxylate oxidase	
		AAC64184	Endopolygalacturonase	
		AAL11503	NADP-isocitrate dehydrogenase	
		CAC00637	Dehydrin-like protein	
		AAF34133	Low molecular weight HSP	
		AAF34133	Low molecular weight HSP	
		AAL32036	Small HSP	
		AAF34133	Low molecular weight HSP	
		AAR99375	Small HSP	
		AAF34133	Low molecular weight HSP	
Decrease during the postharvest at 0°C (6 days)		AAL37492	Pyruvate decarboxylase	
		CAD42909	Catalase	
		AAG53944	Quinone-oxidoreductase QR1	
		ABB78006	Major allergen Pru p 1	
		BAC21261	Glutathione S-transferase	
		NP_186962	ADP-ribosylation factor B1C	

Table 12-1 contd....

Table 12-1 contd....

Condition	Cultivar	NCBI Acc N°	Protein identification	Reference
COLD TREATMENT				
Increase after cold storage at 5°C (3 weeks)	"Oded"	15982950	NADP-isocitrate dehydrogenase	Dagar et al. 2010
		10180029	Putative protein disulfide-isomerase	
		82568691	Oxoacid-dependent dioxygenase	
		108864390	Putative 40S ribosomal protein S5	
		25091405	Thaumatin-like protein 1 precursor	
		30421435	Calmodulin	
		24473796	60 s acidic ribosomal protein	
Increase after cold storage at 0°C vs. 5°C (3 weeks)	"Hongtao"	73762178	Type II SK2 dehydrin	Zhang et al. 2010
		77744891	Temperature-induced lipocalin	
		82492265	Major allergen Pru p 1	
		63192024	Enolase	
Decrease after cold storage at 0°C vs. 5°C (3 weeks)		15238285	Chorismate mutase	
		15235295	Cinnamyl-alcohol dehydrogenase 5	
		109631192	Cinnamyl-alcohol dehydrogenase 1	
		3282505	Polyphenol oxidase precursor	
Increase in 4°C (3 weeks) vs. 20°C	"O'Henry"	BAC98347	Anthocyanidin synthase	Nilo et al. 2010
		O04300	Alpha-1,4-glucan-protein synthase [UDP-forming]	
		CAB57457	Pectin methylesterase	
		AAM00216	Thaumatin-like protein	
		P52877	Phosphoserine aminotransferase chloroplast precursor	
		AAL37492	Pyruvate decarboxylase	
		NP_186962	ADP-ribosylation factor B1C	

Decrease in 4°C (3 weeks) vs. 20°C		AAC64184	Endopolygalacturonase	
		CAC00637	Dehydrin-like protein	
		CAD42909	Catalase	
		AAL11503	NADP-isocitrate dehydrogenase	
		P51615	NADP-malic enzyme	
		AAF34133	Low molecular weight HSP	
		AAL32036	Small HSP	
		AAF34133	Low molecular weight HSP	
HEAT TREATMENT				
Decrease in HT (39°C, 3 days) vs. 20°C	"Dixiland"	3282505	Polyphenol oxidase precursor	Lara et al. 2009
Increase in HT (39°C, 3 days) vs. 20°C		6969974	LMW heat shock protein	
		1235898	17.7 kDa heat shock protein	
		AY500559.1	Small heat shock protein	
		44409451	Major cherry allergen	
		41059801	Small HSP	
		44409451	Major cherry allergen	
		2677828	Cysteine protease	
		2677828	Cysteine protease	
		6969974	LMW heat shock protein	
		82492265	Major allergen Pru p 1	
		1235898	17.7 kDa heat shock protein	
		41059801	Small HSP	
		44409451	Major cherry allergen Pru av 1.0202	
		5257560	Cytosolic class II LMW HSP	
		AY500559.1	Small HSP	
		73762178	Type II SK2 dehydrin	

process in which ROS production and antioxidant machinery (i.e., CAT, SOD and APX) are systematically altered (Jimenez et al. 2002; Sarry et al. 2004). Moreover, during ripening the accumulation of seven low molecular weight and small heat shock proteins (HSP) reveals the importance of these proteins during the postharvest. In this respect, small HSPs from tomato have been involved in the control of pectin depolimerization and juice viscosity during fruit ripening (Ramakrishna et al. 2003). A major allergen Pru p1 and a dehydrin-like protein decreased during fruit ripening (Nilo et al. 2010). Dehydrin proteins belong to the group 2 of late embryogenesis abundant (LEA) proteins. Although they have been associated to several abiotic stresses, their function in the fruit has not been completed elucidated (Close 1996). Three dehydrins have been found in peach and their expressions have been studied in "Loring" fruits (Wisniewski et al. 2006; Bassett el al. 2009). While the expression of *PpDhn1* and 2 declined from early peach fruit development until fully ripe stage; *PpDhn3* was more abundant in ripened fruit (Bassett et al. 2009). In addition, alteration in their expression in response to different postharvest treatment has also been reported (Lara et al. 2009; Zhang et al. 2010). Finally, other proteins differentially expressed during ripening were detected, as they have not been previously related to this process or they have not been identified, their study could give more insight into the molecular events taking place as peach ripens.

12.2.3.2 Postharvest of Peach Fruit

Shipping of fleshy fruits to distant markets and internal marketing requires a lengthy storage period. Since at ambient temperature harvested peach undergo the ripening process and deteriorate quickly, refrigeration of peach fruits has been used to slow these processes, and in consequence to extend fruit market life. In addition, low temperatures not only reduce the enzymatic and microbial activity but also slow down the respiratory rhythm of fruits. However, cold storage affects fruit quality causing physiological disorders collectively termed "chilling injury" (CI, Lurie and Crisosto 2005). In general many tropical and subtropical fruits are sensitive to CI when exposed to low, non-freezing temperatures, leading to serious losses in quality and market value (Salveit and Morris 1990). In peach, CI includes internal and external browning, flesh break-down, lack of juiciness (mealiness or woolliness), reddish discoloration (flesh bleeding), black pit cavity, loss of flavor, loss of ability to ripen and increased incidence of decay. These symptoms are developed when fruit are stored for more than 2–3 weeks at temperatures between 0°C and 8°C (Brummell et al. 2004; Lurie and Crisosto 2005, and references therein). Nevertheless, most of the CI disorders are observed when cold stored fruit are transferred to ambient temperature; that is to say when peach fruit reach their final phase of marketing. At this stage fruit display unattractive organoleptic properties, being rejected by the consumers.

12.2.3.2.1 Proteins involved in development or tolerance of chilling injury symptoms

Exposure to temperature below optimum growth results in damage of specific organs or tissues of cold-sensitive plants (Tomashow 1999). There has been an increasing trend in using proteomic methods in the field of plant stress responses. Proteomic approaches utilizing 2D-PAGE together with mass spectrometry have been successfully used to understand plant responses to a broad range of abiotic stress factors—cold, heat, drought, waterlogging, salinity, ozone treatment, hypoxia and anoxia, herbicide treatments, inadequate or excessive light conditions, desbalances in mineral nutrition, enhanced concentrations of heavy metals, radioactivity and mechanical wounding (Kosová et al. 2011). Regarding peach fruit, our comprehension of complex molecular events that underlie the response to cold treatment of fruit has grown dramatically in the last few years due to adoption of both proteomic and transcriptomic strategies.

Preserving the stability of cell membranes at low temperatures is important for plant resistance to cold stress (Wongsheree et al. 2009). Increased permeability due to changes in membrane composition and structure occurs very early during the chilling stress (Sevillano et al. 2009). Internal-browning symptoms are dependent on membrane integrity. Imbalance between production and elimination of reactive oxygen species (ROS) inside the fruits conducts to the production of brown-colored polymers as a consequence of changes in membrane permeability which allows the interaction of phenolic compounds and polyphenol oxidase (PPO) enzymatic oxidation—normally in different cell compartments—and thus phenolic oxidation (Lurie and Crisosto 2005; Franck et al. 2007). Peach fruit shows less CI incidence and index when refrigerated at 0°C that at 5°C, as accounted by lower electrolyte leakage and lower phenolic compounds (Zhang et al. 2010). Since greater membrane stability was observed in peach stored at 0°C, Zhang and colleagues (2010) searched for proteins differentially expressed in "Hongtao" peaches stored at 0 and 5°C. By using electrospray ionization quadrupole time-of-flight tandem mass spectrometry (ESI-qTOF-MS/MS) 44 proteins differentially expressed were detected (Table 12-1). Most notably, enhanced expression of enolase, temperature-induced lipocalin, major allergen Pru p 1 and Type II SK2 dehydrin in peach fruit stored at 0°C with respect to fruit kept at 5°C was observed. All these proteins have been previously related to the cold acclimation response in other tissues and would contribute to membrane stability in peach (Zhang et al. 2010). Briefly, enolase is up-regulated in response to cold in leaves (Yan et al. 2006) and its mutation conducted to chilling sensitivity of *Arabidopsis* (Lee et al. 2002). Pru p 1 is a lipid transfer protein (LTP) which might be involved in membrane biogenesis and regulation of intracellular fatty acid pools (Miquel et al. 1988). Lipocalins proteins from plants may fulfill a wide range of functions such as protection against photo-oxidative damage and their expression is induced during cold acclimation (Frenette Charron et al. 2002). In addition, dehydrins are also related to abiotic stress tolerance and also proposed as antioxidant and as membrane and proteins stabilizers (Close 1997). On the other hand, decrease levels of cinnamyl-alcohol dehydrogenases 1 and 5, PPO precursor and chorismate mutase were detected by 2-DE coupled to ESI-qTOF-MS/MS in stored peaches at 0° with respect to those kept a 5°C and this response doubtlessly correlates with decrease in flesh browning incidence of the fruit (Zhang et al. 2010).

Another approach to seek for proteins differentially accumulated under conditions that conducts to CI was undertaken by Nilo and colleagues (2010). In this case, proteins extracted from mesocarp of peach fruit "O'Henry" were stored at 4°C for 3 weeks and subsequently kept at 20°C for softening, and compared to those of peach kept at ambient temperature by using a 2D-DIGE approach coupled to LC-MS/MS (Table 12-1). Cold stored peach that softened at ambient temperature showed CI such as browning and mealiness, and increased ethylene production (Nilo et al. 2010). Changes in cell wall metabolism will be discussed in the following section. TLP and dehydrin were found to accumulate in cold stored mesocarp; and their corresponding genes were already reported as cold responsive in previous studies using a transcriptomic approach (Ogundiwin et al. 2008). In addition, differential expression of transcripts encoding two dehydrins was reported in peach bark, leaf and embryo (Wisniewski et al. 2006). Different proteins involved in ROS metabolism (catalase and ferritin) and HSPs (PpsHSP1) were decreased in cold stored fruit suggesting a reduced capacity to cope with oxidative stress due to refrigeration (Nilo et al. 2010). Among proteins overexpressed in fruit exhibiting CI, anthocyanidin synthase could be identified as a factor contributing to the browning development since this enzyme produces red anthocyanidins susceptible of oxidation by PPO and peroxidases (Zhang et al. 2001, 2005). Regarding carbohydrate and organic acid metabolism down-regulation of NADP-malic enzyme and up-regulation of –NADP-isocitrate dehydrogenase in cold fruit (Nilo et al. 2010) indicate altered organic acid and NADPH metabolism in cold stored fruit. On the other hand, increased levels of PDC detected in CI peach have been proposed as very important to cope with the high energy demand of chilled injured fruit (Nilo et al. 2010). Another enzyme probably contributing to deal with cold stress is phosphoserine aminotransferase, through the provision of either serine or intermediates for the synthesis of protective compounds such as Glycine betaine (Nilo et al. 2010). Finally, as mentioned in the previous paragraph, refrigeration affects membrane stability and in consequence trafficking. In this respect, an ADP-ribosylation factor (ARF) was accumulated during cold storage; this factor has been previously associated with membrane remodelation processes and thus its enhanced expression during refrigeration could help to deal with one of the effects of cold stress (Nilo et al. 2010).

12.2.3.2.2 Cell wall metabolism in refrigerated peach fruit

Array technology has contributed to the identification of genes expressed during the postharvest of peach fruits subjected to cold storage (Ogundiwin et al. 2008; Vizoso et al. 2009) and of genes directly involved in woolliness (González-Agüero et al. 2008). By this approach, changes in the expression of genes involved in cell wall metabolism and endomembrane trafficking were observed. More recently, comparative proteomics have provided more information regarding molecular events conducting to chilling injury and softening.

Mealiness (woolliness) is characterized by a lack of juiciness and a mealy texture (Lill et al. 1989). Mealiness is not perceived from the exterior of the fruit, since affected fruit weigh the same, and have a similar firmness to the fingers. This disorder develops in cold stored peach during the subsequent ripening at ambient temperature. It is associated with modification in pectin metabolism due to changes in the relative activities of pectin methylesterase (PME) and polygalacturonase (PG) (Buescher and Furmanski 1978; Brummell et al. 2004 and reference therein). Therefore, the accumulation of high molecular weight pectins together with cell wall calcium may form a gel that binds free water and contributes to the woolliness phenotype. In addition, alteration of cell wall metabolism in mealy fruits is caused by variation in the activity of numerous cell wall-modifying enzymes, as accounted by reduced activity in exo- and endo-polygalacturonase (endoPG), endo-1,4-β-mannase, β-galactosidase, β-arabinosidase and α-arabinosidase enzymes (Zhou et al. 2000b; Brummell et al. 2004). Moreover, decreased expansin content, determined through immunoblot analysis as well as at mRNA level, was detected in "O'Henry" and in "Summer Lady" mealy fruits stored at 5°C with respect to juicy fruits (Obenland et al. 2003).

At a proteomic level, Dagar and colleagues (2010) examined changes in the cell wall proteins and total proteins of peach fruit subjected to cold storage through 2D-PAGE analysis joined with reverse HPLC microspray mass spectrometry (HPLC/MS/MS). Since the resolution capacity of 2D gel electrophoresis is still insufficient to monitor entire protein sets of eukaryotic cells, proteome research often is based on a subset of proteins of eukaryotic cells called "subproteome" (Cordewell et al. 2000; Jung et al. 2000). In plant biology, very successful subproteomic analyses were carried out for the cell wall, the plasma membrane, the thylakoids, the mitochondria (Robertson et al. 1997; Santoni et al. 2000; van Wijk 2000; Kruft et al. 2001). During these investigations, hundreds of proteins were separated, several of which were identified for the first time. Dagar and colleagues (2010) provided the first proteomic analyses of the response of the cell wall proteins in cold-stored peach fruits. To this aim, two cultivars with different cold sensitivity were compared, a CI resistant named "Oded" and a sensitive one, "Hermoza". Proteins extracted from cell wall were used for differential proteomics (Table 12-1). Most of the proteins identified in the cell wall extract from stored fruit belonged to the Thaumatin like protein (TLP) family—a type of pathogenesis-related protein (Hiroyuki and Terauchi 2008) or were similar to unknown proteins already reported in *Prunus* sp. When the response of both cultivars to cold treatment was compared, it was found that even though TLP1 precursor was increased in both cultivars after 3 weeks of storage at 5°C, transcript analysis revealed that the earlier and higher expression in the resistant cultivar might be involved in protection against dramatic changes in cell wall and thus shielding against the development of woolliness (Dagar et al. 2010). On the contrary, in "O'Henry" peaches overexpression of a TLP was detected by 2D-DIGE in fruits that developed CI after refrigeration at 4°C with respect to fruit after harvest or cold stored and softened at ambient temperature (Nilo et al. 2010). In agreement with results of other researchers, Nilo and colleagues (2010) reported enhanced levels of PME and reduced accumulation of endoPG in CI fruits, supporting the relevance of these enzymes in the mealiness development (Table 12-1). In addition, an increased in an α-1,4-glucan-protein synthase—probably involved in arabinose metabolism—in peach exhibiting CI symptoms reinforces the concept that modifications in arabinans polymers are linked to generation of CI signs.

12.2.3.2.3 Treatments used to extend peach postharvest life and to prevent CI

In order to prevent or alleviate CI symptoms, different preharvest and postharvest treatments have been applied to peach, including salicylic acid treatments, exposure to sublethal high temperature or the use of modified atmospheres (Lurie and Crisosto 2005; Budde et al. 2006; Lurie 2006; Wang et al. 2006; Jin et al. 2009; Lara et al. 2009, 2011). Among postharvest handling procedures, exposure to sublethal high temperature constitutes one of the most studied postharvest treatments (Lurie and Klein 1991) applied not only to peach but also to other fruits. Heat treatment (HT) increases the tolerance to subsequent chilling and its use results advantageous in preserving the organoleptic quality such as texture and juicy during the following cold storage of fruits (Cheng et al. 2008). This treatment also reduces pathogen levels and disease development in several fruits (Saltveit 1991; Lurie 1998, 2006; Ferguson et al. 2000; Paull and Chen 2000; Budde et al. 2006; Wang et al. 2006; Jin et al. 2009).

Exposure of "Dixiland" peach fruit to a heat treatment of 3 days at 39°C was proved to be efficient in postharvest technology (Budde et al. 2006). This treatment, delayed the ripening process with softening inhibition and slowdown in ethylene production (Lara et al. 2009). These responses, together with other changes in carbon metabolism and in organoleptic properties were reversed when heated peach were transferred to ambient temperature. Therefore, normal fruit ripening was restored after the treatment. In addition, positive quality features such as increment in the fructose content, largely responsible for the sweetness, and reddish coloration were observed in treated fruit. Nevertheless, the treated fruit displayed high amounts of acetaldehyde and low content of organic acids (Lara et al. 2009).

The rationale for the use of heat treatment (HT) is that exposure to a high temperature triggers physiological responses that allow the fruit to better cope with subsequent cold stress. Although biochemical and molecular changes induced by HT are not yet completely understood, cellular responses due to HT that may be involved in CI protection are starting to emerge. In this respect, differential proteomics is a powerful tool for studying and identifying global changes in structure and abundance of proteins in response to HT and to identify candidate proteins that may prevent some of the CI symptoms as well. Therefore, a 2D-DIGE proteomic approach was used to reveal the response of mesocarp from "Dixiland" peach fruit subjected to 39°C for 72 h (Table 12-1). A total of 57 spots were detected as differentially expressed in the heated vs. unheated peach mesocarp proteome, of which almost a 70% were stress and defence related proteins. As expected from a HT, smHSPs of 17–18 kDa and HSP70 were induced. These proteins may participate in the acquisition of tolerance against some CI symptoms since in other fruits such as tomato and grapefruit, the exposure to a HT prior refrigeration prevented CI, and this protection was endorsed to the increased expression of various HSPs (Sabehat et al. 1996, 1998; Rozenzvieg et al. 2004; Polenta et al. 2007). In addition, the induction of HSPs due to salicylic acid was also correlated with CI alleviation (Wang et al. 2006). On the contrary, reduced levels of HSPs70 and HSP100 were associated in peaches with CI symptoms such as woolliness (González-Agüero et al. 2008). Other stress related proteins were induced in "Dixiland" peach mesocarp in response to HT such as Cys proteases, major allergen, major cherry allergen, and a dehydrin. These proteins may activate different mechanisms which could provide CI tolerance since many of them were reported as responsive to low temperature or other types of stress (Wisniewski et al. 1996; Close 1997; Estelle 2001; Renault et al. 2008). Interestingly, the HT also conducted to a repression of the expression of a polyphenol oxidase (PPO) in treated fruit transferred to ambient temperature in comparison with untreated peaches of the same postharvest age kept at 20°C. Since PPO is a key enzyme of the browning process (Lurie and Crisosto 2005), its repression in heated fruit would reduce this symptom of CI (Lara et al. 2009).

12.2.3.2.4 The use of differential proteomics in other aspects of the postharvest

Two dimensional-PAGE coupled to ESI-qTOF-MS/MS was also applied to study the response of peach fruit exposed to the antagonist yeast *Pichia membranefaciens* and to salicylic acid, both conducting to enhanced resistance and delayed initiation of *Penicillium expansum* infection (Chan et al. 2007). By this approach, different proteins involved in the resistance to the mentioned pathogen were identified.

12.2.3.3 Other Studies in Prunus Including Differential Proteomics

Woody plants such as *P. persica*, present cold hardiness, a dynamic process which changes with time and that involves several contributing factors. The degree of cold hardiness can be modulated by day-length and temperature, and it has an important impact in fruit production (Wisniewski et al. 2003, 2006, 2007 and references therein). In this species, low-temperature induced gene expression was studied (Bassett et al. 2006). The use of 2D-DIGE approach coupled to MALDI-TOF-MS/MS is adequate to enlarge the knowledge and to contrast gene/protein expression (Renaut et al. 2006). Particularly, it was used to investigate the changes in peach bark related to the exposure to low temperature and/or short photoperiod, components that seasonally change together, and to separate photoperiod from temperature effects (Renaut et al. 2008). By this approach several proteins were identified and classified as either "temperature-" or "photoperiod-responsive". This study concluded that low temperature is the factor that most affects the bark proteome, in comparison with day length, and deeply analyzes the relevance of the proteins differentially expressed in the context of cold acclimation (Renaut et al. 2008).

Another interesting exemplification of the value of proteomics research in Prunoids, is provided by the studies conducted by Feng and colleagues (2006, 2009) in which comparative proteomics was conducted to discover proteins differentially expressed in compatible and incompatible interactions between pollen and pistils from apricots (*P. armeniaca*).

12.3 Metabolomics

The term "metabolomics" refers to the systematic quantitative and qualitative study of the multiple chemical constituents found in an organ, tissue or even a single cell. The metabolome as a whole represents the final measurable state of a particular organ or cell and thus, its analysis gives a final view of what is occurring in that organ or cell in a specific situation. Moreover, the high level of responsiveness of the metabolome to different stress situations or developmental signals, and the considerably lower costs of metabolomic experiments in relation to proteomic or transcriptomic studies, are responsible for the current wide application of metabolomic studies in different experimental systems. In this way, metabolomics is currently applied to a large number of research areas, from drug discovery to particular organ function or food quality. In particular, plants possess the unique characteristic of synthesizing a large number of chemical compounds; therefore, it is estimated that up to 200,000 primary and secondary metabolites exist in the plant kingdom (Fiehn et al. 2000; Fiehn 2002; Hall 2005; Oksman-Caldentey and Saito 2005; Fernie 2007). However, among the difficulties inherent in the metabolomic studies are the dynamic changes of metabolites either in time or compartment; as well as the wide disparity of concentrations of the different type of metabolites (Stitt and Fernie 2003).

One of the most metabolite-rich plant organs is the fruit, which constitutes an excellent model for metabolomics studies. During fruit development, from the fertilized ovary to the ripe fruit, a large number of metabolites are synthesized and/or catabolized. Moreover, fruit constitutes an essential part of human diet, not only because of its high levels of nutrients, but also due to the presence of several metabolites with recognized value in human health, such as flavonoids, anthocyanins and phenolics (Macheix et al. 1991; Swanson 1998). Thus, the comprehensive analysis of fruit metabolite content is essential for the goal of obtaining fruit varieties with high levels of nutrients.

Although the content of particular metabolites or structurally or biologically associated compounds have been analyzed in different types of fruits since several years ago, comprehensive metabolic profiling studies have been just recently applied to fruit. These analysis have been mainly focused on tomato (Roessner-Tunali et al. 2003; Carrari et al. 2006; Carrari and Fernie 2006); and more recently, on two non-climacteric fruits, grape (Deluc et al. 2007; Zamboni et al. 2010) and strawberry (Fait et al. 2008; Osorio et al. 2011). With regards to stone fruits, metabolic profiling studies during peach fruit cv. "Dixiland" development and ripening have been recently conducted (Lombardo et al. 2011). The comparison among the metabolic profiling analysis performed in different fruits indicate particular metabolic programs for each fruit, which is in relation to the diverse differences found among fruits in terms of morphology and physiology. Other recent application of metabolomics in peach involves studies during refrigerated storage, in order to analyze metabolites linked to chilling injury in this stone fruit (Lauxmann et al. unpublished results). Here, a brief description of the basic principles and methodologies of the current metabolomics approaches and their application in fruit will be described, with an emphasis on the recent metabolomic studies performed on peach fruit.

12.3.1 Metabolomics: The Basis

A typical metabolomic profiling study includes several steps starting from sample extraction, data processing, statistical analysis and metabolite identification. Following these experimental procedures, the metabolic results are analyzed taking into account the metabolic networks in which the identified metabolites are involved. The several types of technologies for the analysis of metabolites have made an incredible progress in the recent years, especially regarding the number of metabolites that are identified. The choice of the analytical method to be used depends on which types of metabolites will be analyzed. Several reviews about the techniques used in metabolomics and their applications in plants have already been published (Fiehn et al. 2000; Roessner et al. 2001; Hall et al. 2002; Fiehn 2002; Fernie 2003, 2007; Fernie et al. 2004; Lisec et al. 2006; Schauer and Fernie 2006); thus, just a brief overview of the different methodologies most commonly used will be presented.

The most widely used method for metabolome analysis combines chromatographic separation of extracts with detection and identification by mass spectrometry (MS). MS allows the sensitive detection and identification of metabolites through the interpretation of mass spectrum and determination of molecular formulae via accurate mass measurements (Dunn et al. 2005). Gas chromatography mass spectrometry (GC-MS) and liquid chromatography mass spectrometry (LC-MS) are the most widely used techniques for separation of the different compounds in the extracts. GC-MS has high separation efficiency and provides advantage in terms of robustness. For these reasons, it is the principal technique for the analysis of volatile metabolites as well as non-volatile polar metabolites such as amino acids, sugars and organic acids, after their transformation, by chemical derivatization, to volatile and thermostable compounds. The chemical derivatization procedures include trimethylsilylation and reaction with diazomethane. However, although GC-MS is the method of choice for analyzing primary metabolites; this technique can not be applied for thermolabile compounds. On the other hand, LC-MS allows the detection of multiple metabolite classes in a single analysis, even in a very complex matrix such as plant extracts, adding versatility to the methodological range of GC-MS. The introduction of ultra performance LC enables even faster separation. After the data sets are obtained by a MS technique, automated computational tools allows metabolite identification and quantification using several databases (Kopka et al. 2005; Schauer et al. 2005; Moco et al. 2006).

A quite different overall approach to analyze the metabolome of a given sample is the analysis of the extracts by nuclear magnetic resonance (NMR) spectroscopy (Krishnan et al. 2005), which can identify abundant metabolites unambiguously, as each separate resonance from a given spectrum is specific to a particular atom pair. However, it is not useful for analyzing a broad range of compounds due to the low level of resolution (Stitt and Fernie 2003). An alternative platform with a fast, sensitive and broad coverage of a whole metabolome is the capillary electrophoresis-MS (CEMS); and special mention must be given to the Fourier-transform ion cyclotron resonance mass spectrometry (Aharoni et al. 2002).

In any case, after the metabolite content of a given sample have been identified and quantified, several recent developments also allow the construction of metabolic networks, by using diverse mining tools including multivariate analyses and clustering for discovering differences between groups of samples (Stitt et al. 2010). Thus, bioinformatics and biostatistics are among the most essential disciplines utilized in metabolome research.

12.3.2 *Application of Metabolomics to Fruit: Tomato, Grape and Strawberry*

Fruit constitutes a variety of different tissues and represents one of the most complex organs of the plant. During fruit development and ripening, the complex network of metabolites and proteins is dramatically altered. In this regard, metabolomics is an excellent tool for analyzing metabolism in developing fruit, due to its ability to follow a relatively large number of compounds in a single or a few analyses. Moreover, as fruit is one of the most metabolite-rich organs of plants, it constitutes an excellent model for metabolomics studies.

In the case of tomato (Roessner-Tunali et al. 2003; Carrari and Fernie 2006; Carrari et al. 2006), a dramatic change in the pigment composition during development and ripening was detected (Carrari et al. 2006). Sucrose decreased during tomato development, whereas glucose and fructose accumulated in an essentially linear manner. The changes in organic acid content and the levels of amino acids were also highly variable during the time course of tomato development (Carrari et al. 2006). In the case of strawberry fruit, a combination of GC-MS and LC-MS allowed the simultaneous study of the primary and secondary metabolites composition.

The study of these metabolites was performed separately in achene and receptacle during strawberry fruit ripening (Fait et al. 2008). The results obtained indicate that changes in primary and secondary metabolism reflect organ and developmental specificities. In this way, the receptacle was characterized by increases in sugars and their direct derivatives, while the achene was characterized by a major decrease in the levels of carbon- and nitrogen-rich compounds. In the case of grapevine berries, an integration of results obtained at the transcriptomic, proteomic, and metabolomic levels allowed the identification of putative stage-specific biomarkers for berry development and withering (Zamboni et al. 2010).

However, apart from the results on tomato, grape or strawberry, it is essential to extend metabolomics to other fruit species. In this regard, broader use of metabolomics in Rosaceae will provide data sets that would help in modeling metabolic networks related to traits of agronomic interest, as well as in the discovery of human health-promoting metabolites.

12.3.3 *Applications of Metabolomics to Peach*

12.3.3.1 Metabolomics During Peach Development and Ripening

A metabolic profiling analysis has been recently performed during peach cv. "Dixiland" fruit development and ripening (Lombardo et al. 2011). Peach represents a type of fruit clearly different from tomato, strawberry or grape since it is a drupe, with seeds covered by a hard, lignified endocarp—the stone—, being the edible portion a juicy mesocarp. The developmental progress of peach fruit can be described as consisting of four clearly recognized distinct stages (S1–S4) (Tonutti et al. 1997; Lombardo et al. 2011; El-Sharkawy et al. 2007). The first stage (S1) represents the first exponential growth phase, with a rapid cell division and elongation. During the second stage (S2), there is hardly any increase in fruit size. This stage corresponds to the endocarp hardening and the stone is formed (Dardick et al. 2010). In the third stage (S3), a second exponential growth takes place, with a fast increase in fruit size and rapid cell division. At the last stage (S4), the fruit reaches the final full size and comprises the fruit ripening or climacteric stage (Trainotti et al. 2003), which can take place out of the tree (Borsani et al. 2009).

GC-MS applied to peach fruit was able to identify unambiguously 47 different metabolites, which were classified into soluble sugars (14), organic acids (8), amino acids (14), sugar alcohols (4), fatty acids (2) and miscellaneous compounds (5) (Lombardo et al. 2011). After obtaining the whole dataset of the relative changes of these metabolites during the different developmental stages (S1 to S4), harvest (H) and postharvest ripening for 7 days (H3, H5 and H7), associations among the different stages in peach were analyzed by principal component analysis (PCA). This analysis revealed a clear metabolic shift from early stages of development (until S2) to late stages (S2 to S4) and the postharvest ripening process, which represents a separate and distinct group apart from the developmental stages. Figure 12-1 shows a scheme of some of the most relevant metabolic changes during early developmental stages and post harvest ripening of "Dixiland" peach fruit (Lombardo et al. 2011).

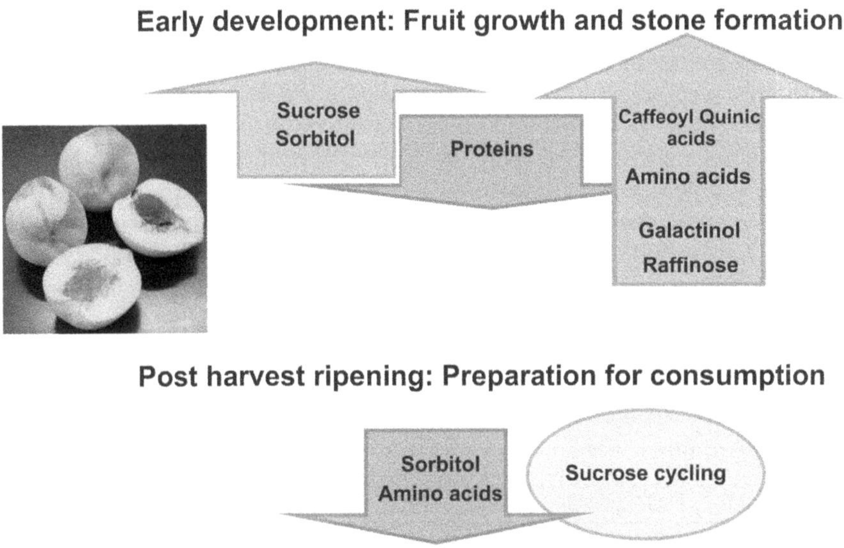

Figure 12-1 Scheme of some metabolic key changes occurring during early development and post harvest ripening of "Dixiland" peach fruit. The scheme highlights key metabolites that increase ↑ or decrease ↓ during development or ripening of peach fruit. During post harvest ripening, sucrose results recycled (Lombardo et al. 2011).

Color image of this figure appears in the color plate section at the end of the book.

During early developmental stages, the metabolic changes observed are in relation to the fruit growth and stone formation. In this period, proteins are used as source of several amino acids, which are present at relatively high levels at early developmental stages. Moreover, important peaks of increase of particular amino acids and caffeoyl quinic acids occur concomitant with the formation of the stone, indicating that amino acids are used as substrates for the phenylpropanoid, lignin and flavonoid pathways that are induced concomitantly with the deposition of lignin in the stone at S2 (Dardick et al. 2010). On the other hand, sugars are drastically increased during early development, in relation to high rate of translocation from the leaves, and correlated with high invertase activity in the fruit during this period (Lombardo et al. 2011). Sorbitol also increases in this period, although to a lower extent than sucrose. On the other hand, high levels of galactinol, *myo*-inositol and raffinose were also detected, which may be related to the seed maturation and to the development of desiccation tolerance (Horbowicz and Obendorf 1994; Bernal-Lugo and Leopold 1995). With regards to organic acids, practically constant levels of quinate and malate were detected at early developmental stages.

On the other hand, during the postharvest ripening process, the peach fruit is transformed into a palatable product for consumers by modifying its color, softening and altering the amounts of sugars, organic acids and volatile compounds (Borsani et al. 2009). Contrary to early stages of development, when high levels of amino acids were detected, several amino acids decrease during ripening, indicating that are used as respiratory substrate during the ripening process. Putrescine levels are drastically reduced during postharvest ripening, which may be related to the potential role of polyamine as a senescence inhibitor. With regards to sugars, practically constant levels of sucrose were found during the ripening process, which was linked to sucrose cycling (Lombardo et al. 2011). With respect to sorbitol, a dramatic decrease was found during the ripening process.

Overall, significant differences were detected when comparing the metabolic profiling during development of peach to other types of fruits, which may be related to the great differences in terms of morphology that are found among fruits. In the case of peach, in which extremely high levels of lignin have been found in the stone (Dardick et al. 2010), it seems that the lignification of the endocarp layer drastically affects primary metabolism to fulfill this process of *Prunus* species.

12.3.3.2 Metabolomics of Peach Fruit During Cold Storage After Harvest

Metabolomic analysis was also performed on "Dixiland" peach fruit stored at 0°C after harvest in comparison to fruit ripening at 20°C (Lauxmann et al. unpublished results). By this approach, significant metabolome changes were detected, some of which may be linked to the induction of fruit defense against chilling. In this way, dramatic increases of some compatible solutes, as galactinol and raffinose were detected. Some amino acids, substrates of the phenylpropanoid pathway, were also increased, as well as putrescine. In addition, changes in the relative levels of saturated and non-saturated fatty acids in relation to peach ripening at 20°C were also detected.

12.3.4 Metabolomics as a Functional Genomic Methodology in Stone Fruits

Metabolomics contributes to the understanding of the complex molecular interactions in different biological systems being currently used in functional genomics research and it is providing also potential targets for the engineering of metabolism in several different plant systems (Hall et al. 2002; Fernie and Stitt 2003; Bino et al. 2004; Trehewey 2004).

In the case of stone fruits, a broader application of this promising tool is needed, with a validation of the metabolomic information obtained by different groups and further analysis of the fluxes of metabolic pathways to decipher the biological relevance of each metabolite. A fundamental step is also the link of the functional metabolomic information to transcriptome and proteome datasets. These issues will enable the application of metabolomics for the prediction of stone fruits metabolism and the identification of essential keys for fruit productivity and nutritional improvement.

12.4 Concluding Remarks

Finally, in the era of omics, major challenges include interpretation and integration of large datasets to understand the principles underlying the regulation of networks and their relevance (Moreno-Risueno et al. 2010). The combination of transcriptomic, metabolomic and proteomic data on peach fruits will facilitate the functional identification of networks controlling development, ripening and response to different postharvest conditions. This information may ultimately be used to understand the different biological processes and be used as model of other stone fruits.

Acknowledgements

MFD and MVL thank financial support from the Argentine National Research Council (CONICET).

References

Aharoni A, de Vos CH, Verhoeven HA, Maliepaard CA, Kruppa G, Bino R, Goodenowe DB (2002) Nontargeted metabolome analysis by use of Fourier transform ion cyclotron mass spectrometry. Omics 6: 217–234.

Aranzana MJ, Abbassi E-K, Howad W, Arús P (2010) Genetic variation, population structure and linkage disequilibrium in peach commercial varieties. BMC Genetics 11: 69.

Baeremfaller K, Grossman J, Grobei MA, Hull R, Hirsch-Hoffman M, Yalovsky S, Zimmermann P, Grossniklaus U, Gruissem W, Baginsky S (2008) Genome-scale proteomics reveals *Arabidopsis thaliana* gene models and proteome dynamics. Science 320: 938–941.

Bassett CL, Wisniewski ME, Artlip TS, Norelli JL, Ranut J, Farrell RE (2006) Global analysis of genes regulated by low temperature and photoperiod in peach bark. J Am Soc Hort Sci 131: 551–563.

Bassett CL, Wisniewski ME, Artlip TS, Richart G, Norelli JL, Farrell RE (2009) Comparative expression and transcript initiation of three peach dehydrin genes. Planta 230: 107–118.

Bernal-Lugo I, Leopold AC (1995) Seed stability during storage: raffinose content and seed glassy state. Seed Sci Res 5: 75–80.

Bienvenut WV (2005) Introduction: Protein analysis using mass spectrometry. In: Bienvenut WV (ed) Acceleration improvement of protein identification by mass spectrometry. Springer, Dordrecht, Netherlands, pp 1–118.

Bino RJ, Hall RD, Fiehn O, Kopka J, Saito K, Draper J, Nikolau BJ, Mendes P, Roessner-Tunali U, Trethewey RN, Lange BM, Wurtele ES, Sumner LW (2004) Potential of metabolomics as a functional genomics tool. Trends Plant Sci 9: 418–425.

Blomberg A, Blomberg L, Norbeck J, Fey SJ, Larsen PM, Larsen M, et al. (1995) Interlaboratory reproducibility of yeast protein-patterns analyzed by immobilized pH gradient 2-dimensional gel-electrophoresis. Electrophoresis 16: 1935–1945.

Borsani J, Budde CO, Porrini L, Lauxmann MA, Lombardo VA, Murray R, Andreo CS, Drincovich MF, Lara MV (2009) Carbon metabolism of peach fruit after harvest: changes in enzymes involved in organic acid and sugar level modifications. J Exp Bot 60: 1823–1837.

Brummell DA (2006) Cell wall disassembly in ripening fruit. Funct Plant Biol 33: 103–119.

Brummell DA, Dal Cin V, Crisosto CH, Labavitch JM (2004) Cell wall metabolism during the development of chilling injury in cold-stored peach fruit: association of mealiness with arrested disassembly of cell wall pectins. J Exp Bot 55: 2041–2052.

Budde CO, Polenta G, Lucangeli CD, Murray RE (2006) Air and immersion heat treatments affect ethylene production and organoleptic quality of "Dixiland" peaches. Postharvest Biol Technol 41: 32–37.

Buescher RW, Furmanski RJ (1978) Role of pectinesterase and polygalacturonase in the formation of woolliness in peaches. J Food Sci 43: 264–266.

Canovas FM, Dumas-Gaudot E, Recorbet G, Jorrin J, Mock HP, Rossignol M (2004) Plant proteome analysis. Proteomics 4: 285–298.

Carrari F, Baxter C, Usadel B, Urbanczyk-Wochniak E, Zanor MI, Nunes-Nesi A, Nikiforova V, Centero D, Ratzka A, Pauly M, Sweetlove LJ, Fernie AR (2006) Integrated analysis of metabolite and transcript levels reveals the metabolic shifts that underlie tomato fruit development and highlight regulatory aspects of metabolic network behaviour. Plant Physiol 142: 1380–1396.

Carrari F, Fernie AR (2006) Metabolic regulation underlying tomato fruit development. J Exp Bot 57: 1883–1897.

Challapalli KK, Zabel C, Schuchhardt J, Kaindl AM, Klose J, Herzel H. (2004) High reproducibility of large-gel two-dimensional electrophoresis. Electrophoresis 25: 3040–3047.

Chan Z, Qin G, Xu X, Li B, Tian S (2007) Proteome approach to characterize proteins induced by antagonist yeast and salicylic acid in peach fruit. J Proteome Res 5: 1677–1688.

Chen J-Y, He L-H, Jiang Y-M, Wang Y, Joyce DC, Ji Z-L, Lu W-J (2008) Role of phenylalanine ammonia-lyase in heat pretreatment-induced chilling tolerance in banana fruit. Physiol Plant 132: 318–328.

Close TJ (1996) Dehydrins: emergence of a biochemical role of a family of plant dehydration proteins. Physiol Plant 97: 795–803.

Close TJ (1997) Dehydrins: a commonalty in the response of plants to dehydration and low temperature. Physiol Plant 100: 291–296.

Corbett JM, Dunn MJ, Posch A, Gorg A (1994) Positional reproducibility of protein spots in 2-dimensional polyacrylamide-gel electrophoresis using immobilized pH gradient isoelectric-focusing in the first dimension—an interlaboratory comparison. Electrophoresis 15: 1205–1211.

Cordewell SJ, Nouwens AS, Verrills NM, Basseal DJ, Walsh BJ (2000) Subproteomics based upon protein cellular location and relative solubilities in conjunction with composite two-dimensional electrophoresis. Electrophoresis 21: 1094–1103.

Dagar A, Friedman H, Lurie S (2010) Thaumatin-like proteins and their possible role in protection against chilling injury in peach fruit. Postharvest Biol Technol 57: 77–85.

Dardick CD, Callahan AM, Chiozzotto R, Schaffer RJ, Piagnani MC, Scorza R (2010) Stone formation in peach fruit exhibits spatial coordination of the lignin and flavonoid pathways and similarity to *Arabidopsis* dehiscence. BMC Biology 8: 13.

Deluc LG, Grimplet J, Wheatley MD, Tillett RL, Quilici DR, Osborne C, Schooley DA, Schlauch KA, Cushman JC, Cramer GR (2007) Transcriptomic and metabolite analyses of Cabernet Sauvignon grape berry development. BMC Genomics 8: 429.

Dunn WB, Bailey NJC, Johnson HE (2005) Measuring the metabolome: current analytical technologies. Analyst 130: 606–625.

El-Sharkawy I, Kim WS, El-Kereamy A, Jayasankar S, Svircev AM, Brown DCW (2007) Isolation and characterization of four ethylene signal transduction elements in plums (*Prunus salicina* L.). J Exp Bot 58: 3631–3643.

Estelle M (2001) Proteases and cellular regulation in plants. Curr Opin Plant Biol 4: 254–260.

Fait A, Hanhineva K, Beleggia R, Dai N, Rogachev I, Nikiforova VJ, Fernie AR, Aharoni A (2008) Reconfiguration of the achene and receptacle metabolic networks during strawberry fruit development. Plant Physiol 148: 730–750.

Feng JR, Chen XS, Wu Y, Liu W, Liang Q, Zhang LJ (2006) Detection and transcript expression of S-RNase gene associated with self-incompatibility in apricot (*Prunus armeniaca* L.). Mol Biol Rep 33: 215–221.

Feng JR, Chen XS, Yuan ZH, Zhang LJ, CI ZJ, Liu XL, Zhang CY (2009) Primary molecular features of self-incompatible and self-compatible F_1 seedling from apricot (*Prunus armeniaca* L.) Katy x Xinshiji. Mol Biol Rep 36: 263–272.

Fenn JB, Mann M, Meng CK, Wong SF, Whitehouse CM (1989) Electrospray ionization for mass spectrometry of large biomolecules. Science 246: 64–71.

Ferguson IB, Ben-Yehoshua S, Mitcham EJ, Mc Donald RE, Lurie S (2000) Postharvest heat treatments. Introduction and workshop summary. Postharvest Biol Technol 21: 1–6.

Fernie AR (2007) The future of metabolic phytochemistry: Larger numbers of metabolites, higher resolution, greater understanding. Phytochemistry 68: 2861–2880.

Fernie AR (2003) Metabolome characterisation in plant system analysis. Funct Plant Biol 30: 111–120.

Fernie AR, Trethewey RN, Krotzky A, Willmitzer L (2004) Metabolite profiling: from diagnostics to systems biology. Nat Rev Mol Cell Biol 5: 763–769.

Fieh O (2002) Metabolomics—the link between genotypes and phenotypes. Plant Mol Biol 48: 155–171.

Fiehn O, Kopka J, Dormann P, Altmann T, Trethewey RN, Willmitzer L (2000) Metabolite profiling for plant functional genomics. Nat Biotechnol 18: 1142–1161.

Franck C, Lammertyn J, Ho QT, Verboven P, Verlinden B, Nicola BM (2007) Browning disorders in pear fruit. Postharvest Biol Technol 43: 1–13.

Frenette Charron JB, Breton G, Badawi M, Sarhan F (2002) Molecular and structural analyses of a novel temperature stress-induced lipocalin from wheat and *Arabidopsis*. Plant Physiol 139: 2017–2028.

Gaskell SJ (1997) Electrospray: Principles and practice. J Mass Spectrom 32: 677–688.

Giovannoni JJ (2004) Genetic regulation of fruit development and ripening. Plant Cell 16: S170–S180.

González-Agüero M, Pavez L, Ibáñez F, Pacheco I, Campos-Vargas R, Meisel LA, Orellana A, Retamales J, Silva H, González, Cambiazo V (2008) Identification of woolliness response genes in peach fruit after postharvest treatments. J Exp Bot 59: 1973–1986.

Gorg A, Postel W, Weser J, Gunther S, Strahler JR, Hanash SM, et al. (1987) Horizontal two-dimensional electrophoresis with immobilized pH gradients in the 1st-dimension in the presence of nonionic detergent. Electrophoresis 8: 45–51.

Hall R, Beale M, Fiehn O, Hardy N, Sumner L, Bino R (2002) Plant metabolomics: the missing link in functional genomics strategies. Plant Cell 14: 1437–1440.

Hall RD (2006) Plant metabolomics: from holistic hope, to hype, to hot topic. New Phytol 169: 453–468.

Heazlewood JL, Millar AH (2003) Integrated plant proteomics—putting the green genomes to work. Funct Plant Biol 30: 471–482.

Henzel WJ, Billeci TM, Stults JT, Wong SC, Grimley C, Watanabe C (1993) Identifying proteins from 2-dimensional gels by molecular mass searching of peptide-fragments in protein-sequence databases. Proc Natl Acad Sci USA 90: 5011–5.

Hillenkamp F, Karas M (1990) Mass spectrometry of peptides and proteins by matrix-assisted ultraviolet laser desorption/ionization. Meth Enzymol 193: 280–295.

Hirano H, Islam N, Kawasaki H (2004) Technical aspects of functional proteomics in plants. Phytochemistry 65: 1487–1498.

Hiroyuki K, Terauchi R (2008) Regulation of expression of rice thaumatin-like protein: inducibility by elicitor requires promoter W-box elements. Plant Cell Rep 27: 1521–152

Horbowicz M, Obendorf RL (1994) Seed desiccation tolerance and storability: dependence of flatulence-producing oligosaccharides and cyclitols. Review and survey. Seed Sci Res 4: 385–405.

Jimenez A, Creissen G, Kular B, Firmin J, Robinson S, Verhoeyen M, Mullineaux P (2002) Changes in oxidative processes and components of the antioxidant system during tomato fruit ripening. Planta 241: 751–758.

Jin P, Zheng Y, Tang S, Rui H, Wang CY (2009) A combination of hot air and methyl jasmonate vapor treatment alleviates chilling injury of peach fruit. Postharvest Biol Technol 52: 24–29.

Jung E, Heller M, Sanchez JC, Hochstrasser DF (2000) Proteomics meets cell biology: the establishment of subcellular proteomes. Electrophoresis 21: 3369–3377.

Kopka J, Schauer N, Krueger S, Birkemeyer C, Usadel B, Bergmuller E, Dormann P, Weckwerth W, Gibon Y, Stitt M, Willmitzer L, Fernie AR, Steinhauser D (2005) GMD@CSB. DB: the golm metabolome database. Bioinformatics 21: 1635–1638.

Kosová K, Vítámvás P, Prásil IT, Renaut J (2011) Plant proteome changes under abiotic stress – Contribution of proteomics studies to understanding plant stress response. J Proteom 74: 1301–1322.

Krishnan P, Kruger NJ, Ratcliffe RG (2005) Metabolite fingerprinting and profiling in plants using NMR. J Exp Bot 56: 255–265.

Krokhin OV, Ens W, Standing KG (2005) MALDI QqTOF MS Combined with Off-line HPLC for Characterization of Protein Primary Structure and Post-Translational Modifications. J Biomol Tech 16: 429–440.

Kruft V, Eubel H, Jänsch L, Werhahn W, Braun H-P (2001) Proteomic approach to identify novel mitochondrial proteins in Arabidopsis. Plant Physiol 127: 1694–1710.

Laemmli UK (1970) Cleavage of structural proteins during the assembly of the head of bacteriophage T4. Nature 227: 680–5.

Lara MV, Borsani J, Budde CO, Lauxmann MA, Lombardo VA, Murray R, Andreo CS, Drincovich MF (2009) Biochemical and proteomic analysis of "Dixiland" peach fruit (*Prunus persica*) upon heat treatment. J Exp Bot 60: 4315–4333.

Lara MV, Budde CO, Porrini L, Borsani J, Murray R, Andreo CS, Drincovich MF (2011) Peach (*Prunus persica*) fruit response to anoxia: Reversible ripening delay and biochemical changes. Plant Cell Physiol 52: 392–403.

Lee H, Guo Y, Ohta M, Xiong L, Stevenson B, Zhu JK (2002) LOS2, a genetic locus required for cold-responsive gene transcription encodes a bi-functional enolase. EMBO J 21: 2692–2702.

Lill RE, O'Donoghue EM, King GA (1989) Postharvest physiology of peaches and nectarines. Hort Rev 11: 413–452.

Lisec J, Schauer N, Kopka J, Willmitzer L, Fernie AR (2006) Gas chromatography mass spectrometry for metabolite profiling in plants. Nat Protoc 1: 387–396.

Loboda AV, Krutchinsky AN, Bromirski M, Ens W, Standing KG (2000) A tandem quadrupole/time-of-flight mass spectrometer with a matrix-assisted laser desorption/ionization source: Design and performance. Rapid Commun Mass Spectrom 14: 1047–1057.

Lombardo VA, Osorio S, Borsani J, Lauxmann MA, Bustamante CA, Budde CO, Andreo CS, Lara MV, Fernie AR, Drincovich MF (2011) Metabolic profiling during peach fruit development and ripening reveals the metabolic networks which underpin each developmental stage. Plant Physiol 157: 1696–1710.

Lurie S (1998) Postharvest heat treatments. Postharvest Biol Technol 14: 257–269.

Lurie S (2006) The effect of high temperature treatment on quality of fruits and vegetables. Acta Hort 712: 165–173.

Lurie S, Crisosto CH (2005) Chilling injury in peach and nectarine. Postharvest Biol Technol 37: 195–208.

Lurie S, Klein JD (1991) Acquisition of low-temperature tolerance in tomatoes by exposure to high-temperature stress. J Am Soc Hort Sci 116: 1007–1012.

MacGillivray AJ, Rickwood D (1974) The heterogeneity of mouse-chromatin nonhistone proteins as evidenced by two-dimensional polyacrylamide-gel electrophoresis and ion-exchange chromatography. Eur J Biochem 41: 181–90.

Macheix JJ, Sapis JC, Fleuriet A (1991) Phenolic compounds and polyphenoloxidase in relation to browning in grapes and wines. Crit Rev Food Sci Nutr 30: 441–486.

Miquel M, Block MA, Joyard J, Dorne AJ, Dubacq JP, Kader JC, Douce R (1988) Protein-mediated transfer of phosphatidylcholine from liposomes to spinach chloroplast envelope membranes. Biochim Biophys Acta 937: 219–228.

Moco S, Bino RJ, Vorst O, Verhoevern HA, de Groot J, van Beek TA, Vervoort J, de Vos CHR (2006) A liquid chromatographymass spectrometry-based metabolome database for tomato. Plant Physiol 141: 1205–1218.

Moreno-Rosueno MA, Busch W, Benfey PN (2010) Omics met networks—using systems approaches to infer regulatory networks in plants. Curr Opin Plant Biol 13: 126–131.

Newton RP, Brenton AG, Smith CJ, Dudley E (2004) Plant proteome analysis by mass spectrometry: principles, problems, pitfalls and recent developments. Phytochemistry 65: 1449–1485.

Nilo R, Saffie C, Lilley K, Baeza-Yates R, Cambiazo V, Campos-Vargas R, González M, Meisel LA, Retamales J, Silva H, Orellana A (2010) Proteomic analysis of peach fruit mesocarp softening and chilling injury using difference gel electrophoresis (DIGE). BMC Genomics 11: 43.

Obenland DM, Crisosto CH, Rose JKC (2003) Expansin protein levels decline with the development of mealiness in peaches. Postharvest Biol Technol 29: 11–18.

O'Farrell PH (1975) High resolution two-dimensional electrophoresis of proteins. J Biol Chem 250: 4007–21.

Ogundiwin EA, Martí C, Forment J, Pons C, Granell A, Gradziel TM, Peace C, Crisosto CH (2008) Development of Chill Peach genomic tools and identification of cold-responsive genes in peach fruit. Plan Mol Biol 68: 379–397.

Oksman-Caldentey KM, Saito K (2005) Integrating genomics and metabolomics for engineering plant metabolic pathways. Curr Opin Biotechnol 16: 174–179.

Osorio S, Bombarely A, Giavalisco P, Usadel B, Stephens C, Aragüez I, Medina-Escobar N, Botella MA, Fernie AR, Valpuesta V (2011) Demethylation of oligogalacturonides by FaPE1 in the fruits of the wild strawberry *Fragaria vesca* triggers metabolic and transcriptional changes associated with defence and development of the fruit. J Exp Bot 62: 2855–2873.

Paull RE, Chen NJ (2000) Heat treatment and fruit ripening. Postharvest Biol Technol 2: 21–37.

Polenta GA, Calvete JJ, González CB (2007) Isolation and characterization of the main small heat shock protein induced in tomato pericarp by thermal treatment. FEBS J 274: 6447–6455.

Rabilloud T (2002) Two-dimensional gel electrophoresis in proteomics: old, old fashioned, but it still climbs up the mountains. Proteomics 2: 3–10.

Rabilloud T, Chevallet M, Luche V, Lelong C (2010) Two-dimensional gel electrophoresis in proteomics: past, present and future. J Proteom 73: 2064–2077.

Ramakrishna W, Deng Z, Ding C-K, Handa AK, Ozminkowski RH (2003) A novel small heat shock protein gene, *vis1*, contributes to pectin depolymerization and juice viscosity in tomato fruit. *Plant Physiol* 131: 725–735.

Renaut J, Hausman J-F, Bassett C, Artlip T, Cauchie H-M, Witters E, Wisniewski M (2008) Tree quantitative proteomic analysis of short photoperiod and low-temperature responses in bark tissues of peach (*Prunus persica* L. Batsch). Tree Genet Genomes 4: 589–600.

Renaut J, Hausman J-F, Wisniewski ME (2006) Proteomics and low-temperature studies: bridging the gap between gene expression and metabolism. Physiol Plant 126: 97–109.

Robertson D, Mitchell GP, Gilroy JS, Gerrish C, Bolwell GP, Slabas AR (1997) Differential extraction and protein sequencing reveals major differences in patterns of primary cell wall proteins from higher plants. J Biol Chem 272: 15841–15848.

Roessner-Tunali U, Luedemann A, Brust D, Fiehn O, Linke T, Willmitzer L, Fernie AR (2001) Metabolic profiling allows comprehensive phenotyping of genetically or environmentally modified plant systems. Plant Cell 13: 11–29.

Roessner-Tunali U, Hegemann B, Lytovchenko A, Carrari F, Bruedigam C, Granot D, Fernie AR (2003). Metabolic profiling of transgenic tomato plants overexpressing hexokinase reveals that the influence of hexose phosphorylation diminishes during fruit development. Plant Physiol 133: 84–99.

Rose JKC, Bashir S, Giovannoni JJ, Jahn MM, Saravana RS (2004) Tackling the plant proteome: practical approaches, hurdles and experimental tools. Plant J 39: 715–733.

Rossignol M (2001) Analysis of the plant proteome. Curr Opin Biotechnol 12: 131–134.

Rozenzvieg D, Elmaci C, Samach A, Lurie S, Porat R (2004) Isolation of four heat shock protein cDNAs from grapefruit peel tissue and characterization of their expression in response to heat and chilling temperature stresses. Physiol Plant 121: 421–428.

Sabehat A, Lurie S, Weiss D (1998) Expression of small heat shock proteins at low temperature: a possible role in protecting against chilling injuries. Plant Physiol 117: 651–658.

Sabehat A, Weiss D, Lurie S (1996) The correlation between heat-shock protein accumulation and persistence and chilling tolerance in tomato fruit. Plant Physiol 110: 531–537.

Saltveit ME (1991) Prior temperature exposure affects subsequent chilling sensitivity. Physiol Plant 82: 529–536.

Salveit ME, Morris LL (1990) Overview on chilling injury of horticultural crops. In: Wang CY (ed) Chilling Injury of Horticultural Crops. CRC Press, Boca Raton, FL, pp 3–16.

Santoni V, Kieffer S, Desclaux D, Masson F, Rabilloud T (2000) Membrane proteomics: use of additive main effects with multiplicative interaction model to classify plasma membrane proteins according to their solubility and electrophoretic properties. Electrophoresis 21: 3329–3344.

Sarry J-E, Sommere N, Sauvage F-X, Bergoin A, Rossignol M, Albagnac G, Romieu C (2004) Grape berry biochemistry revisited upon proteomic analysis of the mesocarp. Proteomics 4: 201–215.

Schagger H, von Jagow G (1991) Blue native electrophoresis for isolation of membrane protein complexes in enzymatically active form. Anal Biochem 199: 223–231.

Schauer N, Fernie AR (2006) Plant metabolomics: toward biological function and mechanism. Trends Plant Sci 11: 508–516.

Schauer N, Steinhauser D, Strelkov S, Schomburg D, Allison G, Moritz T, Lundgren K, Roessner-Tunali U, Forbes MG, Willmitzer L, Fernie AR, Kopka J (2005) GC-MS libraries for the rapid identification of metabolites in complex biological samples. FEBS Lett 579: 332–1337.

Schulze WX (2010) Proteomics approaches to understand protein phosphorylation in pathway modulation. Curr Opin Plant Biol 13: 280–287.

Sevillano L, Sanchez-Ballesta MT, Romojaro F, Flores FB (2009) Physiological, hormonal and molecular mechanisms regulating chilling injury in horticultural species. Postharvest technologies applied to reduce its impact. J Sci Food Agri 89: 555–573.

Shulaev V, Korban SS, Sosinski B, Abbott AA, Aldwinckle H, Folta KM, Iezzoni A, Main D, Aru´ s P, Dandekar AM, Lewers K, Brown SK, Davis TM, Gardiner SE, Potter D, Veilleux RE (2008) Multiple models for Rosaceae genomics. Plant Physiol 147: 985–1003.

Stitt M, Fernie AR (2003) From measurements of metabolites to metabolomics: an on the fly perspective illustrated by recent studies of carbon-nitrogen interactions. Curr Opin Biotechnol 14: 136–144.

Stitt M, Sulpice R, Keurentjes J (2010) Metabolic networks: how to identity key components in the regulation of metabolism and growth. Plant Physiol 152: 428–444.

Swanson CA (1998) Vegetables, fruits, and cancer risk: the role of phytochemicals. In: Bidlack WR, Omaye ST, Meskin MS, Topham DKW (eds) Phytochemicals. A New Paradigm. CRC Press, Boca Raton, FL, pp 1–12.

Thomashow MF (1999) Plant cold acclimation: freezing tolerance genes and regulatory mechanisms. Annu Rev Plant Physiol Plant Mol Biol 50: 571–599.

Tonutti P, Bonghi C, Ramina A (1996) Fruit firmness and ethylene biosynthesis in three cultivars of peach (*Prunus persica* L. Batsch). J Hort Sci 71: 141–147.

Tonutti P, Bonghi C, Ruperti B, Tornielli GB, Ramina A (1997) Ethylene evolution and 1-aminocyclopropane-1-carboxylate oxidase gene expression during early development and ripening of peach fruit. J Am Soc Hort Sci 122: 642–647.

Trainotti L, Bonghi C, Ziliotto F, Zanin D, Rasori A, Casadoro G, Ramina A, Tonutti P (2006) The use of microarray mPEACH1.0 to investigate transcriptome changes during transition from pre-climacteric to climacteric phase in peach fruit. Plant Sci 170: 606–813.

Trainotti L, Tadiello A, Casadoro G (2007) The involvement of auxin in the ripening of climacteric fruits comes of age: the hormone plays a role of its own and has an intense interplay with ethylene in ripening peaches. J Exp Bot 58: 3299–3308.

Trainotti L, Zanin D, Casadoro G (2003) A cell wall oriented genomic approach reveals a new and unexpected complexity of the softening in peaches. J Exp Bot 54: 1821–1832.

Trethewey RN (2004) Metabolite profiling as an aid to metabolic engineering in plants. Curr Opin Plant Biol 7: 196–201.

Unlu M, Morgan ME, Miden JS (1997) Difference gel electrophoresis: a single gel method for detecting changes in protein extracts. Electrophoresis 18: 2071–2077.

van Wijk KJ (2000) Proteomics of the chloroplast: experimentation and prediction. Trends Plant Sci 5: 420–425.

van Wijk KJ (2001) Challenges and prospects of plant proteomics. Plant Physiol 126: 501–508.

Vizoso P, Meisel LA, Tittarelli A, Latorre M, Saba J, Caroca R, Maldonado J, Cambiazo V, Campos-Vargas R, Gonzalez M, Orellana A, Silva H (2009) Comparative EST transcript profiling of peach fruits under different postharvest conditions reveals candidate genes associated with peach fruit quality. BMC Genomics 10: 423.

Wang L, Chen S, Kong W, Li S, Archbold DD (2006) Salicylic acid pretreatment alleviates chilling injury and affects the antioxidant system and heat shock proteins of peaches during cold storage. Postharvest Biol Technol 41: 244–251.

Wisniewski M, Bassett C, Gusta LV (2003) An overview of cold hardiness in woody plants: Seeing the forest through the trees. HortScience 38: 952–959.

Wisniewski M, Bassett C, Norelli JL, Artlip T, Renaut J (2007) Using biotechnology to improve resistance to environmental stress in fruit crops: The importance of understanding physiology. Acta Hort 738: 145–156.

Wisniewski M, Close T, Artlip T, Arora R (1996) Seasonal patterns of dehydrins and 70-kDa heat-shock proteins in bark tissues of eight species of woody plants. Physiol Plant 9: 496–505.

Wisniewski ME, Bassett CL, Renaut J, Farrell R Jr, Tworkoski T, Artlip TS (2006) Differential regulation of two dehydrin genes from peach (*Prunus persica*) by photoperiod, low temperature and water deficit. Tree Physiol 26: 575–584.

Wongsheree T, Ketsa S, van Doorn WG (2009) The relationship between chilling injury and membrane damage in lemon basil (*Ocimum* x *citriodourum*) leaves. Postharvest Biol Technol 51: 91–96.

Yan SP, Zhang QY, Tang ZC, Su WA, Sun WN (2006) Comparative proteomic analysis provides new insights into chilling stress responses in rice. Mol Cell Proteom 5: 484–496.

Zamboni A, Di Carli M, Guzzo F, Stocchero M, Zenoni S, Ferrarini A, Tononi P, Toffali K, Desiderio A, Lilley KS, Pe` ME, Benvenuto E, Delledonne M, Pezzotti M (2010) Identification of putative stage-specific grapevine berry biomarkers and omics data integration into networks. Plant Physiol 154: 1439–1459.

Zhang C, Ding Z, Xu X, Wang Q, Qin G, Tian S (2010) Crucial roles of membrane stability and its related proteins in the tolerance of peach fruit to chilling injury. Amino Acids 39: 181–194.

Zhang ZQ, Pang XQ, Duan XW, Ji ZL, Jiang YM (2005) Role of peroxidase in anthocyanin degradation in litchi fruit pericarp. Food Chem 90: 47–52.

Zhang ZQ, Pang XQ, Ji ZL, Jiang YM (2001) Role of anthocyanin degradation in litchi pericarp browning. Food Chem 75: 217–221.

Zhou H-W, Sonego L, Khalchitski A, Ben-Arie R, Lers A, Lurie S (2000a) Cell wall enzymes and cell wall changes in 'Flavortop' nectarines: mRNA abundance, enzyme activity, and changes in pectic and neutral polymers during ripening and in woolly fruit. J Am Soc Hort Sci 125: 630–637.

Zhou H-W, Ben-Arie R, Lurie S (2000b) Pectin esterase, polygalacturonase and gel formation in peach pectin fractions. Phytochemistry 55: 191–195.

13

Bioinformatics as a Tool for Stone Fruit Research

Sook Jung[1,*] and *Dorrie Main*[1,*]

ABSTRACT

Technological advances continue to generate large amounts of genomic, genetic and breeding data for *Prunus* and related species. These include annotated whole genome sequences, transcriptomic and expression data, proteomic and metabolomic data, genotypic and phenotypic data and genetic and physical maps. Analysis, storage, integration and dissemination of these data using bioinformatic tools and databases are essential to provide utility of the data for basic, translational and applied research. This chapter discusses the currently available bioinformatics resources for *Prunus* species. Sequence databases, comparative genome databases, gene expression databases, protein or metabolome databases and community oriented databases that integrate different types of data in closely related species are presented.

Keywords: *Prunus*, bioinformatics, database, genomic, genetic, breeding, integration

13.1 Introduction

In the last two decades, advances in high-throughput technology have revolutionized biology and turned it into an information-based science. The first spike in data generation occurred in the early 1990s when large scale sequencing became available through the use of Sanger technology for expressed sequence tag (EST) and bacterial artificial chromosome (BAC) sequencing (Adams et al. 1991; Shizuya et al. 1992). As with many other non-

[1]Department of Horticulture and Landscape, Washington State University, Pullman, WA 99164, USA.

*Corresponding authors: *sook jung@wsu.edu; dorrie@wsu.edu*

commodity crops, researchers would have to wait until nearer the end of the decade before funds were available to generate these resources for *Prunus* species. In 2001, the first *Prunus* ESTs became available in the National Center for Biotechnology (NCBI) dbEST repository (http://www.ncbi.nlm.nih.gov/nucest/), rising to over 100,000 Sanger sequenced ESTs by 2011. The same high quality sequencing technology would be used to generate the peach genome sequence, released publicly on April 1, 2010 (http://www.rosaceae.org/peach/genome). In the last few years, the advent of next generation technologies, such as 454 and Illumina, have significantly enhanced the ability to generate large-scale transcriptome and genome sequence at a fraction of the cost of Sanger sequencing. Similar advances in DNA Array, nuclear magnetic resonance (NMR), fourier transform infrared spectroscopy, fourier transform ion cyclotron resonance mass spectroscopy, high performace liquid chromotography, and mass spectrometry have generated large-scale gene expression, proteome and metabolome data for many species. Other data types include molecular marker data along with genetic mapping data and/or large-scale genotyping data of various varieties. More recently, high-throughput phenotypic and genotypic data are also being generated to study the relationship between genotype, phenotype and environment as well as for breeding purposes. All of these large-scale data require proper analysis, storage and integration to enhance our understanding of biology and to be utilized in further research. Bioinformatics tools and methodologies, therefore, have become an essential and integral part of biological research.

Large-scale genome and transcriptome data were initially accumulated for model species, but are now available for a wide range of species. Crop plants with sequenced genomes include rice (*Oryza sativa*) (International Rice Genome Sequencing Project 2005), grapevine (*Vitis vinifera*) (Jaillon et al. 2007), sorghum (*Sorghum bicolor*) (Paterson et al. 2009), cucumber (*Cucumis sativus*) (Huang et al. 2009), maize (*Zea mays*) (Schnable et al. 2009), soybean (*Glycine max*) (Schmutz et al. 2009), and three species in Rosaceae: peach (*Prunus persica*), apple (*Malus domestica*) (Velasco et al. 2010) and strawberry (*Fragaria vesca*) (Shulaev et al. 2010), with several others currently being sequenced.

Prunus belongs to the Rosaceae, a fruit, nut, ornamental and lumber family which contains a variety of crop species that are both biologically and economically important. In addition to the annotated whole genome sequences, a wealth of other genomic and genetic data is available for *Prunus* (Shulaev et al. 2008). These include BAC libraries, peach and apple physical maps, ESTs, numerous genetic maps in various species of *Prunus* and molecular markers that have been used for mapping and genotyping. Large-scale genotyping and phenotyping data are also being generated from various projects including RosBreed (Iezzoni et al. 2010). Currently, there are only limited proteomic and metabolomic data available for *Prunus*.

In this chapter, we discuss bioinformatics resources with a focus on *Prunus* (Table 13-1). Where relevant, resources for other Rosaceae member species are also discussed since both the similarities and differences among these member species present opportunity to increase our understanding in basic biology, as well as to transfer knowledge across the family for crop improvement. Resources available in other model species are also discussed, since these present a valuable tool in conducting and future planning of research in *Prunus*.

13.2 Gene, Genome and other Large-Scale Sequence Database Resources

Researchers seeking peer-reviewed publication of new genomic data are required to submit their genome or gene sequences to one of the three primary sequence databases: GenBank (Benson et al. 2011), European Nucleotide Archive (ENA) of the European Molecular Biology Laboratory (EMBL) (Leinonen et al. 2011); or the DNA Data Bank of Japan (DDBJ) (Kaminuma et al. 2011). These databases synchronize daily, and consequently, they are generally the most up to date source of genomic data for any species. A caveat to this rule is where genome sequences are released ahead of publication through other portals as in the case of the peach, sweet orange and mandarin genomes. GenBank is built and distributed by NCBI. They collect sequence data by submission from authors and bulk submission from high-throughput sequencing centers. The Entrez search site on the home page of NCBI (www.ncbi.nih.gov) allows researchers to search all the different data that it houses. Searching for information on *Prunus* species by typing *Prunus* [ORGN] in the search box, returns a results page categorized by the different databases it houses. As of August 1, 2011, this search returned one genome match for the *Prunus persica* chloroplast sequence, 131 genes, 104,556 ESTs, 7,662 unigenes, 3,740 proteins, 48,261 genome survey sequences, 37,226 nucleotide sequences, ten 3D structures, 200 species, 109 single nucleotide polymorphisms (SNPs), 10 large scale datasets in the short read archives (SRA), 345 probes, 60 gene expression omnibus (GEO) datasets of expression and molecular abundance profiles, 125 population datasets and 2,185 publications on *Prunus*. NCBI, is therefore, an excellent source of genomic data for *Prunus*, although some of the larger datasets such as the SRA are not generally very useful for biologists without access to bioinformatics expertise to help with analysis of data. Researchers can keep up to date on new data submitted to NCBI for their species of interest by creating a My NCBI account which saves search history and filters data, amongst other features. For more information on the data and tools available

Table 13-1 Publicly available Databases for Stone Fruit Research.

Type	Name	URL	Reference
Sequence Database	GenBank	http://www.ncbi.nlm.nih.gov/genbank/	Benson et al. 2011
	European Nucleotide Archive (ENA)	http://www.ebi.ac.uk/ena/	Leinonen et al. 2011
	DNA Data Bank of Japan (DDBJ)	http://www.ddbj.nig.ac.jp/	Kaminuma et al. 2011
	Phytozome	http://www.phytozome.org/	
Comparative Genome Database	CoGe (The Place to Compare Genomes)	http://synteny.cnr.berkeley.edu/CoGe/	Lyons et al. 2008
	The plant genome duplication database (PGDD)	http://chibba.agtec.uga.edu/duplication/	
	Plaza	http://bioinformatics.psb.ugent.be/plaza/	Proost et al. 2009
	GreenPhylD	http://greenphyl.cirad.fr/v2/cgi-bin/index.cgi	Rouard et al. 2011
	SALAD database	http://salad.dna.affrc.go.jp/salad/en/	Mihara et al. 2010
Transcriptome Database	NCBI's Gene Expression Omnibus (GEO)	http://www.ncbi.nlm.nih.gov/geo/	Wheeler et al. 2007
	ESTree DB	http://www.itb.cnr.it/estree/	Lazzari et al. 2008
Protein Database	UniProt	http://www.uniprot.org/	Magrane et al. 2011
	The Worldwide Protein Data Bank (wwPDB)	http://www.wwpdb.org/	Berman et al. 2007
	CATH database	http://www.cathdb.info/	Cuff et al. 2011
	SUPERFAMILY	http://supfam.cs.bris.ac.uk/SUPERFAMILY/	Wilson et al. 2007
	ARAMEMNON	http://aramemnon.uni-koeln.de/	Schwacke et al. 2003
	The Arabidopsis Interactions Viewer	http://bar.utoronto.ca/interactions/cgi-bin/arabidopsis_interactions_viewer.cgi	Geisler-Lee et al. 2007
	The Biomolecular Interaction Network Database (BIND)	http://download.baderlab.org/BINDTranslation/	Isserlin et al. 2011
	The Arabidopsis Subcellular Database (SUBA)	http://suba.plantenergy.uwa.edu.au/	Heazlewood et al. 2007
	The Plant Protein Phosphorylation DataBase (P3DB)	http://p3db.org/	
Metabolome Database	AtMetExpress	http://prime.psc.riken.jp/lcms/AtMetExpress/	Matsuda et al. 2010
	GolmMetabolome Database (GMD)	http://gmd.mpimp-golm.mpg.de/	Hummel et al. 2010
Community Database	Genome Database for Rosaceae (GDR)	http://www.rosaceae.org/	Jung et al. 2008

at NCBI (GenBank), DDBJ or EMBL (ENA), readers should refer to the excellent online tutorials at these sites and the yearly update published in the Nucleic Acids Research Database Addition published every January. The peach genome v1.0 was released by the International Peach Genome Consortium on April 1, 2010 through Phytozome (http://www.phytozome.org/peach.php), the Istituto di Genomica Applicata (IGA, www. http://services.appliedgenomics.org/projects/drupomics/intro/) and the Genome Database for Rosaceae (GDR, http://www.rosaceae.org/peach/genome). Peach v1.0 is a very high quality plant genome. It currently consists of eight pseudomolecules (scaffolds) representing the eight chromosomes of peach, and are numbered according to their corresponding linkage groups. The genome sequencing of the doubled haploid cultivar "Lovell" consisted of approximately 7.7 fold whole genome shotgun sequencing employing the highly accurate Sanger methodology, and was assembled using Arachne (Batzoglou et al. 2002). The assembled peach scaffolds cover nearly 99% of the peach genome, with over 92% having confirmed orientation. To further validate the quality of the assembly, 74,757 *Prunus* ESTs were queried against the genome at 90% identity and 85% coverage, and it was found that only ~2% were missing. While gene prediction and annotation, is an ongoing process, current estimates indicate that peach has 28,689 transcripts and 27,852 genes. Since its release, the peach genome portal on GDR has been visited 14,753 times by researchers from 80 countries over a 16 month period.

Phytozome is a joint project of the Department of Energy's Joint Genome Institute (JGI) and the Center for Integrative Genomics to facilitate comparative genomic studies amongst green plants. Families of orthologous and paralogous genes that represent the modern descendents of ancestral gene sets are constructed at key phylogenetic nodes. These families allow easy access to clade specific orthology/paralogy relationships as well as clade specific genes and gene expansions. As of release v7.0, Phytozome provides access to 25 sequenced and annotated green plant genomes which have been clustered into gene families at 11 evolutionarily significant nodes. Where possible, each gene has been annotated with the Protein Family Database (PFAM) (Finn et al. 2010), the euKaryotic Orthologous Groups (KOG) (Tatusov et al. 2003), Kyoto Encyclopedia of Genes and Genomes (KEGG) (Aoki and Kanehisa 2005), and Protein ANalysis THrough Evolutionary Relationships (PANTHER) (Mi et al. 2010) assignments, and publicly available annotations from NCBI Reference Sequence (RefSeq) (Pruitt et al. 2007), Universal Protein Resource (UniProt) (Magrane et al. 2011), TAIR (Lamesch et al. 2010). As it is still considered in draft status, the full range of comparative genomics tools for peach are not yet available, however, similar to the GDR and IGA peach genome sites, users can view the genome through the genome browser GBrowse, which provides access to transcripts, alternative transcripts, mapped ESTs, aligned plant peptides and genetic markers. User can also search their sequences against the peach genome sequence, transcripts and proteome though the Basic Local Alignment Search Tool (BLAST) and the BLAST-Like Alignment Tool (BLAT) servers, and download all the genome annotation through a bulk data site. The IGA peach site (http://services.appliedgenomics.org/projects/drupomics/intro/) also contains a peach GBrowse and BLAST tools. The IGA peach GBrowse provides access to the same tracks as phytozome but in addition they have RNA-seq Illumina profiles from tissues sequenced from cotyledon and embryo, fruit, leaf and root; repetitive sequences and microsatellites; and a profile of exact genome 20-mers.

Details on the resources and tools for the peach genome in GDR are provided in the Community Database Resources for Stone Fruit section of this chapter.

13.3 Comparative Genomics Database Resources

Comparative genomics is an area of study where the structure and function of genomes from different species or varieties are compared. An important goal of comparative genomics is to obtain an insight on how genomes have evolved. Detection of duplicated regions within species and syntenic regions among species plays an important role in studies of genome evolution. The comparison of genomes and their contents among species allows us to identify genes and other sequence features that are conserved among species and/or genes that are specific to certain clades. These results help us to infer what genes and other sequence features are responsible for the similarities and differences among species. Comparative genomics also allows us to transfer knowledge from well-studied species to less-studied species. With the increasing number of species with the whole genome sequences, several web-based databases are available to compare whole genome data. A few contain the whole genome sequences of *Prunus Persica* and two other publicly available sequenced genomes from the Rosaceae family: *Malus domestica* and *Fragaria vesca*. CoGe (The Place to Compare Genomes) (Lyons et al. 2008), contains whole genome data of 9873 species across all domains of life, including *Prunus persica* and *Malus domestica*. CoGe contains a series of web tools where users can select species of interest to view the genome data and/or perform some comparative analysis. The tools include OrganismView for searching organism, CoGeBlast for blasting sequences against the genomes of interest, FeatView for searching genomic features by name, SynMap for whole genome syntenic dotplot analyses and GEvo for high-resolution sequence analysis of genomic regions. SynMap is the tool for comparative genomics which uses DAGChainer (Haas et al. 2004) as the underlying software for synteny detection.

The plant genome duplication database (PGDD) contains 18 plant genomes including *Prunus persica, Malus domestica* and *Fragraria vesca.* PGDD contains similar tools as those in CoGE where users can explore plant genes in terms of intragenome or cross-genome syntenic relationships. Users can choose two species to view syntenic blocks in Dot Plot, can select a locus to view the regions that contain the locus along with the syntenic regions in multiple species, and can use BLAST to map the sequence of interest to selected genomes. PGDD uses MCscan package (Tang et al. 2008), which uses DAGchainer to predict pairwise segments, as the underlying software for synteny detection.

Plaza (Proost et al. 2009) is another platform where users can access plant genome data to perform comparative analyses. Plaza contains 23 species including *Malus domestica.* Data annotation includes primary gene annotation, gene family, orthologous genes, and functional annotation such as GO terms, InterPro Domains, and Reactome (pathway) data. In the Plaza Synteny Plot tool, users can start from a gene or a gene family to view the gene organization of all homologs of a gene family in selected species. In the Skyline plot, users can enter a gene locus to view an overview of the collinear regions that exist within a set of selected species. In WGDotplot, users can select two species to view syntenic blocks or select the same species twice to view all duplicated blocks within species. The colinear regions within species can also be viewed with Circle Plot. If available, the age of the colinear blocks, determined using Ks, is reported using a color code. From the Skyline plot and WGDotplot, users can access Multiplicon View to view the aligned gene strings of a set of homologous segments. The colinear regions are detected by i-ADHoRe (Simillion et al. 2008). In addition to those described above, Plaza provides other tools for genome evolution and colinearity research. The WGMapping tool allows users to choose all genes or selection of genes to display their location on the chromosomes along with the gene type. The functional clustering visualization tool provides an overview of the location and content of each functional cluster, detected using C-Hunter (Yi et al. 2007), on a chromosome-wide scale. Genomic sequences and genomic features can be viewed using Genome Browser. Tools such as Similarity heatmap, orthologous gene tool, Tree explorer and Gene family finder allows users to further explore gene family evolution.

The GreenPhylDB (Rouard et al. 2011) and Surveyed contained motif Alignment diagram and the Associating Dendrogram (SALAD) databases (Mihara et al. 2010) are related databases but they do not yet contain any *Prunus* or Rosaceae genome data. The difference between these two and the comparative genome databases mentioned above is that GreenPhylDB and SALAD do not contain data or tools for synteny analysis, but its focus is on gene family, phylogeny and ortholog/paralog analyses. GreenPhylD is a web resource for plant comparative and functional genomics. GreenPhylDB v2.0 contains 16 full genomes from the major phylum of plant evolution. The data include various lists of gene families, such as plant specific, phylum specific and species specific list, and tools to facilitate the comparisons. The SALAD database is a web-based resource for genome-wide comparative analysis of annotated protein sequences in plant. In SALAD, users can search for genes and view phylogenetic trees constructed from sequence alignment for a selected single motif or multiple motifs. Another functionality called sequence logo provides users with graphical representation of the sequence conservation of amino acids made from alignment of conserved motif in each node. Users can also compare the sequence logos among members of distinct nodes to evaluate conservation and diversity of amino acid at any sites in the conserved motif.

GDR (Genome Database for Rosaceae) (Jung et al. 2008) uses GBrowse_Syn (McKay et al. 2010), a synteny browser, to show the orthologous regions among the three sequenced genomes of Rosaceae, as detected by the Mercator program (Dewey 2007). GBrowse_Syn is hyperlinked to GBrowse so that users can access various genome annotation data from the conserved syntenic regions shown in GBrowse_Syn. Comparative genomics data made available in GDR thus allow users to explore other data, such as genomic features, anchored trait loci and genetic markers, in the orthologous regions.

13.4 Transcriptome and Gene Expression Database Resources

Initially, transcriptome analyses in *Prunus* focused on generating ESTs to identify candidate genes involved in different stages of plant development and responding to abiotic and biotic stresses. GDR contains the publicly available *Prunus* ESTs and their annotation. The *Prunus* unigene V4 contains 24,307 unigenes with 9,170 contigs and 15,128 singlets, assembled from 92,421 ESTs downloaded from GenBank dbEST. The ESTs were from seven *Prunus* species, 39 libraries, 21 tissues and 20 developmental stages. As an effort to develop an extensive *Prunus* EST database for identification and cloning of genes important to fruit and tree development, a peach transcriptome map have been constructed (Horn et al. 2005). The transcriptome map was constructed using a putative peach unigene set from a peach cDNA library of developing fruit mesocarp. The unigenes were mapped to the general *Prunus* genetic map by hybridizing both ESTs and core markers of the genetic map to BAC clones. The transcriptome data are available from GDR. With the availability of the whole genome sequences of peach, the *Prunus* ESTs, as well as other Rosaceae ESTs, are anchored to the predicted peach gene transcripts (28,692 sequences) from whole genome sequencing.

The best matches between the EST and the predicted genes are available in a spreadsheet with links to the predicted genes in the GDR peach GBrowse. ESTree DB (Lazzari et al. 2008) has a collection of ESTs from *Prunus persica* and *Prunus amygdalus*. The ESTs are from twelve peach libraries and three almond libraries produced in nine different labs. The peach unigene in the sixth release of ESTree DB contains 28,391 unigenes with 7,709 contigs and 20,682 singlets, assembled from 75,404 ESTs. The ESTree mapping project is underway to map about 200 additional ESTs on the peach transcriptome.

High-throughput transcriptome analysis using microarrays have been recently incorporated in various *Prunus* research projects. The microarray platforms developed in these projects include microPEACH1.0, GFNChile_Peach_0.9K_v1.0, and LBEG_Peach_1.5k_v1.0. The microPEACH1.0 consists of 4,806 70mer oligonucleotides designed from *Prunus persica* (peach) EST unigene clusters, mainly in the 3′ end terminal region. GFNChile_Peach_0.9K_v1.0 contains 847 cDNAs from a cv. Loring ripe peach fruit (*Prunus persica*) cDNA library. The cDNAs were selected from this library based on functional properties. The studies using these platforms have mostly been performed to investigate fruit quality traits and other important agronomic traits such as self-incompatibility and dormancy (summarized in Martinez-Gomez et al. 2011). The results are available from NCBI's GEO (Wheeler et al. 2007).

The latest and most efficient tool for transcriptome analysis involves the use of high-throughput sequencing technologies to sequence cDNA. This technology is called whole transcriptome shotgun sequencing (WTSS) or RNA-Seq. RNA-seq allows not only the assessment of the expression level of specific genes but also the detection of less-represented transcripts, allelic-specific expression of transcripts, post-transcriptional mutations and the expression of splice-variants. Even though it has been available for only a couple of years, this technology has been used extensively in humans, mammals and yeast and it is beginning to be used in plant species. *Arabidopsis* was the first plant species to be studied using this technology (Weber et al. 2007). Comparison of deep transcriptome sequencing with the EST database confirmed most of the annotated introns and identified thousands of novel alternatively spliced mRNA isoforms, suggesting at least 42% of intron-containing genes of *Arabidopsis* are alternatively spliced (Filichkin et al. 2010). These RNA-seq data (Lister et al. 2008; Filichkin et al. 2010), along with proteomics data (Baerenfaller et al. 2008; Castellana et al. 2008), have been used to revise the *Arabidopsis* gene models and are incorporated in the TAIR10 release (Lamesch et al. 2010). In *Prunus*, RNA-Seq technology is currently being utilized in projects to study PPV (Plum Pox Virus) resistance, graft incompatibility, fruit quality and flowering time in peach, apricot and plum, as summarized in Martinez-Gomez and coauthors (2011). The results are expected to be incorporated in GDR when the results are publicly available.

13.5 Protein and Metabolome Database

Proteomics and metabolomics refer to the large-scale study of proteins and metabolites. When transcriptomic, proteomic and metabolomic data are integrated, it can help to give us a more complete picture of a living organism in a specific condition. Due to the enormous complexity of the proteome, many different approaches are being taken to generate and catalog proteomics data, such as structural and functional annotation of proteins, protein expression and dynamics, stress and developmental responses, post-translational protein modifications and protein interactions. UniProt (Magrane et al. 2011) provides a comprehensive, high-quality and freely accessible resource of protein sequence and functional information. Currently (searched in July 2011), 2,019 protein entries are available for *Prunus*. Only 68 entries are in Swiss-Prot and the rest (2,951) are in Translated European Molecular Biology Laboratory Nucleotide Sequence Database (TrEMBL).

The Worldwide Protein Data Bank (wwPDB) (Berman et al. 2007) is the main structural protein database. It includes RCSB (A Resource for Studying Biological Macromolecules) PDB in USA, PDBe in Europe (Velankar and Kleywegt 2011) and PDBj in Japan. Currently, PDB contains the coordinates and related information of more than 74,000 structures of proteins, nucleic acids, protein/nucleic acid complexes and other macromolecules that have been determined using X-ray crystallography, NMR and electron microscopy techniques. For *Prunus*, ten structures are currently available from PDB including proteins from *Prunus persica*, *Prunus avium* and *Prunus dulcis*. Other protein structure databases include CATH (Cuff et al. 2011) and SUPERFAMILY (Wilson et al. 2007). The CATH (class, architecture, topology and homology) database provides a hierarchical classification of protein domain structures obtained from PDB. The classification class reflects the amino acid composition, architecture of the general shape of the protein domain and topology the way in which the protein folds into this architecture. SUPERFAMILY provides the prediction of protein domains of known structure in amino acid sequences. The classification of domains is hierarchical, based on nature of the similarity (sequence, evolutionary and structural), into class, fold, superfamily and family, following the structural classification of the protein (SCOP) database (Andreeva et al. 2004). SUPERFAMILY currently includes over 120 meta-genomes from environmental sequencing projects, and has added 2,354 plasmids and over 2,473 viral genomes and their taxonomy (Andreeva et al. 2004). Data from *Prunus* species have not yet been included, but those from *Malus domestica* and

Fragaria vesca, member species of Rosaceae, are available in SUPERFAMILY. Various web-based plant proteome-related databases are summarized in the database section of the proteomics subcommittee of the Multinational Arabidopsis Steering Committee Proteomics Subcommittee (MASCP) Web site (http://www.masc-proteomics.org/). Some of the databases are illustrated below. ARAMEMNON (Schwacke et al. 2003) is a database of plant membrane proteins with *Arabidopsis thaliana* as the reference model plant. Currently, the database holds all putative membrane proteins of five other plant species: grape (*Vitis vinifera*), poplar (*Populus trichocarpa*), rice (*Oryza sativa*), maize (*Zea mays*) and Brachypodium (*Brachypodium distachyon*). The *Arabidopsis* Interactions Viewer (Geisler-Lee et al. 2007) is an interaction database for *Arabidopsis thaliana* predicted from interacting orthologs in yeast (*Saccharomyces cerevisiae*), nematode worm (*Caenorhabditis elegans*), fruitfly (*Drosophila melanogaster*), and human (*Homo sapiens*). The database includes 70,944 predicted and 22,156 confirmed *Arabidopsis* interacting proteins. The confirmed *Arabidopsis* interacting proteins come from BIND, the Biomolecular Interaction Network Database (Isserlin et al. 2011), high-density *Arabidopsis* protein microarrays (Popescu et al. 2007, 2009) and other literature sources. The interactions in BIND were identified using several different methods, such as yeast two hybrid screens, but also via traditional biochemical methods. All subcellular localization data in the *Arabidopsis* Interactions Viewer is from SUBA, the *Arabidopsis* Subcellular Database (Heazlewood et al. 2007). Subcellular localization data in SUBA is brought together from various sources such as studies using chimeric fluorescent fusion proteins, proteomic surveys using mass spectrometry and literature. It also contains precompiled bioinformatic predictions for protein subcellular localizations from a set of 10 different prediction tools. Complex relational queries can be performed between these experimental and predicted datasets to find and collate evidence for the subcellular location of *Arabidopsis* proteins. The Plant Protein Phosphorylation DataBase (P3DB) (Gao et al. 2009) hosts protein phosphorylation data for five species: *Arabidopsis thaliana, Brassica napus, Glycine max, Medicago truncatula* and *Oryza sativa*. AtMetExpress (Matsuda et al. 2010) and the Golm Metabolome Database (Hummel et al. 2007, 2010) are metabolome databases. AtMetExpress contains data from a study of phytochemical accumulation during development of the model plant *Arabidopsis thaliana* using liquid chromatography-mass spectrometry in samples covering many growth stages and organs. The Golm Database contains data of mass spectra from biologically active metabolites quantified using gas chromatography coupled to mass spectrometry. It covers data from mammals, yeast, corynebacterium, model plants, such as crop plants and related wild species, as well as required non-sample controls.

13.6 Community Database Resources for Stone Fruits

Plant community databases provide access to all or most of the datasets for individual or closely related species. Such databases include The Arabidopsis Information Resource (TAIR, Lamesch et al. 2010), the Genome Database for Rosaceae (GDR, Jung et al. 2008), the Solanaceae Genomics Network (SGN, Bombarely et al. 2011), Gramene (Youens-Clark et al. 2011), TreeGenes (Wegrzyn et al. 2008) and MaizeGDB (Schaeffer et al. 2011). Community databases generally store comprehensively integrated data such as annotated sequences of genomes and transcriptomes, genetic data, and molecular and phenotypic diversity data. As a result of data integration, the value of individual types of data increase exponentially, providing essential resources to accelerate the molecular understanding of phenotypic traits and the use of DNA information in crop improvement.

Initiated in 2002, GDR is the sole community database for the Rosaceae family, which includes *Prunus* as well as *Fragaria, Malus, Pyrus, Rosa,* and *Rubus,* the other economically important genera in the family. The *Prunus* research community have generated data for the annotated peach genome sequence, peach physical map, a large collection of ESTs, peach transcriptome map, numerous genetic maps, genetically mapped traits (MTL and QTL), genotypic diversity data, publicly available breeding data with both phenotypic and genotypic data, and cultivar evaluation data for growers. The integration and standardization of the data is crucial for the data to be utilized by different types of users including genomicists, geneticists, molecular biologists, evolutionary biologists, bioinformaticists, breeders and growers. The purpose of GDR is to collect, curate, analyze, and integrate the data and provide efficient interfaces for user access to allow these numerous and complex data to be efficiently utilized. In the sections below, we describe the data and web interface, analysis tools and community tools available from GDR, together with work in progress and future plans. GDR tutorials are available at http://www.rosaceae.org/tutorials.

13.6.1 Data and Web Interface

13.6.1.1 Annotated Whole Genome Sequence

The whole genome sequence of peach v1.0, generated from DNA from the doubled haploid cultivar 'Lovell' by the International Peach Genome Initiative (IPGI) is available on GDR. Users can access the all the annotated peach genome data from the peach genome page. Through the graphic interface GBrowse (Donlin 2001; Stein et al. 2002), users can view various genomic features aligned to the peach genome, such as gene models, repeats, SNPs, as well as alignments of peach ESTs, repeats, genetic markers, and genes from other plant model species. Each feature has hyperlinks that lead to a page with sequences and other information, with further hyperlinks to external databases where applicable. The genetic marker feature in GBrowse is linked to the marker page in GDR where all the detailed marker information is available, such as primers, mapped positions, references, and links to CMap, the comparative map viewer in GDR. The mapped ESTs are also linked to the EST page in GDR, where all the detailed EST information is available. The peach genome page in GDR contains various downloadable files, including the fasta files of predicted peach gene transcripts, CDS (coding sequences), and predicted gene peptides. Excel files of gene transcripts with homologs to *Arabidopsis* genes, Swiss-Prot entries and TrEMBL entries are also available with hyperlinks to external databases. Other Excel downloadable files include various Rosaceae ESTs and *Prunus* genetic markers that map to the predicted peach genes and SNPs for peach and cherry with hyperlinks to GBrowse in GDR.

In addition, GDR contains similar annotated whole genome data for apple (Velasco et al. 2010) and strawberry (Shulaev et al. 2010), two other member species of Rosaceae with sequenced genomes. The orthologous regions among these species, detected by Mercator program (Dewey 2007), are displayed using GBrowse_Syn (McKay et al. 2010).

13.6.1.2 Annotated EST Unigene Data

GDR contains all the publicly available *Prunus* ESTs, as well as other Rosaceae ESTs, downloaded from the dbEST at NCBI (Gibney and Baxevanis 2011). Routine processing in GDR occurs in three stages: sequence filtering and trimming to obtain high-quality sequences, assembly into contigs to reduce the inherent redundancy and build unigene sets from the combined contigs and singlest, and sequence annotation. A unigene is available for *Prunus*, as well as *Malus*, *Fragaria*, *Rosa* and *Pyrus*. The assembled contigs and singlets for the four genera were assembled together to generate a putative unigene set for the entire Rosaceae ESTs. Other annotation includes putative function and Gene Ontology (Ashburner et al. 2000) association to contigs and ESTs by homology with Swiss-Prot, TrEMBL and InterPRO proteins (Mulder et al. 2007). Plant Structure Ontology (Ilic et al. 2007) is also utilized to annotate the ESTs with the tissue from which the ESTs are generated.

The *Prunus* unigene project page is a good starting point for an overview of the various annotated data for the unigenes. It displays the overall results of the project with a side bar containing links to the EST search sites, library details, protocols and downloads, putative homology and microsatellite analysis. A link to the gene ontology (GO) classification is also available for the *Prunus* and other genera assemblies as well as the Rosaceae family assembly. Downloadable data include batch sequence in fasta format, homology results file in Excel format with links into GDR and external databases and SSR/ORF/primers results in Excel format.

The EST search site is for those users who are interested in a subset of ESTs. They can choose to search ESTs of the entire Rosaceae or the genus of interest by selecting the appropriate tabs. Users can also either search ESTs or contigs. In each search page, ESTs or contigs can be searched by their name(s), assembly results, sequence features such as SSR or SNP, taxonomy, tissue type and putative function including match description, match organism and GO term. Users can also perform a batch search by uploading a file with EST names. Previous unigene versions are also available for search to help those who have been using an older version in their research. The results can be downloaded in fasta format or as a tab-delimited file with SWISS-PROT homology results containing hyperlinks back into the data for each sequence retrieved. Instead of displaying all the details on one page, the EST details page initially displays the clone information and the sequence with a side bar containing links to library details, unigene information, sequence homology, SSR/ORF information, map position and anchored BACs when applicable. A unigene information page provides the contig name and hyperlink for both the genus and family unigenes. The contig page gives similar annotation data for the contig with additional links to the SNP results and the comprising ESTs. For the ESTs anchored to peach BACs and/or to Rosaceae genetic maps, the EST detail page provides a link to view the ESTs' map positions using the GDR Map Viewer or CMap.

13.6.1.3 Genetic Map, Peach Physical Map, Transcriptome Map

GDR currently contains data for 25 genetic maps for *Prunus* as well as 23 maps for other Rosaceae member species. GDR uses CMap, the web-based comparative map tool, to allow users to compare maps from different cultivars and species. The comparative mapping facilitates the data transfer from well-studied species to less-studied ones. For example, the GDR map collection includes the TxE map, which is recognized as the reference map for *Prunus*. The TxE map, constructed from an almond x peach F_2 population, contains 826 markers with a total distance of 524 cM (Dirlewanger et al. 2004; Howad et al. 2005). The TxE map contain many markers that are used in the construction of maps of other *Prunus* species such as peach, apricot, sour cherry, plum x almond-peach hybrid and almond x peach, but also other Rosaceae species such as apple and pear. The essential colinearity of the anchored markers in the *Prunus* maps and the presence of large colinear blocks among different genera in Rosaceae, such as *Prunus* and *Malus* (Dirlewanger et al. 2004), enable comparative mapping, an invaluable tool for cross-utilization of data in Rosaceae. In addition to the directly-mapped genetic markers, the TxE map in GDR-CMap displays peach transcriptome map data, major trait loci affecting agronomic characters found in various *Prunus* species, pathogen resistance loci, and Rosaceae Conserved Orthologous Set (RosCOS). The peach transcriptome was established by the hybridization of the putative unique ESTs from a peach cDNA library of developing fruit mesocarp (Horn et al. 2005). The peach ESTs were anchored to the TxE map by hybridization to the genetically anchored peach BACs (Horn et al. 2005). The position of major trait loci have been established in the TxE map using the data from different linkage maps anchored with the TxE reference map (Dirlewanger et al. 2004). Candidate genes representing analogs of major resistance genes have been anchored to this map by the hybridization to the genetically anchored peach BACs (Lalli et al. 2005). 613 out of 3,818 RosCOS rosaceaous unigenes that correspond to single copy *Arabidopsis* genes, are also mapped to TxE reference map (Cabrera et al. 2009). GDR-CMap serves as an integrative tool in the utilization of the data anchored to the TxE reference map in the study of other species in *Prunus* and other genera in Rosaceae. The anchored features, such as marker and ESTs, in the map are also linked to the corresponding GDR sites so that all the relevant information for the features can be viewed. Markers that are anchored to peach genome have hyperlink to the peach GBrowse. Another important resource in GDR is the peach physical map data. The peach physical map (Zhebentyayeva et al. 2008) is constructed from two peach BAC libraries (Georgi et al. 2002) and the physical length of the map is estimated to be 303 cM, which is 104.5% of the peach genome. GDR uses two tools available from the WebAGCoL Package (Pampanwar et al. 2005) to display the current peach physical map.

13.6.1.4 Genetic Markers and Traits

To provide more details of the genetic markers and traits that have been used in genetic map development or genetic diversity studies, GDR contains an extensively annotated molecular marker database. Currently, over 1,400 extensively annotated markers are available from GDR. The marker annotation includes marker aliases, source cultivar, source description, primer sequences, PCR conditions, references, map position, associated ESTs and associated BACs. While annotation of trait data is at an initial stage, the traits are annotated in GDR with aliases, published symbol, curated trait category, taxon, trait description, screening method, map position and references.

The marker search site allows both a simple search by name and an advanced search with various search categories. The search category includes marker type, the species from which the marker is developed, the species to which the marker is mapped, map position, markers with associated BAC clones and markers with associated ESTs. Users can also upload a file of names to get the detailed data. In the trait search site, users can search trait by name, symbol, taxon or curated trait category.

13.6.1.5 DNA Polymorphism

GDR contains DNA polymorphism data from various projects on molecular diversity of Rosaceae species. Currently, data from nine different projects are available, three from *Prunus* diversity studies and the rest from *Malus* and *Pyrus* studies. All the current data are from projects with SSR markers. Users can query by the marker name or species to view the details of diversity projects and the genotype of the varieties used in the analyses. SNP markers from the RosBREED project (www.rosbreed.org) are being added.

13.6.1.6 Breeding Data

One of the newest components of GDR is the search/browse site for breeding data. GDR contains password-protected private breeding data as well as publicly available breeding data. Private breeding data includes data from the Washington Apple Breeding Program and Pacific North West Sweet Cherry Breeding Program. Public data includes data from the federally funded RosBreed project, a program designed to establish a sustainable marker-assisted breeding infrastructure for US Rosaceae crops. The data includes the varieties and their pedigrees, phenotyping and genotyping data, and experimental metadata. The current search interfaces allows users to search by datasets, variety name, trait values, molecular markers that have been used for genotyping, alleles and pedigree. From the result page, users can view detailed results for a variety or download an Excel file with all the phenotyping/genotyping results that user specified. Users can also select a pedigree by selecting a variety and number of ancestral and progeny generations to generate an input file for breeding software such as Pedimap and FlexQTL (Bink and van Eeuwijk 2009).

13.6.2 Analysis Tools

GDR web-based tools include a BLAST server, FASTA server, CAP3 Assembly server and SSR server. The FASTA/BLAST servers allow users to conduct sequence homology analyses against various sequence databases including annotated sequences in GDR. The databases include whole genome nucleotide and protein sequences of peach, strawberry and apple; ESTs of the Rosaceae or each genus from NCBI, genera-specific unigene sets, as well as a family-wide unigene; Rosaceae genomic or protein sequences from NCBI, peach, apple and cherry SNP sequences; *Arabidopsis* protein sequences from TAIR and the ESTs, unigene sets, SSR-containing ESTs from individual cDNA libraries of peach mesocarp, almond, octoploid strawberry and diploid strawberry. Peach mesocarp ESTs that are anchored to the peach BACs are also available for sequence analysis. Batch sequences can be uploaded for analysis and the results are returned as both raw aligned output and parsed out in Excel. The output in Excel has hyperlinks to the GDR and NCBI sites. FASTA formatted library files of both the sequences with or without matches are provided to allow the user to easily conduct further batch searches in GDR and other databases. An EST assembly server using the CAP3 program is also available so that users can assemble their own EST sets. The server returns the raw output, a summary report and a fasta file containing the combined contig sequences and singlet sequences, which are also available as individual files. The contig file lists the contig number and comprising clone names in the comment line for each assembled transcript. Also available is a SSR server that allows user-defined SSRs to be identified in uploaded sequences. Users can also choose to run Primer3 along with the SSR-detection program to generate primer sets for the SSRs. The results are returned in an excel file containing all the SSRs, primers, ORFs in sequence, product size, with summary information on motif number and type.

13.6.3 Community Resources

GDR provides access to community-based news on various pages under the 'community' header bar, such as Rosaceae genomics, USRosEXEC, conferences, meetings, funding, employment, mailing lists and message boards. USRosEXEC stands for US Rosaceae Genomics, Genetics and Breeding Executive Committee, which serves as a communication and coordination focal point for the community. The USRosEXEC page provides the official documents, meeting minutes, membership and subcommittee information. Several mailing lists, in addition to the GDR mailing list, are available to serve the community with information for specific interests or purposes, and the archives can be viewed through the message board sites. All the publications in Rosaceae genomics and genetics are also available in GDR through the publication search site.

13.6.4 Future Directions

With the availability of the whole genome sequences of peach, the future direction of GDR will include further integrating the annotated whole genome data with other genomics, genetics and breeding data to accelerate the usage of DNA information in crop improvement as well as to improve our knowledge on various aspects of *Prunus* and Rosaceae biology. The whole genome data, that is currently available to view and download through GBrowse, will be available to search by various categories so that users can directly access the genes by querying. The genetic markers will also be queried by the anchored genome position and the neighboring trait locus. A PlantCyc database will be constructed for *Prunus* using the predicted genes from the peach genome sequence. The effort toward collecting and curating trait data, including MTL and QTL, will also be continued. When data for genetic markers, DNA polymorphism and trait data are integrated and easily searchable, users will be able to query for markers that are close to the trait of interest and polymorphic between the varieties of interest.

13.7 Conclusion

The recent addition of whole genome sequences of *Prunus* and other two Rosaceae member species to the already existing wealth of genetic data has brought great opportunity for researchers to accelerate *Prunus* research. Multiple whole genome sequences in one family allow genome level comparative analysis to gain evolutionary insight as well as transfer knowledge among species. The second and third generations of high-throughput sequencing technology now allows resequencing of multiple varieties to catalog sequence variations and quantitative gene expression analyses. The proteome and metabolome analyses are still in infancy in *Prunus* but the advances of tools and methodology with other model species will aid greatly in planning and adopting those analyses in *Prunus* research. Integration of different types of data is critical in the interpretation and utilization of these data and hence the role of the community databases to bring together genomics, genetics, breeders and growers data will become increasingly important.

References

Adams MD, Kelley JM, Gocayne JD, Dubnick M, Polymeropoulos MH, Xiao H, Merril CR, Wu A, Olde B, Moreno RF, et al. (1991) Complementary DNA sequencing: expressed sequence tags and human genome project. Science 252: 1651–1656.

Andreeva A, Howorth D, Brenner SE, Hubbard TJ, Chothia C, Murzin AG (2004) SCOP database in 2004: refinements integrate structure and sequence family data. Nucl Acids Res 32: D226–D229.

Aoki KF, Kanehisa M (2005) Using the KEGG Database Resource. Current Protocols of Bioinformatics Chapter 1: Unit 1.12.

Ashburner M, Ball CA, Blake JA, Botstein D, Butler H, Cherry JM, Davis AP, Dolinski K, Dwight SS, Eppig JT, Harris MA, Hill DP, Issel-Tarver L, Kasarskis A, Lewis S, Matese JC, Richardson JE, Ringwald M, Rubin GM, Sherlock G (2000) Gene ontology: tool for the unification of biology. The Gene Ontology Consortium. Nat Genet 25: 25–9.

Baerenfaller K, Grossmann J, Grobei MA, Hull R, Hirsch-Hoffmann M, Yalovsky S, Zimmermann P, Grossniklaus U, Gruissem W, Baginsky S (2008) Genome-scale proteomics reveals *Arabidopsis thaliana* gene models and proteome dynamics. Science 320: 938–941.

Batzoglou S, Jaffe DB, Stanley K, Butler J, Gnerre S, Mauceli E, Berger B, Mesirov JP, Lander ES (2002) ARACHNE: a whole-genome shotgun assembler. Genome Res 13: 91–96.

Benson DA, Karsch-Mizrachi I, Lipman DJ, Ostell J, Sayers EW (2011) GenBank. Nucl Acids Res 2011 Jan; 39(Database iss): D32–37. Epub 2010 Nov 10.

Berman H, Henrick K, Nakamura H, Markley JL (2007) The worldwide Protein Data Bank (wwPDB): ensuring a single, uniform archive of PDB data. Nucl Acids Res 35: D301–D303.

Bink MC, van Eeuwijk FA (2009) A Bayesian QTL linkage analysis of the common dataset from the 12th QTLMAS workshop. BMC Proc 3 Suppl 1: S4.

Bombarely A, Menda N, Tecle IY, Buels RM, Strickler S, Fischer-York T, Pujar A, Leto J, Gosselin J, Mueller LA (2011) The Sol Genomics Network (solgenomics.net): growing tomatoes using Perl. Nucl Acids Res 39(Database iss): D1149–155.

Cabrera A, Kozik A, Howad W, Arus P, Iezzoni AF, van der Knaap E (2009) Development and bin mapping of a Rosaceae Conserved Ortholog Set (COS) of markers. BMC Genomics 10: 562.

Castellana NE, Payne SH, Shen Z, Stanke M, Bafna V, Briggs SP (2008) Discovery and revision of Arabidopsis genes by proteogenomics. Proc Natl Acad Sci USA 105: 21034–21038.

Cuff AL, Sillitoe I, Lewis T, Clegg AB, Rentzsch R, Furnham N, Pellegrini-Calace M, Jones D, Thornton J, Orengo CA (2011) Extending CATH: increasing coverage of the protein structure universe and linking structure with function. Nucl Acids Res 39 (Database iss): D420–426.

Dewey CN (2007) Aligning multiple whole genomes with Mercator and MAVID. Meth Mol Biol 395: 221–36.

Dirlewanger E, Graziano E, Joobeur T, Garriga-Caldere F, Cosson P, Howad W, Arus P (2004) Comparative mapping and marker-assisted selection in Rosaceae fruit crops. Proc Natl Acad Sci USA 101: 9891–9896.

Donlin MJ (2007) Using the Generic Genome Browser (GBrowse). Curr Protoc Bioinformatics. 2007, Chapter 9, Unit 9 9.

Filichkin SA, Priest HD, Givan SA, Shen RK, Bryant DW, Fox SE, Wong WK, Mockler TC (2010) Genome-wide mapping of alternative splicing in *Arabidopsis thaliana*. Genome Res 20: 45–58

Finn RD, Mistry J, Tate J, Coggill P, Heger A, Pollington JE, Gavin OL, Gunasekaran P, Ceric G, Forslund K, Holm L, Sonnhammer EL, Eddy SR, Bateman A (2010) The Pfam protein families database. Nucl Acids Res 38 (Database iss): D211–222.

Gao J, Agrawal GK, Thelen JJ, Xu D (2009) P3DB: a plant protein phosphorylation database. Nucl Acids Res 37(Database iss): D960–962.

Geisler-Lee J, Caldwell C, Gallie DR (2010) Expression of the ethylene biosynthetic machinery in maize roots is regulated in response to hypoxia. J Exp Bot 61: 857–71.

Georgi LL, Wang Y, Yverggniaux D, Ormsbee T, Inigo M, Reighard GL, Abbott AG (2002) Construction of a BAC library and its application to the identification of simple sequence repeats in peach [*Prunus persica* (L.) Batsch]. Theor Appl Genet 105: 1151–1158.

Gibney G, Baxevanis AD (2011) Searching NCBI Databases using Entrez. Current Protocols of Bioinformatics Chapter 1: Unit 1.3.

Haas BJ, Delcher AL, Wortman JR, Salzberg SL (2004) DAGchainer: a tool for mining segmental genome duplications and synteny. Bioinformatics 20: 3643–3646.

Heazlewood JL, Verboom RE, Tonti-Filippini J, Small I, Millar AH (2007) SUBA: the Arabidopsis Subcellular Database. Nucl Acids Res 35 (Database iss): D213–218.

Horn R, Lecouls AC, Callahan A, Dandekar A, Garay L, McCord P, Howad W, Chan H, Verde I, Main D, Jung S, Georgi L, Forrest S, Mook J, Zhebentyayeva T, Yu Y, Kim HR, Jesudurai C, Sosinski B, Arús P, Baird V, Parfitt D, Reighard G, Scorza R, Tomkins J, Wing R, Abbott

AG (2005) Candidate gene database and transcript map for peach, a model species for fruit trees. Theor Appl Genet 110: 1419–1428.
Howad W, Yamamoto T, Dirlewanger E, Testolin R, Cosson P, Cipriani G, Monforte AJ, Georgi L, Abbott AG, Arús P (2005) Mapping with a few plants: using selective mapping for microsatellite saturation of the *Prunus* reference map. Genetics 171: 1305–1309.
Huang S, Li R, Zhang Z, Li L, Gu X, Fan W, Lucas WJ, Wang X, Xie B, Ni P, Ren Y, Zhu H, Li J, Lin K, Jin W, Fei Z, Li G, Staub J, Kilian A, van der Vossen EA, Wu Y, Guo J, He J, Jia Z, Ren Y, Tian G, Lu Y, Ruan J, Qian W, Wang M, Huang Q, Li B, Xuan Z, Cao J, Asan, Wu Z, Zhang J, Cai Q, Bai Y, Zhao B, Han Y, Li Y, Li X, Wang S, Shi Q, Liu S, Cho WK, Kim JY, Xu Y, Heller-Uszynska K, Miao H, Cheng Z, Zhang S, Wu J, Yang Y, Kang H, Li M, Liang H, Ren X, Shi Z, Wen M, Jian M, Yang H, Zhang G, Yang Z, Chen R, Liu S, Li J, Ma L, Liu H, Zhou Y, Zhao J, Fang X, Li G, Fang L, Li Y, Liu D, Zheng H, Zhang Y, Qin N, Li Z, Yang G, Yang S, Bolund L, Kristiansen K, Zheng H, Li S, Zhang X, Yang H, Wang J, Sun R, Zhang B, Jiang S, Wang J, Du Y, Li S (2009) The genome of the cucumber, *Cucumis sativus* L. Nat Genet 41: 1275–81. http://www.ncbi.nlm.nih.gov/pubmed/19881527
Hummel J, Strehmel N, Selbig J, Walther D, Kopka J (2010) Decision tree supported substructure prediction of metabolites from GC-MS profiles. Metabolomics 6: 322–333.
Hummel J, Selbig J, Walther D, Kopka J (2007) The Golm Metabolome Database: a database for GC-MS based metabolite profiling. In: Nielsen J, Jewett MC (eds) Metabolomics. Springer, Berlin, Heidelberg, New York, pp 75–96.
International Rice Genome Sequencing Project (2005) The map-based sequence of the rice genome. Nature 436: 793–800.
Ilic K, Kellogg EA, Jaiswal P, Zapata F, Stevens PF, Vincent LP, Avraham S, Reiser L, Pujar A, Sachs MM, Whitman NT, McCouch SR, Schaeffer ML, Ware DH, Stein LD, Rhee SY (2007) The plant structure ontology, a unified vocabulary of anatomy and morphology of a flowering plant. Plant Physiol 143: 587–599.
Isserlin R, El-Badrawi RA, Bader GD (2011) The Biomolecular Interaction Network Database in PSI-MI 2.5. Database 2011: baq037.
Jaillon O, Aury JM, Noel B, Policriti A, Clepet C, Casagrande A, Choisne N, Aubourg S, Vitulo N, Jubin C, Vezzi A, Legeai F, Hugueney P, Dasilva C, Horner D, Mica E, Jublot D, Poulain J, Bruyère C, Billault A, Segurens B, Gouyvenoux M, Ugarte E, Cattonaro F, Anthouard V, Vico V, Del Fabbro C, Alaux M, Di Gaspero G, Dumas V, Felice N, Paillard S, Juman I, Moroldo M, Scalabrin S, Canaguier A, Le Clainche I, Malacrida G, Durand E, Pesole G, Laucou V, Chatelet P, Merdinoglu D, Delledonne M, Pezzotti M, Lecharny A, Scarpelli C, Artiguenave F, Pè ME, Valle G, Morgante M, Caboche M, Adam-Blondon AF, Weissenbach J, Quétier F, Wincker P (2007) French-Italian Public Consortium for Grapevine Genome Characterization. The grapevine genome sequence suggests ancestral hexaploidization in major angiosperm phyla. Nature 449: 463–467.
Jung S, Staton M, Lee T, Blenda A, Svancara R, Abbott A, Main D (2008) GDR (Genome Database for Rosaceae): integrated web-database for Rosaceae genomics and genetics data. Nucl Acids Res 36 (Database iss): D1034–1040.
Kaminuma E, Kosuge T, Kodama Y, Aono H, Mashima J, Gojobori T, Sugawara H, Ogasawara O, Takagi T, Okubo K, Nakamura Y (2011) DDBJ progress report. Nucl Acids Res. 2011 Jan; 39(Database iss): D22–27. Epub 2010 Nov 9.
Lalli DA, Decroocq V, Blenda AV, Schurdi-Levraud V, Garay L, Le Gall O, Damsteegt V, Reighard GL, Abbott AG (2005) Identification and mapping of resistance gene analogs (RGAs) in *Prunus*: a resistance map for *Prunus*. Theor Appl Genet 111: 1504–1513.
Lazzari B, Caprera A, Vecchietti A, Merelli I, Barale F, Milanesi L, Stella A, Pozzi C (2008) Version VI of the ESTree db: an improved tool for peach transcriptome analysis. BMC Bioinformatics 9 (Suppl 2): S9.
Lamesch P, Dreher K, Swarbreck D, Sasidharan R, Reiser L, Huala E (2010) Using the Arabidopsis information resource (TAIR) to find information about Arabidopsis genes. Current Protocols of Bioinformatics. Chapter 1: Unit1.11.

Leinonen R, Akhtar R, Birney E, Bower L, Cerdeno-Tárraga A, Cheng Y, Cleland I, Faruque N, Goodgame N, Gibson R, Hoad G, Jang M, Pakseresht N, Plaister S, Radhakrishnan R, Reddy K, Sobhany S, Ten Hoopen P, Vaughan R, Zalunin V, Cochrane G (2011) The European Nucleotide Archive. Nucl Acids Res 39 (Database iss): D28–31.

Lister R, O'Malley RC, Tonti-Filippini J, Gregory BD, Berry CC, Millar AH, Ecker JR (2008) Highly integrated single-base resolution maps of the epigenome in Arabidopsis. Cell 133: 523–36.

Martínez-Gómez P, Crisosto CH, Bonghi C, Rubio M (2011) New approaches to *Prunus* transcriptome analysis. Genetica 139: 755–69.

McKay SJ, Vergara IA, Stajich JE (2010) Using the Generic Synteny Browser (GBrowse_syn). Current Protocols of Bioinformatics Chapter 9: Unit 9.12.

Mi H, Dong Q, Muruganujan A, Gaudet P, Lewis S, Thomas PD (2010) PANTHER version 7: improved phylogenetic trees, orthologs and collaboration with the Gene Ontology Consortium. Nucl Acids Res 38 (Database iss): D204–210.

Mihara M, Itoh T, Izawa T (2010) SALAD database: a motif-based database of protein annotations for plant comparative genomics. Nucl Acids Res 38 (Database iss): D835–842.

Mulder NJ, Apweiler R, Attwood TK, Bairoch A, Bateman A, Binns D, Bork P, Buillard V, Cerutti L, Copley R, Courcelle E, Das U, Daugherty L, Dibley M, Finn R, Fleischmann W, Gough J, Haft D, Hulo N, Hunter S, Kahn D, Kanapin A, Kejariwal A, Labarga A, Langendijk-Genevaux PS, Lonsdale D, Lopez R, Letunic I, Madera M, Maslen J, McAnulla C, McDowall J, Mistry J, Mitchell A, Nikolskaya AN, Orchard S, Orengo C, Petryszak R, Selengut JD, Sigrist CJ, Thomas PD, Valentin F, Wilson D, Wu CH, Yeats C (2007) New developments in the InterPro database. Nucl Acids Res 35: D224–228.

Pampanwar V, Engler F, Hatfield J, Blundy S, Gupta G, Soderlund C (2005) FPC Web tools for rice, maize, and distribution. Plant Physiol 138: 116–126.

Paterson AH, Bowers JE, Bruggmann R, Dubchak I, Grimwood J, Gundlach H, Haberer G, Hellsten U, Mitros T, Poliakov A, Schmutz J, Spannagl M, Tang H, Wang X, Wicker T, Bharti AK, Chapman J, Feltus FA, Gowik U, Grigoriev IV, Lyons E, Maher CA, Martis M, Narechania A, Otillar RP, Penning BW, Salamov AA, Wang Y, Zhang L, Carpita NC, Freeling M, Gingle AR, Hash CT, Keller B, Klein P, Kresovich S, McCann MC, Ming R, Peterson DG, Mehboob-ur-Rahman, Ware D, Westhoff P, Mayer KF, Messing J, Rokhsar DS (2009) The *Sorghum bicolor* genome and the diversification of grasses. Nature 457: 551–6.

Popescu SC, Popescu GV, Bachan S, Zhang Z, Seay M, Gerstein M, Snyder M, Dinesh-Kumar SP (2007) Differential binding of calmodulin-related proteins to their targets revealed through high-density Arabidopsis protein microarrays. Proc Natl Acad Sci USA 104: 4730–5.

Popescu SC, Popescu GV, Bachan S, Zhang Z, Gerstein M, Snyder M, Dinesh-Kumar SP (2009) MAPK target networks in *Arabidopsis thaliana* revealed using functional protein microarrays. Genes Dev 23: 80–92.

Proost S, Van Bel M, Sterck L, Billiau K, Van Parys T, Van de Peer Y, Vandepoele K (2009) PLAZA: a comparative genomics resource to study gene and genome evolution in plants. Plant Cell 21: 3718–3731.

Pruitt KD, Tatusova T, Maglott DR. NCBI reference sequences (RefSeq): a curated non-redundant sequence database of genomes, transcripts and proteins (2007) Nucl Acids Res 35(Database iss): D61–65.

Rouard M, Guignon V, Aluome C, Laporte MA, Droc G, Walde C, Zmasek CM, Périn C, Conte MG (2011) GreenPhylDB v2.0: comparative and functional genomics in plants. Nucl Acids Res 39 (Database iss): D1095–1102.

Schaeffer ML, Harper LC, Gardiner JM, Andorf CM, Campbell DA, Cannon EK, Sen TZ, Lawrence CJ (2011) MaizeGDB: curation and outreach go hand-in-hand. Database 2011: bar022.

Schnable PS, Ware D, Fulton RS, Stein JC, Wei F, Pasternak S, Liang C, Zhang J, Fulton L, Graves TA, Minx P, Reily AD, Courtney L, Kruchowski SS, Tomlinson C, Strong C, Delehaunty K, Fronick C, Courtney B, Rock SM, Belter E, Du F, Kim K, Abbott RM, Cotton M, Levy A,

Marchetto P, Ochoa K, Jackson SM, Gillam B, Chen W, Yan L, Higginbotham J, Cardenas M, Waligorski J, Applebaum E, Phelps L, Falcone J, Kanchi K, Thane T, Scimone A, Thane N, Henke J, Wang T, Ruppert J, Shah N, Rotter K, Hodges J, Ingenthron E, Cordes M, Kohlberg S, Sgro J, Delgado B, Mead K, Chinwalla A, Leonard S, Crouse K, Collura K, Kudrna D, Currie J, He R, Angelova A, Rajasekar S, Mueller T, Lomeli R, Scara G, Ko A, Delaney K, Wissotski M, Lopez G, Campos D, Braidotti M, Ashley E, Golser W, Kim H, Lee S, Lin J, Dujmic Z, Kim W, Talag J, Zuccolo A, Fan C, Sebastian A, Kramer M, Spiegel L, Nascimento L, Zutavern T, Miller B, Ambroise C, Muller S, Spooner W, Narechania A, Ren L, Wei S, Kumari S, Faga B, Levy MJ, McMahan L, Van Buren P, Vaughn MW, Ying K, Yeh CT, Emrich SJ, Jia Y, Kalyanaraman A, Hsia AP, Barbazuk WB, Baucom RS, Brutnell TP, Carpita NC, Chaparro C, Chia JM, Deragon JM, Estill JC, Fu Y, Jeddeloh JA, Han Y, Lee H, Li P, Lisch DR, Liu S, Liu Z, Nagel DH, McCann MC, SanMiguel P, Myers AM, Nettleton D, Nguyen J, Penning BW, Ponnala L, Schneider KL, Schwartz DC, Sharma A, Soderlund C, Springer NM, Sun Q, Wang H, Waterman M, Westerman R, Wolfgruber TK, Yang L, Yu Y, Zhang L, Zhou S, Zhu Q, Bennetzen JL, Dawe RK, Jiang J, Jiang N, Presting GG, Wessler SR, Aluru S, Martienssen RA, Clifton SW, McCombie WR, Wing RA, Wilson RK (2009) The B73 maize genome: complexity, diversity, and dynamics. Science 326: 1112–5.

Schwacke R, Schneider A, van der Graaff E, Fischer K, Catoni E, Desimone M, Frommer WB, Flügge UI, Kunze R (2003) ARAMEMNON, a novel database for Arabidopsis integral membrane proteins. Plant Physiol 131: 16–26.

Shizuya H, Birren B, Kim UJ, Mancino V, Slepak T, Tachiiri Y, Simon M (1992) Cloning and stable maintenance of 300-kilobase-pair fragments of human DNA in *Escherichia coli* using an F-factor-based vector. Proc Natl Acad Sci USA 89: 8794–8797.

Shulaev V, Sargent DJ, Crowhurst RN, Mockler TC, Folkerts O, Delcher AL, Jaiswal P, Mockaitis K, Liston A, Mane SP, Burns P, Davis TM, Slovin JP, Bassil N, Hellens RP, Evans C, Harkins T, Kodira C, Desany B, Crasta OR, Jensen RV, Allan AC, Michael TP, Setubal JC, Celton JM, Rees DJ, Williams KP, Holt SH, Ruiz Rojas JJ, Chatterjee M, Liu B, Silva H, Meisel L, Adato A, Filichkin SA, Troggio M, Viola R, Ashman TL, Wang H, Dharmawardhana P, Elser J, Raja R, Priest HD, Bryant DW Jr, Fox SE, Givan SA, Wilhelm LJ, Naithani S, Christoffels A, Salama DY, Carter J, Lopez Girona E, Zdepski A, Wang W, Kerstetter RA, Schwab W, Korban SS, Davik J, Monfort A, Denoyes-Rothan B, Arus P, Mittler R, Flinn B, Aharoni A, Bennetzen JL, Salzberg SL, Dickerman AW, Velasco R, Borodovsky M, Veilleux RE, Folta KM (2010) The genome of woodland strawberry (*Fragaria vesca*). Nat Genet 43: 109–116.

Simillion C, Janssens K, Sterck L, Van de Peer Y (2008) i-ADHoRe 2.0: An improved tool to detect degenerated genomic homology using genomic profiles. Bioinformatics 24: 127–128.

Stein LD, Mungall C, Shu S, Caudy M, Mangone M, Day A, Nickerson E, Stajich JE, Harris TW, Arva A, Lewis S (2002) The Generic Genome Browser: A building block for a model organism system database. Genome Res 12: 1599–1610.

Tang H, Bowers JE, Wang X, Ming R, Alam M, Paterson AH (2008) Synteny and collinearity in plant genomes. Science 320: 486–488.

Tatusov RL, Fedorova ND, Jackson JD, Jacobs AR, Kiryutin B, Koonin EV, Krylov DM, Mazumder R, Mekhedov SL, Nikolskaya AN, Rao BS, Smirnov S, Sverdlov AV, Vasudevan S, Wolf YI, Yin JJ, Natale DA (2003) The COG database: an updated version includes eukaryotes. BMC Bioinformatics 4: 41.

Velasco R, Zharkikh A, Affourtit J, Dhingra A, Cestaro A, Kalyanaraman A, Fontana P, Bhatnagar SK, Troggio M, Pruss D, Salvi S, Pindo M, Baldi P, Castelletti S, Cavaiuolo M, Coppola G, Costa F, Cova V, Dal Ri A, Goremykin V, Komjanc M, Longhi S, Magnago P, Malacarne G, Malnoy M, Micheletti D, Moretto M, Perazzolli M, Si-Ammour A, Vezzulli S, Zini E, Eldredge G, Fitzgerald LM, Gutin N, Lanchbury J, Macalma T, Mitchell JT, Reid J, Wardell B, Kodira C, Chen Z, Desany B, Niazi F, Palmer M, Koepke T, Jiwan D, Schaeffer S, Krishnan V, Wu C, Chu VT, King ST, Vick J, Tao Q, Mraz A, Stormo A, Stormo K, Bogden R, Ederle D, Stella A, Vecchietti A, Kater MM, Masiero S, Lasserre P, Lespinasse Y, Allan

AC, Bus V, Chagné D, Crowhurst RN, Gleave AP, Lavezzo E, Fawcett JA, Proost S, Rouzé P, Sterck L, Toppo S, Lazzari B, Hellens RP, Durel CE, Gutin A, Bumgarner RE, Gardiner SE, Skolnick M, Egholm M, Van de Peer Y, Salamini F, Viola R (2010) The genome of the domesticated apple (*Malus* x *domestica* Borkh.). Nat Genet 42: 833–839.

Velankar S, Alhroub Y, Alili A, Best C, Boutselakis HC, Caboche S, Conroy MJ, Dana JM, van Ginkel G, Golovin A, Gore SP, Gutmanas A, Haslam P, Hirshberg M, John M, Lagerstedt I, Mir S, Newman LE, Oldfield TJ, Penkett CJ, Pineda-Castillo J, Rinaldi L, Sahni G, Sawka G, Sen S, Slowley R, Sousa da Silva AW, Suarez-Uruena A, Swaminathan GJ, Symmons MF, Vranken WF, Wainwright M, Kleywegt GJ (2011) PDBe: Protein Data Bank in Europe. Nucl Acids Res 39 (suppl 1): D402–D410.

Weber APM, Weber KL, Carr K, Wilkerson C, Ohlrogge JB (2007) Sampling the Arabidopsis transcriptome with massively parallel pyrosequecing. Plant Physiol 144: 32–42.

Wheeler DL, Barrett T, Benson DA, Bryant SH, Canese K, Chetvernin V, Church DM, DiCuccio M, Edgar R, Federhen S, Geer LY, Kapustin Y, Khovayko O, Landsman D, Lipman DJ, Madden TL, Maglott DR, Ostell J, Miller V, Pruitt KD, Schuler GD, Sequeira E, Sherry ST, Sirotkin K, Souvorov A, Starchenko G, Tatusov RL, Tatusova TA, Wagner L, Yaschenko E (2007) Database resources of the National Center for Biotechnology Information. Nucl Acids Res 35 (Database iss): D5–12.

Wilson D, Madera M, Vogel C, Chothia C, Gough J (2007) The SUPERFAMILY database in 2007: families and functions. Nucl Acids Res 35(Database iss): D308–313.

Yi G, Sze SH, Thon MR (2007) Identifying clusters of functionally related genes in genome. Bioinformatics 23: 053–1060.

Jill L, Wegrzyn, Jennifer M, Lee, Brandon R Tearse, David B Neale (2008) TreeGenes: A Forest Tree Genome Database. Int J Plant Genom, vol 2008, Article ID 412875, 7 pages: doi: 10.1155/2008/412875.

Youens-Clark K, Buckler E, Casstevens T, Chen C, Declerck G, Derwent P, Dharmawardhana P, Jaiswal P, Kersey P, Karthikeyan AS, Lu J, McCouch SR, Ren L, Spooner W, Stein JC, Thomason J, Wei S, Ware D (2011) Gramene database in 2010: updates and extensions. Nucl Acids Res 39 (Database iss): D1085–94.

Zhebentyayeva T, Swire-Clark G, Georgi L, Garay L, Juns S, Forrest S, Blenda A, Blackmon B, Mook J, Horn R, Howad W, Arus P, Main D, Tomkins J, Sosinski B, Baird W, Reighard G, Abbott A (2008) A framework physical map for peach, a Model *Rosaceae* Species. Tree Genet Genomes 4: 745–756.

14

Retrospect and Prospects

Albert G. Abbott

Of the fruiting plant species, stone fruits are among the most important crops worldwide and play a significant role in small farm agriculture and rural economies. They are comprised of the species in the genus *Prunus* in the family Rosaceae and are valued for their fruit, lumber and ornamental uses. Prior to the advent of molecular genetic technologies, our understanding of the genetics of the important traits in fruit tree breeding was extremely limited, however, with the application of molecular marker and genomics technologies, these species are rapidly becoming models for the study of fruit and forest tree genetics and genomics and for the discovery of genes critical to tree development and sustainability.

In this volume, we describe the history, traditional breeding and many of the recent advances in genomics research in these key species that are uncovering the genes and characters that govern agriculturally relevant traits. One of the central advantages to gene discovery that these species have over others is the relatively small size of their genomes. This fact coupled with fairly short generation times for some species has enabled rapid deployment of modern genomic technologies to substantially advance our understanding of the regions of the stone fruit genome that significantly impact many traits not only important to fruit tree growth and sustainability but to forest tree species as well.

Traditionally, many agricultural breeding programs have focused on increasing yields, without incorporating other desirable characteristics in the food crop such as the food quality. However, many consumers have grown more health and diet conscious. Foods have now assumed the status of 'functional' foods, which should be capable of providing additional benefits, such as preventing or delaying the onset of chronic diseases. Phytochemicals

Department of Genetics and Biochemistry, Clemson University, 116 Jordan Hall, Clemson, SC 29634, USA; e-mail: *aalbert@clemson.edu*

in fruits and vegetables have attracted a great deal of attention due to their role in preventing diseases caused by oxidative stress. Stone fruits are excellent sources of these health beneficial compounds and thus stone fruit breeders and researchers are interested in manipulating through traditional and transgene approaches the gene networks responsible for production of these compounds. Current and future stone fruit breeding efforts are being directed at pyramiding fruit quality traits with other traits of importance for varietal improvement and sustainability.

Until recently, the genetic understanding of characters important to stone fruit agriculture lagged well behind that of the large commodity crops, (maize, rice, soybean, etc.). This was primarily due to lack of significant public investment and the refractory nature of the plants for achieving high-resolution genetic manipulations (juvenility periods in years, requirements for large amounts of labor and space). However, with the shift in the marketplace focus in many countries from large-scale commodity crops to smaller specialty crops of increased nutritional value, the development of genetic resources in the stone fruit crops came to the forefront. In the last 8-10 years, significant investment by the USDA in the United States and the European Union (EU) granting agencies in Europe and state of the art molecular genetics technologies, have substantially contributed to progress in development of the genetic and genomic resources for key *Prunus* species. In *Prunus*, peach [*Prunus persica* (L.) Batsch] is currently the most highly genetically characterized species. The recent completion of whole genome sequences for the peach has now paved the way for linking candidate genes in marked intervals with actual traits. Thus peach serves as a reference genome for identification and characterization of genes important to *Prunus* agriculture and more broadly to other species in the Rosaceae as well.

Predominantly through the EU *Prunus* mapping communities' efforts to integrate their mapping data through a common core set of SSRs many important genomic intervals influencing specific characters have been identified and are publicly available through the Genome Database for Rosaceae (GDR). All of these trait mapping programs to date have identified the trait containing intervals through use of single parent mapping crosses; F_1, F_2,or backcrosses,with limited numbers of progeny in the analysis (<300). This has enabled low resolution of trait containing intervals for both simple and complex traits. Additionally using the physical map resources in peach anchored on the *Prunus* general genetic map, potential gene candidates were previously placed in trait containing regions to develop specialized gene maps such as resistance gene maps that could be integrated into this *Prunus* genetic framework to serve as a resource for identification of resistance gene containing regions of the *Prunus* species genomes.

Although significant reproducible QTL intervals have been identified for specific traits (e.g., chilling requirement, bloom time, sucrose, etc.), resolution of these regions is severely limited by progeny numbers. It is clear that for MAS more tightly linked markers need to be developed for many of the important traits.

Due to the physical restrictions (time, space, and labor) on molecular mapping in fruit trees, high resolution of QTLs through single or multiple family crosses is extremely difficult. An alternative to the use of individual segregating families for mapping QTLs is association mapping. Association mapping takes advantage of the phenomenon of linkage disequilibrium for markers and traits in diverse established populations. This alternative QTL mapping method has been applied to analysis of human genetic characters and is currently being explored in plants as a way of resolving QTLs in particular species where requisite populations exist. It would seem that this approach would have great application for fruit tree mapping since many species of fruiting trees have broadly diverse existing populations maintained in germ plasm centers around the world. Additionally perennial species provide opportunities to continually score different characters over multiple years in old populations to mark multiple phenotypes on a single population, a distinct advantage over traditional strategies that follow only a few polymorphic characters in a single population where most other phenotypic loci are fixed.

A basic strategy for linkage disequilibrium mapping in fruiting trees could be as follows: 1. Establish and screen a genetically diverse population, population diversity established using programs such as Structure (Prichard et al. 2000a) for particular characters of interest; 2. Individual trees are then scored for the traits of interest; 3. Utilizing highly polymorphic marker types (SNPs, or SSRs) score the trees across the genome for allelic composition at the various loci; and 4. Utilizing statistical analysis packages such as Tassel (*www.maizegenetics.net*) to search for linkage associations of specific traits and specific alleles at marker loci. Markers obtained in such a manner can then be tested in field material and crosses for utilization in MAS. Of course this approach relies on adequate populations and significant numbers of polymorphic markers. For species in the Rosaceae, a number of initiatives worldwide are providing the gene sequence resources for robust marker development (whole genome sequences of peach, apple, strawberry are completed; SNP resources are available through RosBREED; extensive EST/marker map resources; GDR).

An adaptation of association mapping called Pedigree Genotyping has been employed in apple QTL analysis (Van de Weg *et al.*, 2004). In this approach, pedigree information is coupled with phenotypic information to follow the segregation of founder alleles for specific traits through breeding pedigrees. When alleles of molecular markers, such as SSRs, are found to associate with the phenotype through the pedigree, one can assign potential regions of the genome governing the trait. These regions are then targeted for further study. This system should, in theory work with any species that has: 1) a sufficiently saturated molecular genetic map; and 2) breeding materials of phenotyped, pedigreed individuals.

Many of the important characters of interest in *Prunus* breeding programs (e.g., chilling requirement and bloom time, resistance to diseases, fruit nutritional value) are quantitative in genetic nature and thus present significant challenges to rapid genetic manipulation for varietal improvement. Genomics resources (genes, markers etc.) offer the tools to substantially augment the traditional breeders toolbox to improve the speed and efficiency of the character selection process. However, in order to capitalize on the use of the *Prunus* genomics resources and translate them into breeding programs the genetic resolution of the trait containing intervals in the genome needs to be substantially improved. This is no simple task as the physical limitations (time, space) required for tree species genetics complicates studies to refine QTL intervals into marker selectable units. Thus new approaches to QTL resolution (e.g., association studies and candidate gene approaches) are rapidly augmenting more traditional approaches (F_1 or F_2 mapping populations) with the promise of higher resolution of QTL for marker-assisted breeding purposes. However, even with increased marker resolution, the time in tree breeding required to test a marker-assisted breeding scheme for successful varietal improvement is a severe bottleneck to realizing the potential of translational approaches.

In stone fruit genetics and genomics, collaborative decisions to combine map information from many laboratories to produce a genera wide linkage map created a model system for integration of related species genome information enabling comparative genomics studies and a broad access to markers for applications in all *Prunus* breeding programs. The ability to move seamlessly from the genome of peach to apricot, almond, etc. is currently driving advances of marker-assisted breeding schemes for selection of abiotic and biotic stress resistance, phenological traits, fruit quality traits, tree architecture and others. All these achievements have been highlighted in this volume. We are now moving into the period where detailed genetics and genomics information is finally becoming part of the fruit tree breeders' toolbox.

Still some of the remaining hurtles to breeding in fruit trees (generation time, space, marginalization of lands, preservation of germplasm) all present significant challenges to improving and sustaining stone fruits. Recent advances that integrate molecular breeding and transgenic lines with reduced or no juvenility hold great promise to eliminate some of these barriers, however to date in stone fruits these approaches are limited to plums. In plums, Dr. R. Scorza's group at the AFRS-ARS has demonstrated that altered expression of at transgenic *FT* gene produces a tree lacking any juvenility period and exhibiting indeterminate flowering. This continuously flowering plum can be used to significantly shorten the breeding time for generation of new varieties. Subsequent outcrosses can eliminate the transgene yielding new non-transgenic varieties. Significant investment into development of such approaches for the other species would seem a likely path for the future. Having the marked portions of the genome controlling the traits of interest and methods to rapidly collect these into elite breeding materials will create a completely new landscape in fruit tree breeding and presumably provide the breeders with the tools needed to combat many challenges faced by their fruit tree breeding programs.

In the application of new genomics technologies to our understanding of basic biology of trees, stone fruit researchers are uniquely positioned. The genome size and simplicity of the stone fruit tree genomes makes them ideal candidates to test association genetics approaches based on whole-genome sequence genotyping for gene discovery and genome-wide selection in QTL breeding. Having a high quality reference genome sequence for peach coupled with whole genome re-sequencing technologies for species such as apricot with its extensive germplasm diversity afford great opportunities for testing these approaches. Indeed laboratories worldwide are applying whole-genome re-sequencing as a means to uncover candidate genes for a myriad of important characters (e.g., tree architecture, flowering time, chilling requirement, disease resistance) in *Prunus* species.

Indeed it is likely that stone fruit breeding in the future will capitalize on many of these recent technologies as important tools for advancing the breeding process however, we still face many challenges of which perhaps the most important one today is the cataloging and preserving the extensive germplasm for these species.

Germplasm is the primary resource for identification and introgression of many important gene variants that breeding programs need to face the challenges to sustain and advance breeding lines and cultivars. In stone fruits, much of the germplasm exists in remote areas of countries where lack of resources and political forces greatly impact our access and ability to preserve these important materials for future breeding applications. Additionally, breeding programs must continually balance investment of funds for development of their elite breeding lines and maintenance of exotic germplasm materials often leading to loss of important germplasm variants. Coupled with land development in the countries harboring the wild germplasm and tighter restrictions on release of germplasm from these countries, the future preservation of the wild diversity of stone fruit species is increasingly in doubt. Thus, although we may have the tools to expeditiously advance breeding materials to address industry problems, we may not have the genetic variants available to achieve our goals. Solutions to these problems will likely rely heavily on identifying and understanding the extent of wild germplasm diversity available around the globe, significant investment in germplasm repositories, politically acceptable ways to protect and distribute wild germ plasm materials, and advanced mutagenesis and transgenic technologies for gene targeted development of specific gene variants.

Finally, researchers in stone fruit do not operate in a genomics informational vacuum. Public databases such as the "Genome database for Rosaceae" (GDR, www.rosasceae.org), Phytozome (www.phytozome.org) and others continually highlight the significant advances in genetics and genomics of these tree fruit crops. This fact has led to the increasing awareness of the importance of these simple compact genomes to the broader understanding of tree biology and genetics. Stone fruit species genomes, such as that of the peach, are rapidly coming to the forefront of comparative genomics in trees as they represent some of the most streamlined genomes and possibly the most preserved ancestral genome states. These species span the human utility spectrum from food to ornamental to lumber applications and thus can provide important genetic and genomic information to these different industries.

The integration of stone fruit genetics and genomics knowledge into the broader spectrum of tree research offers many exciting and future collaborative avenues to continue to advance the breeding of these plant species while at the same time demonstrating the relevance of these species to the foundations of our understanding of basic tree biology. The future lies in the fruits of our efforts.

Index

M

N

O

P

Q

R

Color Plate Section

Chapter 6

Figure 6-1

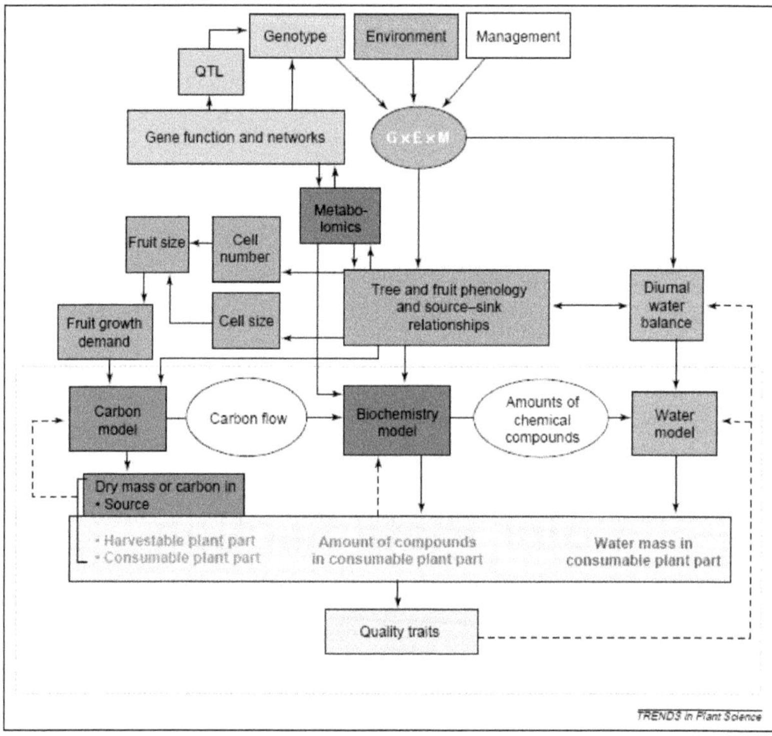

Figure 6-2

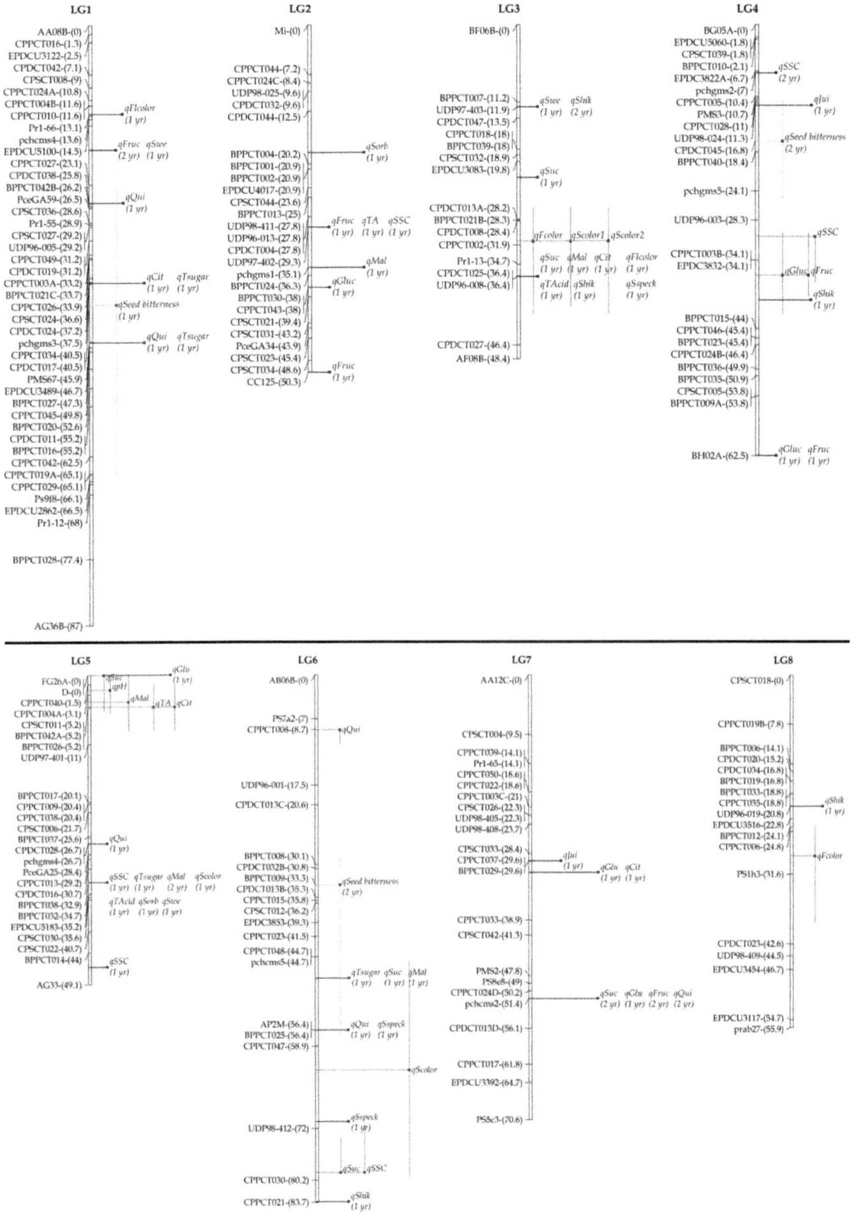
LG1
AA08B-(0)
CPPCT016-(1.3)
EPDCU3122-(2.5)
CPDCT042-(7.1)
CPSCT008-(9)
CPPCT024A-(10.8)
CPPCT004B-(11.6)
CPPCT010-(11.6)
Pr1-66-(13.1)
pchcms4-(13.6)
EPDCU5100-(14.5)
CPPCT027-(23.1)
CPDCT038-(25.8)
BPPCT042B-(26.2)
PceGA59-(26.5)
CPSCT036-(28.6)
Pr1-55-(28.9)
CPSCT027-(29.2)
UDP96-005-(29.2)
CPPCT049-(31.2)
CPDCT019-(31.2)
CPPCT003A-(33.2)
BPPCT021C-(33.7)
CPPCT026-(33.9)
CPSCT024-(36.6)
CPDCT024-(37.2)
pchgms3-(37.5)
CPPCT034-(40.5)
CPDCT017-(40.5)
PMS67-(45.9)
EPDCU3489-(46.7)
BPPCT027-(47.3)
CPPCT045-(49.8)
BPPCT020-(52.6)
CPDCT011-(55.2)
BPPCT016-(55.2)
CPPCT042-(62.5)
CPPCT019A-(65.1)
CPPCT029-(65.1)
Ps9f8-(66.1)
EPDCU2862-(66.5)
Pr1-12-(68)
BPPCT028-(77.4)
AG36B-(87)
qFlcolor (1 yr)
qFruc qSuc (2 yr) (1 yr)
qQui (1 yr)
qCit qTsugar (1 yr) (1 yr)
qSeed bitterness (1 yr)
qQui qTsugar (1 yr) (1 yr)
LG2
Mi-(0)
CPPCT044-(7.2)
CPPCT024C-(8.4)
UDP98-025-(9.6)
CPDCT032-(9.6)
CPDCT044-(12.5)
BPPCT004-(20.2)
BPPCT001-(20.9)
BPPCT002-(20.9)
EPDCU4017-(20.9)
CPSCT044-(23.6)
BPPCT013-(25)
UDP98-411-(27.8)
UDP96-013-(27.8)
CPDCT004-(27.8)
UDP97-402-(29.3)
pchgms1-(35.1)
BPPCT024-(36.3)
BPPCT030-(38)
CPPCT043-(38)
CPSCT021-(39.4)
CPSCT031-(43.2)
PceGA34-(43.9)
CPSCT023-(45.4)
CPSCT034-(48.6)
CC125-(50.3)
qSorb (1 yr)
qFruc qTA qSSC (1 yr) (1 yr) (1 yr)
qMal (1 yr)
qGluc (1 yr)
qFruc (1 yr)
LG3
BF06B-(0)
BPPCT007-(11.2)
UDP97-403-(11.9)
CPDCT047-(13.5)
CPPCT018-(18)
BPPCT039-(18)
CPSCT032-(18.9)
EPDCU3083-(19.8)
CPDCT013A-(28.2)
BPPCT021B-(28.3)
CPDCT008-(28.4)
CPPCT002-(31.9)
Pr1-13-(34.7)
CPDCT025-(36.4)
UDP96-008-(36.4)
CPDCT027-(46.4)
AF08B-(48.4)
qSuc qShik (1 yr) (2 yr)
qSuc (1 yr)
qFcolor qScolor1 qScolor2
qSuc qMal qCit qFlcolor (1 yr) (1 yr) (1 yr) (1 yr)
qTAcid qShik qSspeck (1 yr) (1 yr) (1 yr)
LG4
BG05A-(0)
EPDCU5060-(1.8)
CPSCT039-(1.8)
BPPCT010-(2.1)
EPDC3822A-(6.7)
pchgms2-(7)
CPPCT005-(10.4)
PMS3-(10.7)
CPPCT028-(11)
UDP98-024-(11.3)
CPDCT045-(16.8)
BPPCT040-(18.4)
pchgms5-(24.1)
UDP96-003-(28.3)
CPPCT003B-(34.1)
EPDC3832-(34.1)
BPPCT015-(44)
CPPCT046-(45.4)
BPPCT023-(45.4)
CPPCT024B-(46.4)
BPPCT036-(49.9)
BPPCT035-(50.9)
CPSCT005-(53.8)
BPPCT009A-(53.8)
BH02A-(62.5)
qSSC (2 yr)
qQui (1 yr)
qSeed bitterness (2 yr)
qSSC
qGluc qFruc
qShik (1 yr)
qGluc qFruc (1 yr) (1 yr)
LG5
FG26A-(0)
D-(0)
CPPCT040-(1.5)
CPPCT004A-(3.1)
CPSCT011-(5.2)
BPPCT042A-(5.2)
BPPCT026-(5.2)
UDP97-401-(11)
BPPCT017-(20.1)
CPPCT009-(20.4)
CPPCT038-(20.4)
CPSCT006-(21.7)
BPPCT037-(25.6)
CPDCT028-(26.7)
pchgms4-(26.7)
PceGA25-(28.4)
CPPCT013-(29.2)
CPDCT016-(30.7)
BPPCT038-(32.9)
BPPCT032-(34.7)
EPDCU5183-(35.2)
CPSCT030-(35.6)
CPSCT022-(40.7)
BPPCT014-(44)
AG33-(49.1)
qGlu (1 yr)
qSuc
qpH
qMal
qTA
qCit
qQui (1 yr)
qSSC qTsugar qMal qScolor (1 yr) (1 yr) (2 yr) (1 yr)
qTAcid qSorb qSuc (1 yr) (1 yr) (1 yr)
qSSC (1 yr)
LG6
AB06B-(0)
PS7a2-(7)
CPPCT008-(8.7)
UDP96-001-(17.5)
CPDCT013C-(20.6)
BPPCT008-(30.1)
CPDCT032B-(30.8)
BPPCT009-(33.3)
CPDCT013B-(35.3)
CPPCT015-(35.8)
CPSCT012-(36.2)
EPDC3853-(39.3)
CPPCT023-(41.5)
CPPCT048-(44.7)
pchcms5-(44.7)
AP2M-(56.4)
BPPCT025-(56.4)
CPPCT047-(58.9)
UDP98-412-(72)
CPPCT030-(80.2)
CPPCT021-(83.7)
qQui
qSeed bitterness (2 yr)
qTsugar qSuc qMal (1 yr) (1 yr) (1 yr)
qQui qSspeck (1 yr) (1 yr)
qScolor
qSspeck (1 yr)
qSuc qSSC
qShik (1 yr)
LG7
AA12C-(0)
CPSCT004-(9.5)
CPPCT039-(14.1)
Pr1-65-(14.1)
CPPCT050-(18.6)
CPPCT022-(18.6)
CPPCT003C-(21)
CPSCT026-(22.3)
UDP98-405-(22.3)
UDP98-408-(23.7)
CPSCT033-(28.4)
CPPCT037-(29.6)
BPPCT029-(29.6)
CPPCT033-(38.9)
CPSCT042-(41.3)
PMS2-(47.8)
PS8e8-(49)
CPPCT024D-(50.2)
pchcms2-(51.4)
CPDCT013D-(56.1)
CPPCT017-(61.8)
EPDCU3392-(64.7)
PS5c3-(70.6)
qQui (1 yr)
qGlu qCit (1 yr) (1 yr)
qSuc qGlu qFruc qQui (2 yr) (1 yr) (2 yr) (2 yr)
LG8
CPSCT018-(0)
CPPCT019B-(7.8)
BPPCT006-(14.1)
CPDCT020-(15.2)
CPDCT034-(16.8)
BPPCT019-(16.8)
BPPCT033-(18.8)
CPPCT035-(18.8)
UDP96-019-(20.8)
EPDCU3516-(22.8)
BPPCT012-(24.1)
CPPCT006-(24.8)
PS1h3-(31.6)
CPDCT023-(42.6)
UDP98-409-(44.5)
EPDCU3454-(46.7)
EPDCU3117-(54.7)
prab27-(55.9)
qShik (1 yr)
qFcolor

Figure 6-3

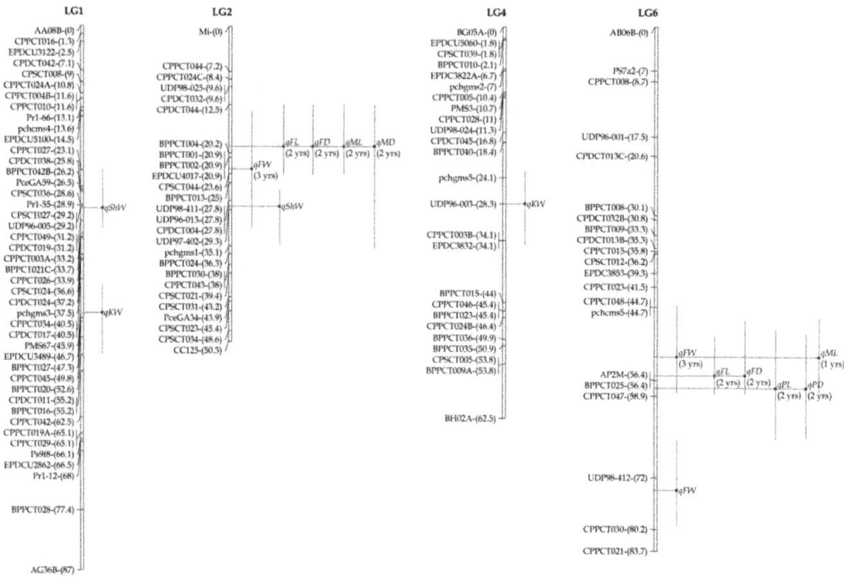

Figure 6-4

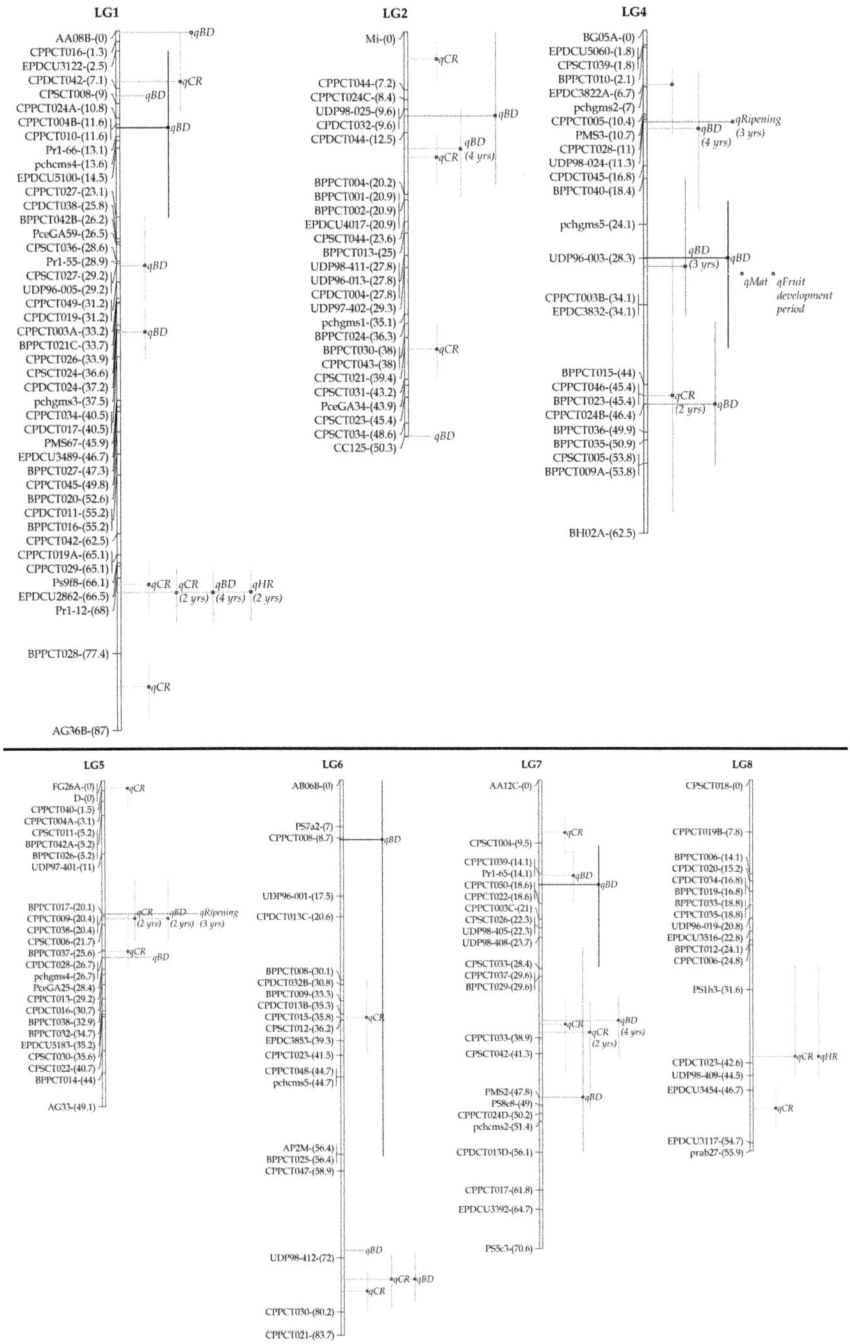
LG1
AA08B-(0)
CPPCT016-(1.3)
EPDCU3122-(2.5)
CPDCT042-(7.1)
CPSCT008-(9)
CPPCT024A-(10.8)
CPPCT004B-(11.6)
CPPCT010-(11.6)
Pr1-66-(13.1)
pchcms4-(13.6)
EPDCU5100-(14.5)
CPPCT027-(23.1)
CPDCT038-(25.8)
BPPCT042B-(26.2)
PceGA59-(26.5)
CPSCT036-(28.6)
Pr1-55-(28.9)
CPSCT027-(29.2)
UDP96-005-(29.2)
CPPCT049-(31.2)
CPDCT019-(31.2)
CPPCT003A-(33.2)
BPPCT021C-(33.7)
CPPCT026-(33.9)
CPSCT024-(36.6)
CPDCT024-(37.2)
pchgms3-(37.5)
CPPCT034-(40.5)
CPDCT017-(40.5)
PMS67-(45.9)
EPDCU3489-(46.7)
BPPCT027-(47.3)
CPPCT045-(49.8)
BPPCT020-(52.6)
CPDCT011-(55.2)
BPPCT016-(55.2)
CPPCT042-(62.5)
CPPCT019A-(65.1)
CPPCT029-(65.1)
Ps9f8-(66.1)
EPDCU2862-(66.5)
Pr1-12-(68)
BPPCT028-(77.4)
AG36B-(87)
qBD
qCR
qBD
qBD
qBD
qBD
qCR
qCR (2 yrs)
qBD (4 yrs)
qHR (2 yrs)
qCR
LG2
Mi-(0)
CPPCT044-(7.2)
CPPCT024C-(8.4)
UDP98-025-(9.6)
CPDCT032-(9.6)
CPDCT044-(12.5)
BPPCT004-(20.2)
BPPCT001-(20.9)
BPPCT002-(20.9)
EPDCU4017-(20.9)
CPSCT044-(23.6)
BPPCT013-(25)
UDP98-411-(27.8)
UDP96-013-(27.8)
CPDCT004-(27.8)
UDP97-402-(29.3)
pchgms1-(35.1)
BPPCT024-(36.3)
BPPCT030-(38)
CPPCT043-(38)
CPSCT021-(39.4)
CPSCT031-(43.2)
PceGA34-(43.9)
CPSCT023-(45.4)
CPSCT034-(48.6)
CC125-(50.3)
qCR
qBD
qBD
qCR (4 yrs)
qCR
qBD
LG4
BG05A-(0)
EPDCU5060-(1.8)
CPSCT039-(1.8)
BPPCT010-(2.1)
EPDC3822A-(6.7)
pchgms2-(7)
CPPCT005-(10.4)
PMS3-(10.7)
CPPCT028-(11)
UDP98-024-(11.3)
CPDCT045-(16.8)
BPPCT040-(18.4)
pchgms5-(24.1)
UDP96-003-(28.3)
CPPCT003B-(34.1)
EPDC3832-(34.1)
BPPCT015-(44)
CPPCT046-(45.4)
BPPCT023-(45.4)
CPPCT024B-(46.4)
BPPCT036-(49.9)
BPPCT035-(50.9)
CPSCT005-(53.8)
BPPCT009A-(53.8)
BH02A-(62.5)
qRipening (3 yrs)
qBD (4 yrs)
qBD (3 yrs)
qBD
qMat
qFruit development period
qCR (2 yrs)
qBD
LG5
FG26A-(0)
D-(0)
CPPCT040-(1.5)
CPPCT004A-(3.1)
CPSCT011-(5.2)
BPPCT042A-(5.2)
BPPCT026-(5.2)
UDP97-401-(11)
BPPCT017-(20.1)
CPPCT009-(20.4)
CPPCT038-(20.4)
CPSCT006-(21.7)
BPPCT037-(25.6)
CPDCT028-(26.7)
pchgms4-(26.7)
PceGA25-(28.4)
CPPCT013-(29.2)
CPDCT016-(30.7)
BPPCT038-(32.9)
BPPCT032-(34.7)
EPDCU5183-(35.2)
CPSCT030-(35.6)
CPSCT022-(40.7)
BPPCT014-(44)
AG33-(49.1)
qCR
qCR (2 yrs)
qBD (2 yrs)
qRipening (3 yrs)
qCR
qBD
LG6
AB06B-(0)
PS7a2-(7)
CPPCT008-(8.7)
UDP96-001-(17.5)
CPDCT013C-(20.6)
BPPCT008-(30.1)
CPDCT032B-(30.8)
BPPCT009-(33.3)
CPDCT013B-(35.3)
CPPCT015-(35.8)
CPSCT012-(36.2)
EPDC3853-(39.3)
CPPCT023-(41.5)
CPPCT048-(44.7)
pchcms5-(44.7)
AP2M-(56.4)
BPPCT025-(56.4)
CPPCT047-(58.9)
UDP98-412-(72)
CPPCT030-(80.2)
CPPCT021-(83.7)
qBD
qCR
qBD
qCR
qBD
qCR
LG7
AA12C-(0)
CPSCT004-(9.5)
CPPCT039-(14.1)
Pr1-65-(14.1)
CPPCT050-(18.6)
CPPCT022-(18.6)
CPPCT003C-(21)
CPSCT026-(22.3)
UDP98-405-(22.3)
UDP98-408-(23.7)
CPSCT033-(28.4)
CPPCT037-(29.6)
BPPCT029-(29.6)
CPPCT033-(38.9)
CPSCT042-(41.3)
PMS2-(47.8)
PS8c8-(49)
CPPCT024D-(50.2)
pchcms2-(51.4)
CPDCT013D-(56.1)
CPPCT017-(61.8)
EPDCU3392-(64.7)
PS5c3-(70.6)
qCR
qBD
qBD
qCR
qBD (4 yrs)
qCR (2 yrs)
qBD
LG8
CPSCT018-(0)
CPPCT019B-(7.8)
BPPCT006-(14.1)
CPDCT020-(15.2)
CPDCT034-(16.8)
BPPCT019-(16.8)
BPPCT033-(18.8)
CPPCT035-(18.8)
UDP96-019-(20.8)
EPDCU3516-(22.8)
BPPCT012-(24.1)
CPPCT006-(24.8)
PS1h3-(31.6)
CPDCT023-(42.6)
UDP98-409-(44.5)
EPDCU3454-(46.7)
EPDCU3117-(54.7)
prab27-(55.9)
qCR
qHR
qCR

Figure 6-5

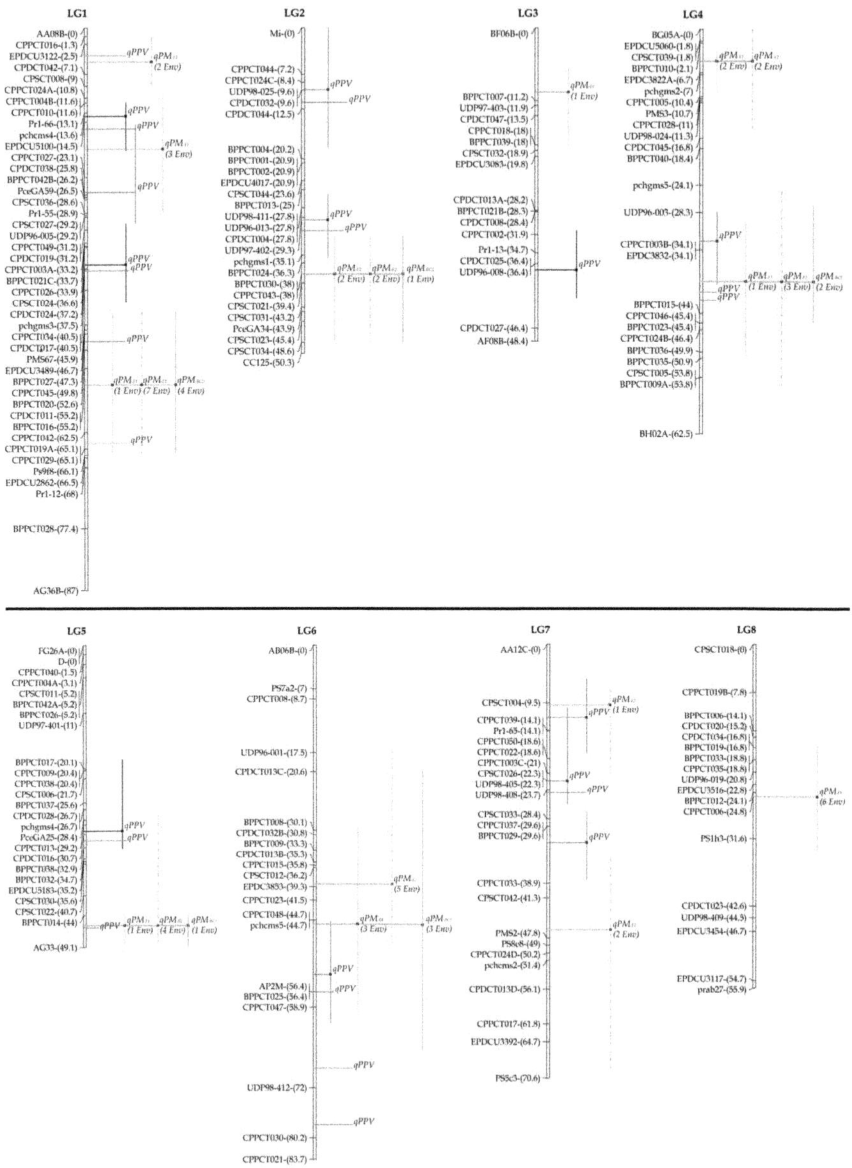

Chapter 9

Figure 9-1

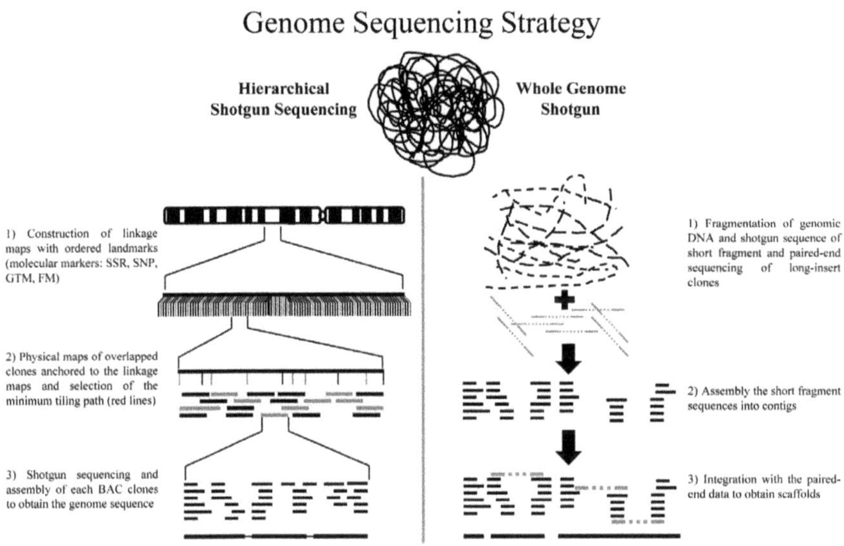

Figure 9-2

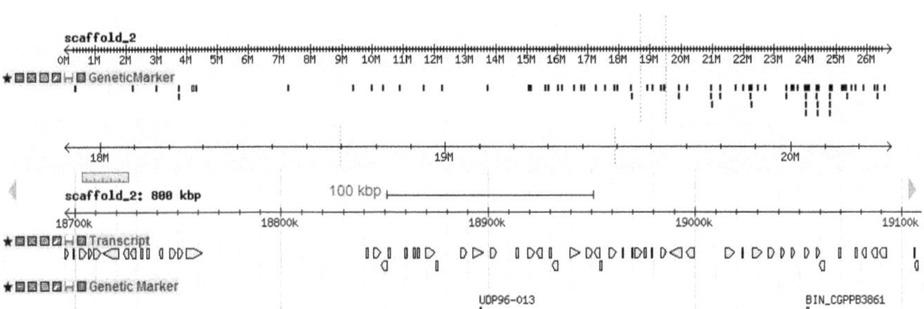

Figure 9-3

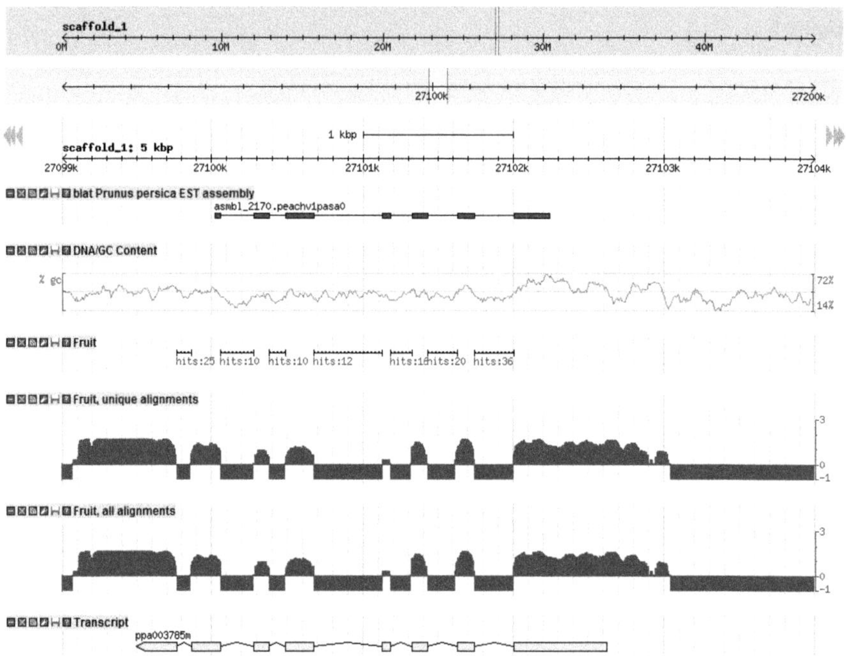
scaffold_1
0M
10M
20M
30M
40M
27100k
27200k
1 kbp
scaffold_1: 5 kbp
27099k
27100k
27101k
27102k
27103k
27104k
blat Prunus persica EST assembly
asmbl_2170.peachv1pasa0
DNA/GC Content
% gc
72%
14%
Fruit
hits:25
hits:10
hits:10
hits:12
hits:16
hits:20
hits:36
Fruit, unique alignments
3
0
-1
Fruit, all alignments
3
0
-1
Transcript
ppa003785m

Chapter 10

Figure 10-1

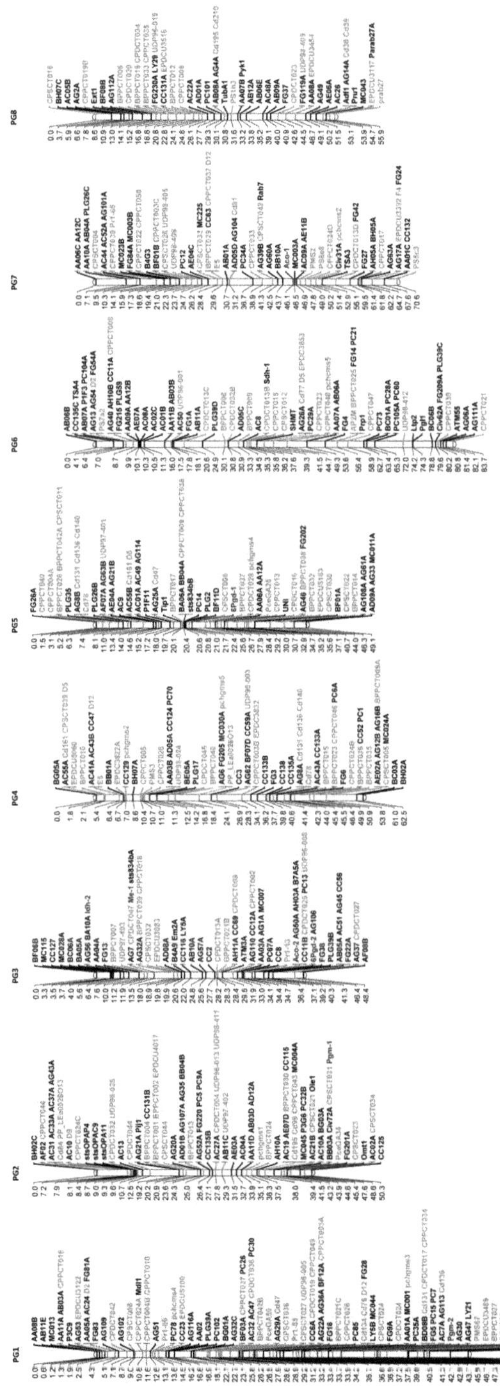

Figure 10-2

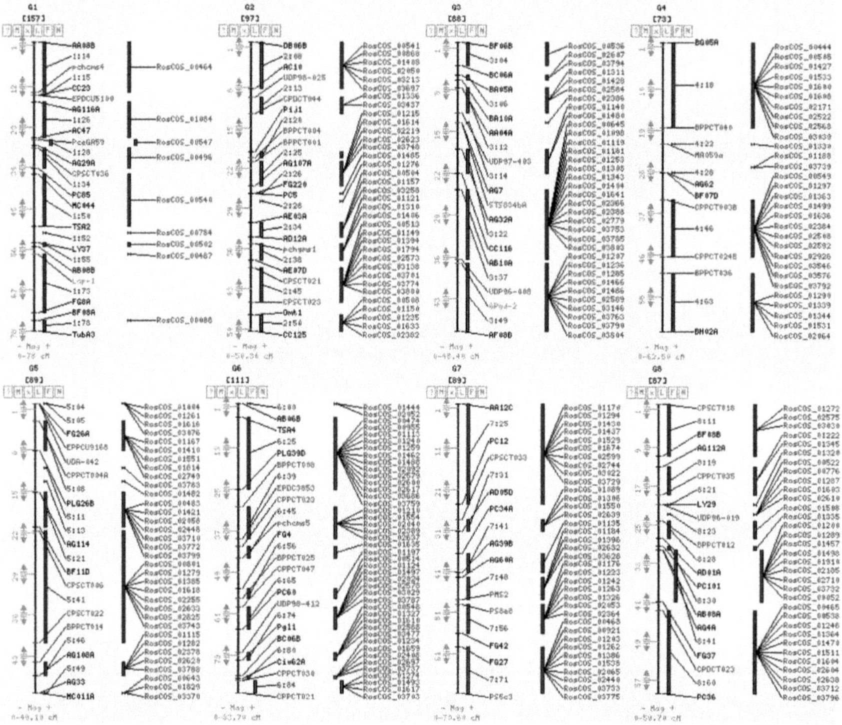

Figure 10-3

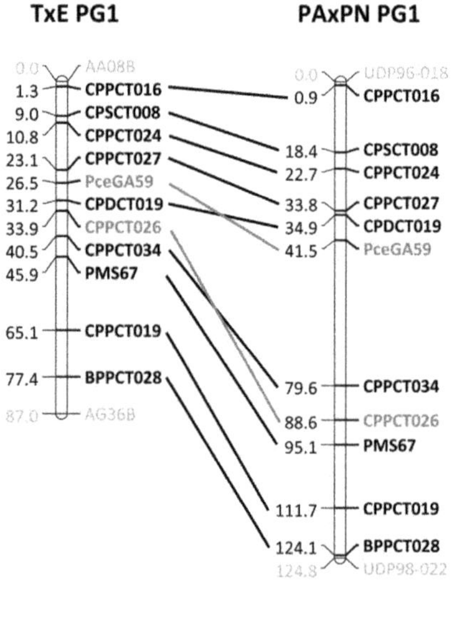

Figure 10-6

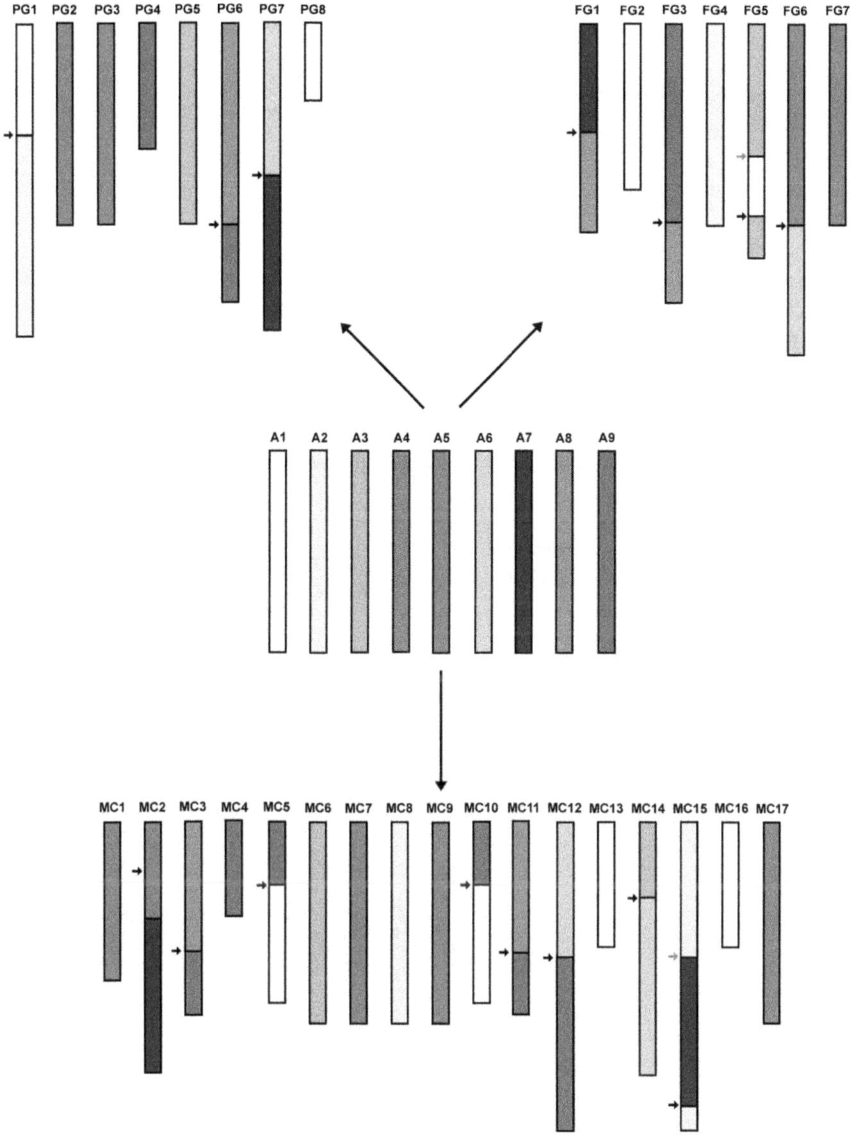
PG1 PG2 PG3 PG4 PG5 PG6 PG7 PG8
FG1 FG2 FG3 FG4 FG5 FG6 FG7
A1 A2 A3 A4 A5 A6 A7 A8 A9
MC1 MC2 MC3 MC4 MC5 MC6 MC7 MC8 MC9 MC10 MC11 MC12 MC13 MC14 MC15 MC16 MC17

Chapter 11

Figure 11-1

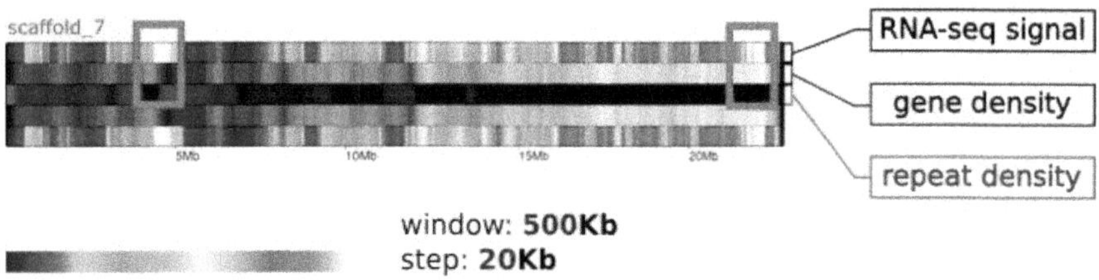

Chapter 12

Figure 12-1

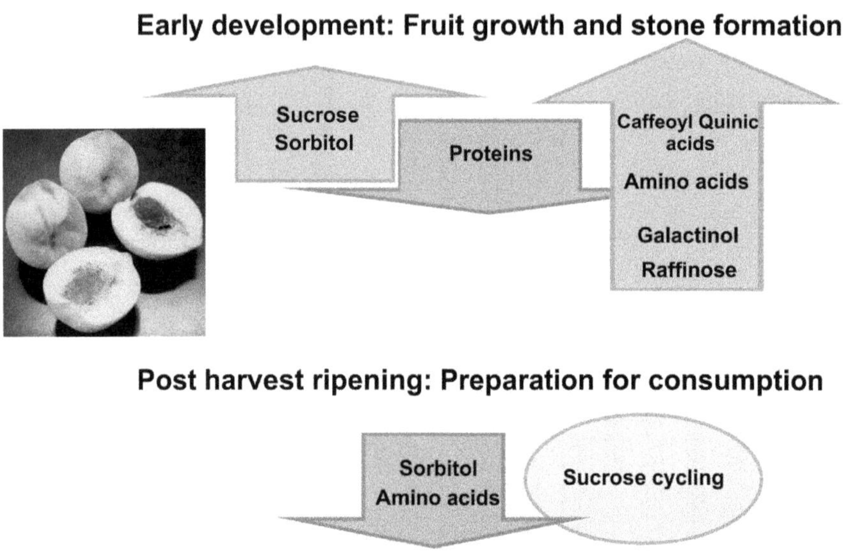

For Product Safety Concerns and Information please contact our EU
representative GPSR@taylorandfrancis.com
Taylor & Francis Verlag GmbH, Kaufingerstraße 24, 80331 München, Germany

www.ingramcontent.com/pod-product-compliance
Ingram Content Group UK Ltd.
Pitfield, Milton Keynes, MK11 3LW, UK
UKHW022213190926
475702UK00003B/337
* 9 7 8 1 5 7 8 0 8 8 0 1 0 *